Resilience in Historical Landscapes

Resilience in Historical Landscapes

Guest Editors

Angelo Castrorao Barba
Pilar Diarte-Blasco
Manuel Castro-Priego
Giuseppe Bazan

MDPI
Basel • Beijing • Wuhan • Barcelona • Belgrade • Novi Sad • Cluj • Manchester

Guest Editors
Angelo Castrorao Barba
Escuela de Estudios Árabes
Consejo Superior de
Investigaciones Científicas
Granada
Spain

Pilar Diarte-Blasco
Instituto de Historia
Consejo Superior de
Investigaciones Científicas
Madrid
Spain

Manuel Castro-Priego
Department of Archaeology
University of Alcalá
Alcalá de Henares
Spain

Giuseppe Bazan
Department STEBICEF
University of Palermo
Palermo
Italy

Editorial Office
MDPI AG
Grosspeteranlage 5
4052 Basel, Switzerland

This is a reprint of the Special Issue, published open access by the journal *Land* (ISSN 2073-445X), freely accessible at: https://www.mdpi.com/journal/land/special_issues/resilience_historical_landscapes.

For citation purposes, cite each article independently as indicated on the article page online and as indicated below:

Lastname, A.A.; Lastname, B.B. Article Title. *Journal Name* **Year**, *Volume Number*, Page Range.

ISBN 978-3-7258-7957-1 (Hbk)
ISBN 978-3-7258-7958-8 (PDF)
https://doi.org/10.3390/books978-3-7258-7958-8

Cover image courtesy of Angelo Castrorao Barba

Contents

About the Editors

Angelo Castrorao Barba

Angelo Castrorao Barba (born in Palermo, 1983) is currently a tenure-track researcher (Ramón y Cajal Fellow) at the School of Arabic Studies (EEA) of the Spanish National Research Council (CSIC) in Granada. He has directed excavations, surveys, and interdisciplinary research in the hinterlands of Palermo and Málaga, aiming to understand the transformations in Islamic settlements, landscapes, and material culture during Late Antiquity and the Early Middle Ages. After earning his PhD in Medieval Archaeology from the University of Siena in 2013, Castrorao Barba has worked as a researcher and fellow at various prestigious institutions. These include the University of Palermo, the Royal Netherlands Institute in Rome (KNIR), VU Amsterdam, University of Tübingen, University of Hamburg, School of Arabic Studies (EEA-CSIC), and the Institute of Archaeology and Ethnology/Centre for Late Antique and Early Medieval Studies of the Polish Academy of Sciences (IAE-PAN). He has published over 50 scientific papers and a monograph on the end of Roman villas in Italy and has recently edited volumes on Late Antique, Byzantine, and Islamic archaeology for Archaeopress Archaeology, Brepols, University Press of Florida, and De Gruyter.

In recent years, Castrorao Barba has been PI of research projects on Late Antique, Byzantine and Islamic archaeology in Sicily and Andalusia, funded by the Barakat Trust, the Gerda Henkel Stiftung, Marie Skłodowska-Curie Actions, CSIC and Fundación Palarq.

Pilar Diarte-Blasco

Pilar Diarte-Blasco is a Tenured Researcher at the Department of Archaeology and Social Processes at the Instituto de Historia, part of the Spanish National Research Council (CSIC). She is a specialist in Late Antique and Early Medieval Archaeology and landscape studies, with extensive experience in survey techniques, geophysics, and the management and visualisation of archaeological data. She earned her European PhD in 2011 from the Universidad de Zaragoza (Spain) and has since held several prestigious postdoctoral fellowships in Italy, Spain, and the United Kingdom. With a strong publication record, she has led and coordinated numerous multidisciplinary projects focused on the post-classical transformation of landscapes and townscapes. These include her participation in various research initiatives and her leadership of several international projects in Spain and Italy. Pilar is also the co-director of the Reccopolis Archaeological Project, which explores the only archaeologically attested Visigothic city foundation in the Iberian Peninsula. Founded by King Leovigild in A.D. 576, Reccopolis played a significant role in the region's landscape, serving as a territorial focal point until the 8th century.

Manuel Castro-Priego

Manuel Castro-Priego is currently an Assistant Professor of Archaeology. Since 2010, he has co-directed the research project on the Visigothic city of Reccopolis. In recent years, he has developed two regional research projects for the study of the material culture of Reccopolis and its dissemination and inclusivity (Reccult and ACCARQ). He also manages the excavations in the Andalusi medina of Uceda (Guadalajara, Spain). He has been awarded research stays and competitive scientific contracts in the USA and Latin America, where he has developed projects on the first European presence in Ecuador. His field of study is the economy and medieval ceramics, as well as the European impact on the Pacific coast of South America during the 16th century. He currently has three books and more than forty publications in high-impact journals (*Radiocarbon*, *Nature Communications*, *Medieval Archaeology*, among others).

Giuseppe Bazan

Giuseppe Bazan (Italy) is a Professor of Environmental and Applied Botany at the University of Palermo. He teaches courses in Landscape Ecology as part of the master's programme in "Spatial Planning," Historical Landscape Ecology in the Master's course in Archaeology, and Botany in the bachelor's programme in "Biological Sciences" at the University of Palermo. His research focuses on landscape ecology, with a primary interest in understanding how plant communities contribute to the formation and evolution of land mosaics. Combining field studies on plant biodiversity with GIS techniques, his work addresses complex aspects of landscapes at both the spatial and temporal scales. Since 2000, he has been continuously engaged in scientific research, resulting in over 150 publications. His research spans topics such as ornamental and wild plants, the relationship between vegetation and landscape, and vegetation mapping using GIS. Giuseppe Bazan is an active member of scientific societies such as IAVS and IALE. He served as the Principal Investigator for the University of Palermo team in the EU FP7 Project Mediterranean Mountainous Landscapes: An Historical Approach to Cultural Heritage Based on Traditional Agrosystems (MEMOLA). In 2015, the MEMOLA project was selected by the EU Humanities Scientific Committee as one of the top five innovative research projects in the humanities. Within this interdisciplinary initiative, Giuseppe Bazan emphasised the role of vegetation science in understanding landscape history. Additionally, he was the PI for the research project Harvesting Memories: Ecology and Archaeology of Monti Sicani Landscapes (Central-Western Sicily), a collaboration between Bona Furtuna LLC (the project's funder), the University of Palermo, and the Superintendency BB.CC.AA of Palermo.

land

Editorial

Approaching Resilience in Historical Landscapes

Angelo Castrorao Barba [1,*], Pilar Diarte-Blasco [2], Manuel Castro-Priego [3] and Giuseppe Bazan [4]

1 Escuela de Estudios Árabes (EEA), Consejo Superior de Investigaciones Científicas (CSIC), 18010 Granada, Spain
2 Department of Archaeology and Social Processes, Instituto de Historia (IH), Consejo Superior de Investigaciones Científicas (CSIC), 28037 Madrid, Spain; pilar.diarte@cchs.csic.es
3 Department of Archaeology, University of Alcalá, C/Trinidad, 1, 28801 Alcala de Henares, Spain; manuel.castro@uah.es
4 Department STEBICEF, University of Palermo, Viale delle Scienze, Ed. 17, 90128 Palermo, Italy; giuseppe.bazan@unipa.it
* Correspondence: castroraobarba@eea.csic.es

1. Introduction

In recent years, the concept of resilience has become increasingly prominent across a wide range of disciplines, offering a valuable lens through which to explore how landscapes evolve over the long term. Far from being static entities, landscapes are shaped by the ongoing interactions between human activities and natural processes. These interactions give rise to patterns of change, resistance, and adaptation that unfold over time. As a result, the resilience of historical landscapes has emerged as a key topic of interdisciplinary investigation, drawing together insights from both the natural sciences and the humanities.

This Special Issue, 'Resilience in Historical Landscapes', sets out to explore these dynamics by focusing on the interplay between environmental forces and human responses across different historical contexts. Our premise is that resilience cannot be fully understood without considering both its ecological and cultural dimensions. Natural phenomena such as climate shifts, vegetation changes, and geomorphological processes intersect with human practices—ranging from land use and settlement to social and political organisation—producing landscapes that are constantly in flux.

The contributions gathered in this volume approach these questions from multiple disciplinary perspectives, aiming to illuminate how landscapes have responded to, and been reshaped by, episodes of disruption and transition. We aim to address the following questions: What forms has resilience taken in different historical settings? How have natural systems reacted to human interventions, and how have past societies adapted to environmental challenges? What indicators allow us to recognise a "resilient landscape" in the archaeological and historical record?

By combining methods and frameworks from landscape ecology, environmental and vegetation sciences, and geomorphology with those of archaeology, history, and human geography, the articles in this Special Issue provide new insights into the long-term dynamics of landscape transformation and persistence.

2. Structure of the Studies

The structure of the volume reflects an intentionally open and flexible editorial approach. Rather than adopting a rigid thematic framework, the Special Issue includes a wide variety of case studies and methodological perspectives, each addressing the broader question of historical resilience from distinct disciplinary angles. The twelve articles presented here cover a broad geographical spectrum, ranging from the Netherlands, Italy,

Spain, and France to Egypt, Uzbekistan, and China. Contributors engage with the theme of resilience through landscape archaeology, vegetation science, geology, and remote sensing. The contributions collected in this Special Issue are organised into four thematic sections, each addressing a core dimension of resilience in historical landscapes: (1) Water, aridity and the management of marginal landscapes; (2) Vegetation and land-use legacies: plantscapes, terraces and ecological strategies; (3) Tools and scales of landscape analysis: remote sensing, cartography and environmental archaeology; and (4) Settlement persistence and change: rural and urban adaptation across the time. This structure is designed not as a rigid classification, but rather as a flexible framework that reflects the diversity of the case studies, methods, and temporal scales explored throughout the volume. While each section highlights specific analytical perspectives, all contributions share a common concern with understanding how past landscapes responded to disruption, adapted over time, and continue to inform present and future approaches to land-use and heritage.

The first section 'Water, aridity and the management of marginal landscapes' brings together studies situated in hydrologically sensitive or climatically stressed environments, including coastal wetlands, arid zones, and river deltas. Throughout history, water has shaped the very possibility of life, settlement, and production. In marginal environments—such as wetlands, arid zones, and deltaic plains—hydrological conditions have constituted both opportunity and constraint, often pushing human communities to develop innovative strategies for survival, continuity, and adaptation. The three contributions in this section examine how the long-term management of water and soil underpinned the resilience of past landscapes, exploring the dynamic interface between environmental stress and cultural response. The wetlands of northern Apulia (Italy), the arid plateau of Eastern Marmarica (north-western Egypt), and the geomorphologically complex Schelde Delta (Belgium and the Netherlands) represent radically different ecologies, yet they all share a common feature: they have demanded intentional, situated, and often long-term planning in order to remain habitable and productive. Resilience, in these cases, is not a passive condition but the result of material interventions, infrastructural ingenuity, and cultural knowledge transferred across generations. The contribution on northern Apulia (Contribution 1) examines the historical trajectory of Salpia vetus and its Roman successor, Salapia, within a lagoon environment that required continuous adaptation to hydrological shifts. By reassessing the ancient urban centres of Salpia vetus and Roman Salapia, the authors reconstruct a diachronic narrative of environmental transformation and settlement adaptation, drawing on a broad interdisciplinary toolkit that includes archaeological data, remote sensing, and paleoenvironmental analysis. Rather than presenting decline and relocation as signs of failure, the study foregrounds the capacity of human communities to reorganise spatially and infrastructurally in response to ecological pressures. The case offers a compelling example of how urban resilience in wetland environments was historically built through the integrative management of water systems across lagoon, river, and coastal interfaces.

In contrast, the arid landscapes of Eastern Marmarica (Contribution 2) offer a model of social-ecological resilience based on mobility, decentralised resource use, and embedded knowledge systems. Here, the archaeological and geoarchaeological records reveal a flexible strategy of land use, where water scarcity was mitigated through small-scale interventions and adaptive settlement practices extending from the Bronze Age to the Graeco-Roman period.

Lastly, the Schelde Delta contribution (Contribution 3) situates the question of resilience within a deep-time geological framework. By synthesising the formation of seven major landforms and their influence on cultural and economic developments, the authors demonstrate how the delta's environmental complexity shaped long-term human responses.

https://doi.org/10.3390/land15071155

Projecting into the Anthropocene, the article explores future scenarios for sustainable management, positioning historical resilience as a foundation for biodiversity recovery and geopark designation. Together, these three studies underscore the diversity of ways in which resilience was historically produced and maintained in hydrologically marginal environments. They also remind us that landscape resilience emerges not from environmental stability but from human capacities to engage, manage, and adapt to ecological uncertainty over the longue durée.

The section 'Vegetation and land-use legacies: plantscapes, terraces and ecological strategies' focuses on the historical management of vegetation and cultivated landscapes, exploring how biotic resources—such as native plant species and terraced slopes—were structured, combined, and valued across time. These contributions approach vegetation not simply as a biological layer but as a dynamic materialisation of human–environment relations. They demonstrate how specific plant combinations, spatial arrangements, and agricultural forms reflect historical land-use strategies and simultaneously condition the potential for ecological continuity and future sustainability. From imperial China to the Mediterranean mountains, the articles collected here address both the material structure and the symbolic dimensions of cultivated landscapes.

The first case (Contribution 4), focusing on the Qing Dynasty suburbs of Beijing, reconstructs historical plantscapes through the analysis of classical texts and spatial data. Drawing on a rich combination of geospatial modelling and textual interpretation, the authors identify and classify over 650 plant species mentioned in literary sources, mapping their distribution and evaluating their aesthetic and ecological roles. This method allows for the reconstruction of historically authentic vegetation patterns, providing a valuable tool for contemporary landscape restoration and urban greening. The plantscape here is shown to be a medium through which cultural values, ecological functions, and spatial organisation intersect—demonstrating the potential of historical vegetation studies to inform present-day environmental strategies.

In contrast, the second study (Contribution 5) addresses the long-term evolution of terraced landscapes in the Roya valley located in the Maritime Alps (France) and Imperia Province (Italy). Using GIS analysis and historical maps, the authors track land-use changes across five centuries, identifying demographic and environmental shifts that led to the partial abandonment of terraced agriculture after the mid-nineteenth century. Despite this decline, the terraces have persisted physically in the landscape and recently demonstrated structural resilience in the face of an extreme climatic event. The research illustrates how historical agro-technical infrastructure, even in a degraded or inactive state, continues to play an environmental and symbolic role in sustaining the landscape. It also highlights the possibilities for future revitalisation grounded in heritage-based land management. Together, these two contributions reveal how vegetation and land-use legacies serve as critical components of landscape resilience. Whether through the intentional design of plantscapes or the enduring presence of agricultural infrastructure, the human shaping of plant ecologies leaves traces that transcend individual periods and offers foundations for sustainable futures.

The third section, 'Tools and scales of landscape analysis: remote sensing, cartography and environmental archaeology', presents methodological reflections on how historical landscapes are identified, interpreted, and modelled. Understanding historical landscapes requires not only archaeological evidence and environmental knowledge but also the tools to capture, compare, and represent change across time and space.

The first contribution (Contribution 6) provides a concise yet comprehensive overview of aerial and remote sensing techniques in archaeology, with a particular focus on the use of unmanned aerial systems (UASs) equipped with multispectral sensors. Drawing on case studies from northern Greece, the authors show how crop and soil marks—often invisible to the naked eye—can be enhanced through spectral indices to detect buried archaeological features. The paper discusses both the theoretical underpinnings and the practical applications of this method, underlining its value as a non-destructive tool for large-scale landscape prospection.

The second paper (Contribution 7) examines the role of historical cartography in landscape archaeology through a case study from the province of Guadalajara in inland Spain. By combining archaeological survey data, textual sources, and digitised pre-modern maps within a GIS environment, the study explores diachronic transformations in land use and connectivity. Employing a longue durée perspective, the authors assess the potential and limitations of historical mapping for reconstructing both gradual and disruptive changes in the rural landscape. The article argues for the critical use of cartographic sources in understanding the interplay between spatial representation and environmental history.

The third contribution (Contribution 8) turns to environmental archaeological and spatial data from the hillfort of Tossal de la Vila in eastern Spain to analyse biotic resource management across two widely separated occupation phases: the Late Bronze Age and the early Islamic period. Through radiocarbon dating and archaeobotanical analysis, the authors reconstruct contrasting subsistence strategies that reflect shifting functional roles of the site over time. The study shows how long-term environmental exploitation and resource configuration can be detected through material proxies, enriching our understanding of local resilience and economic adaptation across millennia.

These studies underscore the methodological breadth and interpretative potential of landscape archaeology today. By integrating spatial technologies, historical sources, and archaeobiological evidence, they offer new ways of engaging with past landscapes and highlight the importance of multi-scalar analysis in reconstructing long-term socio-environmental processes.

The final section, 'Settlement persistence and change: rural and urban adaptation acróss the time', explores the long-term dynamics of settlement organisation and reorganisation, from protohistoric hillforts to Islamic rural systems and post-Roman transformations. Drawing on archaeological, spatial, and ecological data, these studies analyse how human groups responded to political, environmental, and economic pressures by adapting, relocating, or reconfiguring their relationship with the landscape. This section also incorporates theoretical contributions—such as resilience thinking and panarchic frameworks—which offer interpretive models for understanding settlement systems as complex adaptive responses to shifting conditions.

The first contribution (Contribution 9) focuses on the Sicani Mountains in central-western Sicily, tracing how rural settlement patterns evolved from the Roman through the Islamic period. Using GIS analysis and ecological variables such as topography, vegetation, and access to water, the authors reconstruct shifting site hierarchies and land-use logic across centuries. Their findings point to a complex interplay of continuity and transformation, in which Islamic-era configurations emerged within, and in response to, long-standing territorial frameworks.

A second study (Contribution 10) compares two rural landscapes in southeastern Iberia, analysing why some areas were occupied continuously over nearly a millennium, while others experienced only brief and episodic use. Drawing on intensive survey data and excavations, the authors reflect on the social and environmental conditions that fostered either resilience or abandonment, and critically assess whether consistent methodologies allow meaningful comparison across cases. The research raises broader questions about how archaeological evidence can reveal the long-term sustainability—or fragility—of historical settlement systems.

https://doi.org/10.3390/land15071155

The third contribution (Contribution 11) turns to the urban site of Ancient Termez in southern Uzbekistan, a key node along Eurasian trade routes. Spanning over 1500 years of occupation, from the Greco-Bactrian to the Islamic period, the study uses a multidisciplinary approach—combining excavation, survey, aerial imagery, radiocarbon dating, and archaeobotanical analysis—to explore how urban forms and resource use adapted over time. The analysis reveals cycles of contraction, reorganisation, and reoccupation that reflect the capacity for urban resilience in an arid riverine environment.

The final paper (Contribution 12) in this section presents a theoretical and empirical investigation of settlement dynamics in the Basentello Valley of southern Italy between the fourth and eighth centuries CE. Engaging explicitly with resilience thinking and complexity theory, the authors propose a panarchic model of landscape change that frames settlement systems as complex adaptive responses to ecological and social stressors. Their application of the adaptive cycle offers a conceptual toolkit for reinterpreting settlement reorganisation not as collapse, but as a systemic transformation embedded in multi-scalar processes.

This section illustrates the value of combining empirical landscape data with interpretative models that foreground adaptability, reconfiguration, and historical contingency. By examining how settlements endured, shifted, or re-emerged in response to changing pressures, these studies provide compelling insights into the resilience of human landscapes in the face of uncertainty and transformation.

To conclude, this Special Issue brings together a wide range of case studies, temporal frameworks, and analytical methods that highlight the material, spatial, and ecological dimensions of resilience. It encourages critical reflection on the long-term consequences of land-use strategies, the legacies of adaptation, and the relevance of historical landscape knowledge for addressing contemporary environmental and heritage challenges. In this sense, the volume advocates for a dynamic and relational understanding of resilience. Landscapes are not passive settings but active historical constructs, marked by negotiation between people and their environments across generations. Through this perspective, the contributions not only enhance our understanding of past societies but also offer meaningful insights into how we might imagine and inhabit resilient landscapes in the present and future.

Funding: Angelo Castrorao Barba's work is supported by the Ramón y Cajal programme (RYC2022-035404-I), funded by MCIN/AEI/10.13039/501100011033—FSE+ (PI Angelo Castrorao Barba), and by the project SUR-TABISA (Ayudas de Excelencia RYC-MaX-CSIC, 20241MAX005) (PI Angelo Castrorao Barba). Pilar Diarte-Blasco's work is supported by the SUBRECC project (PID2022-136708NA-I00), funded by MCIN/AEI/10.13039/501100011033/ and the European Union through the "ERDF A way of making Europe" programme; and by CONURBAN project (CNS2023-144171), funded by MICIU/AEI/10.13039/501100011033 and by the European Union NextGenerationEU/PRTR.

Conflicts of Interest: The authors declare that they have no conflicts of interest.

List of Contributions:

1. Goffredo, R.; Totten, D.M. Building Resilience through Territorial Planning: Water Management Infrastructure and Settlement Design in the Coastal Wetlands of Northern Apulia (*Salpia vetus-Salapia*) from the Hellenistic Period to Late Antiquity. *Land* **2024**, *13*, 1550. https://doi.org/10.3390/land13101550
2. Rieger, A.-K. Water as a Problem and a Solution in Arid Landscapes: Resilient Practices and Adapted Land Use in the Eastern Marmarica (NW-Egypt) between the 2nd Millennium BCE and the 1st Millennium CE. *Land* **2023**, *12*, 1109. https://doi.org/10.3390/land12051109
3. Kluiving, S.; Waterman, R. The Anthropocene in the Aspiring UNESCO Global Geopark Schelde Delta Area: Geological History, Human Resilience and Future Landscape Management. *Land* **2023**, *12*, 990. https://doi.org/10.3390/land12050990

4. Xu, D.; Zhu, J.; Chen, Z.; Hu, N.; Wang, P.; Li, Y. Identifying the Authenticity of Plantscapes through Classics: A Case Study of Beijing Suburbs in the Qing Dynasty. *Land* **2024**, *13*, 1171. https://doi.org/10.3390/land13081171
5. Le Vot, T.; Cohen, M.; Nowak, M.; Passy, P.; Sumera, F. Resilience of Terraced Landscapes to Human and Natural Impacts: A GIS-Based Reconstruction of Land Use Evolution in a Mediterranean Mountain Valley. *Land* **2024**, *13*, 592. https://doi.org/10.3390/land13050592
6. Kaimaris, D. Aerial Remote Sensing Archaeology—A Short Review and Applications. *Land* **2024**, *13*, 997. https://doi.org/10.3390/land13070997
7. Diarte-Blasco, P.; Castro-Priego, M. Is Historical Cartography a Useful Tool for Landscape Analysis? A Perspective from Inland Spain (Zorita de los Canes, Guadalajara) from the Middle Ages to the Present. *Land* **2023**, *12*, 1627. https://doi.org/10.3390/land12081627
8. Pérez-Polo, M.; Negre, J.; Falomir, F.; Pérez-Jordà, G.; de Haro, S.; Aguilella, G. Landscape Exploitation and Biotic Resource Management at the Tossal de la Vila Hillfort through the Long Durée. *Land* **2023**, *12*, 1033. https://doi.org/10.3390/land12051033
9. Castrorao Barba, A.; Aleo Nero, C.; Battaglia, G.; Zambito, L.; Virga, L.; Messina, A.; Cangemi, M.; Bazan, G. Continuity, Resilience, and Change in Rural Settlement Patterns from the Roman to Islamic Period in the Sicani Mountains (Central-Western Sicily). *Land* **2024**, *13*, 400. https://doi.org/10.3390/land13030400
10. Sarabia-Bautista, J. Exploring the Dynamics of Occupation between Resilience and Abandonment in Two Post-Classic Rural Landscapes on the Iberian Peninsula. *Land* **2023**, *12*, 768. https://doi.org/10.3390/land12040768
11. Ariño, E.; Uribe, P.; Angás, J.; Piqué, R.; Portero, R.; Martínez-Ferreras, V.; Gurt, J.M. Adaptive Dynamics of Settlement Models in the Urban Landscape of Termez (Uzbekistan) from c. 300 BCE to c. 1400 CE. *Land* **2023**, *12*, 1550. https://doi.org/10.3390/land12081550
12. Munro, M. Resilience Thinking and Landscape Complexity in the Basentello Valley (BA, MT), c. AD 300–800. *Land* **2023**, *12*, 651. https://doi.org/10.3390/land12030651

land

MDPI

Article

Building Resilience through Territorial Planning: Water Management Infrastructure and Settlement Design in the Coastal Wetlands of Northern Apulia (*Salpia vetus-Salapia*) from the Hellenistic Period to Late Antiquity

Roberto Goffredo [1,*] and Darian Marie Totten [2]

[1] Dipartimento di Studi Umanistici, Università di Foggia, 71122 Foggia, Italy
[2] Department of History and Classical Studies, McGill University, Montréal, QC H3A 2T7, Canada; darian.totten@mcgill.ca
* Correspondence: roberto.goffredo@unifg.it

Abstract: This Gulf of Manfredonia has, for millennia, been the primary water feature of the coastal wetland of Northern Apulia, Italy, although modern reclamation works make writing its long-term history challenging. Our recent paleoenvironmental research has reconstructed the evolution of the southern half of this lagoon since the Neolithic period. Here, we write a history of water management and environmental change in this landscape from the perspective of two key urban sites: pre-Roman *Salpia vetus* and Roman *Salapia.* The Roman architectural historian Vitruvius recounts the abandonment of *Salpia vetus* and the refoundation of *Salapia.* We employ his narrative as a frame for a more complex environmental history, starting from a historiography of this landscape's study and a summary of our interdisciplinary research agenda, which unifies environmental, topographical, remote sensing, and archaeological approaches. Resilience in this changeable wetland environment was only possible through an integrated and intentional management of water among rivers, the lagoon, and the Adriatic Sea. While *Salpia vetus* exploited this wetland and thrived for centuries, the settlement eventually collapsed due to human and environmentally impelled factors. Roman *Salapia* subsequently emerged with a different approach, new infrastructure, and a new location. This blueprint would sustain urban life in this wetland for six centuries and lay the groundwork for the Medieval town.

Keywords: Italy; Northern Apulia; wetland management; landscape archaeology; environmental history; interdisciplinary approaches; pre-Roman and Roman urbanism

1. Introduction

Resilience is a theme richly explored when considering how ancient cities confronted environmental change and challenges. This assertion is further strengthened when those cities were situated in wetland environments, prone to sometimes radical and deleterious changes resulting from natural trends and human activities. In coastal *Apulia,* in southeastern Italy, the histories of pre-Roman *Salpia vetus* and Roman *Salapia* were already known in antiquity, a cautionary tale and colonial success story, respectively, employed by the architectural theorist and urban planner Vitruvius.

In his treatise *De Architectura*, Vitruvius dedicated part of his first book to choosing the most salubrious place for the foundation of a new city (Vitr. 1, 4, 11–12)[1] [1]. He recalls how, unlike *Altinum, Ravenna,* and *Aquilcia*, the *oppidum vetus* of *Salpia* was built near a stagnant marsh (*palus*). This location was the cause of pestilential vapors that led the exasperated inhabitants to entrust Hostilius, probably the patronus of the community, with the task of finding a more suitable site to relocate the settlement. After thoroughly surveying the surrounding area, Hostilius purchased more salubrious land four miles from the old town (*oppidum vetus*), not far from the sea, on the shores of a coastal *lacus*, now called the Salpi Lagoon. There, he refounded *Salapia*. Finally, to preserve the salubriousness of the area and avoid new swamping, Hostilius opened the basin of the coastal lagoon toward the sea, transforming it into a port to serve the new city.

The leading players in this history, more complex than his narrative divulges, were multiple: the *oppidum vetus* of *Salpia*, a site in severe urban decline; the refounded city of *Salapia*, a town in ascent due to favorable, even if heavily managed, circumstances; and the coastal wetland, with which the two *Salapia* related over centuries and whose evolution affected the settlement choices of local communities.

We start with a definition of resilience in a literal sense, based on its Latin root, the ability of a community to "bounce back" when faced with environmental stress and disturbance, both naturally conditioned and human-impelled [2,3]. The ecological systems model of Hollings and Gunderson provides a useful baseline, acknowledging the cyclicality of changes while not presuming that all occur at the same pace or involving the same variables in all contexts [4,5]. However, this theory poses some problems and encourages questions. Past and recent theorizing has often presumed the static location of settlements or systems that enable a clear path to tracking adaptability and transformations. Even the short-scale movement seen in the wetland under investigation here must be interrogated: do they maintain the function, structure, and identity required of the panarchy model [5–7]?

Additionally, treating systems as internally homogenous [8,9] allows little flexibility in interpreting the multi-variate, political, economic, and social character of human interactions, presupposing a predetermined narrative or "rationality" to problem-solving [2,10]. Indeed, what would balance be when seen from a distinctly political, economic, or social perspective? And balance for whom? Can imbalances exist in one arena but not in others and still produce stability? The challenges of charting a model in a wetland context become increasingly apparent because it is predicated on seasonal, annual, and even longer changes. The marsh is not necessarily consistent from day to day, let alone year-to-year or century-to-century.

Our focus on the relationship of human settlements around the Salpi Lagoon to water makes this slipperiness ever more apparent (no pun intended!). This was not just a negotiation with the lagoon itself. Our context being permeable, this marsh was impacted by relationships to rivers, their outflows, and the sea. In his model of climate change archaeology, Van der Noort encourages us to delve into the study of local contexts to chart the strategies of communities to confront this changeability [10,11]. Marshlands can be risky environments: they bring great opportunities to inhabitants and ruinous stresses. However, the weight of each over time can be complex and not easily predicted. Resilience is highly contextual, in other words.

To chart the resilience of these towns over nearly one thousand years, a multidisciplinary team of archaeologists, bioarchaeologists, geomorphologists, and palynologists, coordinated by the Department of Humanities of the University of Foggia and McGill University in Montreal, has been reconstructing the long-term, co-evolutionary processes shared between these anthropic settlements and the environment in the coastal plain of northern *Apulia* (Figure 1). With the resurrection of myriad variables and tracking patterns over centuries, we have written this complex history that was not linear and surely not predetermined. Indeed, wetlands can be highly productive, meaningful contexts for urban development. While Vitruvius and his evocative story of settlement transition in a wetland are the starting point for this exploration, environmental and material data must be studied and interpreted to flesh out more fully the complexities of change through time, the waxing and waning and waxing again of urban life along and around the Salpi Lagoon. In this study, we chart what might be considered a single "adaptive cycle": the growth, floruit, decline, and demise of pre-Roman *Salpia vetus* that led to the growth and floruit of Roman *Salapia*. These stories were spatially proximate, and resilience was communicated and imparted environmentally and materially.

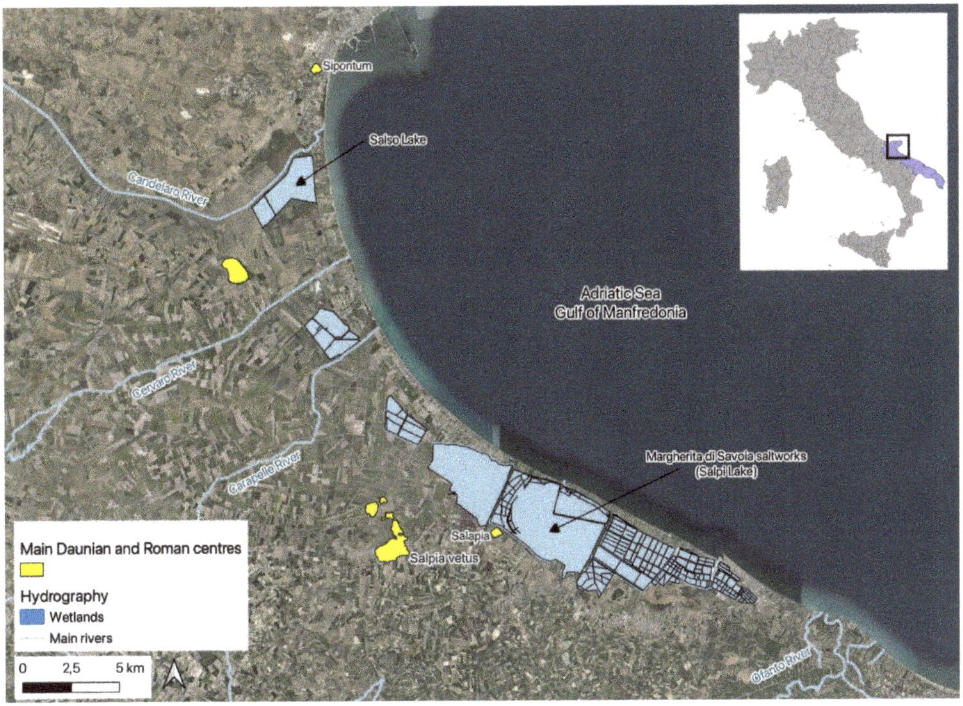

Figure 1. Coastal Daunia (Northern Apulia): Location of the Daunian and Roman settlements mentioned in the text (elab. by R. Goffredo).

First, we present a contextualization of the *Salapia* lagoon landscape, from a geographical, historical, and archaeological point of view. What did we already know about it before our research project began? What are the already available sources for its study? How had previous historical and archaeological research addressed the reconstruction of this habitat? From there, we present the research agenda designed by our group to resurrect the history of this wetland, whose past boundaries and character have been obscured by 19th and 20th century reclamations works. Finally, we evaluate the relationship of *Salpia vetus* and *Salapia* with the surrounding wetlands and how water management strategies were crucially important to both sites.

2. The Study Area and Background

The coastal plain of Northern Apulia (Figure 2), in southeastern Italy, has long been defined by the presence of a wide lagoon. Prior research has registered diachronic evolution prompted by climatic changes, sea level fluctuations, river floods, and adaptation/containment/exploitation strategies influenced by local communities and their everyday activities over centuries [12,13]. At the beginning of the 20th century, the growing demand for cultivable land and the progressive degeneration of human health in this zone pushed the State to reclaim most of the coastal wetlands and lagoons spread along this gulf [14]. Alluvial deposits, purposefully transported, obliterated almost every trace of the ancient morphology of the shoreline, saving only two derivations of the original coastal lagoon: the *Palude Frattarolo*, near Manfredonia, and a part of the *Lago di Salpi*, which was transformed into saltpans, and are still in use. Therefore, the modern image of this lagoon is likely radically different from its ancient form and appearance.

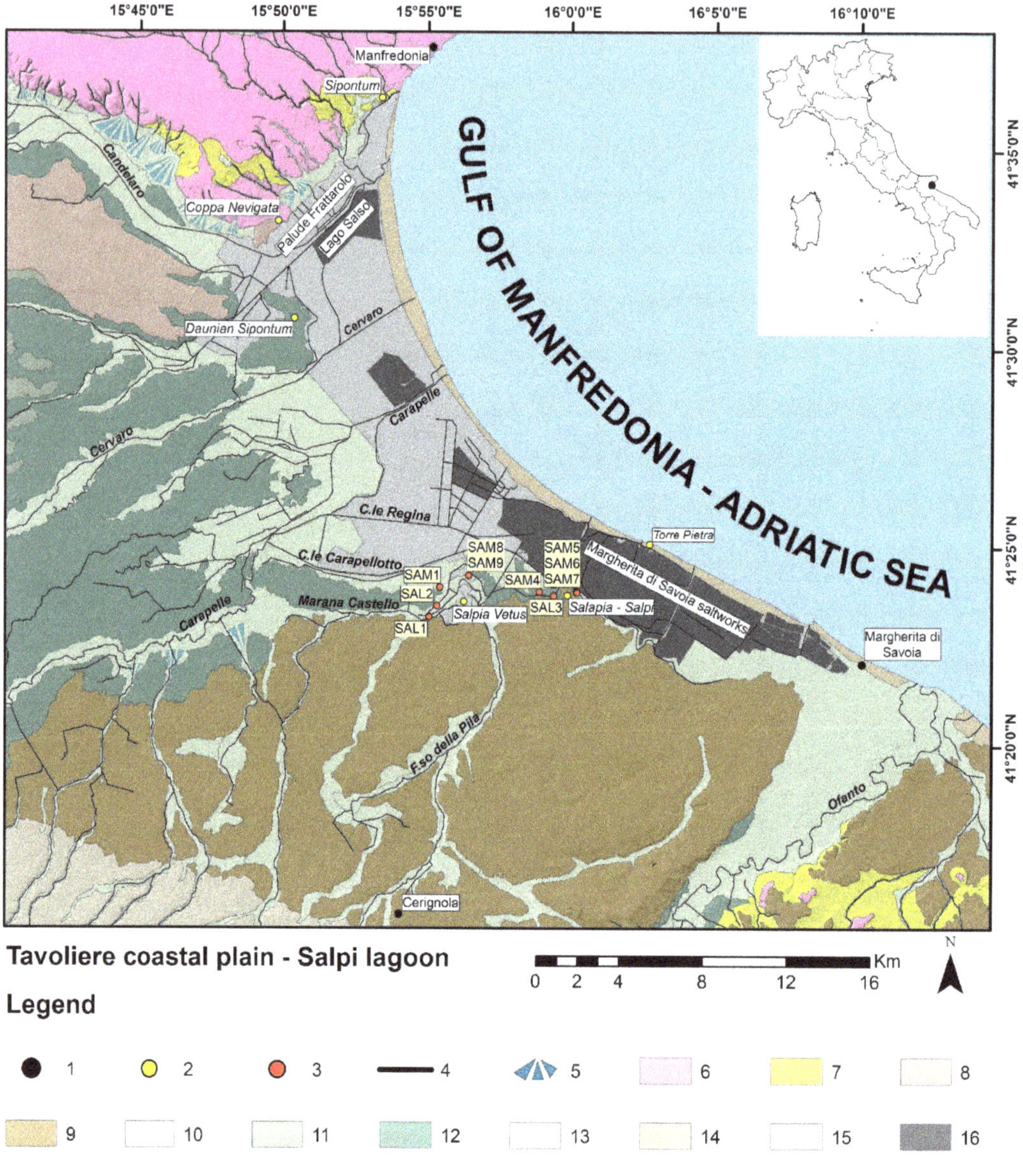

Figure 2. Geological and geomorphological map of the study area (from Susini et al., 2023 [13]). Legend: (1) main towns; (2) archaeological sites; (3) boreholes; (4) water stream (natural/artificial channel); (5) alluvial fan; (6) carbonate Units (Mesozoic); (7) carbonate Units (Pliocene); (8) marine deposit—Middle Pleistocene; (9) alluvial deposit—Middle Pleistocene; (10) infralittoral deposit—Middle Pleistocene; (11) alluvial deposit—Middle/Upper Pleistocene; (12) alluvial deposit—Upper Pleistocene; (13) alluvial plain—Upper Pleistocene/Holocene; (14) beach deposit—Holocene; (15) reclamation deposit—Recent time; (16) artificial reservoir—Recent time (for interpretation of the references to color in this figure legend, the reader is referred to the web version of this article, as for all figures henceforth).

Ancient textual sources have helped to write part of this history or at least provide glimmers of past engagement with this coastal context. The most ancient source on the presence of a lake or lagoon along the Adriatic coast of Northern Apulia is a passage of the *Alexandra* of Lychóphron (Lyc. *Alex.* 1126–1140), dated to the end of the 4th century BCE. In his words, a temple dedicated to Cassandra was built by the local inhabitants (the Dauni) on the shores of a *limne* near *Salpe*. Two centuries later, Strabo, in his description of coastal Daunia (Strabo 6, 3, 9), presented *Salapía* as the *epíneion* of *Arpi*, 140 *stadia* from *Sipontum*, situated near to a navigable river and a *stomalímne megale* where the grain coming from *Arpi* and *Sipontum* was traded. Even if some doubts remain about the meaning of the term *stomalímne* (a lagoon? an estuary? an estuary flowing into a lagoon?), we have more certainty in identifying Strabo's *Salapía* with the settlement identified by Giulio Schmiedt in 1964 in the area of Lupara-Torretta dei Monaci (see *infra*), a site with phases of occupation dating back to the 10th–9th centuries BCE. That was the site directly involved in the Hannibalic War (Liv. 24, 20, 16; 26, 38, 6; 27, 28, 3–13) and also active in the Social War until 89 BCE, the year of its conquest and destruction (App. *BC.* 1, 52).

A few decades later, in 63 BCE, in his oration against the Agrarian Law of Servilius Rullus, Cicero evoked an image of the pestilential nature of the land inhabited by the *Salapini* (Cic. *leg. agr.* 2, 71). This is a testimony of great importance: on the one hand, it allows us to understand the difficulties in which the inhabitants of the lagoon city lived at that time; on the other hand, it foreshadows the event that, shortly afterward, would be described by Vitruvius [1]: the founding of the 'second' *Salapia* and the migration of the local community from the *oppidum vetus* to the new one.

The available written sources of the Roman period do not mention the Salapia lagoon. Even in medieval and late medieval documents, it is barely mentioned. The *Pantanum Salparum* is cited rarely in documents of the 12th and 13th centuries CE [14]. In these texts, there is no reference to topographic features, only a precise listing of landholdings, rural houses, or churches in the territory of the medieval town of *Salpi*, whose buildings occupied part of the area of the pre-existing Roman city (see *infra*). The meaning of *pantanum* evokes an image of a stagnant, unhealthy place unsuited to settlement, but we cannot exclude the possibility that the term might have been used in a more "neutral" way, to indicate portions of closed water. Also, immediately adjacent to the *Pantanum Salparum* was the probable location of one of the most important *domus solatiorum* constructed by Frederick II in Northern *Apulia*. Thus, in the mid-13th century, the area was still an accessible place, productive, and not lacking natural appeal [14] (p. 96).

If, as it seems possible from a few notations, the *Pantanum Salparum* was full of water, rich in fish still in the late 15th–16th century [15,16], we are confronted with a very different picture than that recorded in travel narratives of the 17th–18th century and in some official documents written between the 19th and the beginnings of the 20th century, in which the Salpi Lagoon is described as an unproductive marsh, in an area deserted by people because of its pestilential nature [17] (pp. 23–24).

Writing the history of this wetland has posed considerable challenges, as until recently, information from written sources, historical cartography, and historical aerial photographs were the only sources available to reconstruct this coastline and changes through time. Historical maps visually represent the *Pantanum Salparum*/Salpi Lagoon and the coast from the 15th century to the 1950s. Depending on the scale adopted, they can provide precious details and are very descriptive, offering images of landscapes that have completely disappeared or have been radically transformed (Figure 3). However, the high complexity inherent in coastal environments, subject to profound and sometimes rapid changes over time, does not allow wholesale or even piecemeal retrojection of this information without a critical approach. Instead, these maps visually represent the more recent evolution of the territorial context under examination.

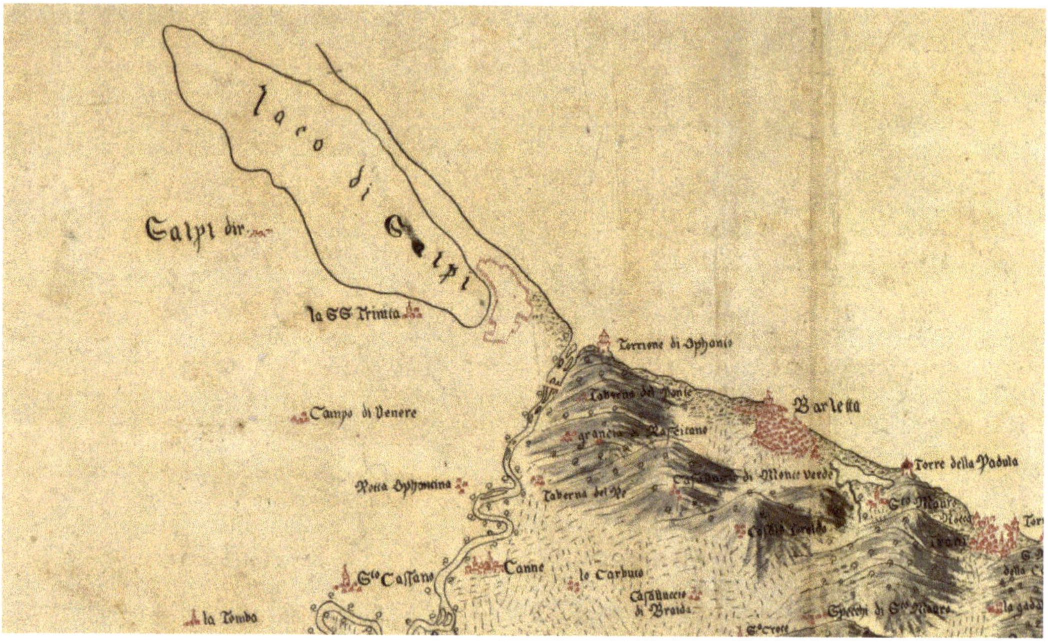

Figure 3. Excerpt from the Aragonese Map of the Province of Bari and part of Basilicata, copied by Ferdinando Galiani (1767). Paris, National Library of France, Maps and Plans, Ge AA 1305/3.

Aerial photographs are predominantly vertical photos taken by the Italian Military Geographic Institute (IGM-Instituto Geografico Militare) in 1954–1955, which captured how the area looked before the development of intensive agriculture, which has played a commanding role in changing local morphology (Figure 4).

Using the IGM 1954–1955 photos, Giulio Schmiedt first hypothesized the setting and limits of the urban centers of *Salpa vetus*, *Salapia*, and *Salpi*, and of the areas allegedly covered by the recently reclaimed lagoon system [18–20]. Thanks to a complex network of cropmarks and damp marks, he suggested a series of hypotheses, identifying correspondences with descriptions in the ancient literary sources.

First, he identified the center of *Salpia vetus* with the widespread settlement located in Torretta dei Monaci-Lupara, 8 km from the coast (Figures 5 and 6). As suggested by the photo interpretation, the settlement rose up on slightly elevated mounds (the so-called *penisole*), and it was reached via multiple roads from the countryside, visible as cropmarks and grassmarks; a semi-circular ditch also protected the city. Excavations conducted in the area between 1967 and 1979, by Fernanda Tinè Bertocchi, proved this theory and dated the occupation of the settlement from the 10th–9th century BCE to at least the end of the 2nd century BCE [21].

Second, Schmiedt identified the plateau where the new *Salapia* was founded, along the present-day saltpans of Margherita di Savoia, approximately 6 km (the 4 Roman miles mentioned by Vitruvius) from the *oppidum vetus* located at Lupara-Torretta dei Monaci. He also recognized the *Monte di Salpi*, at that time 10–12 m above sea level and still surrounded by a defensive moat, as the site of the medieval city of *Salpi*. Moreover, he identified many roads originating from and leading to both the Roman and Medieval towns, connecting them to other main centers of the coast and the hinterland (Figure 7).

Figure 4. IGM aerial photographic coverage of the northern Apulia coast (elab. by R. Goffredo).

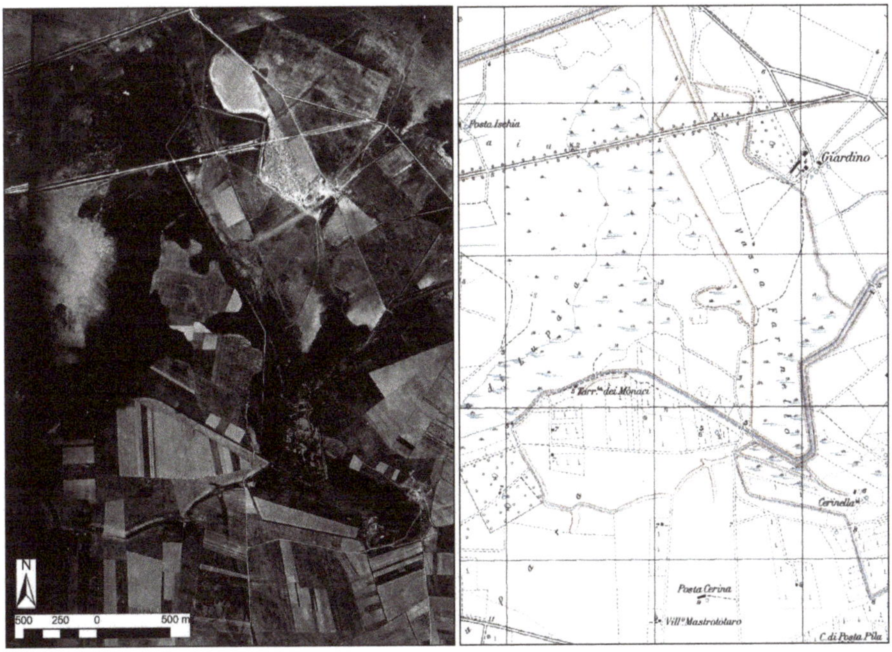

Figure 5. The site of *Salpia vetus* and the wetland area of Marana di Lupara (on the left: 1955 aerial photograph; on the right: topographic map by IGM).

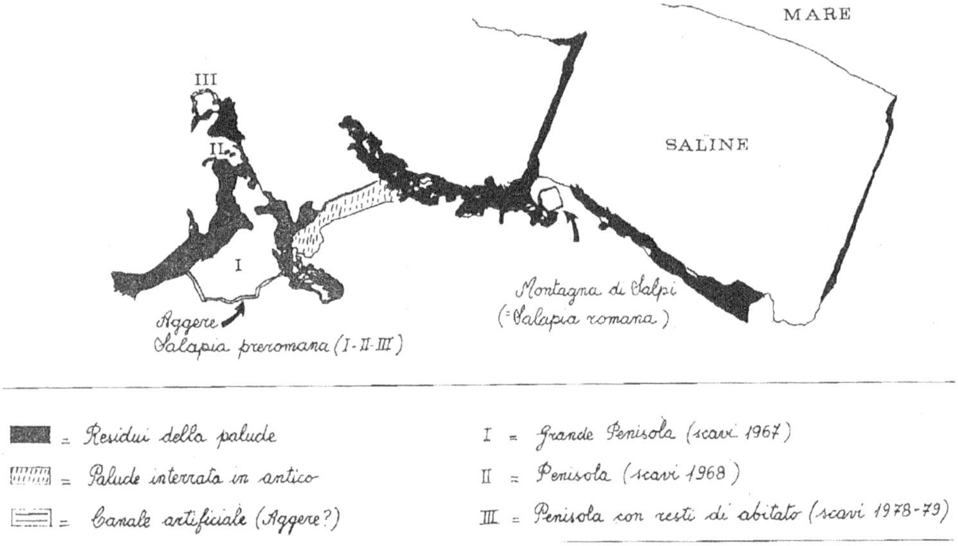

Figure 6. The three *Peninsulas* of *Salpia vetus* and the lagoon in the reconstruction proposed by Fernanda Tiné Bertocchi [21]. Aggere = embarkment; Salapia preromana = pre-roman Salapia; Montagna di Salpi = Salpi Mountain; Salapia romana = Roman Salapia; saline = saltworks; mare = sea; residui della palude = remnants of the swamp; palude interrata in antico = swamp buried in ancient times; canale artificiale = artificial canal; grande penisola (scavi 1967) = large peninsula (excavations 1967); penisola (scavi 1968) = peninsula (excavations 1968); penisola con resti di abitato (scavi 1978-79) = peninsula with ruins (excavations 1978-79).

Third, thanks to dumpmarks, Schmiedt designed the ancient lagoon extent, stretching from the foothills of the Gargano Mountain to the Ofanto River. The Bronze Age settlement of Coppa Nevigata [22], the pre-Roman site of *Sipontum* (Masseria Cupola) [23], the Roman and medieval city of *Sipontum* [24], *Salpia vetus* (Torretta dei Monaci-Lupara), *Salapia*, and *Salpi* (Monte di Salpi) were therefore located on its shores, defining the limits of the water basin.

Schmiedt identified the sites and demonstrated why an in-depth analysis of the area was necessary, as ancient landscapes were affected by the drastic changes to local geomorphology, to reconstruct the history of Daunia. Nevertheless, he stretched the data to confirm the ancient sources. Geoarchaeological analyses would not adequately support his coastal environmental reconstruction. In the early 2000s, the first integrated program of geoarchaeological and paleoenvironmental analyses was conducted by a team of geologists and geomorphologists of the University of Bari, focused on the northern part of the ancient wetlands, near the Neolithic and Bronze Age site of Coppa Nevigata [22]. Such evidence had been lacking for the southern portion of the lagoon until our research campaign.

Therefore, at the start of our research on the *Salpi* Lagoon, we had a few certainties and many questions: how did this coastal wetland evolve from the Neolithic period to the Middle Ages in relation to interior watersheds and the sea? What trends can we reconstruct in the environmental history of this wetland? How has this transitional landscape been organized and managed over the centuries, particularly throughout the long period from the emergence of *Salpia vetus* (9th–8th c. BCE) to the abandonment of medieval *Salpi* (15th c. CE)?

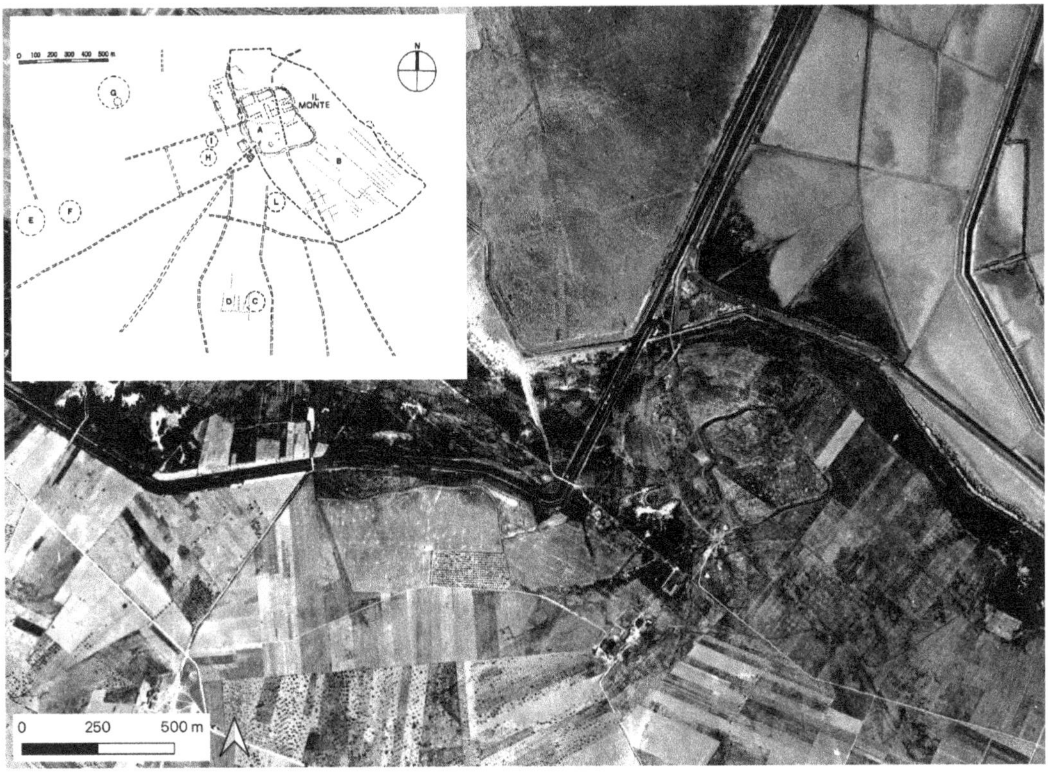

Figure 7. The site of Roman *Salapia* and medieval *Salpi* in a historical aerial photograph by IGM (1954–1955). In the upper left corner, the interpretation of aerial photographic anomalies by G. Schmiedt [19]. A: Monte di Salpi, site of medieval *Salpi*, which Schmiedt believed to be the acropolis of the Roman city; B: urban perimeter during the Roman period, which subsequent research has shown to be much smaller; C-F-G-H-I-L: traces of Neolithic villages; D: traces of Roman farms.

3. Materials and Methods

To address these questions, the research project *Life on the Lagoon: Reconstructing the Biography of Human-Landscape Dynamics on the Salpi Lagoon, Italy* defined and tested a set of best practices for a *wetLandscape archaeology*. The research program, from the macro to microscale, included the following:

- *Dedicated off-site coring*: our project gathered and analyzed sedimentological, palynological, and micropaleontological data coming from a total of 10 drilled boreholes (depth between 9 and 16 m), collected in the area between the sites of *Salpia vetus*, *Salapia*, and *Salpi* (Figure 8). Accelerator Mass Spectrometry ^{14}C dates were obtained from selected charcoals, ostracod shells, and organic sediments (processed by Beta Analytic Radiocarbon Dating—Miami, FL, USA) [13] (in part. pp. 40–41 for details on the calibration methods used). Calibrated ages have been obtained from SAL3, SAM5, SAM6, SAM7, and SAM9 cores, from which datable material was recovered. Apart from the SAM9 core, the relative scarcity of plant remains and organic sediments did not allow the recovery of a consistent chronological sequence for the SAL3, SAM5, SAM6, and SAM7 cores. Despite this, calibrated ages from all dated cores show a good chronological coherence, with an interval spanning from the Late Northgrippian (6206–6104 cal BP = 4256–4154 BCE, SAM9, Figure 8) to the Late Meghalayan (231–164 cal BP = 1719–1786 CE, SAL3, Figure 8). The age-depth model and sedimentation rate of the SAM9 core were built using the Bchron R package (v. 4.7.6), which fits a stochastic linear interpolation in the Compound Poisson–Gamma model. Such multi-proxy paleoenvironmental analyses allowed us to outline better the history of the evolution of this coastal ecosystem over seven millennia.

- *Systematic collection, georeferencing, and GIS management* of all available historical maps and data on known sites, infrastructural systems, field systems, and historical road networks, with particular reference to the wetland area of *Salpia / Salapia / Salpi* (the coastal area between the salt pans of Margherita di Savoia, and the courses of the Ofanto and Carapelle Rivers) (Figure 9);
- *Aerial photo interpretation*: analysis of vertical black-and-white IGM aerial photographs (years 1954–1955 and 1976), AGEA orthophotos from 1997 and 1999; Regione Puglia orthophotos from 2000, 2004, 2005, 2006, 2013, and 2015; oblique aerial photographs taken in the early 2000s, all as part of the systematic aerial reconnaissance campaigns of Northern Puglia conducted by the Department of Humanities of the University of Foggia. Examining this extensive and varied documentation within a GIS framework and considering its quality, resolution, and the years and seasons of capture allowed for the identification of approximately 1700 traces (weedmarks, cropmarks, soilmarks). These traces, primarily located in the hinterlands of *Salpia vetus*, *Salapia*, and *Salpi*, are attributable to settlements, infrastructure, agrarian divisions, and cultivation that no longer relate to the contemporary landscape [25] (Figure 10). With regard to linear features in particular, a series of basic spatial analyses (azimuth calculation, trace extrapolation by orientation classes) were also applied in a GIS environment to verify the possible existence of pre-modern land divisions;
- *Systematic field survey*: the archaeological field survey, conducted between 2017 and 2019, covered an area of approximately 29 km^2. This area, corresponding to the southeastern part of the *ager Salapinus*, is bordered to the north and east by the inner banks of the coastal lagoon (Saline di Margherita di Savoia), to the west by the Fosso della Pila (a tributary of the Carapelle River), and to the south by the route of the *Regio Tratturello Cerignola-Casale della Trinità*, which seems to run along the 'boundary' between the historical jurisdiction of *Salpia-Salapia* and the (much larger) one of *Canusium* (Figure 11);
- *Extensive geophysical surveys* recovered the urban core, including street grid, and city walls of the Roman city of *Salapia* and medieval *Salpi* [26–28], as well as suburban and artisanal quarters, the extra-urban road network, and landscape management infrastructure beyond the walls of these towns;
- *Systematic, open-area archaeological excavations* in 20 × 20 m grid squares, expanded as needed to expose complete architecturally defined spaces, to register contexts of daily life and associated artifacts over long time scales, and demonstrate settlement transformations [29,30];
- *On-site archaeobotanical analyses* of samples recovered from stratigraphically excavated sealed contexts dated by artifact analyses from *Salapia* and *Salpi* (predominantly midden deposits, cooking surfaces, floor surfaces, and burial fills); determined anthracological and carpological remains are compared with the diachronic environmental trends reconstructed through analysis of natural sedimentary sequences obtained from off-site coring [31–33].

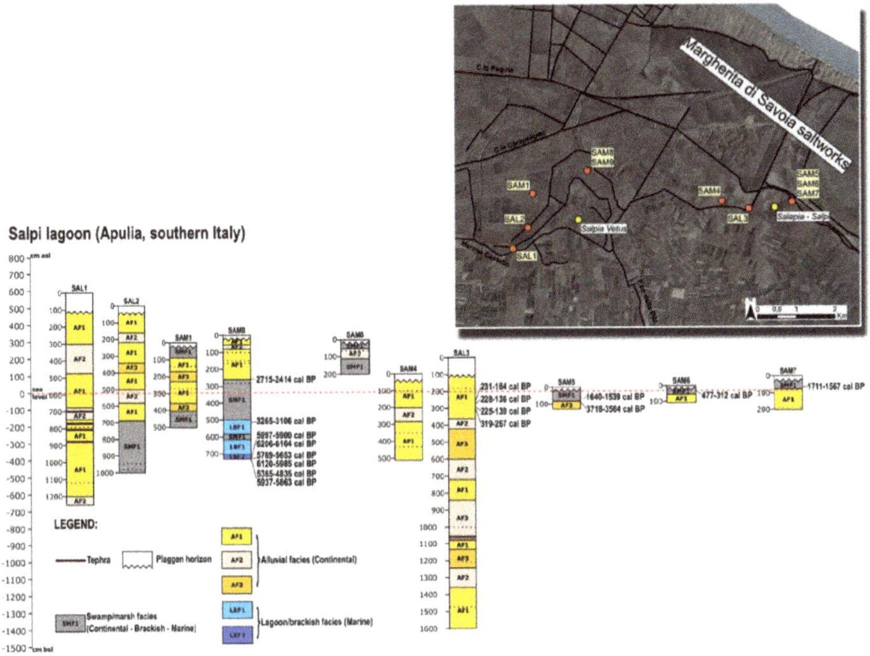

Figure 8. Map of the 10 drilled boreholes, in the area between the sites of *Salpia vetus*, *Salapia*, and *Salpi* (from Susini et al., 2023 [13]).

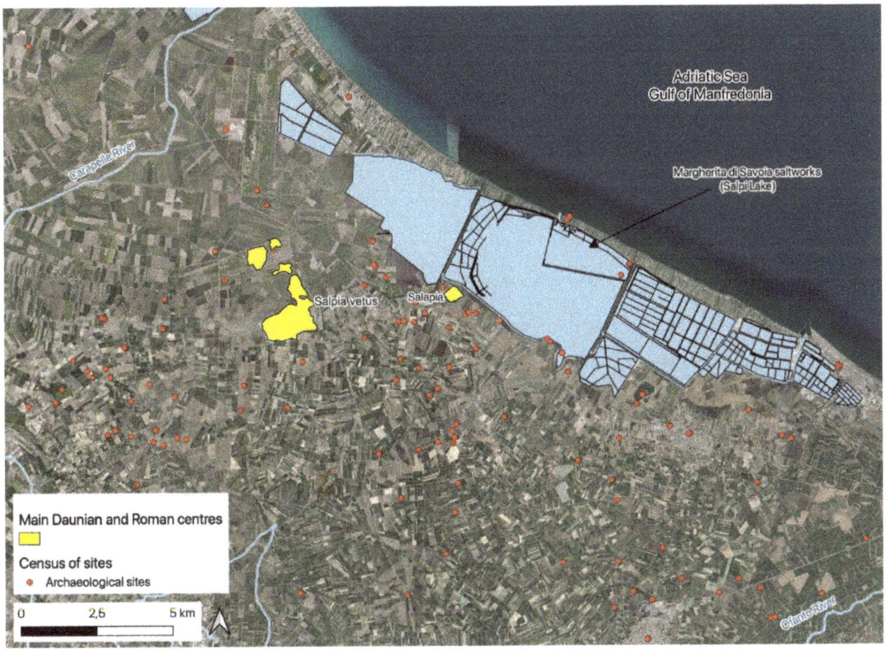

Figure 9. Georeferenced survey in a GIS environment of all the archaeological sites already known before the start of our research in the *Salpia vetus*-*Salapia* area (elab. by R. Goffredo).

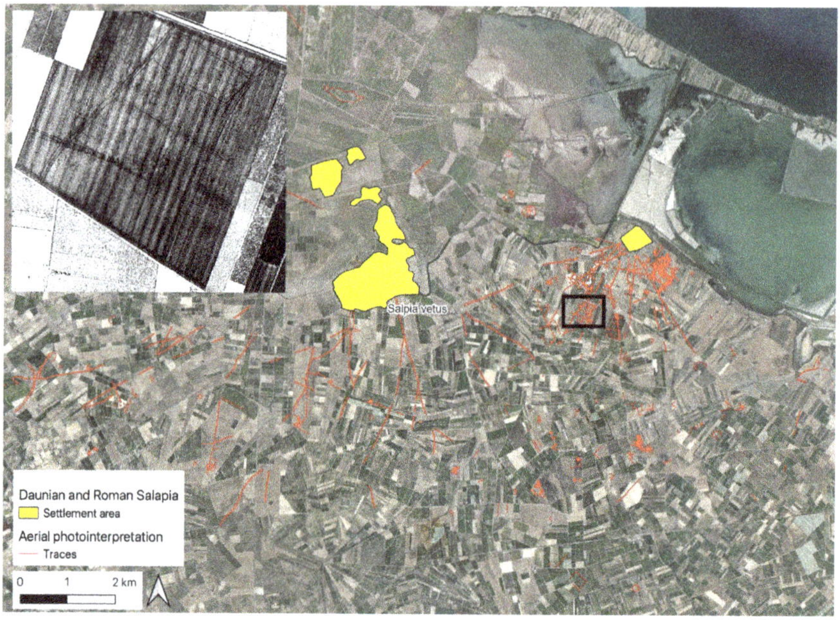

Figure 10. Georeferenced mapping of anomalies detected through aerial photo interpretation. In the upper left: detail of an analyzed sector (within the black square), where the complex stratigraphy of traces related to ancient road networks and agricultural plots is clearly visible (elab. by R. Goffredo).

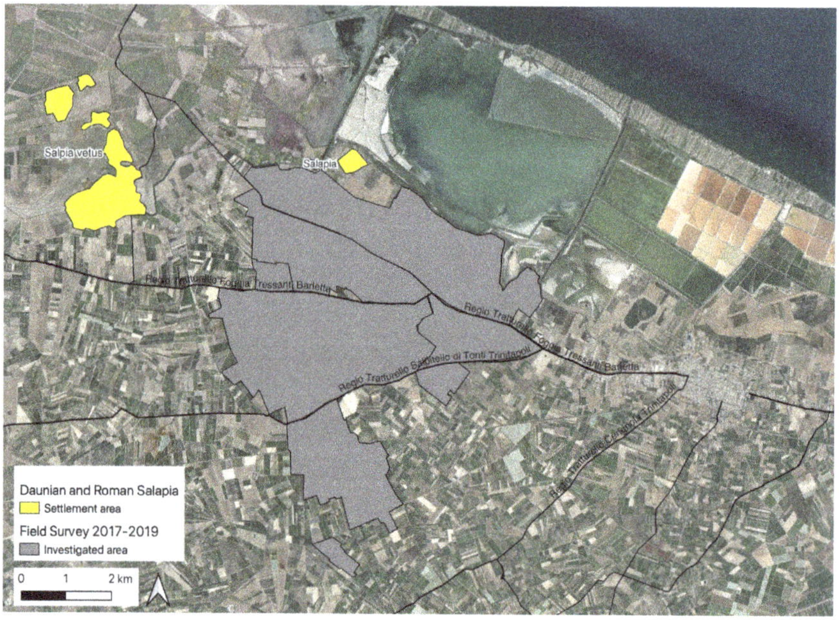

Figure 11. Map of the territory between the settlements of *Salpia vetus-Salapia-Salpi* and the present-day salt pans of Margherita di Savoia, indicating the area subjected to systematic field survey (in gray) and the routes of the modern *tratturi* (drove-roads) crossing the study area.

4. Results

4.1. Salpia vetus

4.1.1. A Wrong Choice of Location?

During the Neolithic period (6500–3000 ca BCE), most of the coastal plain of Northern Apulia was occupied by a large lagoon, partially opened to the sea, extending for about 40 km from the Gargano to the mouth of the Ofanto River (Figure 12).

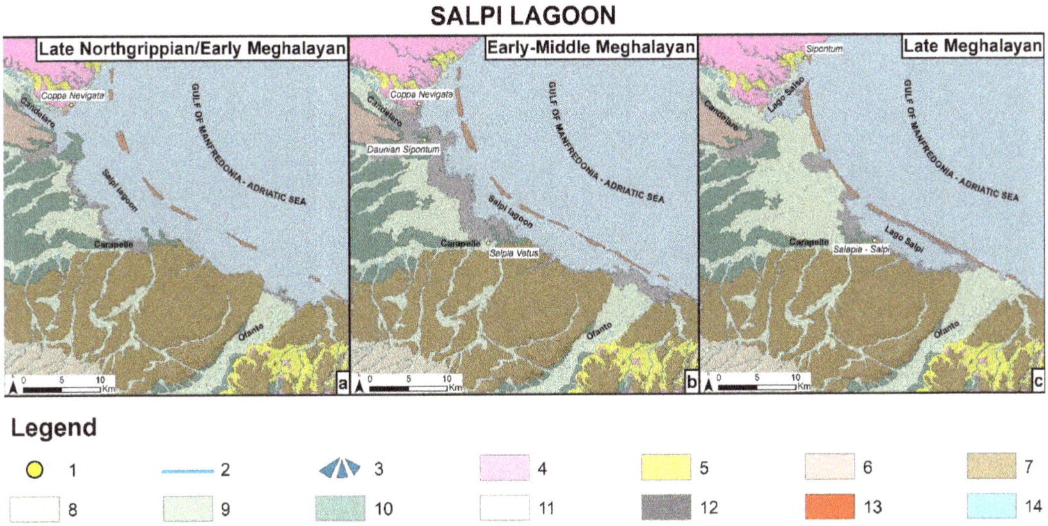

Figure 12. Evolutionary model of the Salpi lagoon during the Holocene (from Susini et al., 2023 [13]). (**a**) Late Northgrippian (4.250–2.250 ca BCE); (**b**) Early Meghalayan (2.250–550 ca BCE); (**c**) Late Meghalayan (550 BCE—Present). Legend: (1) archaeological sites; (2) water streams; (3) alluvial fan; (4) carbonate Units—Mesozoic; (5) carbonate Units—Pliocene; (6) marine deposit—Middle Pleistocene; (7) alluvial deposit—Middle Pleistocene; (8) infralittoral deposit—Middle Pleistocene; (9) alluvial deposit—Middle/Upper Pleistocene; (10) alluvial deposit—Upper Pleistocene; (11) alluvial deposit—Upper Pleistocene/Holocene; (12) swamps deposit—Holocene; (13) Dune belts—Holocene; (14) Lagoon—Holocene.

Over the centuries, this body of water experienced a progressive reduction in size due to the northward progradation of a system of coastal dunes. This development of sand barriers favored the incremental isolation of the lagoon from the sea, shifting from a marine-brackish environment into a more sheltered freshwater environment [13] (pp. 45–49).

During the Bronze Age (2000–1000 ca BCE) and throughout the Iron Age (1000–700 ca BCE) and the Daunian period (600–300 ca BCE), brackish marshland and freshwater swamps began to occupy most of the area of the southern part of the coastal plain [13] (pp. 49–50).

What do we know about the place where, already at the beginning of the 1st millennium, the settlement of *Salpa vetus* took shape?

The area in question is the one that the topographic maps of the Military Geographic Institute (IGM) identified as the wetland area of Marana di Lupara, with traces still visible in the mid-20th century: a depression subject to the stagnation of freshwater, following floods of the Carapelle River. In maps of the late 16th to 19th centuries, the area was drawn as a swamp or lake (the so-called *Lago Giardino*) (Figure 13).

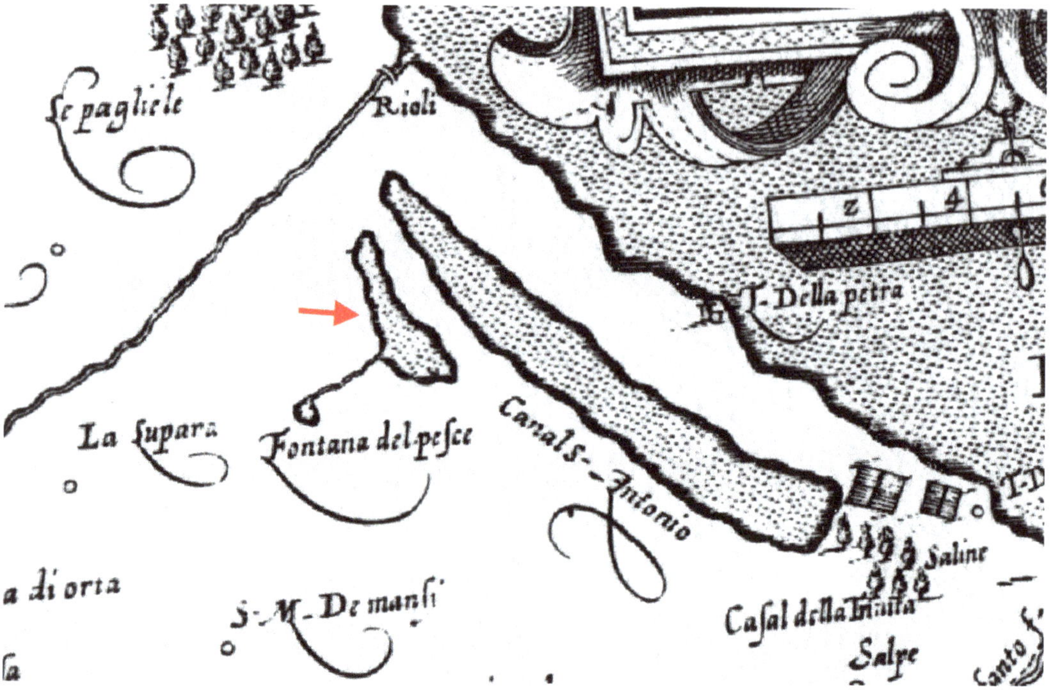

Figure 13. The *Lago Giardino* on the map *Capitanata. Olim Melapia, et Japigia pars* by Giovanni Antonio Magini (16th century).

Based on the shape of the place, Giulio Schmiedt [19] and Catherine Delano Smith [34] hypothesized that the settlement of *Salpia vetus* had developed on strips of land surrounded by the waters of a peripheral branch of the great coastal lagoon (Figure 14). How accurate is this reconstruction?

Evidence from recent cores drilled in the area surrounding this urban center offers data to reflect on the relationship between *Salpia vetus* and its environment.

The analyzed sediments of the SAL1 and SAL2 cores, located immediately west of the settlement (*supra*, Figure 8), do not provide any evidence of a lagoon environment. On the contrary, starting from the Final Neolithic period (3500–3000 ca BCE) and until quite recently, the area was configured as a floodplain, affected by the periodic overflowing of the waters of the Carapelle River, which, in ancient times, flowed immediately north of the settlement of *Salpia*.

The sediments analyzed from the SAM9 core (*supra*, Figure 8) also show how the brackish lagoon environment connected to the sea, documented in the immediate north-east area from the settlement during the Early Neolithic (6500–4500 ca BCE) and the initial phases of the Bronze Age (2000–1500 ca BCE). It was replaced, between the Middle Bronze Age and the mid-5th century BCE, by an alluvial plain characterized by the presence of marshes and swamps influenced, above all, by the freshwater inputs of the Carapelle.

Finally, starting from the beginning of the 4th century BCE and throughout the Roman period, but also afterward, the *oppidum vetus* of *Salpia* and its hinterland definitively lost any relationship with the coastal lagoon. Both to the east and to the west, the SAL2, SAM8, and SAM9 cores clearly attest to the presence of floodplains periodically affected by the formation of shallow freshwater or very shallow saline water pools (Figure 15).

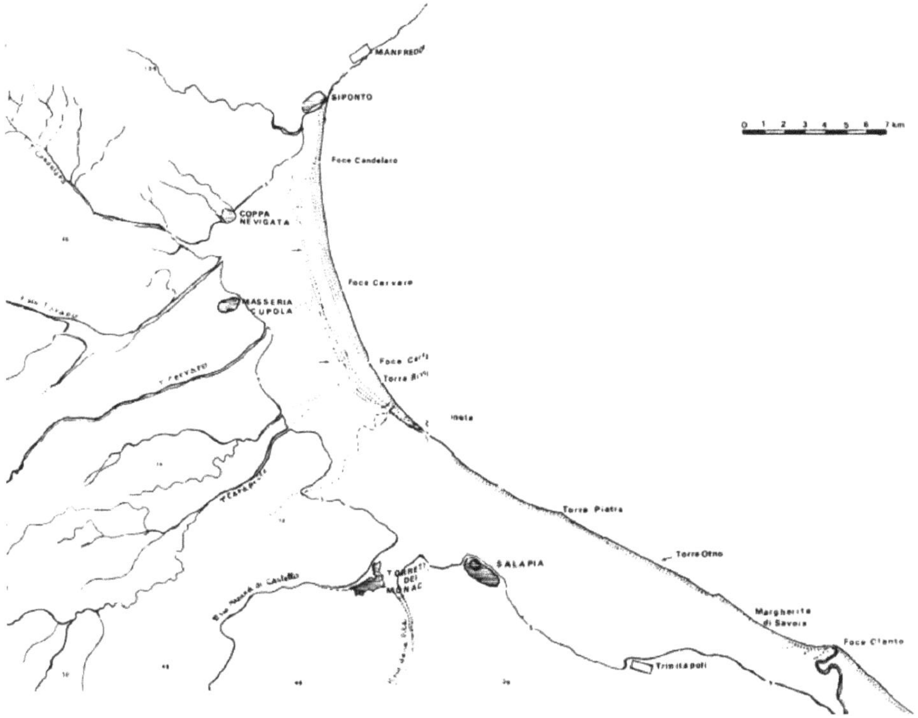

Figure 14. Reconstructive hypothesis of the coastal strip between *Sipontum* and *Salpa-Salapia* according to C. Delano Smith [34].

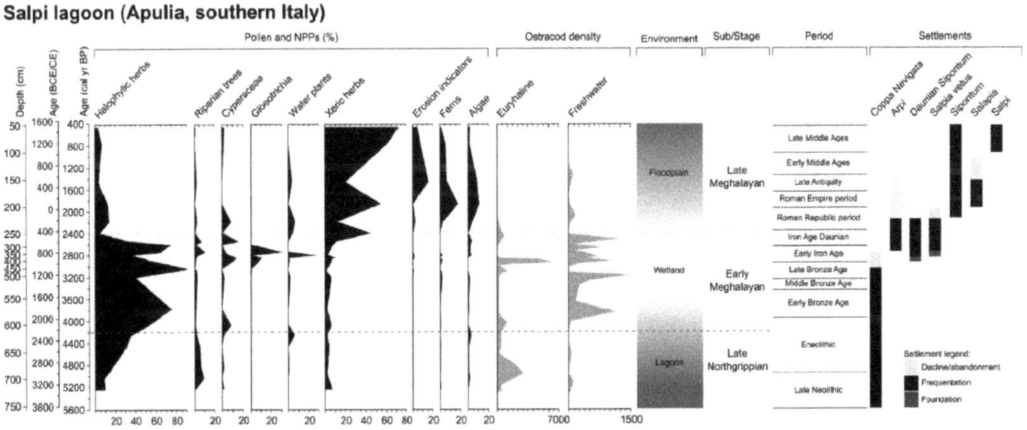

Figure 15. Comparison between pollen and micropaleontological data from the SAM9 core. The ecological groups of pollen, Non-Pollen Palynomorphs (NPPs), and ostracods are aligned with the main paleoenvironmental changes of the Salpi Lagoon and with archaeological data from the Tavoliere plain. Halophytic herbs: Amaranthaceae; Riparian trees: *Alnus*, *Populus*, *Tamarix*; Xeric herbs: Cichorieae; Erosion indicators: *Glomus*, *Pseudoschizaea*; Algae: *Botryococcus*, *Pediastrum*, Zygnemataceae (from Susini et al., 2023 [13]).

An overall assessment of the data presented thus far, therefore, leads us to believe that, starting from the 10th century BCE, the *Salapini* community had begun within a territorial context whose environmental characteristics were unstable, but not necessarily unfavorable to the development of a complex socio-economic system [35,36] (Figure 16).

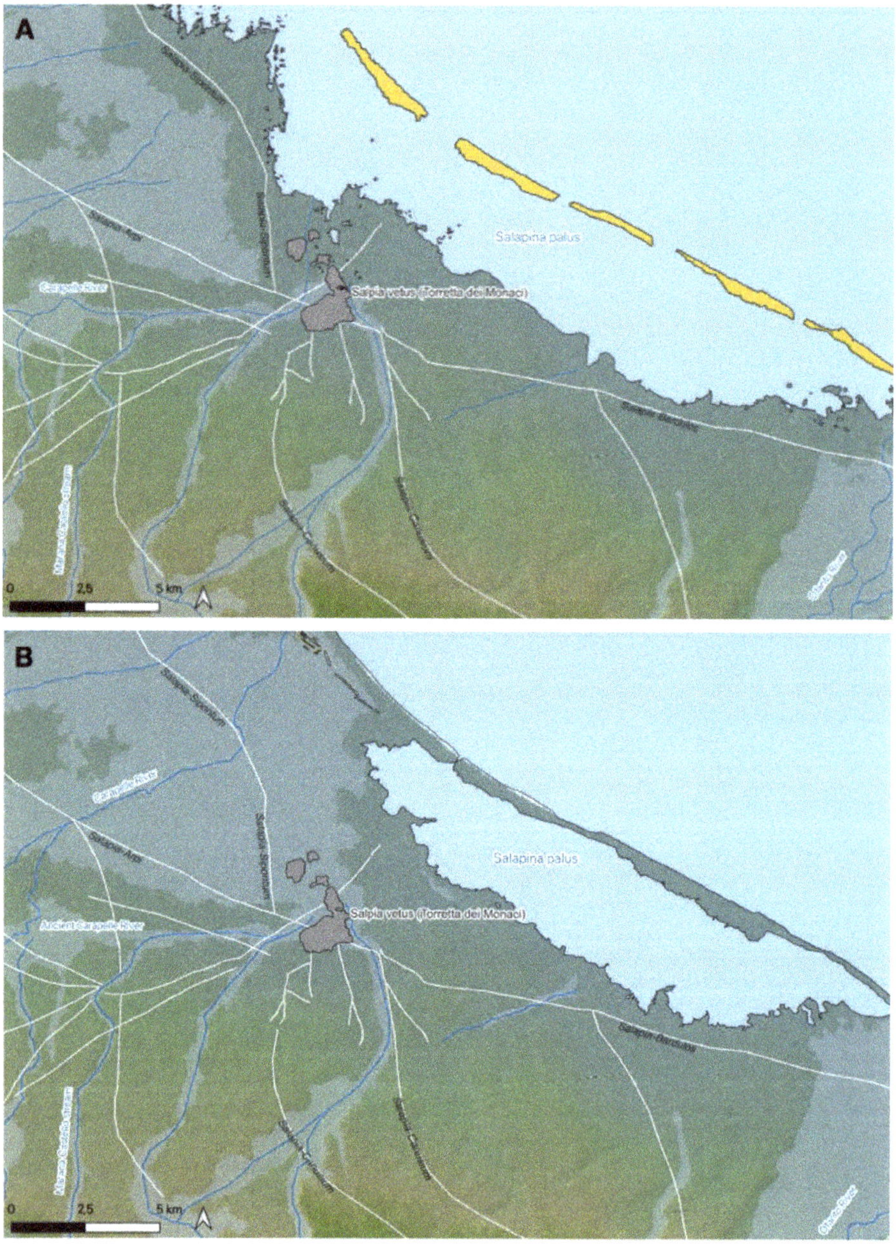

Figure 16. The evolution of the coastal lagoon near *Salpa vetus* from the 7th–6th centuries BCE (**A**) to the 4th–2nd centuries BCE (**B**), indicating the main and secondary road system (elab. by R. Goffredo).

As Aristotle had already observed well before Vitruvius (Arist. *Pol.* 7, 11), a city could arise and develop optimally only if it was well connected to both the mainland and the sea, had an abundant water supply, was sheltered from the winds, and could be easily defended from enemy attacks.

All archaeological data and literary sources available allow us to understand that, over the long term, the *oppidum vetus* of *Salpia* benefited from all these favorable conditions, eventually becoming, especially between the 6th and 4th centuries BCE, with more than 300 hectares of land designated for human occupation, one of the main proto-cities of Daunia, along with *Canusium* and *Arpi*.

First, the proximity of the Carapelle River ensured the supply of fresh water for human and animal life and agriculture. It also allowed convenient access to the hinterland, the coastal lagoon, and the Adriatic Sea. It is worth mentioning again the testimony of the Greek geographer Strabo on the existence, between *Salapia* and the coastal settlement of *Sipous/Sipontum* (Cupola-Beccarini), of a navigable river and a large lagoon estuary, through which goods from *Sipontum*, especially grain, were transported.

Despite the progressive retreat of the lagoon from the Bronze Age onwards, the large coastal basin and its precious resources (in particular salt [37,38], fish [39][2], game and fowl[3] [40]) remained easily accessible, enabling *Salpia vetus* to act, from the late Iron Age onwards, as an *emporium* both for the main settlements of inland Daunia (Strabo presents the settlement as the *epineion* of Arpi) and for Adriatic trade [41].

Furthermore, arboreal pollen from the SAM9 core reveals that, since the final stages of the Eneolithic period, an almost treeless landscape with scattered Mediterranean shrubs and trees stretched all around the settlement [13]. This open and, at times, wet countryside offered extensive pastures for sheep and goat farming and especially for horse breeding [42], while from the late 5th century BCE, bioarchaeological indicators point to the spread of cereal cultivation, horticulture, and olive growing (Figure 17). From the late 4th century BCE, see the increase of anthropogenic pollen indicators in SAM9 core, such as: Cichorieae, Asteroideae, Poaceae, Plantago (areas used for grazing); Apiaceae and Brassicaceae (horticultural crops); walnut, chestnut, olive. The cultivation of cereals is indirectly attested by pollen grains of weedy plants such as *Centaurea* (cereals are known to be very low-pollen producers) [13] (p. 44, Figure 5).

Finally, the presence of waterways, canals, and temporary swamps in the site's immediate vicinity made the *oppidum* more defensible [43]. The analysis of the IGM historical aerial photographs clearly shows that only the southern side of the so-called *Peninsula I* of the settlement was protected by a massive fortification, consisting of a 2 km long ditch associated with an embankment toward the interior of the settlement [21] (p. 63) (Figure 18). This defensive structure, likely constructed between the 6th and 5th centuries BCE, roughly similar to that of the settlement of Arpi[4] [44], faced an inland area farther from the course of the Carapelle River, where the lands were generally drier. Consequently, access to the site was easier. All the other sides of *Peninsula I*, just like all the other parts of the settlement (the so-called *Peninsulas II and III*), were closer to the wet area crossed by the river, and perhaps were not devoid of defenses but certainly were not protected by fortifications like that of the southern side. Wetlands and swamps themselves probably served as a deterrent to potential attackers.

4.1.2. Salubrity and Water Management at *Salpia vetus*: Challenging Roman Exceptionality

In Aristotle's reflection on the optimal conditions for a city, the philosopher underscores the salubrity of air and water. *Salubritas* was an "foundational" value for the survival of human beings. Vitruvius reaffirms a similar attachment to *salubritas* as a central component to the juridical–institutional organization of urban settlements. The choice of the most suitable place for the establishment and construction of a city should be front of mind (*electio loci saluberrimi*: Vitr. 1, 4).

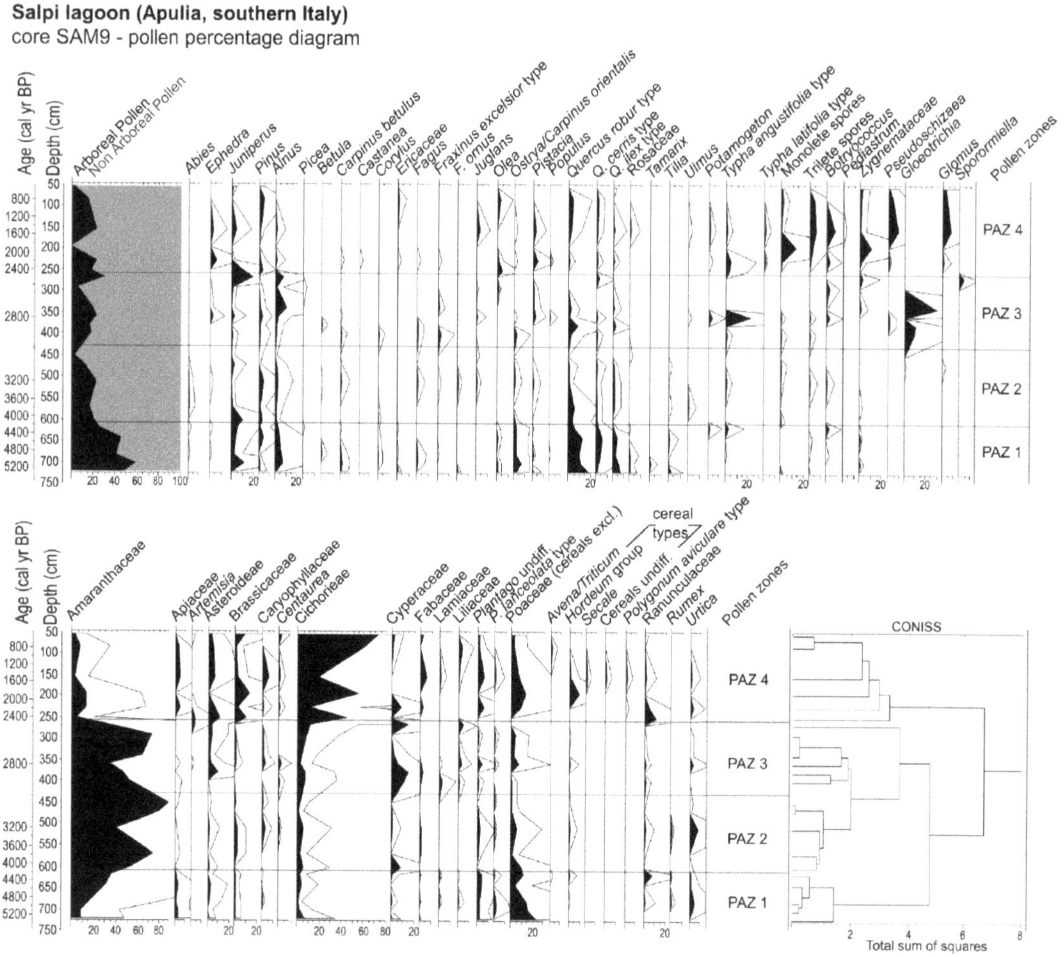

Figure 17. Core SAM9. Percentage diagram of selected pollen and Non-Pollen Palynomorph (NPP) taxa. Curves are exaggerated by a factor of 5× (from Susini et al., 2023 [13]).

In Vitr. 1, 4, 11–12 [1], he explains the nature of *salubritas* through the true opposition between the "positive examples" of *Aquileia*, *Altinum*, and *Ravenna* and the "negative examples" of the Pontine Plain and *Salpia vetus*. It is therefore worth revisiting this opposition and delving into its meaning, with the aim of then returning to examine the archaeological data at our disposal from *Salpia vetus* for a more precise contextualization of the written sources.

Vitruvius' discourse deserves evaluation on the empirical and technical level. There should be an adequate altimetric difference between the areas where lagoons and marshes are found; the internal waters should not be "dead", but regenerated by river inputs, and strong tides should impel a flow seawater inland. Therefore, *salubritas* is primarily the result of geomorphological and environmental conditions "naturally" suitable for life. Therefore, it is nature that makes the difference between the healthy *Gallicae paludes*, where all these favorable circumstances occurred, and the wet plains of the Pontine region or the area surrounding *Salpia vetus*, significantly depressed below sea level and crossed by streams with torrential, seasonal, or sporadic character.

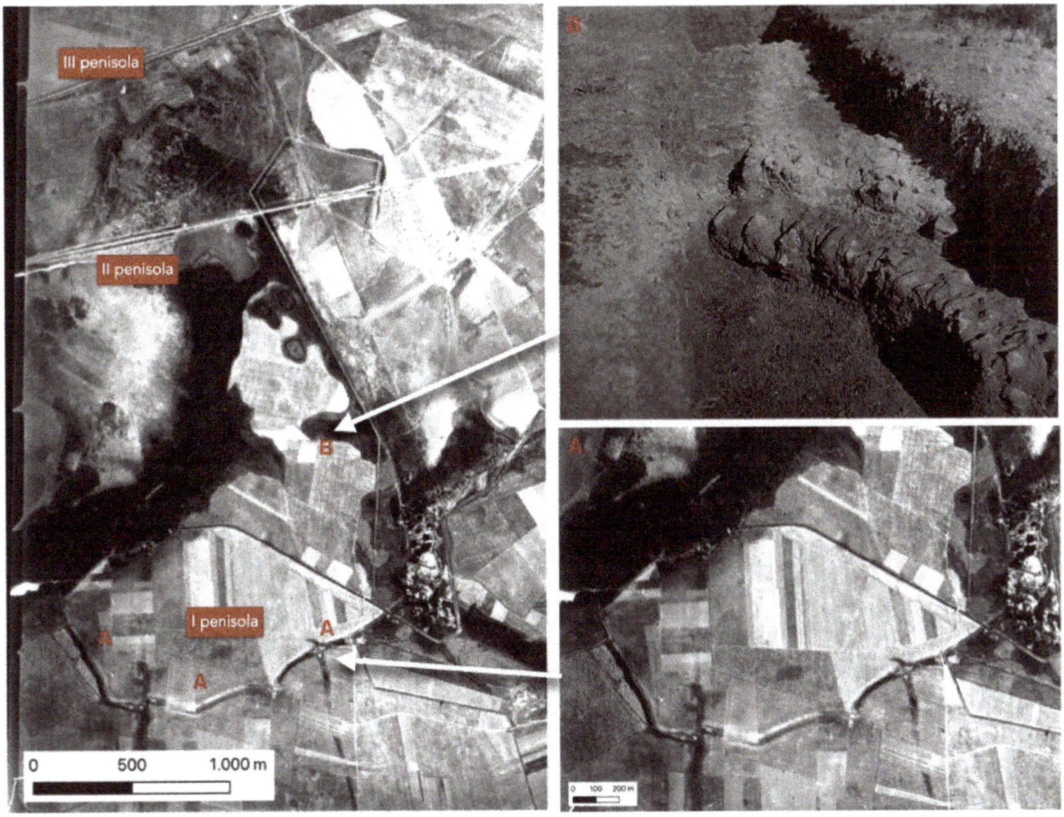

Figure 18. The fortification of the *I Peninsula* and the probable dam (from Fernanda Tiné Bertocchi [21]).

At the same time, in Vitruvius' reflection, the simple empirical–technical observation takes on a much stronger significance: human hands can strengthen what nature already offers or they can realize what nature itself does not allow. Therefore, *salubritas* is achieved through the construction of canals and water drainage/management works, and is the result of culture, knowledge, and skill.

As is well known, Vitruvius wrote his treatise during the age of Augustus, pouring into this work his expertise as well as experiences, testimonies, and knowledge of places presumably gathered during his service following Caesar, *proconsul* of Cisalpine Gaul [45]. By that time, the Romans had conquered the Po Valley regions, accompanied by the construction of new road infrastructures, the establishment of extensive land allotment systems for *viritim* assignments, hydraulic regulation, and the founding of new colonies, often established in coastal or near-coastal contexts with problematic hydrogeological settings [46].

At *Aquileia*, 60 *stadia* from the open sea and connected to the north-Adriatic gulf by the *Natiso cum Turro*, which bordered the eastern side of the city, the foundation of the Latin colony (181 BCE) already had a massive environmental impact, not only for the creation of the new urban layout but also for the extensive reclamation works carried out: canals that flanked the northern and western sides of the city along with centuriation, initiated at the beginning of the 2nd century BCE and completed by the first half of the 1st century BCE. Moreover, to protect the western and southern suburbs of the city, both depressed areas prone to flooding, the contemporaneous excavation of the *Canale Anfora* was undertaken: this water collector started outside the western stretch of the urban walls toward the lagoon, with a straight path of 6 km, thus allowing the drainage of surface waters from the hinterland to the sea and the influx of brackish lagoon waters inland, brought by the push of the tides [47–50].

At *Altinum*, a city-island among the main endo-lagoon *emporia* of the *Veneti*, the establishment of the Roman *urbs* was marked at first, perhaps already by the time of Caesar, by the construction of the walls and, subsequently, by the excavation of the *Canale Sioncello*, which connected the Sile River with the pre-existing ring of watercourses that already enclosed the Altinate settlement by an internal navigable route. Therefore, a complete circuit of canals directly connected to the coastal lagoon was, promoting the ingress and outflow of tides near the settlement [51].

A city of governed waters, at least since the Augustan age, was also *Ravenna*. It directly overlooked a large internal lagoon, closed and crossed by a system of artificial canals and hydraulic works so sophisticated and complex that it was the center one of the most strategic military and commercial ports on the Adriatic coast [52].

It is the contrast with these territories and cities expertly shaped by Roman technical and engineering skill that Vitruvius made the case that the *oppidum vetus* of *Salpia*, founded not by Rome but (in his words) by the Greek hero Diomedes or by the oikist *Elpias* of Rhodes [1], was a paradigmatic negative example. The *rationes* that the Romans had at their disposal for some time, that is, that body of theoretical principles and skills matured precisely through the territorial planning carried out with the conquest of the peninsula, would never have led to an urban foundation in a depressed area, surrounded by stagnant marshes and devoid of waterways or canals connected to the sea.

The opposition is, therefore, ideological and cultural: it is no coincidence that, in the narrative construct, Marcus Hostilius's intervention and the re-foundation (this time entirely Roman) of the second *Salapia* are presented as solutions to remedy the error of the mythical (Greek) founders.

Vitruvius' narrative serves as an example of the "Roman exceptionality thesis" [53], persisting in scholarship until very recent times. However, archaeological evidence from pre-Roman settlements such as *Salpia vetus* provides the counterweight for such models. Elsewhere, recent, fruitful studies that have led to a revision of the paradigm that Roman conquest (and colonization in particular) introduced a structurally new territorial and socio-economic organization in conquered territories. The results of increasingly numerous and in-depth projects have now clearly demonstrated how many Italic communities of the peninsula, well before coming into contact with Rome, were capable of expressing forms of territorial organization and planning, creating scattered rural settlements, and innovating agricultural and natural resource exploitation. It is unsurprising then that Italic peoples already had a culture of managing local hydrogeology and, in particular, communities settled in transitional environments such as coastal wetlands.

At *Salpia vetus*, reference has already been made to the impressive moat with an embankment of the so-called *I Peninsula*, whose construction, beyond its defensive and symbolic value, was also functional for draining lands periodically flooded by the overflow of the nearby Carapelle River and the flow of floodwaters toward the lagoon. Along the eastern edge of the same peninsula, a dam or levee was built, extending about 10 m in length with a foundation of large stones (*supra* Figure 18). However, the most striking evidence of land management interventions comes from the immediate hinterland surrounding the *Salpia* settlement.

The integrated analysis of the extensive dossier of historical and recent aerial photographs considered in our study (see above) allowed us to recognize and map a complex and coherent system of linear traces immediately south of the settlement, within a maximum radius of about 7–8 km from the site (Figures 19 and 20). (1) at least five main axes, with widths ranging between 13 and 20 m, that depart from the moat closing the southern part of the *I Peninsula* and fan out toward the hinterland; (2) a series of "secondary" axes arranged to connect the main axes or as further branches of them.

Figure 19. Detail of a pair of IGM aerial photographs from 1955 to 1954, in which the linear traces moving from *Salpia vetus* toward the settlement's hinterland are clearly visible (elab. by R. Goffredo).

Unfortunately, reliable excavation data cannot yet confirm this interpretation. Giulio Schmiedt [19] and Giovanna Alvisi [54] had previously partially identified and mapped some of these traces, which they argued were evidence of the network of main and secondary roads connecting *Salpia vetus* to the surrounding territory.

This proposal is undoubtedly plausible, and it offers the most immediate and straightforward reading of the network of anomalies. However, two factors must be considered: (1) the traces are crop marks and damp marks have considerable widths and could well refer to tracks partially dug into the ground: thus ditches; (2) the traces closely interact with the intricate network of geomorphological incisions through which spring and runoff waters from the interior terraces of Daunia flowed toward the Carapelle River.

Therefore, we cannot exclude the possibility that this capillary system of dug or semi-dug paths, radiating from *Salpia vetus*, responded not only to the need to allow mobility in and out of the settlement but also to the need to organize the living space surrounding the *oppidum*, through water management and drainage strategies that ensured the area's healthiness and facilitated agricultural practices.

As far as dating, the establishment of the road/ditch system might have been implemented concurrently or shortly after the construction of the large moat delimiting the *I Peninsula*. If so, the territorial infrastructure would have been the result of a socio-economic and cultural growth process that *Salpia vetus* experienced between the 6th and 5th centuries BCE, with the emergence of prominent élites capable of projecting their interests toward land control and managing alliances, reciprocity, and the exchange of goods. It is also worth noting that the sedimentary sequences of the SAM9 core, immediately northeast of the settlement, recorded a significant increase in the local sedimentation rate (c. 0.31 cm/yr) between 1250 and 450 BCE [13] (p. 48). Thus, the implementation of a general plan to protect the site from water and manage the territory's hydrogeological arrangements fit well in a period of enhanced hydrological runoff.

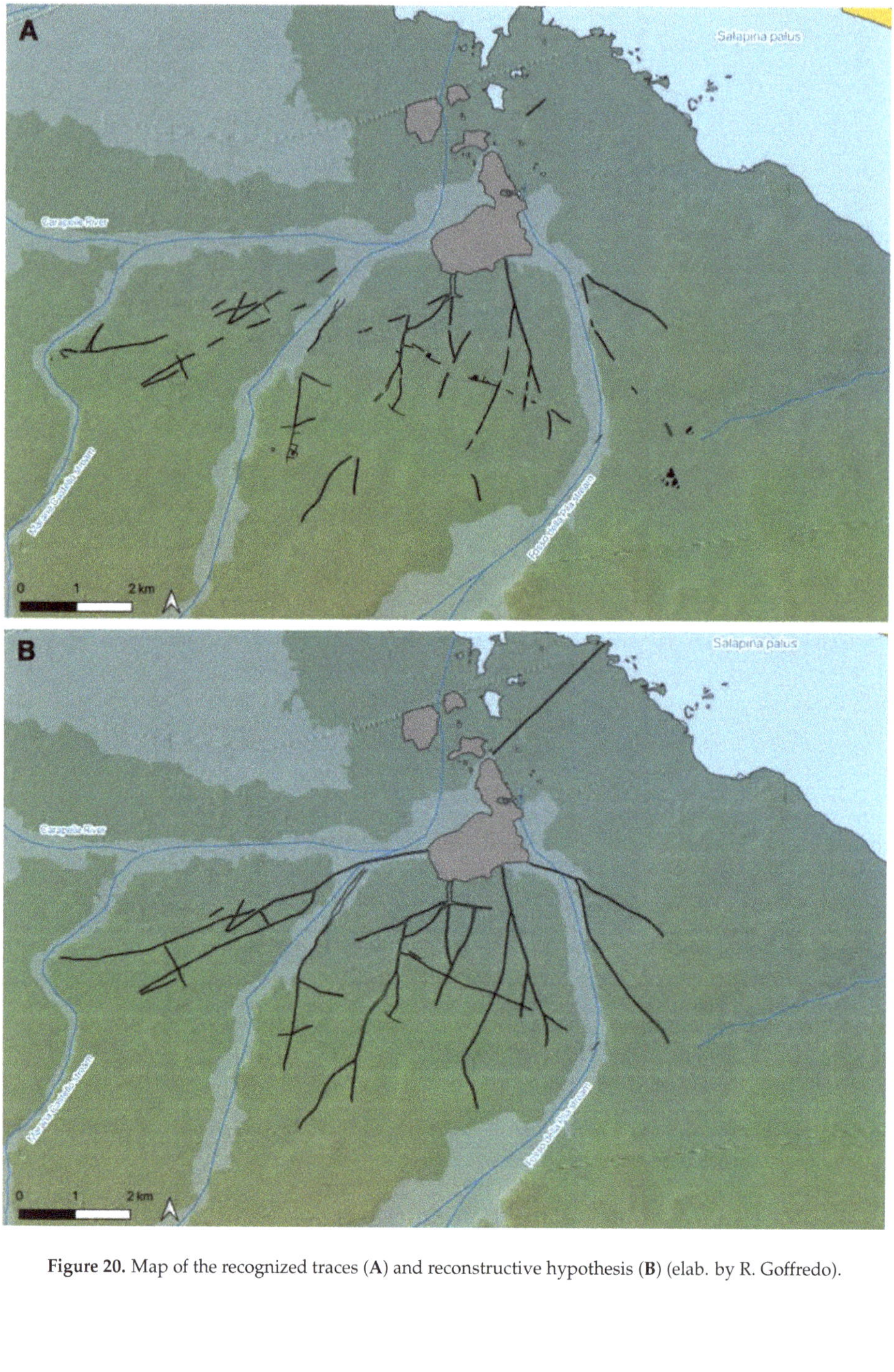

Figure 20. Map of the recognized traces (**A**) and reconstructive hypothesis (**B**) (elab. by R. Goffredo).

4.1.3. The End of Resilience?

What has been outlined so far, based on the scarce archaeological evidence available from *Salpia vetus* and, above all, the most recent paleoenvironmental findings, encourages us to re-examine Vitruvius' account and to consider with more robust critical tools the unsuitability of the place chosen by the *Salapini* for their original settlement.

Wetlands are ecosystems characterized by delicate, fragile balances, strongly conditioned in their expansion or contraction, by the impact of climatic *conjunctures* and by the outcomes of economic choices. Likewise, the same strategies deployed by humans in interacting with these 'fragile' ecosystems have fluctuated over time, moving from phases characterized by greater attention to water control, and more balanced relations between agriculture and the economic exploitation of fallow marshland, to phases increasingly marked by greater aggressiveness in the exploitation of local resources, with diminishing care for the land.

The relocation of the *Salapini* was a drastic response to a situation that had emerged after centuries of balanced coexistence and had gradually become unsustainable over time (*quotannis aegrotando laborantes*, the source reports) [1]; a situation that Vitruvius traces back to the presence of undrained internal marshes incapable of finding outlets to the sea (were these the freshwater marshes created by the Carapelle waters flowing into the alluvial plains surrounding *Salpia vetus*, as seen above?); and the consequent formation of miasmas harmful to the health of the city's inhabitants.

Which events had an impact directly on the local 'water culture'? What caused the failure of the territorialization efforts that had successfully transformed inevitably unstable spaces into permanently habitable places for centuries?

To answer these questions, we need to consider multiple factors, starting with the wetter climatic trends, that seem to have occurred at a peninsular level and, in particular, at a regional level between the 2nd and 1st centuries BCE [55] (pp. 19–22), [56]. This circumstance may have favored the recurrence of seasons with more intense rainfall, such as to increase the flow of the Carapelle River and its tributaries, and facilitate the release, in the plain, at the mouth, or within the lagoon, of sediment transported by river water, especially during floods and overflows.

This period of climatic instability coincided with a political, economic, and social crisis at *Salpia vetus* with the conclusion of the Roman–Carthaginian conflict. In fact, during the Hannibalic War, *Salpia vetus* was among the centers that betrayed its alliance with Rome and, at the end of the conflict, paid for its disloyalty with the punishment, if not the complete replacement, of its ruling class [57] (pp. 153–200). Diminishing grave goods from the necropolis of the *II Peninsula*, which ceased to be frequented precisely at the end of the 3rd century BCE, communicate social and economic impoverishment associated with the settlement.

At the same time, *Salpia* was deprived of vast portions of its territory. The *Liber coloniarum* lists sectors of the *Sipontum* and *Salpia agri* to Roman public lands, which then had policies of *limitation* imposed alongside allocation to settlers, probably in the wake of Gracchan land reform measures (*Lib. Col.* I, 210–211 L). Some faint linear aerial photographic anomalies oriented roughly 64–71° east, identified to the south of the settlement and immediately southwest of the site later occupied by the second *Salapia*, might at first glance suggest the presence of traces of the division intervention mentioned by the *Liber coloniarum*. Extreme caution is necessary, however, due to the following: (1) regular modules with recurring measurements (e.g., modules of 20 *actus*) do not seem to be recognizable; (2) the axial traces are numerically sparse.

To verify this hypothesis, we attempted to trace the persistence of orientations compatible seen in the aerial photographs by analyzing contemporary land organization patterns over an area much larger than the probable ancient territory of *Salpia vetus*. A surface of about 850 km^2 has been scanned, which also includes part of the current territories of Trinitapoli and Cerignola and extends to the valley of the Ofanto River. The outcome of the investigation has highlighted how, in several sectors of the examined area, there are numerous orientations compatible with the range between 60° and 75° east; therefore, the data is very problematic (Figure 21).

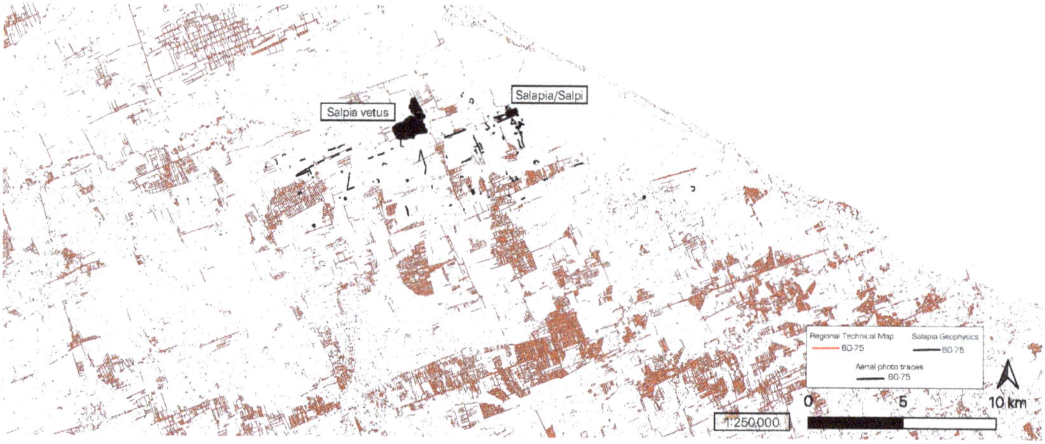

Figure 21. Extrapolation of orientations 60–75° east. Cartographic basis: Regional Technical Map. In black, the traces of axes with a linear development and an orientation of 64–71° east visible in aerial photography (courtesy of Valeria Volpe).

Written evidence of centuriation was exacted, as is also well attested in other areas of Daunia in the Republican period, to extend cultivable areas (also by virtue of the draining function performed by the centurial *limites*) by intensifying specialized vine and olive cultivation, alongside the more established cereal cultivation. From this point of view, all the indicators of anthropization gathered so far, for the periods after the 4th–3rd century BCE, point in the direction of an intensification of anthropic pressure on the area surrounding the city, with a persistent and intense activity of landscape clearing pursued by humans, also through the reiteration of fires, to the detriment of the woodland and to extend the spaces allocated to agriculture and grazing [13]. In fact, beyond agriculture, and not infrequently in conflict with it, the areas closest to the coastal brackish lagoon and salt-growing sites might also have been affected by the phenomenon of transhumant sheep-breeding, whose origins can be traced back to the post-Hannibalic period, with an increase in the 1st century BCE [58] (pp. 212–216).

Lastly, at the beginning of the 1st century BCE, the city took an active part in the Social War and was brought back to order by Cosconius, who nevertheless burnt it down in 89 BCE (App. *BCE*. 1, 52).

After the Roman pacification, therefore, the *oppidum vetus* of *Salpia* no longer found favorable conditions for a revival: the downsizing of the city induced by the prolonged conflict, the probable lack of coordination of the community in the management/maintenance of drainage systems and embankments, beds/swales and mouths of canals, streams, and the lagoon itself, in the face of the wetter climate mentioned above, contributed to the collapse of the entire local socio-ecological system. Aided by deforestation and the spread of agricultural areas and pastures unprotected from erosive processes, the formation of 'dead water' stagnation in the immediate surroundings of the settlement could be traced back to the difficulties encountered in containing or draining the masses of water poured into the depressed *Salpia vetus* plain by rainfall, the flooding of the nearby Carapelle or the overflowing of the coastal lagoon itself; finally, as is well known, humid summers and still, sun-soaked waters, whether fresh or brackish, are the ecological settings of choice for the reproduction of what has been, from antiquity to very recent times, the main vector of malaria in the Mediterranean area, the *Anopheles* mosquito [59] (pp. 43–114, 265–267).

The ironic tone with which, in 63 BCE, Cicero dismissed *Servilius Rullus*'s proposal to send new settlers to *Salapinorum pestilentiae fines* (Cic. *leg. agr.* 2, 71) is indicative of the difficult living conditions in which the city now found itself. The lengthy period of resilience of *Salpia vetus* was now over (the "release phase" of Holling's adaptive cycle). This was the prelude to a necessary reorganization between the community and surrounding environment, in another place but still within the same precious coastal wetland.

4.2. *Roman Salapia (1st c. BCE to Late 5th c. CE): Foundation, Growth, Floruit*

4.2.1. Vitruvius on (Re)founding Salapia: Creating a Healthy Site in a Productive Wetland

The case study of *Salpia vetus*' demise and *Salapia*'s refounding culminates in Chapter 4; through this account, he must summarize the chapter's key themes. He provides many details of location, logistics, and infrastructure, whose confluence made the new site a superior one to the old [1]. Vitruvius places rhetorical emphasis through his repeated, but shifting, use of *locus* as a throughline for the reader: position becomes crucially important to the success or failure of any settlement. *Salapia*'s refoundation and transformation are also tracked by the evolving adjectives and indirect pronouns connected to *locus*. Section Twelve begins with a reference to the Pontine marshes, *in his locis*, where standing waters "putrefy" and emit unhealthy air because of the marshy water's stasis. *Salpia vetus* was put in a place "of that kind" (*in eiusmodi loco*), i.e., pestilential. The *Salapini*'s request to Marcus Hostilius is for an appropriate place (*idoneum locum*); his intervention, by selecting a site adjacent to the sea, was made in a healthy place (*loco salubri*). The transformations did not stop there, as we see the former *oppidum* become a *municipium*, with the imposition of Roman walls, a street grid with *insulae*, and the sale of lots. However, this urban organization was not all that made the *municipium*. In fact, Vitruvius lingers over the idea of the lagoon's remaking, which is thoroughly done (*perfecit*, with *per-* as intensifier), by creating a port for the again-emphasized *municipium*. In *itaque*, which concludes this chapter, we are reminded that the *Salapini*, only four miles distant from the old town (*ab oppido veteri*) finally lived in a healthy place (*salubri loco*). In other words, the discovery of a healthful site did not require a distant move, but it needed to be well-researched for selection within the broader wetland zone.

As Hostilius is said to have purchased the land, we might conjecture that its healthful position had already been uncovered by Roman centuriation in this zone. While pre-Roman peoples surely knew how to apply adept technological strategies to settlement problems, sometimes Roman exceptionality was born of being at the right place and time with the correct tools, resources, investment, and acquired knowledge. In fact, *Altinum* was a Roman refoundation, similarly not far removed from the pre-Roman site, while *Aquileia* and *Ravenna* were new urban foundations within long-exploited wetland contexts.

It is worth investigating each of these details in turn, as we have augmented them through the evidence recovered in our paleoenvironmental and archaeological investigations and through consideration of modern historical sources about the Salpi Lagoon. While *Salapia* was founded on a more healthful site, we might trace the interventions and infrastructure—indeed, the decisions made to modify and continually engage with this wetland environment—that solidified the settlement's resilience and enabled it to thrive. We can contribute to understanding why Roman urbanism ensured long-term stability in such a context and brought this wetland in a new direction.

4.2.2. A Roman Urban Foundation: "Laying Out Lots"

Our geophysical prospection and field survey in 2013 [26,27,60], followed by six years of excavation (2014–2019) have confirmed the position of the Roman town [29,61]. We have resurrected its plan, along with details of its chronological development. The layout of city walls, with its urban grid, is clear from our magnetometry investigations, along with extra-urban structures, infrastructure, and the road network (Figure 22). The town's orientation toward the north and east conforms to Vitruvius's advice to eschew a western orientation, therefore avoiding the strongest of the sun's rays (aspect). Turned north and east, the town would have also, in Vitruvius's estimation, evaded the strongest of seaborne winds (I, 4, 1).

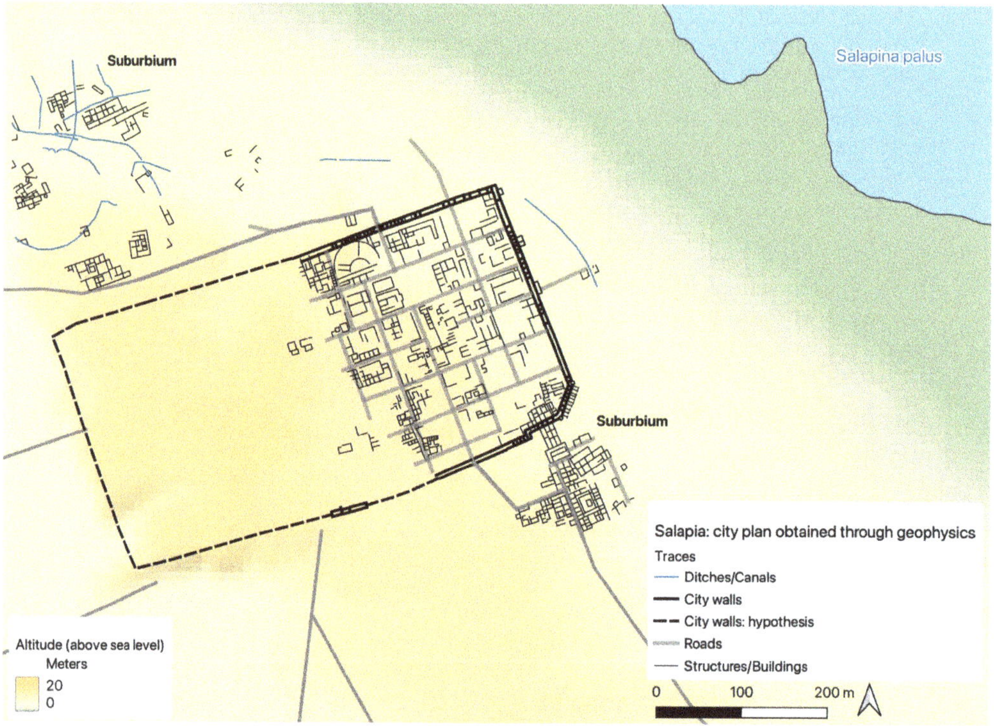

Figure 22. Magnetometry results of the urban plan, city walls, and extra-urban structures. Note the relief gradient of the Roman town and the surrounding area (elab. R. Goffredo).

While Vitruvius advised against foundations in marshlands, he admitted that proper topographic conditions might be found. Such marshy deltas were difficult to avoid in a Mediterranean context, as they provided protected and profitable landings for ports (again *Altinum*, *Aquileia*, and *Ravenna* might serve as examples). Vitruvius emphasized that those marshes that are higher (by position) than the sea's coast (*eaeque paludes excelsiores fuerint quam litus marinum*) should be preferred [1] (I, 4, 11), to facilitate drainage. The refounded *Salapia*, it should be made clear, had an elevation marginally elevated (by perhaps one meter), if not equal to the former *Salpia vetus* in what is an extremely low-lying coastal landscape. *Salapia*'s advantage, instead, was born of the confluence of its elevation, its position, and its geomorphological profile. The SAL3 and SAM4 cores, sunk to the northwest of *Salapia*, have revealed alluvial facies, stable since the Bronze Age, when compared to the marsh-like quality of the SAM8 core, proximate to *Salpia vetus*. These alluvial facies provided a land-based buffer for this town, protecting it from the vicissitudes and pooling of the inland marshlands afflicting *Salpia vetus* [13] (p. 49). Additionally, the new town was positioned on an outward facing lagoon, a large and extensive saltwater body.

Remote sensing, field survey, and excavation work provide an opportunity to reconstruct the material impacts of this new town on the surrounding landscape and to understand how water was managed. Its intramural area covered nearly 20 hectares, a typical size for towns in Apulia, while evidence of a suburban quarter to the south and artisanal activities to the north nearly double the surface area of this settlement [26,27].

As customary, its grid plan was organized around a north-south cardo and an east-west *decumanus*. We might conservatively say that the street grid was laid out around in the second half of the 1st c. BCE to the early 1st c. CE, based on the distribution of limited preliminary architectural features confirmed by residual ceramics and the long-lived position of streets we have uncovered. Our excavations have also revealed that the town's bedrock foundation was undulating and uneven, of limited porosity, and likely slow draining. Evidence for water management in the town is patchy but might speak to broader patterns. An arched drain was found as part of a likely 1st c. BCE phase of construction, running west to east, with a downward slope toward a circular tank, was closed probably before the 2nd c. CE, with the imposition of the tannery [62] (p. 116) (Figure 23). However, it confirms the downward slope we registered in the floor and street surfaces oriented west–east. As Kolowski-Ostrow has noted for *Pompeii* [63] and observed in *Herculaneum*, streets and open drains had low-gradient slopes to ensure water—usually from heavy rainfall or ever-flowing street fountains—traveled down and out toward the sea. Most small towns lacked full and even partial coverage sewer systems. *Salapia* would not then have been out of place with such "low tech" intra-urban water management that brought water down and away toward the lagoon.

Figure 23. Photos of remains of W–E running drain from the early phase of *Salapia*. (**A**) in profile, (**B**) in extent. (G. De Venuto, 2014).

Small snatches of water-themed infrastructure are also evident in the late antique phases of the town. Sometime in the 4th c., a modest bath complex was built to the northeast of a well-appointed Roman *domus*. Water drainage for this structure, confirmed by a channel with piping, flowed from north to south toward a cesspool, in part dug deeply into the bedrock and in part constructed of masonry (Figure 24). We did not excavate to its base for reasons of safety, but its dimensions as documented would have held no fewer than 15,000 L, but likely more [64] (pp. 187–189). It is possible that elsewhere in this town, the drainage of wastewater was managed in this way, with an agglomerated urban "patch" of dampness in this wetland. The *caldarium* was constructed above ground level, on hypocausts (poorly preserved in our case due to their proximity to the surface of the modern plough soil), when more regularly, this radiant heating technology was constructed below floor level. Was this floor surface raised with the aim to protect the hot range from flooding, which would have been ruinous for this technology? Whether from the overflow of the adjacent cesspool, or intermittent flooding in the town, or both, this limited datapoint provides an inkling that mitigating flood risk was front of mind for this community [64], (pp. 189–194), [65], (p. 142, and Figure 8.22) and the architecture central to Roman life was built to adapt to its context.

Figure 24. Cesspool of the 4th c. CE bath complex at *Salapia*. (**A**) full feature from above, (**B**) detail of drainpipe, masonry, and line of bedrock (photo R. Goffredo, 2014).

These details can be contextualized among other indicators of water management from immediately beyond the town walls. Magnetometry investigations have brought forth long and narrow anomalies that might represent drainage canals running along the town's northern edge and skirting its eastern side, traversing the proposed artisanal quarter and then moving between the city walls and the present-day lagoon edge [26] (Figure 25). This easternmost anomaly approximates the width of a street artery, which was perhaps 3–5 m wide. The SAM6–7 cores were either aligned with it, while the SAM5 core was immediately beyond the path of this canal. These cores have brought forth very limited evidence of sedimentation, with the SAM7 core's sediment dated by ^{14}C from the mid-3rd to early 4th c. CE, and the SAM5 core to the mid-4th to early 5th c. CE. SAM6 rather had sediment dated to the 15th c. [13] (Figure 8). Possible and interwoven interpretations present themselves. The first is that this drainage infrastructure processed increased sedimentation in a somewhat similar period (4th c.–5th CE) to when De Santis et al. charted increased sediment outputs in the Carapelle watershed to the north of *Salapia*, a result of infrequent but heavy flooding events [66]. Since *Salapia* was within the same watershed, such sediments might have also been transported downstream toward the town. Second, the town was subject to consistent, low accumulation sedimentation from its 1st c. BCE founding and beyond Late Antiquity, which might have been largely removed in earlier and later phases to manage the flow of water close to the site. The canals that were fundamental to the health of *Altinum*, *Aquileia*, and *Ravenna* have been discussed above, skirting around the outer edge of these towns, even defining them spatially, while negotiating the relationship between the marshes and sea. Each infrastructural system shared a similar chronology dated to the 2nd c. BCE–1st c. CE, in which span *Salapia*'s refounding occurred. The brackish sediments might be attributed to backflows from the lagoon toward the town, due to their proximity to the lagoon's edge. The limited stratigraphy in the cores is intriguing because unlike *Altinum* and *Aquileia*, there appears no evidence so far of urban trash dumping into this canal, even though it might have separated the town from possible port infrastructure, as discussed below [47] (pp. 188–193), [67]. This attests to its likely continuous and crucial role in managing water around the site over centuries, especially the eastern flank, which would have encountered the town's street-borne runoff.

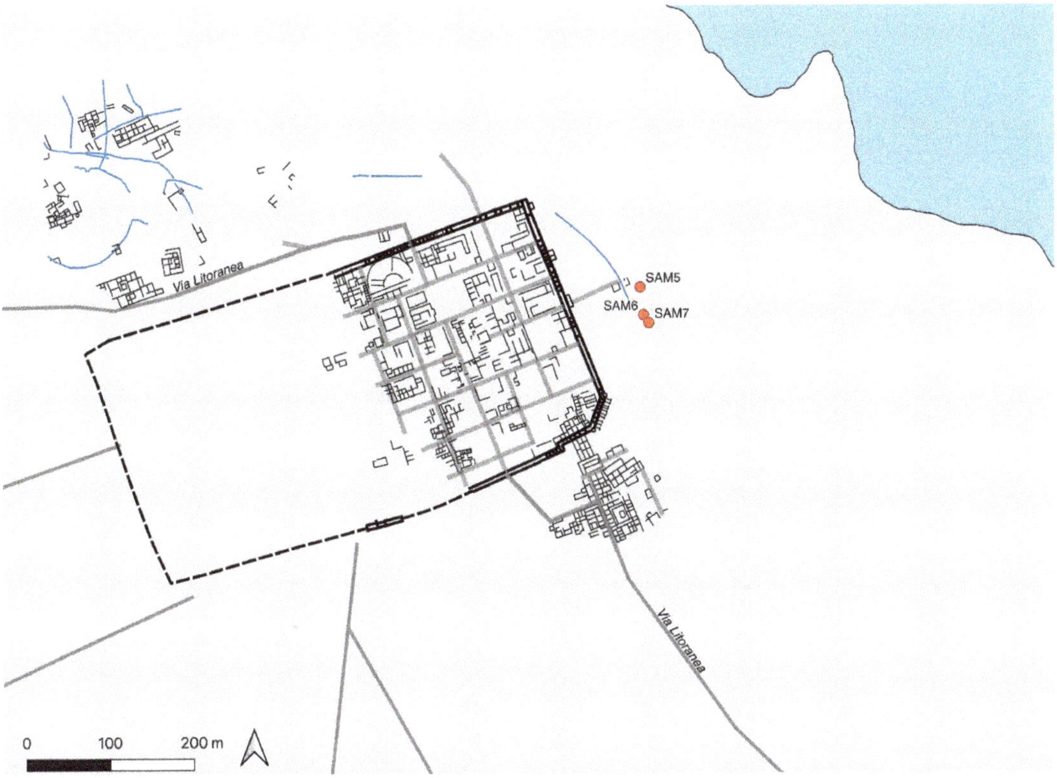

Figure 25. Plan of *Salapia*, with position of the canals revealed by magnetometry investigations, and the positions of the SAM5, 6, and 7 cores (elab. R. Goffredo).

The movement of these canals from the north, from the direction of *Salpia vetus*, might also speak to their role in draining the broader territory, instilling overall health in the hinterland. The eastern flank ran parallel to the lagoon edge as it moved south into a hinterland zone that our surveys have revealed to be somewhat unsettled over the imperial and late antique floruit of the town itself (Figure 9). We lack the evidence to say with certainty that this zone was dedicated to further drainage infrastructure. However, we must consider this datapoint, along with that of urban drainage, as part of an integrated system of waters, including lagoon to sea, as part of Vitruvius' narrative.

4.2.3. "Making a Harbor of the Lake"

Although Vitruvius recounts that the town was situated adjacent to the sea (*secundum mare*), this was not the reality on the ground. A few phrases later, he reported that the lagoon was opened to the sea, making clear that the town was separated from the outer coastline that formed the lagoon's eastern edge. This closed lagoon had only been formed a few centuries before *Salapia*'s founding, in the 4th c. BCE at the latest, but after *Salpia vetus*'s establishment and growth. The inner lagoon edge was already likely formed, but changeable, due to the sedimentation carried by the Carapelle River watershed at the north, and the Ofanto River watershed at the south. Vitruvius tells us of a cut made to provide this lagoon access to the sea, confirming the solidity and contiguity of the outer dunes to the east. With future coring, we hope to identify whether other naturally formed openings were also present.

Again, the SAM5, 6, and 7 cores uniformly reveal a brackish environment from about the 3rd c. CE to the 16th c. CE, confirming that the lagoon was salty, likely due to its connection to the Adriatic Sea. Although we lack specific data dated to *Salapia*'s refounding, the brackish environment, recorded for centuries later, might be evidence of a continued careful curation of this relationship, not only as a space for trade, but also for the general health of the lagoon, and therefore the sustainability of the town itself. Similar modifications have been registered at Cosa's Lagoon and at the *Acherusia palus* in Campania dated to the Roman period [68] (p. 82).

Naturally protected approaches like the Salpi Lagoon were not uncommon in the antique Mediterranean. On the Tyrrhenian Sea, Pisa was such a port, as was Lixus in Morocco [69]. Contemporary examples include the Nador Lagoon, again in Morocco, and the Mar Menor, in southeastern Spain [70], among myriad others in North Africa [71,72]. From these examples, we can list the demands placed on such water features, which include the management of port traffic, population stress from their urban centers, and demands to accommodate river outflow, in addition to natural progradation that can influence the depth and spatial extent of a lagoon. Apparent in all cases is the delicate balance needed over time so that the lagoon's ecosystem, its form, and interrelationship to land, seas, and rivers, might be maintained.

The Salpi Lagoon, while not subject to progradation in the Holocene period [73], was rather influenced by its position between the Carapelle and Ofanto Rivers and their watersheds. These patterns are sustained by the agglomeration of alluvial deposits, as evidenced by all cores studied by our research team, over a very long timespan up to and including the ancient period that is our focus. From the Roman period onward, the lagoon was likely impacted by agricultural activities not only in the immediate hinterland of the site, but also by those upstream of these two river systems, should we take recent survey work as indicative of intensive farming activities well into Late Antiquity [74,75]. Tending to the health of the lagoon would have therefore been tantamount for the successful evolution of the urban settlement, as well as to bolster its economic potential.

Early modern sources help provide a sense of the complexity attendant for this mediating body of water, between freshwater rivers and the Adriatic Sea. Saverio Russo details the conflicts created by divergent resource interests, which when coupled with natural processes, destabilized the lagoon environment [76,77]. On the one hand, there was the push to create industrial-scale salt production, requiring the necessary salty inflows from the Adriatic [78][5]. These interests were at odds with those invested in large-scale fishing, which was dependent on freshwater inputs and an enclosed basin, so that the fish could breed and grow within this closed system of limited movement. Freshwater inputs, especially in large volumes, can destabilize salt production. Interchange with the Adriatic was therefore limited, while riverine outputs were halted. In very hot summers, this lagoon was prone to rapid evaporation and became quite shallow, leading to the formation of small pools which crystalized salt, enough to collect contraband salt produced and sold outside of the rigorously regulated monopoly. These pools also increased the potential for malarial breeding in the summers [79][6]. Management of the lagoon's relationship to the sea and to freshwater inputs was influenced by myriad human interests, prompted debates, and conflict, and could only be resolved by institutional decision-making and an enormous investment of capital in infrastructural changes related to both economic development and land reclamation for sanitary reasons, some of which were realized and others not.

Vitruvius's small notation, therefore, underscores, at least in his telling, the unilateral decision of Hostilius, perhaps representative of the Roman interests in this region. A harbor would have actualized trade and connections with a broader Mediterranean network, in a Roman mode. The remains of structures at Torre Pietra, still visible on the shore in the form of rough masonry (Figure 26), stones bound with a mortar matrix in loosely horizontal courses, today imperfectly perpendicular to the coastline. The work of Cocchi et al., employing remote sensing technology, has also recovered the evidence of small blocks (50–60 cm in size) similarly oriented off the coast at a depth of about 10 m, which have been interpreted as piers related to this harbor infrastructure, perhaps part of a dragged-ship channel [80] (p. 99). Lopez also interprets a sizeable concrete mass, found a few kilometers from the shore, as a defensive post [81]. On a typical summer day, the wind can be quite strong standing on the remains of these walls, an indication that *Salapia*'s lagoon, while protected, did not have the easiest approach for sailors. The remains of shipwrecks immediately beyond these dunes might attest both to the level of harbor traffic and the necessity of finding an adequate entrance to the harbor itself [82]. However, this managed opening, in which there was clearly great infrastructural investment [83], both of money and labor, would have also served to open the lagoon more regularly, and likely continually, to the sea.

Figure 26. Drone photographs of the remains of masonry walls at Torre Pietra (R. Ragno).

We are lacking the material evidence for the curation of the dune lands at the interior of the lagoon, beyond the harbor entrance at Torre Pietra. Modern reclamation works and the construction of both the Bourbon saltworks in the 18th c. and the contemporary salt installations have likely destroyed the material traces of ancient modifications to the interior lagoon and its shores, including compromising the remains of any interior harbor infrastructure. However, material evidence and the shape of the lagoon argue for certain orientations. The town of *Salapia* and this entrance appear to face each other across the lagoon, signaling a nearly straight approach to the opposite shore (Figure 27). We have hypothesized the lagoon basin as a single open pocket, oriented north-south, without any further inlets or dunes dividing it. Further to the south, two additional landings, one at the villa of San Vito and the other at the village of Mattoni, were likely positioned sequentially along the shore, conditioning the new harbor perhaps to have grown in the more immediate shores around *Salapia* and further north [82] (pp. 174–182)[7] [84]. In viewing the plan of *Salapia* as resurrected by magnetometry, the town itself is set somewhat inland from the current coastline. In the strip of land between these walls and the inner shore, our surveys have revealed a heavy concentration of ceramic finds, including numerous amphorae, that might point to spaces of temporary storage, or offloading, for this merchandise arriving into the harbor [85] (Figure 28). Ravenna's relationship to Classe might provide a point of contrast, where a lagoon composed of various inlets saw its commercial and naval harbor separated by topography from the shores immediately adjacent to the town [86]. The Claudian infrastructure at *Portus* was a harbor complex designed to receive ships, offload, and store cargoes that assumed a form similar to the pocket harbor of *Salapia*, where offloading infrastructure was positioned opposite the entrance points along the shore [87] (pp. 71–103). The final piece is the position of the river systems, which would have discharged sediment most directly into the lagoon's northern (Carapelle) and southern (Ofanto) ends. At *Portus*, too, considerations were taken to locate the harbor at the median space between these offloads, reducing the impacts of sedimentation [88]. Therefore, *Salapia*'s position might have been fortuitous in easing harbor management in this urbanized coastal space.

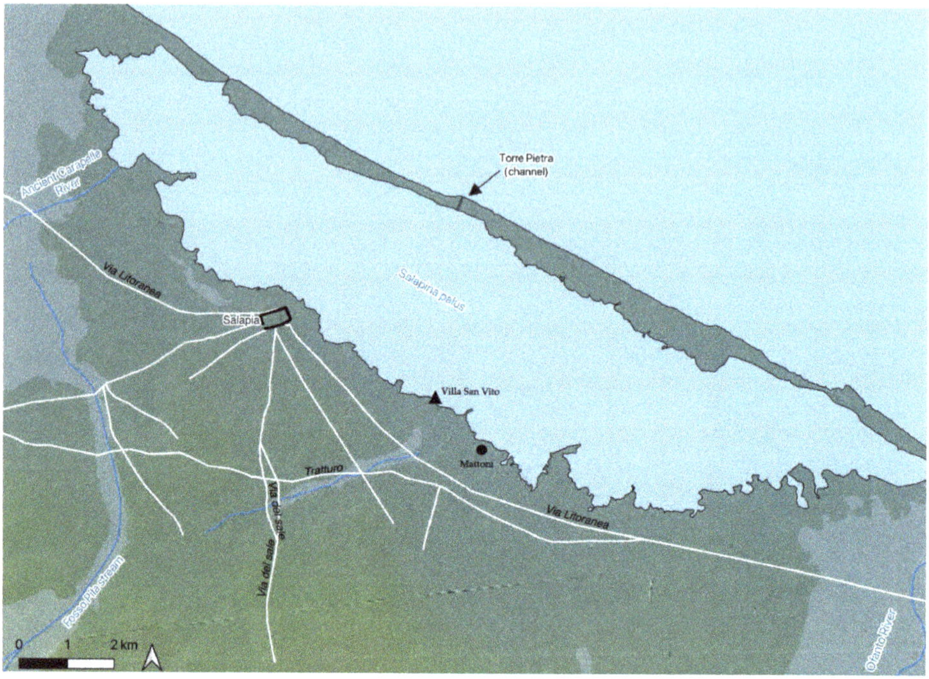

Figure 27. Map of the Salpi Lagoon, with the positions of *Salapia*, the villa of San Vito, and Mattoni (elab. R. Goffredo).

We might envision that docks were built for ships and wharves for unloading cargoes, cranes for unloading, and potentially spaces to perform ship repair and maintenance on this interior shore [89,90]. Underwater archaeology in the 1980s, as well as the ad hoc presence of sea-worn amphora sherds along the beaches of Margherita di Savoia, corroborate trading in this harbor, and potentially its scale [81,82]. *Salapia* itself was awash in amphorae, as noted by their reuse in urban buildings, such as in the foundations of walls (used mostly whole) and in fragments for earthen wall uprights. The ample presence of Italian, Aegean, and eastern Mediterranean amphorae, dated from the 1st c. BCE to the late 5th c. CE, demonstrates the activity of the port, with products consumed in the town itself [91].

With such activity and its supporting material culture, most harbors had to contend with the negative environmental impacts that prompted their own short- and long-term problems [92][8]. Beyond this infrastructure, sailing activity would have affected aquatic animal and plant life, causing constant movement and introducing material pollution into the ecosystem. When added to the potential for wrecks of approaching ships and the role harbor infrastructure played—by virtue of its shape and position—in attracting waste deposited at the interior, the environmental effects cannot be understated [93] (pp. 199–202); [94][9]. Therefore, these logistical challenges needed to be balanced alongside the economic growth such harbor traffic enabled, as well as the open contact it encouraged between the lagoon and the sea. Indeed, such continuous interchange would have ensured consistent water levels in the lagoon, crucial for maintaining the depths necessary to receive ships. Freshwater inputs, which could be variable, would not guarantee the same consistency.

Figure 28. Amphora finds from the 2013 urban survey of the *Salapia* site. Note the concentration between the eastern flank of the city walls and the present-day shore (elab. R. Goffredo).

As noted above, the SAM5 and 7 cores had limited evidence of sediment accumulation, and we might find support by proxy for the idea of a heavily curated lagoon environment to maintain the functionality of the harbor as well. The ancient epigraphic sources allude to dredgers responsible for removing sand from port basins to enable docking in other contexts [95], logistics that perhaps were also present in the *Salapia* context. The notable example of the *Fossa Augusta*, also referenced above, maintained clear channels and, therefore, lightened the effects of siltation in Ravenna's lagoon, sustaining the myriad sites of docking that have been hypothesized at various scales [86]. Such harbor curation might have set the tone for a broader tending of the lagoon, not only its waters, but also the surrounding land.

Again, modern comparisons might be informative. In both the context of the Salpi Lagoon and the nearby Lesina Lake [77] (p. 78), coastal farmers insisted on the interchange between the water of these lakes and the sea, not only to prevent inundation from river waters, but also to encourage "movement" that would have unsettled problematic marsh plants and sediments hosting breeding insects, including mosquitos. While we have not conducted survey work along these dunes, the present-day small farms that inhabit them have a vested interest in maintaining their pest- and weed-free lands, including the removal of marsh plants to mitigate insect breeding; these plants could be for resource extraction (reeds and thrushes for building and fires), to liberate uncultivated land for other uses, or to keep the boundaries of one's property clean [96–98][10]. The transformation in land use, from pasture to garden and small farm cultivation, coincided progressively with the early 20th c. reclamation works, ensuring that the land was curated [99] (Figure 2). Flowing water with interchange surely hinders the egg-laying and gestation of anopheles mosquitos [90]. We might also conjecture a culling of unwanted plants to maintain and keep the harbor clean. The concomitant effects, then, of curating the lagoon space with a specific purpose would have aided water management and tended to adjacent areas. To this we might add again evidence from the cores (SAM9) and archaeobotanical data from our excavated contexts: the surrounding hinterland was a coordinated one as well, with a range of grain and vegetable crops typical of Roman farming regimes in Italy. These data have been supported by the archaeobotanical assemblage from the 5th c. CE town, where wheat, barley, and emmer were all cultivated, along with garden plants like cucumbers and melons [32]. Although the challenges posed by agricultural sedimentation were likely present, charcoal data from this same 5th c. CE context points to a mix of riparian, deciduous, and conifer forests nearby that would have helped to slow sedimentation and erosion and maintain a balance of moisture in the system [100,101][11].

Taken together, these data point to a vested interest by the town of *Salapia* in tending to its harbor and adjacent lands, which impacted water circulation and quality, at least into Late Antiquity.

4.2.4. Beyond Production: Salt-Making Infrastructure as Water Management Infrastructure

However, to understand fully *Salapia*'s relationship to the lagoon and its management of water, we must move beyond the account of Vitruvius and consider our only other textual sources linked to the Imperial and Late Antique town. Dating to the 3rd and 4th c. CE, when *Salapia* was still a vibrant town, the evidence of the Antonine Itinerary and the Peutinger Table directly reference *Salapia*'s role in salt production. *Salapia* does not appear in these itineraries, but the appellation *salinis* is in its place, literally "from the saltpans." Although Meluta Marin, the renowned scholar of this territory, argued in 1989 that *salinis* referred specifically to the storage sheds for this salt [102], we know from other ancient contexts and "best practice" that cultivated sea salt should be stored outside, where the wind and sun help it to dry and solidify [103][12]. Therefore, there were likely no dedicated structures for the storage of salt, but the infrastructure to produce it was surely visible [104][13]. Building on the harbor's position proposed at the center of the lagoon, we hypothesize that the ancient pans were found somewhere to the south of the town, along the inner shores of the lagoon, possibly as far distant as the village site of Mattoni. This would put the pans near where they appear now in the southern half of the lagoon (Figure 11). Although we cannot determine the scale of this production, the replacement of *Salapia* in these itineraries with *salinis* as a key landmark might underscore what was its relevance in the landscape to both local inhabitants and those traveling through the region [105]. While the mountains of salt, depending on size, might have been the first visual marker, as with the case of Utica in North Africa, we cannot discount the possibility that the infrastructure for salt-making itself stood out in this landscape.

While we have no material evidence for this production to date, it is likely, based on more recent archaeological studies and the weight the literary texts placed on evaporative technology in the Mediterranean that this was the infrastructure employed at *Salapia* [106–108]. In this case, in framing the site's resilience vis-à-vis water, the pans themselves could be a highly efficient tool for water management. The technology of these pans would feed salt water through increasingly shallower and smaller pans as it increases in density via evaporation under the hot summer sun and with the help of the wind. The pans, built of compacted clayey earth, at times lined with stones or outlined by reused amphorae [106], regimented the marshes where they were placed, providing a certain regularity to the water and its flow through the system. While we do not know how large the pans at *Salapia* would have been, even a small installation would have required a certain level of attention and curation of the broader lagoon space. As attested by the modern sources, giving both space and time to fishing and salt production required plans for barrier walls to ensure that especially freshwater would not contaminate the system as well as sluices to control outflows of freshwater from rivers for this reason [76–78]. As we know from the examples of excavated remains, as well as from the modern example of pans of Barletta from the 18th and 19th c., these infrastructures required careful planning and knowledge of the lagoon environment, to be placed in an efficacious spot that would facilitate the entry of salt water into the system, and movement through the structure [109]. In the case of the Barletta pans, water screws and other mechanisms were employed to overcome the flatness of the zone and ensure the movement of increasingly salty water from one pan to the next.

What stands out in these early modern accounts is the cyclical nature of the production and operation of the pans in coordination with the seasons. As understood in greater detail from modern sources, the "spring cleaning" phase for these pans was an important one that required a significant number of workers to remove the marshy overgrowth and detritus that had accumulated over the winter months when the pans were not in operation [109]. This period of overgrowth fortuitously coincided with colder weather. Even if hibernating mosquitos and other pests passed their winters there, their homes would be dislodged each spring when the marshes were prepared for the new season [79][14]. The "growing" of salt occurred over the summer months and into the early Fall, when the sun was strongest and the air hottest, to facilitate the evaporative technology. This would have helped confront the peak breeding season for mosquitos and other pests, reducing plant overgrowth through the concentration of activity and its product, as high salinity is not conducive to sustaining either plant or animal life in these pans.

One contends that the required attention to this wetland via salt production meant a consistent and regulated engagement with another swath of the lagoon. Balance was needed between the port and the pans; *Portus* and its own Stagno Maccarese saw that interrelated hydraulic works were part of a larger system of complete landscape organization [110] (p. 264). The example of *Ficocle*/Cervia and its own pans [111], likely situated to the north of the settlement, which was built on an island at the center of a natural lagoon, might also speak to the symbiosis of urban centers and salt production that extended beyond economic concerns and perhaps were part of a complementary strategy to manage the wetland. Such activity ensured attention and tending, while keeping threats from stagnant waters and insects at bay.

By prioritizing, as proposed here, the port and the salt infrastructure, the site of *Salapia* could maximize the advantages of its well-drained floodplain and its position in this marshy zone. Predictably, however, the maintenance of such logistically intensive infrastructures, which required both the investment of materials and labor owed to Roman administration and coordination over centuries, were probably not sustainable, at the same scale or intensity, as *Salapia*'s urban institutions waned with the dawn of the 6th c. CE. While beyond the scope of this argument, it is worth noting that settlement did persist at the site of *Salapia* for another three centuries. The community there, although declining in population with each passing century, continued a complex agricultural regime [31,33], saw the growth of olive cultivation beyond its scale detected even in the Roman period [13] (p. 49), and continued to produce salt, potentially still at scale into the 8th c. CE, as it was integrated into the new Lombard economic and social networks of exchange and obligation (*CSS* I, 11 cod. Vat. Lat 4939). The patrimony of the Roman city's infrastructure and water management regime likely produced concomitant effects and influenced the long-term cultural engagement with the lagoon, in varying degrees, for many subsequent generations.

5. Discussion and Conclusions

In sketching the settlement history of *Salpia vetus* to its abandonment and then *Salapia*'s refounding and *floruit*, some key trends are worth emphasizing, as we have elaborated the complexity of human–environment interactions in this coastal ecosystem.

These reconstructed histories demonstrate how decisions taken and actualized around water and broader territorial management could be adaptive, dynamic, and responsive to environmental conditions on the ground, provided that actors had the resources and logistical capabilities at their disposal [112] (pp. 8–10).

Salpia vetus's approach to environmental resilience grew with the expansion of the settlement, most visible in the 6th and 5th c. BCE. Strategies included reinforcing the natural topography of the peninsulas on which the settlement was built through excavating moats. This was likely coupled with a dendritic road network that would have doubled as a set of drainage channels to funnel water away from the settlement and toward its hinterland.

However, as detailed here, over time, these strategies did not evolve as needed alongside both the changing environmental and geopolitical situation as contact and conflict with Rome increased. It is quite possible that the augmented agricultural exploitation detailed across the 4th and 3rd c. BCE was incompatible with the low-lying landscape, prone to pooling stagnant water, and putting the nearby urban ecosystem at risk.

With the waning resilience of *Salpia vetus* in the wake of greater external pressure, we are reminded that conversations between people and their environments are not just a singular, intimate dialogue but involve interruptions and, at times, new interlocutors. *Salpia vetus* was not only challenged by its deteriorating environmental situation but also exasperated by external political and military stressors. Viewed from the perspective of Holling and Gunderson's panarchy model, adaptation was only feasible to a point before the system broke, represented by the abandonment of the town itself [4]. What might have become of *Salpia vetus* had it been able to regroup in a different political and cultural context not predicated on conflict with Rome?

In Vitruvius' telling, the *Salapini* asked to be moved to a new site, but can we trust such an account when the Romans had so much to gain with their broad-scale settlement transformations and adaptive responses to environmental challenges? Their perspective was an imperialist one aimed at standardizing landscape infrastructure and organization, a plan being sold up and down the corridor of the Adriatic coastal wetlands. Given the Romans' ability to mobilize resources, the outcome was that the *Salapini*, whether willfully or under duress, abandoned their town after the progressive collapse of their elite networks and their failing interchange with an increasingly waterlogged and destabilized landscape of inland ponds and lakes. We should not take such a break lightly, even if the new town of *Salapia* arose in the same coastal wetland. We are left to ask: Did resilience requiring such profound changes really work for this Daunian community? For only certain members, or perhaps no one at all? And at what social and cultural cost to their community identity?

If we are to fit these cases to a model of panarchy, should we consider the connection between *Salpia vetus* and *Salapia* as a continuity or a rupture of the system? If resilience is meant to entail a community that can absorb disruptive forces without altering function, processes, and identity [4–7,113], much depends on whether *Salapia* was a refounding of the old model, or a settlement made anew.

The founding of the Roman *municipium* at *Salapia* required the incorporation of new administrative structures and, at the material level, a new urbanism with formal differences to *Salpia vetus*, which included a port with seaward-facing infrastructure, city walls with a grid plan, and baths as leisure sites, for example. There was also the incorporation of new water management features and agricultural regimes, changes that surely insisted on new cultural modes and human connections. Coupled with a new context on a lagoon made to be outward-facing, with greater expectations for the scale of trade and salt production, a community now saltwater-based and not fresh-water-based, might lead us to see a new trajectory for this wetland rather than a remaking of the old.

As is apparent, in a way that panarchy does not wholly account for, local contexts and experiences matter in this case. Van der Noort's model for climate change archaeology calls for particular attention to local conditions and customs to understand the human–environment relationship fully, and to counteract deterministic narratives [3,10,113]. Indeed, by any measure, the story of *Salpia vetus* and its long-lived and vibrant settlement is worthy of discussion for the particularly local and specific variables attendant in its life cycle (birth, growth, conservation, collapse). The same might be said for the birth, growth, and persistence of *Salapia*, if we accept the diversity of cultural, social, and material parameters involved, and put aside the flattening required of the panarchy model.

What links these settlements intimately, however, is born of their shared local context and the questions of how to manage a dynamic wetland. The Romans not only had the advantages of military might and resources, but they also had the knowledge born of observing *Salpia vetus*' struggles directly. A sense of this prior history was worn in the material remains of the settlement, including its drainage works, its earthen barrier walls, and its former territorial organization. Rome therefore had the ability to cannibalize *Salpia vetus'* past and miscues at adaptation. This co-option gets at the heart of Vitruvius' message: search nearby to find a healthful place and try again, but do not make the same mistakes. Sharing a name, from *Salpia* to *Salapia*, might not have been a wholesale lifting of identity, but a nod toward this recognized, but troubled, past.

Salapia's resilience was born of the selection of a more healthful site, in a well-drained floodplain on a lagoon they intentionally connected to the sea. There was careful attention to managing the outflows and interchange of fresh and saltwater in this system. Canal works drained to the lagoon, as did the intra-urban street grid. Salt-making infrastructure could have played double duty, in both producing a necessary resource and ensuring that parts of the lagoon not englobed by the port and its carefully tended infrastructure also received similar manicuring to tame changeable flora and fauna in the marsh over the year.

Our juxtaposition of *Salpia vetus* and *Salapia* furthers the charge for resilience models that bolster local histories and their particularities. Acknowledging breaks in the system rather than "remaking" are also crucial for the sake of accuracy. Failure matters as much as success, as does acknowledging the diversity of experience. Assuming continuity and ease of transformation due to a shared context or similar challenges risks flattening interpretations of the outcomes.

Jumping ahead, when we consider the end of *Salapia*, sometime between the late 8th and early 9th c. CE, we are confronted with a subsequent narrative of abandonment. There is no perceptible evidence of human life on the plain adjacent to the lagoon, and burials dated to the 9th c. CE, through ^{14}C dating and stratigraphic interpretation, might signal that here too, resilience was not possible. Instead, a wholesale separation was required. Unlike the *Salpia vetus–Salapia* connection, the break was not caused by space, but by time. For nearly two centuries, there was no community to speak of. Were they regrouping? And where?

It was perhaps as early as the 10th or early 11th century CE that the settlement core of *Salapia* began to recompose itself in the western half of the ancient city of Roman foundation, further away from the waters of the lagoon and more elevated with respect to the surrounding landscape [28]. To maintain the strategic relationship with the wetland, therefore, the same floodplain was inhabited but made more suitable and protected by elevating itself in space, the chosen strategy to revitalize the settlement. Furthermore, the sedimentary sequences of the SAM5 and SAM6 cores indicate the marshes, already discernible in the 3rd century CE immediately east of *Salapia*, persisted into the 16th century. Links could be found in a shared sense of a saltwater lagoon and its potential.

It was the advent of Norman control, at the end of the 11th century, that marked the relaunch of the center of *Salpi* (this was to be the name of the Medieval city) in the geopolitical arena of northern Apulia; the settlement was redesigned on a large embankment surrounded by a deep moat, reworking a strategy that helped the interior *Salpia vetus* manage its own relationship to freshwater, now realized on the shores of the salty lagoon. While sharing space with its predecessor, the medieval town cannibalized much of its material legacy through its literal stones, reassembled on the elevated *castrum* and a loose reworking of the Roman urban plan. However, this was a new wetland community, with new concerns and institutions. In fact, the settlement thrived with such an organization regaining the dignity of *civitas* and bishopric; the settled community was reorganized with new roles and hierarchies; the ruling class was renewed. A new but different life cycle began, cognizant of the past lives lived on this lagoon.

Author Contributions: Conceptualization, R.G. and D.M.T.; methodology, R.G. and D.M.T.; data curation, R.G. and D.M.T.; writing—original draft preparation, R.G. (Sections 1–3, 4.1.1–4.1.3 and 5) and D.M.T. (Sections 1, 3, 4.2.1–4.2.4 and 5)); writing—review and editing, R.G. and D.M.T.; visualization, R.G.; funding acquisition, D.M.T. All authors have read and agreed to the published version of the manuscript.

Funding: The project Life on the Lagoon: Reconstructing the Biography of Human–Landscape Dynamics on the Salpi Lagoon (Italy) benefited from the NEH National Endowment for the Humanities Collaborative Research Grant RZ-249965-16 (P.I. Darian Marie Totten—Davidson College-McGill University). This research also forms part of the project PRA-HE 2021 at the Università di Foggia Wetlands. Humans and Environment in transition (P.I. Roberto Goffredo).

Data Availability Statement: Data from the research project *Life on the Lagoon: Reconstructing the Biography of Human–Landscape Dynamics on the Salpi Lagoon (Italy)* included published and unpublished sources.

Acknowledgments: We will write the acknowledgments if this article passes the peer review.

Conflicts of Interest: The authors declare no conflicts of interest.

Notes

1 Vitr. 1, 4, 12: *Quibus autem insidentes sunt paludes et non habent exitus profluentes neque per flumina neque per fossas, uti Pomptinae, stando putescunt et umores graves et pestilentes in is locis emittunt. Item in Apulia oppidum Salpia vetus, quod Diomedes ab Troia rediens constituit sive, quemadmodum nonnulli scripserunt, Elpias Rhodius, in eiusmodi locis fuerat conlocatum, ex quo incolae quotannis aegrotando laborantes aliquando pervenerunt ad M. Hostilium ab eoque publice petentes impetraverunt, ut his idoneum locum ad moenia transferenda conquireret eligeretque. Tunc is moratus non est, sed statim rationibus doctissime quaesitis secundum mare mercatus est possessionem loco salubri ad senatuque populoque Romano petit, ut liceret transferre oppidum, constituitque moenia et areas divisit nummoque sestertio singulis municipibus mancipio dedit. His confectis lacum aperuit in mare et portum e lacu municipio perfecit. Itaque nunc Salpini quattuor milia passus progressi ab oppido veteri habitant in salubri loco.* ("Those places, however, which have stagnant marshes, and lack flowing outlets, whether rivers or by dykes, like the Pontine marshes, by standing become foul and send forth heavy and pestilent moisture. Also in Apulia, the town of Old Salpia, which Diomede returning from Troy established, or, as some have written, Elpias of Rhodes, was situated in such place. As a result, the inhabitants suffered every year from various ailments and they soon came to M. Hostilius, and they beseeched him publicly to seek out and choose an appopriate site for transferring their walls. He did not delay and immediately, after fully ascertaining all the conditions, bought a site in a healthy place, and obtained permission from the senate and Roman people to remove the town. He established the walls and divided the lots and gave formal possession to the individual townsmen for a sesterce each. When this was done, he opened the lake into the sea, and made a harbor out of the lake for the municipality. And so the people of Salpia now dwell on a healthy site at a distance of four miles from the old town") [1].

2 The familiarity of Daunian coastal communities with the practice of fishing is suggested by the stelae of the Archaic period (8th-7th c. BCE) from the area of *Sipontum* and *Salpia*, where the depiction of stylised fishes and more articulate fishing scenes (both onshore and offshore) is recurrent: see [40] (nn. 592A, 613A; 615B; 616B; 737B; 742B; 775B).

3 From Neolithic and until very recent times, hunting activity, although variously regulated, has been an important part of the local economy. Among the fillers of the Daunian s'ipontine' and s'alapine' stelae, large birds with open wings in flight or marshy birds are abundant, as are hares chased by dogs, while more complex hunting scenes with spears, on foot or on horseback, of deer and other large animals are noteworthy. See [40] (nn. 592A, 613A; 615B; 616B; 737B; 742B; 775B).

4 The fortifications on the southern side of *Salpia vetus* have never been excavated, so the dating is based on the analysis of ceramic artifacts found on the surface. It is certain that the *oppidum* was already fortified during the Hannibalic War (late 3rd century BCE): Livy (27.28) mentions that the city was equipped with walls, towers, and a gate with a sliding shutter system. A precise comparison for the fortifications of *Salpia vetus* is represented by the defensive system of *Arpi*, one of the main Daunian cities, located about 8 km northeast of the current city of Foggia. The defensive system enclosed the settlement on only three sides, while the fourth side was protected by the course of the Celone Stream. The structure, built during the 7th century BCE, consisted of a deep moat, 15 m wide, followed on the inside by an embankment on which a wall of raw clay was erected. Between the 5th and 4th centuries BCE, the fortifications underwent renovations. See [44] (pp. 931–951).

5 This source registers impressions and perspectives of the primary architect of the 18th c. saltpans in the Salpi Lagoon, Luigi Vanvitelli, via his ample correspondence [78].

6 Although originally thought to be connected to the salty character of the pools [77], p. 79), recent research has made plain that anopheles mosquitos can breed in both salty and freshwater. The lack of movement in these pools likely encouraged heightened propagation. All species currently present in Europe of the *anopheles* genus will breed in either fresh or brackish water (*an. Atroparvus; an. Labranchiae; an. Plumbeus; an. Sacharovi*) [79].

7 The proposed village of Mattoni has been discovered (and disturbed) by clandestine activity. We systematically resurveyed the site in 2017, and its finds are currently under study [84].

8 The authors demonstrate in Eastern Mediterranean ports in Palestine that heavy siltation was prompted by the scale of Roman seaborne trade when compared to previous periods [92].

9 Tuck notes the censors at Rome were responsible for maintaining the channel of the Tiber River, which included its cleaning [94].

10 Pickel provides a reading of the Roman agronomists that sees efforts to reduce malarial breeding in agricultural field [96].

11 The benefits of this forest management might have been realized by the town, as it is at variance with agricultural landscapes from elsewhere in Italy, such as in central Tuscany in the imperial period. This landscape was largely cleared, probably through exploitation of forest resources and to reclaim land for agriculture [100]. Downstream, in Rome, the effects of this clearing were noted in the ancient sources [101].

12 Pliny the Elder records the sight of *Utica*'s salt mounds, visible when approaching by ship to the harbor [103]. The Atisale installation, visible across the lagoon from the site of *Salapia* is defined by the mountain of salt, left exposed to the elements, before it is cleaned and packaged. The present-day production at Cervia serves as another example. Rainwater will not melt exposed salt, which forms an intractable crust when exposed to the sun. Only water infiltration from below can melt these mountains. One should also note that in early-modern and present-day Cervia, the salt "market" is positioned in town, not in the pans themselves.

13 Notations of the *saccarii salarii*, or "porters who transported sacks of salt" might provide indication of how this product was packaged and transported from the mound [104].

14 All species of anopheles mosquito currently present in Europe benefit from low to no water flow, and myriad crevices (hollowed out trees and roots, nooks in plant leaves, etc.) to lay their eggs [79].

References

1. Vitruvius. *On Architecture, Volume I: Books 1–5*; Edited from the Harleian Manuscript 2767; English translation modified from that of Frank Granger, Loeb Classical Library 251; Harvard University Press: Cambridge, MA, USA, 1955; pp. 44–47.
2. *Oxford Latin Dictionary, Fascicle VII*; Glare, P.G.W., Ed.; Oxford University Press: Oxford, UK, 1980.
3. Brown, K. Global environmental change I: A social turn for resilience? *Prog. Hum. Geogr.* **2014**, *38*, 107–117. [CrossRef]
4. Holling, C.S. Resilience and stability of ecological systems. *Annu. Rev. Ecol. Syst.* **1973**, *4*, 1–23. [CrossRef]
5. Gunderson, L.H.; Holling, C.S. *Panarchy: Understanding Transformations in Human and Natural Systems*; Island Press: Washington, DC, USA, 2002.
6. Daems, D.; Vandam, R. Tracing adaptive cycles and resilience strategies within the Sagalassos settlement record, SW Türkiye. *Holocene* **2024**, 09596836241259771. [CrossRef]
7. Walker, B.; Holling, C.S.; Carpenter, S.R.; Kinzig, A. Resilience, adaptability and transformability in social–ecological systems. *Ecol. Soc.* **2004**, *9*, 5. [CrossRef]
8. Davis, D.S. Past, present, and future of complex systems theory in archaeology. *J. Archaeol. Res.* **2023**, 1–48. [CrossRef]
9. Bradtmöller, M.; Grimm, S.; Riel-Salvatore, J. Resilience theory in archaeological practice—An annotated review. *Quat. Int.* **2017**, *446*, 3–16. [CrossRef]
10. Van de Noort, R. *Climate Change Archaeology: Building Resilience from Research in the World's Coastal Wetlands*; Oxford University Press: Oxford, UK, 2013.
11. Walsh, K.; Attema, P.; de Haas, T. The Pontine Marshes (Central Italy): A case study in wetland historical ecology. *Babesch* **2014**, *89*, 27–46.
12. Caldara, M.; Simone, O. L'Ambiente Fisico nell'area dell'Insediamento di Coppa Nevigata. In *Coppa Nevigata e L'area Umida alla Foce del Candelaro Durante l'età del Bronzo*; Cazzella, A., Moscoloni, M., Recchia, G., Eds.; Grenzi Editore: Foggia, Italy, 2013.
13. Susini, D.; Vignola, C.; Goffredo, R.; Totten, D.M.; Masi, A.; Smedile, A.; De Martini, P.M.; Cinti, F.R.; Sadori, L.; Forti, L.; et al. Holocene palaeoenvironmental nd human settlement evolution in the southern margin of the Salpi lagoon, Tavoliere coastal plain (Apulia, Southern Italy). *Quat. Int.* **2023**, *655*, 37–54. [CrossRef]
14. di Biase, P. *Puglia Medievale e Insediamenti Scomparsi. La Vicenda di Salpi*; Schena: Fasano, Italia, 1985.
15. Biblioteca Provinciale De Gemmis—Bari Mss. B/4, vol. 28, fol. 102.
16. Biblioteca Provinciale De Gemmis—Bari Mss. B/3, vol. 54, fol. 56.
17. Goffredo, R. La Laguna, la Città. In *Salapia-Salpi 1. Scavi e Ricerche 2013–2016*; De Venuto, G., Goffredo, R., Totten, D.M., Eds.; Edipuglia: Bari, Italia, 2022; pp. 19–34.
18. Schmiedt, G. *Contributo Della Fotointerpretazione alla Ricostruzione Della Situazione Geografico-Topografica degli Insediamenti antichi Scomparsi in Italia*; Istituto Geografico Militare: Firenze, Italy, 1964.
19. Schmiedt, G. *Atlante Aerofotografico Delle Sedi Umane in Italia. Parte II: Le Sedi Antiche Scomparse*; Istituto Geografico Militare: Firenze, Italy, 1970.
20. Schmiedt, G. Contributo della fotografia aerea alla ricostruzione della antica laguna compresa fra Siponto e Salapia. *ArchStorPugl* **1973**, *26*, 160–172.
21. Tiné Bertocchi, F. Gli Scavi del 1967–1968 e del 1978–1979. In *Salpia Vetus. Archeologia di Una Città Lagunare*; Lippolis, E., Gianmatteo, T., Eds.; Osanna Edizioni: Venosa, Italy, 2008; pp. 77–94.
22. Cazzella, A.; Moscoloni, M.; Recchia, G. *Coppa Nevigata e L'area Umida Alla Foce del Candelaro Durante l'età del Bronzo*; Grenzi Editore: Foggia, Italy, 2013.
23. Nava, M.L. I precedenti insediativi: L'area di Cupola-Beccarini. In *Siponto Antica*; Mazzei, M., Ed.; Grenzi Editore: Foggia, Italy, 1999; pp. 45–69.
24. Mazzei, M. *Siponto Antica*; Grenzi Editore: Foggia, Italy, 1999.

25. Cardone, A.; Volpe, V. Tra Persistenze e Discontinuità: Analisi Spaziali a *Salapia-Salpi* e nel Suburbio. In *Oltre La Convenzione. Pensare, Studiare, Costruire il Paesaggio Vent'anni Dopo;* Castiglioni, B., Puttilli, M., Tanca, M., Eds.; Società Studi Geografici: Firenze, Italy, 2021; pp. 1066–1082.
26. Cerri, L. La Prospezione Magnetica. La Città e il Suburbio. In *Salapia-Salpi 1. Scavi e Ricerche 2013–2016;* De Venuto, G., Goffredo, R., Totten, D.M., Eds.; Edipuglia: Bari, Italy, 2022; pp. 63–70.
27. Cerri, L. La Prospezione Magnetica a Salapia-Salpi 2017–2019. In *Salapia-Salpi 2: Scavi e Ricerche 2017–2019;* Goffredo, R., Totten, D.M., Eds.; Edipuglia: Bari, Italy, *in Progress.*
28. Goffredo, R.; Cardone, A. Salpi in Capitanata. Fonti, dati archeologici, analisi spaziali per la restituzione di un paesaggio urbano sepolto. *Archeol. Mediev.* **2021**, *48*, 301–322.
29. De Venuto, G.; Goffredo, R.; Totten, D.M. (Eds.) *Salapia-Salpi 1. Scavi e Ricerche 2013–2016;* Edipuglia: Bari, Italy, 2022.
30. Goffredo, R. Salpi tra Medioevo ed Età moderna. Nascita, sviluppo e scomparsa di una città. *Mélanges De L'école Française De Rome-Moyen Âge* **2021**, *133*, 421–465. [CrossRef]
31. Della Penna, V.; D'Aquino, S.; Totten, D.M.; Goffredo, R.; Fiorentino, G. Paesaggi, Sistemi Agricoli e Alimentazione tra Continuità e Cambiamento. Analisi Archeobotaniche dal Sito di *Salapia-Salpi* tra Tarda Antichità e Medioevo. In *IX Congresso Nazionale di Archeologia Medievale, Alghero 2022;* Milanese, M., Ed.; All'insegna del Giglio: Firenze, Italy, 2022; Volume 2, pp. 415–420.
32. Fiorentino, G.; D'Aquino, S.; Della Penna, V. Dalla Laguna all'Entroterra: Il Contributo dell'Archeobotanica alla Ricostruzione dei Paesaggi e dei Sistemi Agricoli a Salapia tra Tardo Antico e Alto Medioevo. In *Salapia-Salpi 1. Scavi e Ricerche 2013–2016;* De Venuto, G., Goffredo, R., Totten, D.M., Eds.; Edipuglia: Bari, Italy, 2022; pp. 573–587.
33. Fiorentino, G.; Della Penna, V.; D'Aquino, S. Paleoambiente e Paesaggi, Sistemi Agricoli e Alimentazione Vegetale a Salapia-Salpi: Continuità e Cambiamento tra Tarda Antichità e Basso Medioevo. In *Salapia-Salpi 2. Scavi e Ricerche 2017–2019;* Goffredo, R., Totten, D.M., Eds.; Edipuglia: Bari, Italy, *in Progress.*
34. Delano Smith, C. *Daunia Vetus. Terra, Vita e Mutamenti Sulle Coste del Tavoliere;* Tipografia Laurenziana: Napoli, Italy, 1978.
35. Lippolis, E. Introduzione. In *Salpia Vetus. Archeologia di Una Città Lagunare;* Lippolis, E., Gianmatteo, T., Eds.; Osanna Edizioni: Venosa, Italy, 2008; pp. 33–40.
36. Traina, G. *Paludi e Bonifiche del Mondo antico: Saggio di Archeologia Geografica;* L'Erma di Bretschneider: Roma, Italy, 1988; p. 93.
37. D'Ercole, M.C.; Goffredo, R. Entre Cités et Campagnes: Le sel et les Économies Territoriales de l'Apulie, de l'Antiquité au Haut Moyen Âge. In *Sel et Société, Tome 2: Santé—Croyances—Économie;* Hoet-van Cauwenberghe, C., Masse, A., Prilaux, G., Eds.; XLIIe Colloque International d'HALMA (Lille 2017); Septentrion: Lille, France, 2020; pp. 275–295.
38. Goffredo, R. Da Salpia Vetus a Salapia: La Città di Età Romana. In *Salapia-Salpi 1. Scavi e Ricerche 2013–2016;* De Venuto, G., Goffredo, R., Totten, D.M., Eds.; Edipuglia: Bari, Italy, 2022; pp. 618–622.
39. Goffredo, R.; Totten, D.M.; Fiorentino, G.; Mazzini, I.; Susini, D. Diachronic Human-Wetland Interactions on the Salpi Lagoon (Apulia, Italy): From the Daunian Period to the Early Middle Ages. In *Waterscape Archaeology: Multi-scalar Human-Environment Interactions in Coastal Lagoons;* Bivolaru, A., Cottica, D., Morhange, C., Eds.; Archeopress: Oxford, UK; *in print.*
40. Nava, M.L. Stele Daunie; Firenze, Italy. 1980. Available online: https://www.cambridge.org/core/journals/journal-of-roman-studies/article/abs/m-l-nava-stele-daunie-i-studi-e-materiali-di-etruscologia-e-antichita-italiche-xviii-florence-sansoni-1980-pp-vi-225-31-illus-incl-2-folding-maps-395-plates/9E3679D91028833C9DF56043E1B6A493 (accessed on 17 September 2024).
41. D'Ercole, M.C. *Importuosa Italiae Litora. Paysage et Échanges Dans l'Adriatique Méridionale à L'époque Archaïque;* CJB: Napoli, Italy, 2002.
42. Grelle, F. Allevamento Equino, Transumanza e Agricoltura Nella Puglia Romana, Fra Quarto e Primo Secolo a.C. *MEFRA* 2016, 128.2. Available online: http://journals.openedition.org/mefra/3746 (accessed on 22 June 2016).
43. Santagati, E. Paludi e Fortificazioni nelle Fonti Storiografiche: Una Prima Ricognizione. In *Limnai. Archeologia delle Paludi e delle Acque Interne;* Caliò, L.M., Lepore, G., Raimondi, G., Todaro, S.V., Eds.; Edizioni Quasar: Roma, Italy, 2022; pp. 141–149.
44. Pouzadoux, C.; Munzi, P.; Santoriello, A.; Amato, A.; Leone, M.; Muntoni, I.M. Vivere ad Arpi: Le Trasformazioni di una Città Daunia tra Alessandro il Molosso e Annibale. In *Roma e il Mondo Adriatico dalla Ricerca Archeologica alla Pianificazione del Territorio;* Atti del Convegno Internazionale (Macerata 18–20 Maggio 2017), Perna, R., Carmenati, R., Giuliodori, M., Eds.; Edizioni Quasar: Roma, Italy, 2022; II.2; pp. 931–951.
45. Gros, P. *Vitruvio. De Architectura;* Einaudi: Torino, Italy, 1997.
46. Bandelli, G. Considerazioni Storiche sull'Urbanizzazione Cisalpina di Età Repubblicana (283–89 a.C.). In *Forme e tempi dell'urbanizzazione nella Cisalpina (II secolo a.C.-I secolo d.C.), Atti delle Giornate di Studio (Torino, 4–6 maggio 2006);* Brecciaroli Taborelli, L., Ed.; All'Insegna del Giglio: Firenze, Italy, 2007; pp. 15–28.
47. Strazzulla, M.J. In paludibus moenia constituta: Problemi urbanistici di Aquileia in età repubblicana alla luce della documentazione archeologica e delle fonti scritte. *AAAd* **1989**, *35*, 187–228.
48. Maselli Scotti, F.; Mandruzzato, L.; Tiussi, C. Primo Impianto Coloniario di Aquileia: L'Area Tra foro e Macellum. In *Forme e tempi dell'urbanizzazione nella Cisalpina (II secolo a.C.-I secolo d.C.), Atti delle Giornate di Studio (Torino, 4–6 maggio 2006);* Brecciaroli Taborelli, L., Ed.; All'Insegna del Giglio: Firenze, Italy, 2007; pp. 35–40.
49. Bonetto, J.; Furlan, G.; Ghiotto, A.R.; Missaglia, I. Il Canale Anfora e il centro urbano di Aquileia: Osservazioni cronologiche alla luce di nuovi dati. *JAT* **2020**, *XXX*, 175–202.
50. Maggi, P.; Maselli Scotti, F.; Pesavento Mattioli, S.; Zulini, E. (Eds.) *Materiali per Aquileia. Lo scavo di Canale Anfora (2004–2005);* Edizioni Quasar: Roma, Italy, 2018.

51. Cresci Marrone, G.; Tirelli, M. Altino Romana: Limites e Liminarità. In *Forme e Tempi dell'Urbanizzazione nella Cisalpina (II secolo a.C.-I secolo d.C.)*; Atti delle Giornate di Studio (Torino, 4–6 maggio 2006), Brecciaroli Taborelli, L., Eds.; All'Insegna del Giglio: Firenze, Italy, 2007; pp. 61–66.
52. Gelichi, S. Ravenna, ascesa e declino di una capitale. In Ripoll, G., Ed., Sedes regiae (annis 400–800). *Mem. De La Real Acad. De Buenas Let. Barc.* **2000**, *25*, 109–134.
53. On this theme, see Perlgrom, J. The Roman rural exceptionality thesis revisited. *Mélanges De L'école Française De Rome-Antiq.* **2018**, *130*, 69–103.
54. Alvisi, G. *La Viabilità Romana della Daunia*; Tipografia del Sud: Bari, Italy, 1970.
55. Bernard, S.; McConnell, J.; Di Rita, F.; Michelangeli, F.; Magri, D.; Sadori, L.; Masi, A.; Zanchetta, G.; Bini, M.; Celant, A.; et al. An Environmental and Climate History of the Roman Expansion in Italy. *J. Interdiscip. Hist.* **2023**, *54*, 1–41. [CrossRef]
56. Zanchetta, G.; Baneschi, I.; Magny, M.; Sadori, L.; Termine, R.; Bini, M.; Vannière, B.; Desmet, M.; Natali, S.; Luppichini, M.; et al. Insight into summer drought in southern Italy: Palaeohydrological evolution of Lake Pergusa (Sicily) in the last 6700 years. *J. Quat. Sci.* **2022**, *37*, 1280–1293. [CrossRef]
57. Grelle, F.; Silvestrini, M. *La Puglia nel Mondo Romano. Storia di Una Periferia. Volume I. Dalle Guerre Sannitiche Alla Guerra Sociale*; Edipuglia: Bari, Italy, 2013.
58. Grelle, F.; Silvestrini, M.; Volpe, G.; Goffredo, R. *La Puglia Nel Mondo Romano. Storia di Una Periferia. L'Avvio dell'Organizzazione municipale*; Edipuglia: Bari, Italy, 2017.
59. Sallares, R. *Malaria and Rome. A History of Malaria in Ancient Italy*; Oxford University Press: Oxford, UK, 2002.
60. De Venuto, G.; Goffredo, R.; Totten, D.M.; Ciminale, M.; De Mitri, C.; Valenzano, V. Salapia: Storia e Archeologia di Una Città Tra Mare e Laguna. *Mélanges De L'école Française De Rome Antiq.* **2015**, *127*, 97–130. Available online: http://mefra.revues.org/2719 (accessed on 17 September 2024). [CrossRef]
61. Goffredo, R.; Totten, D.M.; Loprieno, S. *Salapia* Romana. *Salpi* Medievale (Ceringnola, FG): Notizie Dagli Scavi 2017. *Fold&R J. FastiOnline* **2018**, *426*. Available online: https://www.fastionline.org/docs/FOLDER-it-2018-426.pdf (accessed on 17 September 2024).
62. De Venuto, G. L'Area della Conceria dell'Isolato XII. In *Salapia-Salpi 1: Scavi e ricerche 2013–2016*; De Venuto, G., Goffredo, R., Totten, D.M., Eds.; Edipuglia: Bari, Italy, 2022; pp. 115–140, in particular p. 116.
63. Kolowski-Ostrow, A.O. *The Archaeology of Sanitation in Roman Italy: Toilets, Sewers and Water Systems*; UNC Press: Chapel Hill, NC, USA, 2016.
64. Totten, D.M. The Excavation of *Insula XII*: Architectural and Archaeological Features North and East of the *Domus* (*Edificio A and Edificio B*). In *Salapia-Salpi 1: Scavi e ricerche 2013–2016*; De Venuto, G., Goffredo, R., Totten, D.M., Eds.; Edipuglia: Bari, Italy, 2022; pp. 173–218.
65. Crowson, A.; Giles, O.J. The Archaeology of the Vrina Plain: An Assessment. In *Roman Butrint: An Assessment*; Hansen, I.L., Hodges, R., Eds.; Oxbow Books for the Butrint Foundation: Oxford, UK, 2007; pp. 119–164.
66. De Santis, V.; Scicchitano, G.; Scardino, G.; Mele, D.; Sulpizio, R.; Colangelo, G.; Zigaro, M.; Antonio, N.L.; Tomaselli, V.; Caldara, M. Construction of a deltaic strandplain during the Roman period in the Tavoliere di Puglia plain and palaeoclimatic implications. *Front. Earth Sci.* **2023**, *11*, 1278105. [CrossRef]
67. Rogers, A. We Might also Consider the Complexity of Canals and Drainage Infrastructure Instituted in the Urban Sites of Roman Britain: Five Urban Waterscapes. In *Water and Roman Urbanism: Towns, Waterscapes, Land Transformation and Experience in Roman Britain*; Brill: Leiden, The Netherlands, 2013.
68. Frassine, M. *Palus in Agro: Aree Umide, Bonifiche e Assetti Centuriali in Epoca Romana."Agri Centuriati" Supplementa 1*; Fabrizio Serra Editore: Pisa/Roma, Italy, 2013; p. 82.
69. Morhange, C.; Marriner, N.; Blot, M.L.; Bony, G.; Carayon, N.; Carmona, P.; Flaux, C.; Giaime, M.; Goiran, J.-F.; Kouka, M.; et al. Dynamiques géomorphologiques et typologie géoarcheologique des ports antiques en contextes lagunaires. *Quaternaire* **2015**, *26*, 117–139. [CrossRef]
70. García-Allyón, S. Integrated management in coastal lagoons of highly complexity environments: Resilience comparative analysis for three case-studies. *Ocean. Coast. Manag.* **2017**, *143*, 16–25. [CrossRef]
71. Ayache, F.; Thompson, J.R.; Flower, R.J.; Bourjarra, A.; Rouatbi, F.; Makina, H. Environmental characteristics, landscape history and pressures on three coastal lagoons in the Southern Mediterranean Region: Merja Zerga (Morocco), Ghar El Melh (Tunisia) and Lake Manzala (Egypt)'. *Hydrobiologia* **2009**, *622*, 15–43. [CrossRef]
72. El Mahrad, B.; Abalasana, S.; Newton, A.; Icely, J.D.; Snoussi, M.; and Kacimi, I. Socio-Environmental Analysis of the Management of Coastal Lagoons in North Africa. *Front. Environ. Sci.* **2020**, *8*, 37. [CrossRef]
73. De Santis, V.; Caldara, M.; de Torres, T.; Ortiz, J.E. Stratigraphic units of the Apulian Tavoliere plain (Southern Italy): Chronology, correlation with marine isotope stages and implications regarding vertical movements. *Sediment. Geol.* **2010**, *228*, 255–270. [CrossRef]
74. Goffredo, R. *Aufidus. Storia, Archeologia e Paesaggi della Valle dell'Ofanto*; Edipuglia: Bari, Italy, 2011.
75. Goffredo, R.; Ficco, V. Tra *Ausculum* e *Herdonia*: I Paesaggi di Età Daunia e Romana della Valle del Carapelle. In *Faragola 1: Un insediamento rurale nella Valle del Carapelle. Ricerche e studi*; Volpe, G., Turchiano, M., Eds.; Edipuglia: Bari, Italy, 2009; pp. 25–56.
76. Russo, S. La bonifica del Lago Salpi, in Capitanata. *L'Ambiente Storico* **1985**, *8–9*, 119–135.
77. Russo, S. La bonifica come conflitto, il caso del lago Salpi. *Storia Urbana* **1997**, *80*, 63–79.

78. Poso, R. Gli Autografi Vanvitelliani Sulla Bonifica Delle Saline di Barletta. Paone, M., Ed.; Studi di Storia Pugliese in Onore di Giuseppe Chiarelli IV; Galatina Editore: Lecce, Italy, 1976; pp. 25–138.
79. Available online: https://www.ecdc.europa.eu/en/disease-vectors/facts/mosquito-factsheets (accessed on 12 August 2024).
80. Cocchi, L.; Stefanelli, P.; Carmisciano, C.; Caratorti, F.; Taramaschi, L.; Cipriani, S. Marine Archaeogeological Prospection of Roman *Salapia* Settment (Puglia, Italy): Detecting Ancient Harbour Remains. *Archaeol. Prospect.* **2012**, *9*, 89–101. [CrossRef]
81. Lopez, S. *La Città delle Saline Più Grandi D'Europa: Margherita di Savoia e i Suoi Antichi Insediamenti Sommersi: Cenni Topostorici*; Santobuono: Margherita di Savoia, Italy, 1981.
82. Volpe, G. *La Daunia Nell'età della Romanizzazione*; Edipuglia: Bari, Italy, 1990.
83. Leidwanger, J. Opportunistic Ports and Spaces of Exchange in Late Roman Cyprus. *J. Marit. Archeol.* **2013**, *8*, 221–243. [CrossRef]
84. Grelle, F.; Silvestrini, M.; Volpe, G.; Goffredo, R. *La Puglia nel Mondo Romano. Storia di Una Periferia. Vol. II. L'avvio Dell'organizzazione Municipale*; Edipuglia: Bari, Italy, 2017; pp. 282–283.
85. Goffredo, R.; Totten, D.M.; De Mitri, C.; Valenzano, V. La Ricognizione in Superficie (Campagna 2013). In *Salapia-Salpi 1: Scavi e ricerche 2013–2016*; De Venuto, G., Goffredo, R., Totten, D.M., Eds.; Edipuglia: Bari, Italy, 2022; pp. 37–61.
86. Cirelli, E. Roma sul Mare e il Porto Augusteo di Classe. In *Ravenna e l'Adriatico dalle Origini All'età Romana*; Boschi, F., Ed.; Ad Quem: Bologna, Italy, 2013; pp. 109–121.
87. Keay, S.; Millet, M.; Strutt, K. The Survey Results. In *Portus*; Keay, S., Millet, M., Parioli, L., Strutt, K., Eds.; Archaeological Monographs of the British School at Rome: London, UK, 2005; Volume 15, pp. 71–172.
88. Salomon, F.; Strutt, K.; Mladenovic, D.; Goiran, J.-P.; Keay, S. Management of fluvio-coastal dynamics in the Tiber delta during the Roman period: Using an integrated waterways system to cope with environmental challenges at Ostia and Portus. *Water Hist.* **2023**, *15*, 105–123. [CrossRef]
89. Oleason, J.P. The technology of Roman harbours. *Int. J. Naut. Archaeol. Underw. Explor.* **1988**, *70*, 147–157. [CrossRef]
90. Tuck, S.L. Ports. In *A Companion to the Roman Republic*; DeRose Evans, J., Ed.; Blackwell: Hoboken, NJ, USA, 2013; pp. 323–334.
91. Loprieno, S. I Contenitori da Trasporto. In *Salapia-Salpi I: Scavi e Ricerche*; De Venuto, G., Goffredo, R., Totten, D.M., Eds.; Edipuglia: Bari, Italy, 2022; pp. 371–402.
92. Marriner, N.; Morhange, C.; Kaniewski, D.; Carayon, N. Ancient Harbour Infrastructure in the Levant: Tracking the birth and rise of new anthropogenic pressure. *Sci. Rep.* **2014**, *4*, 5554. [CrossRef] [PubMed]
93. Blackman, D.J. Ancient harbours in the Mediterranean. Part 2. *Int. J. Naut. Archaeol. Underw. Explor.* **1982**, *11*, 185–211. [CrossRef]
94. Tuck, S.L. Ports. In *A Companion to the Roman Republic*; DeRose Evans, J., Ed.; Blackwell: Hoboken, NJ, USA, 2013; p. 231.
95. Robinson, D.; Rice, C.; Schorle, K. Ship Loses and the Growth of Roman Harbour Infrastructure. *J. Mediterr. Archaeol.* **2020**, *33*, 102–125. [CrossRef]
96. Pickel, D. Villa Estates and Malaria Risk in Roman Central Italy. *Am. J. Archaeol.* **2024**, *128*, 243–277. [CrossRef]
97. Roiz, D.; Ruiz, S.; Soriguer, R.; Figuerola, J. Climatic effects on mosquito abundance in Mediterranean wetlands. *Parasites Vectors* **2014**, *7*, 333. [CrossRef]
98. Kant, R.; Pandey, S.D.; Sharma, S.K. Mosquito Breeding in Relation to Aquatic Vegetation and Some Physico-Chemical Parameters in Rice Fields in Central Gujarat. *Indian J. Marariology* **1996**, *33*, 30–40.
99. Bissanti, A.A. Le Saline di Margherita di Savoia: Localizzazione e Cambiamenti nell'Organizzazione del Lavoro. In *Sale e Saline nell'Adriatico (secc. XV–XX)*; di Vittorio, A., Ed.; Gianni Editore: Naples, Italy, 1981; pp. 31–50, in particular Figure 2.
100. Arnoldus, A. Agriculture and Land Use. In *The Roman Peasant Project 2009–2014: Excavating the Roman Rural Poor*; Bowes, K., Ed.; University of Pennsylvania Press: Philadelphia, PA, USA, 2020; Volume 2, pp. 471–516.
101. Keenan-Jones, D. Large-Scale Water Management Projects in Roman Central-Southern Italy. In *The Ancient Mediterranean Environment between Science and History*; Harris, W.V., Ed.; Columbia Studies in the Classical Tradition, V. 39; Brill: Leiden, The Netherlands, 2013; pp. 233–256.
102. Marin, M.M. Problemi Riguardanti la Stazione di *Salinis* della *Tabula Peutingeriana*. In *Saline e Sale nell'Antichità, Atti del Convegno sul Basso Tavoliere (Margherita di Savoia)*; Edipuglia: Bari, Italy, 1989; pp. 81–90.
103. Pliny. *Historia Naturalis*. XXXI.29.73–83.
104. Marzano, A. Marine Salt Production in the Roman world: The *saline* and their ownership. *Quat. Sci. Rev.* **2024**, *336*, 108776. [CrossRef]
105. Totten, D.M. Life in the Late Antique Town: *Salapia* in the 4th–6th c. CE. In *Salapia-Salpi 1: Scavi e Ricerche 2013–2016*; De Venuto, G., Goffredo, R., Totten, D.M., Eds.; Edipuglia: Bari, Italy, 2022; pp. 629–660.
106. Grossi, M.C.; Sivili, S.; Arnoldus-Huyzendveld, A.; Facciolo, A.; Rinaldi, M.L.; Ruggieri, D.; Morelli, C. A Complex Relationship between Human and Natural Landscape: A Multi-Disciplinary Approach to the Study of the Roman Saltworks in "Le Vignole-Interporto" (Maccarese, Fiumicino-Roma). In *Archaeology of Salt: Approaching an Invisible Past*; Brigand, R., Weller, O., Eds.; University Press: Leiden, The Netherlands, 2015; pp. 83–101.
107. Curràs, B.X. The Salinae of O Areal (Vigo) and Roman Salt Production in NW Iberia. *J. Rom. Archaeol.* **2015**, *30*, 325–349. [CrossRef]
108. Villalobos, C.A.; Gracia Prieto, F.J.; Ménanteauy, L. Las Sailinas de la Bahía de Cádiz durante la antigüedad: Visión geoarcqueológica de un problema histórico. *SPAL* **2013**, *12*, 317–332. [CrossRef]
109. Russo, S. *Le Saline di Barletta tra Sette e Ottocento*; Grenzi Editore: Foggia, Italy, 2001.
110. Morelli, C.; Carbonara, A.; Forte, V.; Grossi, M.C.; Arnoldus-Huyzedveld, A. La Topografia Romana dell'Agro Portuense alla Luce delle Nuove Indagini. In *Portus and Its Hinterland*; Monographs of the British School at Rome 18; Oxbow: London, UK, 2011.

111. Augenti, A.; Bondi, M.; Cavalazzi, M.; Fiorini, A.; Sericola, M. Archeologia dei Paesaggi nel territorio ravennate: Il Progetto Cervia. *Archeol. Mediev.* **2020**, *47*, 115–139.
112. Van Limbergen, D.; Hoffelinck, A.; Taelman, D. Pathways to Reframing the Roman Economy: From Uniformity to Diversity? In *Reframing the Roman Economy. New Perspectives on Habitual Economic Practices*; Van Limbergen, D., Hoffelinck, A., Taelman, D., Eds.; Palgrave: Cham, Switzerland, 2022; pp. 1–22.
113. Erdkamp, P.; Manning, J.G.; Verboven, K. (Eds.) Introduction. In *Climate Change and Ancient Societes in Europe and the Near East: Diversity in Collapse and Resilience*; Palgrave: Cham, Switzerland, 2021; pp. v–xxv.

land

MDPI

Article

Water as a Problem and a Solution in Arid Landscapes: Resilient Practices and Adapted Land Use in the Eastern Marmarica (NW-Egypt) between the 2nd Millennium BCE and the 1st Millennium CE

Anna-Katharina Rieger

Institute of Classics, University of Graz, Universitätsplatz 3/II, 8010 Graz, Austria; anna.rieger@uni-graz.at

Abstract: Arid environments are suitable for researching the resilience of landscapes, since their ecological conditions pose continuous water stress to plants, animals, and humans living there. It is not only water, but also soil that is a limited resource. The arid landscape of the Eastern Marmarica (NW-Egypt) serves as an example for studying the resilience in and of a past landscape and its inhabitants from the 2nd millennium BCE to the 1st millennium CE, which is conceptualised as a 'social arid landscape'. The adapted life strategies and resilient practices to make a living in the arid environment are reconstructed from (geo-) archaeological evidence, discussing the applicability of the concept of resilience for ancient (landscape) studies. Resilience is an etic concept, depending on the perspective on and scale of a system. With the categories of 'event', 'practice' and 'knowledge', however, various scales can be bridged; life strategies can be defined as communities of practice and dichotomies be solved. Niche dwellings in the ancient Marmarica, where exposure to stress was normal, functioned because of an elaborate water management and the mobility of the people living there. The resilience of the arid social landscape is based on mixed life strategies, where only a multi-factored crisis (economic and climatic) or a series of smaller shocks (many dry years) could have destructive impacts.

Keywords: resilience; aridity; landscape archaeology; socio-ecological system; ancient Marmarica (NW-Egypt); bronze age; Graeco-Roman period; resource management (water and soil); knowledge; communities of practice

1. Arid Landscapes and Human Agents

For ages, arid climatic conditions have formed very specific landscapes—steppe, grassland, and tundra as well as stony or sandy deserts. The key feature from the point of view of natural conditions is the absence or scarcity of water. However, many more parameters factor in the formation and development of arid landscapes: meteorological conditions, geology of the underground, geographical location and topography, relief, and geomorphology. Arid landscapes are a complex ecological system.

If human agents use landscapes and live in them, a process of adaptation on both sides begins. Inhabitants adapt to the extreme natural conditions by establishing practices of livelihood, which might change the landscape, whereas the landscape and its natural conditions are altered by the anthropogenic impact. These are mutually reciprocal and dynamic processes in which landscapes and humans are both agents and reactants. Vegetation, depending on soil and water conditions, as well as the animal species having a natural habitat in the arid landscape are part of such adaptation processes of a human population, which itself had and has to adapt to the arid conditions.

The interrelations and interdependencies of arid landscapes, human agents, and fauna and flora form a complex system of a socio-ecological environment.

Resilience in and of such a system and how resilience as a descriptive and analytical concept can be applied to past arid landscapes and their inhabitants will be explained using the example of the Eastern Marmarica on the northern fringes of the Sahara in NW-Egypt in the time from the 2nd millennium BCE to the middle of the 1st millennium CE (Figure 1).

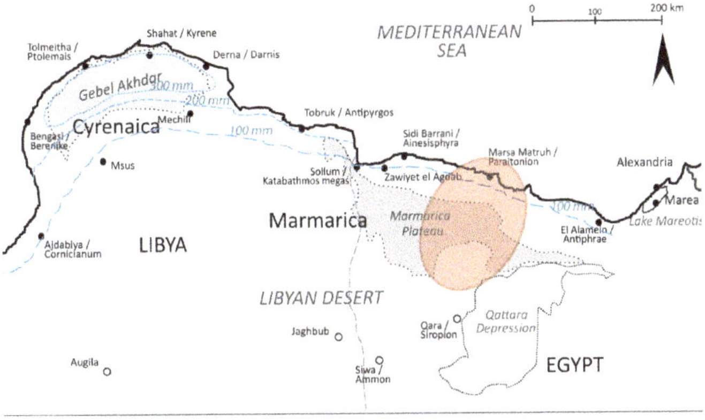

Figure 1. Northeastern Africa with the Graeco-Roman regions Cyrenaica, Marmarica, and Aegyptus in today's northwestern Egypt and northeastern Libya is a semi-arid and arid landscape on the fringes of the Sahara as the isohyets parallel to the shoreline show. The study area in the Marmarica is marked in red. A.-K. Rieger.

1.1. *Arid Social Landscapes and Archaeology*

Since resources—first of all water but also soil—are limited, the carrying capacities of arid landscapes in premodern times can be assumed as being even smaller than today. They are and were marginal landscapes not only in ecological but also social regards, which is reflected in lower population densities in arid regions, since not as many people can and could live on the scarce resources in arid regions compared to more humid regions—if resource availability is not drastically enhanced, which is facilitated through technological developments in modern times. In the past, people had to adapt to the arid environment and they transformed the landscape in order to improve resource availability and minimize risks and uncertainties regarding their livelihoods and subsistence. The mutual influences of (wo)men and landscapes happen on various scales and in different realms of the socio-ecological system, such as space and connectivity or social practices to inhibit the stressors. Hence, the long-term utilisation of arid landscapes in premodern times has formed specific landscapes with specific livelihoods, such as mobile life strategies [1]. Using larger spaces through mobile life strategies means having access to more resources in the resource-poor environment as base of people's subsistence.

The scarceness of resources leads to low population numbers and specific life strategies. For archaeological records, this means that they are also not numerous in arid landscapes—neither in quantity nor category. The lack of data has to be compensated by multi-disciplinary, even transdisciplinary, approaches in the study of past arid landscapes [2].

Instead of using socio-ecological systems used in environmental studies or in environmental sociology [3], in the context of this paper, I combine the notion of the 'social' with landscape, which strengthens the 'human side' in landscapes, which is implicit in landscape but rarely explicated. Even though in landscape studies, the term in Germanic languages 'Landschaft' or 'landscape', in Romanic languages 'paysage' or 'paesaggio', is understood rather as a (wo)man-made surface and habitat of the earth as it prevails also in its etymological origin—land that is 'scaped', formed or made, in German 'geschaffen', deriving from the Middle English or Old Dutch 'scap' or 'skap', meaning quality, condition, constitution, shape, but also act, power, skill. The term social landscape is understood as 'encultured and enearthed' [4]. It covers, for example, aspects of vegetation as well as of socio-economic practices, of geological forms as well as of route systems. The term landscape implies an etic, human perspective on a natural land surface, which is an historical archive and not only a set of ecological parameters. This anthropogenic aspect of landscape is emphasised by the addition of 'social' to the landscape [5]. A social landscape is an enlivened landscape, a dynamic system which forces interaction, embracing co- and counteracting between the more natural and biophysical pole as well as the more human, social pole [6] (Figure 2).

Figure 2. Scheme of an arid social landscape and its entangled components on the nature and the human side. The human and social side has agency and intentionality, whereas the side of biophysical parameters has only agency. Events are the impacts on the social landscape from both humans and nature; practices and knowledge impact the landscape and the biophysical parameters only from the human side. A.-K. Rieger adapted from [7] (Figures 1 and 3).

For the special case of arid environments, which are central to my argument, the social landscape is characterised by

1. A regime of risks of 'too much' or 'too little' water ('gluts' and 'dearth' as extreme forms of availability or unavailability),
2. A distinct logic in livelihoods and strategies of subsistence as a response to risk and stress,
3. A topographical fragmentation or "patchiness" of productive, less productive areas of the region,
4. A distinctive concept of space and a distinct regime of communication—large spaces, few people, but a high degree of organisation in order to function.

These conditions, prerequisites and resulting life strategies of arid landscapes as marginal landscapes make social landscapes either more resilient or more fragile when climatic, social, or economic parameters change, which would put stress on the system or parts of it [8]. Stressors to plants, animals, and humans in an arid environment are first of all water scarcity and variability, so this element will be the focus of the following arguments on the resilience of an arid social landscape.

1.2. Approaches to an Arid Landscape in Archaeology—The Eastern Marmarica as a Case Study

In order to analyse and study an arid social landscape, a number of methods have to be applied and adopted—geoarchaeology, pedology, survey and excavation archaeology were the basic set of methods and approaches, complemented by the material analyses (mainly pottery) used in the Eastern Marmarica Survey.

The data collection of climatic and environmental parameters in an arid landscape focuses mainly on hydrological and pedological conditions—humidity, fluvial geomorphology, vegetation geography, sedimentation and Aeolic impacts. Historical and recent surveys, for example, of precipitation or soil conditions of the present-day landscape, have to be integrated into the analysis of these data to enable a reconstruction of the palaeoenvironment [9]. In the course of the project, stratified and C14-dated archaeobotanical evidence as well OSL-dated sediment fills form the base of the palaeonvironmental reconstruction; beyond this, analogies and comparisons complement the image, whereas further geo-archaeological sampling and analysis was not possible in the course of the project. Once past environmental factors are understood, one can assess the availability of resources and the technologies people applied in order to reconstruct livelihoods, social organisation and socio-economic relationships. For historical periods, at this point, written sources also have to be considered, since they may contain glimpses of life strategies and people's socio-economic and -cultural connections.

Despite the fact that the comparison of the practices of groups living later or recently in the same environment can mislead our 'reading' of the conditions and influence our interpretation of processes in antiquity, presupposing a historical continuity [1] and ref. [10] (for a criticism of this methodology), inquiring into anthropological material and travel accounts on agricultural methods or economic relations from the 19th or earlier 20th century can support a better understanding of the material from Graeco-Roman times [11–14]. When comparisons of socio-environmental parameters are vigilantly applied to the reconstruction of past societal systems, a more detailed picture of social and socio-economic behaviour and practices can be obtained, as for example the organisation of mobility and movements, the frequentation of places, the economic production and connectivity [15].

To this written and documented evidence belongs historical cartographic material which offers a record of now lost places, visible traces, or courses of routes and places. Even though this approach enters the field of ethno-archaeological or historical anthropological comparison, it offers access to the study of organisation of space and mobility patterns in an arid social landscape [16] and ref. [17] (pp. 460–463).

Since a resource-poor environment correlates with a rather find-poor environment, it is not only anthropology, but also prehistory and methods from natural science (OSL dating, micro-morphology of utilisation horizons etc.), which are useful, because they develop methodologies and techniques to analyse phenomena such as periodical movements, the frequentation patterns of places, the utilisation of material objects, etc. Much of the evidence in arid social landscapes consists of surface finds, since stratigraphies are shallow and horizons of utilisation scarce and not very thick [18] and ref. [19] (23–25 on the assessment of surface finds) (Figures 3–5). In many cases, this leads to a non-stratigraphically build-up body of material. Secondly, and unlike groups in prehistoric times, even in historical periods (Graeco-Roman times in this case), multi-sited communities or nomadically living groups did not live with a large quantity of objects and relied on temporary camp sites. Only through the combination of records, data, and material as well as methodologies can results be obtained, which can be difficult in Egypt because of sample export bans and limited laboratory facilities.

Figure 3. Examples of the archaeological evidence on the surface in the Eastern Marmarica from Graeco-Roman times: fireplaces in the depression of the cistern Bir Abu Mukhayat on the Marmarica Plateau (see the cistern map below in Figure 20) with small numbers of Graeco-Roman pot shards (**left**). A.-K. Rieger; the embankment of a tableland field in Wadi Umm el-Ashdan on the Northern tableland with a concentration of Graeco-Roman pot shards between the fieldstones of the embankment (**right**). S. Valtin.

Figure 4. Examples of the archaeological evidence on the surface in the Eastern Marmarica from Roman times: the cistern site Bir Qattrani on the Marmarica Plateau (see the cistern map below in Figure 20), which has traces of Roman up to recent frequentation. The person in the photograph walks along the channelling bund. A. Nicolay; a pile of shards containing the remains of the firing process of a Roman pottery production site at Alam el-Rom (2nd to 4th century CE) [20]. A.-K. Rieger.

Figure 5. Examples of the archaeological evidence on the surface and subsurface in the Eastern Marmarica from Roman times: fieldstone walls of buildings in the settlement of Wadi Umm el-Ashdan. A.-K. Rieger; a trench in Wadi Umm el-Ashdan reaching the natural bedrock 60 cm below the surface. A.-K. Rieger.

Starting from the described approaches which cover the more data-related documentation and interpretation of the geo-and biophysical as well as the human and social phenomena and their material reflection in an arid region, I first have to introduce the concept of resilience and explain how resilience as a modern concept in the field of ecology to psychology and economy [21] can be applied to an "natureculture"-complex, a landscape, of the past [22]. Then the social landscape of the Eastern Marmarica will be introduced, in order to research the resilience of arid landscapes, and applied as a descriptive and analytical concept in interdisciplinary studies of landscapes and in landscape archaeology [23]. For the purpose of the analysis of the material from the past social landscape of the Eastern Marmarica, I will work along three categories that allow a connection of geo-archaeological and socio-archaeological data and approaches. These categories are events [24], practices [25], and knowledge [26] (see Figure 2), that correspond—not fully, but almost—to the characteristics or arid social landscapes mentioned above. They may help to overcome the dichotomy in the data of the eco-facts and arti-facts that obscures the complexity of the systems in order to understand the socio-ecological system. All categories are reflected in archaeological records and physical-geographical material and data. Events such as rainfall and practices such as constructing fieldstone walls to obtain fields are linked by people's knowledge. In this combination of methods, records, and approaches, which will not be described in full detail, the resiliencies in and of the system will be assessed2. Approaches to the Concept of Resilience in Social Landscapes and LandscapeArchaeology.

What can be resilient and what might resilience of an arid social landscape look like? How can the adaptation of people to the specific conditions of a water-scarce environment and its climatic, biological, and geological conditions and the impacts and changes in the landscape be seen through the lens of resilience?

2. Definition and Problems

To start with the common definition of resilience, as understood and applied in either ecological, economic or psychological contexts, it is the capacity of a system or individual "to absorb disturbance and reorganize while undergoing change so as to still retain essentially the same function, structure, identity, and feedbacks" [21] and [27] (p. 6). Resilience means the ability of a socio-political or ecological system, of an organisation or of individuals to recover after a shock, to keep going even though they have experienced changes they were able to absorb and adapt practices and ways of living to a changed state of being. Resilience includes both an aspect of stability or conservatism and the aspect of transformation and adaptation, as long as some characteristics of the former state are still recognisable (often defined by an etic entity, such as researchers, or by emic groups, such as elites). Further, the ability to marginalise the losses, which might have resulted from stress or crisis, belongs to a reaction leading to or enhancing resilience.

Resilience as a paradigm in research started off as looking at larger systems (ecosystem, economic system etc.), but in the wake of socio-constructivist and science-technology studies from the 1980s [22,28], there was a turn in resilience studies to inquire into socio-psychological and socio-ecological resilience, including multi-level perspectives, which allow the consideration of subsystems or co-occurring systems [29,30]. These differentiated approaches to resilience help overcome the nature/culture split which runs through many socio-ecological and archaeological studies [31–33].

Recent research and projects try to connect the individual and the societal-collective layers when looking for resiliencies on the one hand, and to bridge the disciplinary borders researchers have to transgress on the other hand [34–36]. Resilience is not a property but emerges in relations between and dependencies of agents and factors and is a highly relational concept. Resilience turns out to be a feasible descriptive and analytical concept to examine the interactions and relations of the environment and humans [37]. Multi-level perspectives as well as a historicising approach are necessary to make use of resilience as an analytical concept, in order to know about the scale the study looks at on the one hand (micro-, meso-, or macro-scale, corresponding, for example, in the case study with a wadi catchment, the Marmarica Plateau, or the fringe of Saharan Desert), and about the interpretative powers on the other hand [38].

In research, resilience tends to be often essentialised for the purpose of explaining adversities and catastrophes while a system keeps 'stable' [39] (pp. 201–202). However, if it is used as a normative term, it needs a clear definition of [31] (pp. 4–8) and ref. [40] (pp. 87–90 with a contradiction to normative aspects of resonance pp. 104 and 106). When it comes to the application of resilience, the determination of the areas and scale in which one looks for 'change' or 'sameness' as well as the definition of what was the 'state before' has not to be essentialised but is discursively negotiable [41,42], ref. [43] (call the system also "socio-ecological-techn(olog)ical") and [44]. If resilience is about the reaction to disturbances, from long-term stresses to acute shocks and disruption aiming at restoring the previous state, we also have to clarify what a stressor or disruptive impact is and how differently a system, people, a landscape or parts of them, react to it [45]. It is an etic and not fixed concept, whose meaning and content depends on ascriptions of what is resilient towards which stressors [46] and ref. [47] (partial perspective of humans). Ascriptions bear the risk of giving resilience a normative character, where only in a strictly historicising perspective, deciphering interpretative powers in a system and its entities looked at [48]. The same is true for continuities and discontinuities, important in historical research, which can also be normatively loaden, depending on the perspective [49].

2.1. Resilience in Systems of the Past

Resilience as a modern concept cannot be transferred directly to ancient studies. It is not an emic term or concept in the mindset of people in the past. However, many activities, practices, institutions and technologies show that people in ancient times applied material, narrative, or performative practices to explain (retrospectively and prospectively, e.g., in funerary speeches), to find relief (e.g., memorials, religious festivals) or to re-arrange shaken systems (redundancy, prevention, e.g., grain storage, building techniques) the changed conditions, so that we can call it a resilient reaction or resilience of the 'systems' and co-occurring systems [50], such as populations, economies, trade networks, marriage alliances. However, we need to clarify the standpoint of what to understand as resilience in a certain case and ascriptions of identity and sameness to a state before. Resilience makes only sense if the human perspective is explicated, especially when it comes to landscape studies. So, the concept turns out to be a good descriptive and analytical tool for the study of marginal social landscapes where human beings are exposed to more stressors than in more temperate regions.

A system which is not fragile and not easily breaking up but can resist shocks and changes can be called resilient. Hence, an arid social landscape does not seem to belong to such a socio-ecological system. However, as mentioned above, resilience is an etic term, and should not be used in a normative way implying that 'not breaking up' is considered 'better'. From a cultural-anthropological view, the concept of resilience is able to overcome the nature-culture split. However, it should be used strictly descriptively and analytically instead of normatively in order not to interpret the human or climatic impacts on a landscape as improving or worsening. Since a landscape is a dynamic system on a temporally very extended scale, it is only in the combination with the human agent that resilience becomes an issue [39,51]. Only the people inhabiting a landscape had (and have) an understanding of what a landscape, animals, plants, and the humans look like, how they conceptualise it, how they live their lives, and what use they make of the landscape. Humans act with agency and intentionality (see Figure 2), whereas the biophysical landscape enacts only agency ('events') [52,53]. I argue that a social landscape has resilient features and the people using it can cope with stressors. A landscape cannot be resilient, a landscape exists. Even though this seems to be an essentialising statement, the landscape does not care about the changes.

Resilience as an analytical and descriptive category in historical research has to be used context- and perspective-related. It is productively applied—also to historical social landscapes—when it is used as a descriptive tool to grasp the tension between system and agency, between stability and ongoing change in societies, their entities and institutions, or individuals as well as their interrelations, of ecosystems or economic networks [30]. Dissonances and irritations can result from this tension and have an influence on landscapes, societies, technologies, or religions. In the worst cases, catastrophes and shocks can disrupt. How strong this influence or these disruptions are depends on their intensity, the scale which we look at, and the relationships of the various elements that take part, as for example climatic impacts, markets, administrations, social groups, or religious concepts. Parts and subsystems can be influenced differently.

Resilience is also a useful analytical concept in historical research to examine how frictions were reduced, opportunities enhanced, or a status quo of a regime or interrelated regimes maintained (such as climatic and economic regimes or two conflicting communities). It is not synonymous with adaption, resistance, or flexibility and should not just replace these terms. Its strength lies rather in opening our perspective to uncertainty and variabilities in systems and parts of them. Its analytical power coincides with its focus on processes as well as on action and reaction—in human–nature relations as well as social relations.

To apply resilience in combination with a multi-level perspective on a past social landscape is productive because it allows to consider the different scales in the systems' entity and parts. Different scales apply to time and space, since an eco-system can be much larger and long lived than the groups inhabiting it. Social and individual time differs from the ones of plants, animals, climatic and geomorphological changes (ecological time). Groups might be resilient but not an individual and vice versa. Parts of an ecosystem or social system may survive unchanged, others may not. Stability on a community level transgresses also the individual and generational life spans which are in most cases shorter compared to ecological time spans. Only a natural catastrophe might have the same immediate effect on the social system as well as on the ecological system, where it leaves archaeological and geo-archaeological traces. However, the difficulty for (geo-)archaeology is that often the fine resolution of chronologies in the archaeological and palaeoenvironmental record is missing or technically unrecognisable, so that the differentiation of short-term variabilities is rarely possible [54–56].

2.2. *Categories to Approach Resilience in an Arid Social Landscape*

Arid social landscapes induce a specific interaction of humans and the geo- and biophysical parameters [57]. This interplay is closely linked to the resource water. Precipitation is variable, its absence poses a risk to plants, animals, and people; dry years and droughts are common. 'Gluts of water' in the rain season can vice versa cause extreme damage because of the accumulation of rainwater in the wadis. These are the events, related to water and soil, to which as another meteorological impact wind could be added influencing soils, morphology, plants, animals, and humans (Aeolic and fluvial regime) [58].

People develop specific practices to adjust and adapt to the environment as well as to the socio-economic domains and networks to which they belong or in which they are embedded. The practices of people in the ancient Marmarica which we can detect archaeologically comprise agricultural activities, dwellings, burial practices, pottery making and utilisation, diet, as well as trading.

The practices are based on knowledge, including the awareness and knowledge about the unpredictability, the uncertainties; knowledge about how to improve water availability and how to balance its variabilities. Detailed knowledge is necessary to use and navigate the landscape where water and human occupation are only scarcely distributed. The collective and individual factor comes into play here, since (part of the) knowledge lies in each individual; yet it is also a collective knowledge, passed on from generation to generation. Since knowledge, which can be a tacit and embodied knowledge, is closely related to practices, the two will be dealt with together.

Social landscapes reflect and are outcomes of events, practices, and knowledge and they are also a part of it. Hence, landscapes are assemblages of human (individual/social), natural (animated/unanimated) and technological factors [59,60] and ref. [61] (p. 1094 the strength of assemblage "lies precisely in their capacity to deal with coexisting complexities, keeping open their multiplicities without reducing them to singularities"). Looking at landscapes as assemblages of events, practices and knowledge allow for a systematic analysis of data sets, while at the same time the option to adjust or enlarge the data sets or to connect them to another (part of the) 'system' can be kept open [62]. The assemblage allows to easier switch from a micro-scale to a meso- or even macro-scale as well as including a meta-level in the study and adjust the design accordingly.

For a detailed analysis of the applicability and the value of resilience in a historical landscape, these three categories allow to bridge the human and the natural aspects of it. Using the example of the Eastern Marmarica in NW-Egypt from the Bronze Age to Byzantine times, I will exemplify the range of factors, agents, and their relations, be it land surface, morphology, water, soil, climate, humans, productivity, plants, or animals.

3. The Arid Social Landscape of the Eastern Marmarica and Its Resiliencies—Long-Term Continuities and Short-Term Variabilities

An arid social landscape can be investigated against the backdrop of resilience according to the main characteristics described above—(i) scarceness of resources, (ii) adaptive strategies for production, (iii) patchiness of utilised areas, (iv) space, communication, and mobility. We can ask about its resiliencies while being under permanent stress, mainly induced by the scarceness of water. Animals, plants, and people live at risk, mostly, by not having enough water and land to put down roots, by not finding food to reproduce or to grow crops and breed livestock.

A clear-cut set of questions about the resilience of a social landscape has to address the skills, options, and techniques which adapted species have and had developed to survive in arid environments, and in what ranges and magnitudes they are successful in adjusting themselves (reorganisation in the adaptive cycle [29] (Figures 1 and 2) to hardening or adverse conditions, influences and changes (release in the adaptive cycle). Moreover, the periods and time spans in which they adapt need to be assessed historically.

3.1. *The Ancient Marmarica*

Research was conducted on the Eastern Marmarica Plateau in northwestern Egypt between 2004 and 2011, aiming at a reconstruction of the interrelation of landscape, economic potentials, life strategies, and habitational patterns as well as route networks in this semi-arid to arid area between the Mediterranean coast and the Qattara Depression in Graeco-Roman times (Figures 1 and 6). The design was interdisciplinary and geoarchaeological, and the study area covered an area 50 km east and west of Marsa Matruh, the ancient Paraitonion, as well as 150 km south of the coast in the direction of Siwa.

North of the Marmarica Plateau, three main ecological zones can be distinguished (Premarmarican Plain, Northern Tableland, Coastal zone) with environmental conditions comparable to today. Only the conditions for people's livelihood changed from approx. the early 20th century onwards, because of technological developments in, for example, cistern construction, motorised farming, and adapted crops). Life, survival, and livelihood depended in antiquity on the successful management of local resources and adaptation to natural uncertainties. Regional ancient water and soil harvesting systems that were developed over decades or even centuries are a testament to the long-term experience of their builders and triggered special water and sediment dynamics.

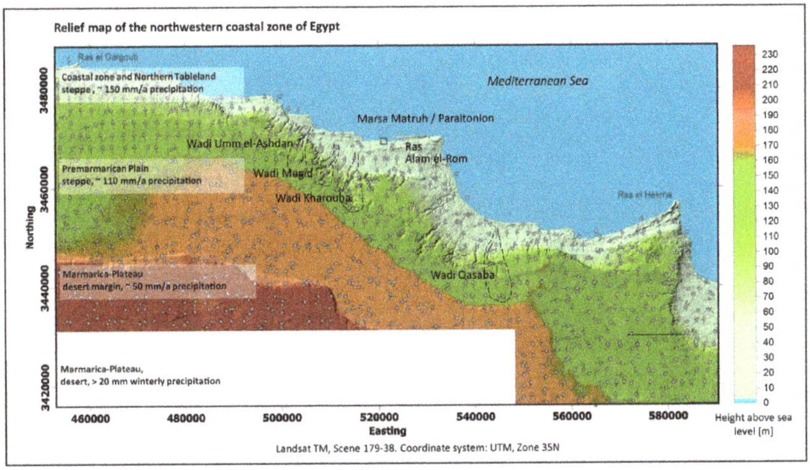

Figure 6. Relief map of the Eastern Marmarica with the ecological zoning and wadi and places mentioned in the text; the arrows show the direction of the overland flow. A. Nicolay, based on Landsat TM Scene 179-38.

The following characterisation of the ancient Marmarica will be cursory, but it mentions the basic parameters of the socio-ecological system.

3.2. Geology, Pedology, and Relief

The Marmarica is an arid landscape on the northern fringes of the Libyan Desert, the Sahara. The region is characterised by a sequence of northward sloping plains, interrupted by three scarps. The areas in between form a sequence of almost parallel ecological zones. The coastal zone, which is approx. 5 km wide, has a Mediterranean climate and is formed by the sediments brought down from the escarpment emerging south of the coastal zone raising approx. to 100 to 120 m asl. It leads up to a Table land formation with an inclination of less the 2%. Numerous wadis are cut into this tableland and the escarpment, which drain into the Mediterranean Sea. Their catchment areas reach 20 to 25 km south on the tableland (Figure 6). With the Pliocene Marmarican Cliff at a distance of approx. 40 km from the coast, the Marmarica Plateau begins, continuing on a mean altitude of approx. 200 m asl until it steps down to the Qattara Depression with the Oasis of Siwa. It is constituted of limestone, which is the prevailing geological material [63]. The extension of the Plateau reaches to the west to the Golf of Sollum and to east to the bay of Marina el-Alamein.

The soils of these plains are generally shallow, loamy, calcareous, and stony. In some places, wind-blown sands occur and can form shallow sand-sheets [64]. All fine sediments, including aeolian deposits, can be secondarily relocated because of the dynamic overland flow regime of the fringes of the Marmarica Plateau, as geomorphological evidence proved frequently, for example, on valley shoulders with a pavement of residual soil skeleton on bedrock outcrops.

3.3. Rainfall

Precipitation is the primary determining ecological factor influencing fauna and flora as well as land use by the inhabitants. Based on modern data, maximum seasonal rainfall in the coastal zone does not exceed approx. 150 mm (Figure 7). However, 5 km inland, 160 mm was measured. In the south on the Premarmarican plain and on the Marmarica Plateau, the annual rainfall drops from 80 mm to less than 20 mm. The main wet season is December to February, with high annual variabilities—which are characteristic for arid environments. We have chosen S'de Boqer in Israel as a comparison with less mean annual rainfall, however, variabilities there have very similar patterns and amplitudes.

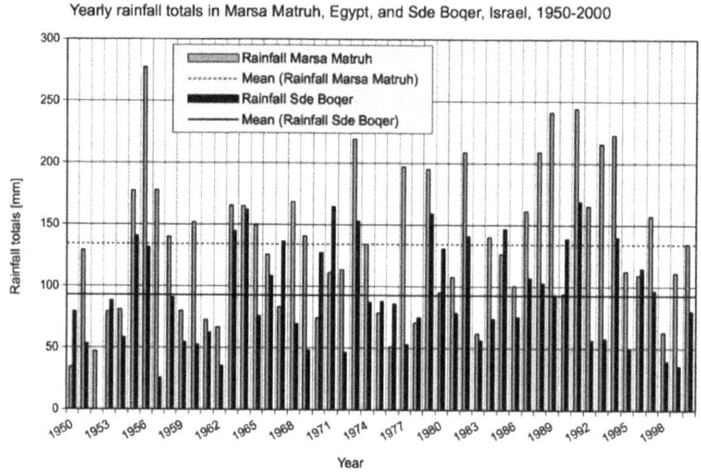

Figure 7. Diagram showing the precipitation of Marsa Matruh compared to S'de Boqer in Israel from 1950 to 2000 to exemplify the magnitudes and variabilities (both aggregated January to December). Database from Meteorological Service of Egypt and Israel Meteorological Service), diagram Th. Vetter [65] (Figure 2).

The entire region lies below the limit of rain-fed agriculture at 200 mm/a and only allows dry farming, which according to palaeoenvironmental studies in the Eastern Sahara and the results from our research, mainly based on the archaeobotanical micro- and macro-remains, but also on archaeozoological evidence, was not different in the time period covered in this study [18,66], ref. [67] (pp. 42–43 with references) and ref. [68,69], so that agricultural utilisation of the region is only possible through water harvesting methods. An attestation confirming the meteorological regime from Roman times exists in the form of a graffito scratched into the fresh plaster of a cistern close to Paraitonion, which is now lost: A certain Isalas has imprinted his hand and his name in the fresh plaster, as well as an altar palm branch and the date of August of the year 6 BCE. This could only have happened when the cistern was empty [70] (Figure 8).

Figure 8. Plaster from a cistern southeast of Paraitonion/Marsa Matruh, where Isalas had imprinted his hand in the freshly plastered cistern in August of the year 6 BCE. This indicates dry summers [70] (Figure 1).

Today, water supply is—as studies of the correlation of income of the rural population and precipitation show—the critical factor for agricultural yields in the arid environment of the Eastern Marmarica and can be assessed for Graeco-Roman antiquity. Water stress was the main issue.

3.4. Overland Flow, Wadi Runoff and Soil

Morphology, topographical features, and conditions of the land surface play a big role for water availability as they influence the overland flow of precipitation. On the one hand the overland flow was (and is) technically altered to enhance water availability (see below Section 3.6). On the other hand, flood protection is an issue whereever overland flow accumulates in wadis, as this can have devastating effects on agricultural areas and plants as well as animals and people. Therefore, both control and redistribution of the overland flow on the slightly northward sloping Tableland and Marmarica Plateau which form huge catchment areas, and of wadi runoff play a role for the water supply (Figure 6). The water and soil harvesting measurements are suitable for adding up to a surplus of water flow over the tableland and into the valleys, from where the tableland, the valleys, and the coastal plains, which already receive the highest rainfall, receive harvested water, that can add up to another 150 mm to reach the minimum threshold for cropping cereals such as wheat or for tree cultures.

The water harvests which the land-units receive through runoff are sufficient to fill the supply gap for agricultural use, which would in more humid areas be filled by precipitation. However, this is only possible through another factor—soil. The water storage capacities of soil, which naturally accumulates in depressions or which is artificially accumulated behind terrace walls or embankments, increases the amount of water available to plants (Figures 9 and 10).

Figure 9. Men-made dams, terraces and bunds direct water and sediment carried with the overland flow. The water and soil harvesting structures (**left**: Wadi Senab, Roman?; **right**: Wadi Umm el-Ashdan, modern) enhance considerably the availability of these two resources, which naturally accumulate in depressions or slow down in the wadi beds. A. Nicolay, Ch. Schill.

In the Eastern Marmarica, the soil depth in the valleys and on the tableland was significantly increased by ancient water harvesting interventions from the 2nd millennium BCE onwards (Figure 11 and ref. [71]). The deepest soils occur in valleys, on alluvial fans, in coastal plains and endorheic depressions. Tableland soils are generally shallow, rarely deeper than 0.3 m. Deeper soils or relics of deep soils are frequently related to ancient water harvesting schemes. Today, the old wadi fills turn increasingly into badlands due to lacking water management. In some cases, wadi fills are almost completely removed, leaving behind barren rocky beds [72].

Figure 10. The sediment fills provide the basis for agricultural terraces and plots, which are at risk of being severely destroyed by gully erosion, which is a sign for the devastating impact of heavy rainfalls: eroded fills in the bed of Wadi Kharouba (**left**), accumulated at the latest from Graeco-Roman to recent times according to the ancient settlements and pottery production in the area. A.-K. Rieger; Wadi Megid (**right**), fills of lateral terraces from the 1st half of the 1st millennium BCE, dated via OSL dating, see [63,71]. A. Nicolay.

Runoff area / Yermic Calcisol (BP 8)

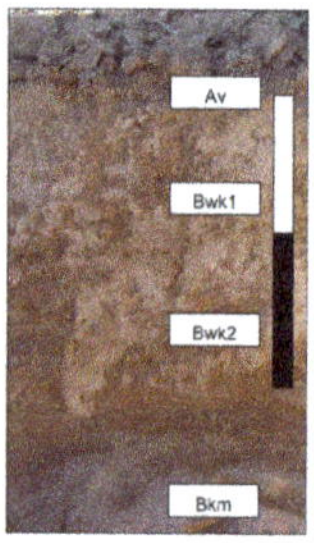

Depth [cm]	Description
0 - 1	very pale brown (10 YR 7/4); Lu; silt loam, vesicular crust, strongly to extremely calcareous; few fine gravel
- 8	very pale brown (10 YR 7/4); Lu; silt loam (SiL); massive (coherent) / subangular blocky structure, extremely calcareous; few fine gravel
- 29	very pale brown (10 YR 8/4); Tu3; silty clay loam (SiCL); massive (coherent) / subangular blocky structure, extremely calcareous; very few fine gravel

Runin area / Anthrosol or Haplic Calcisol (BP 1)

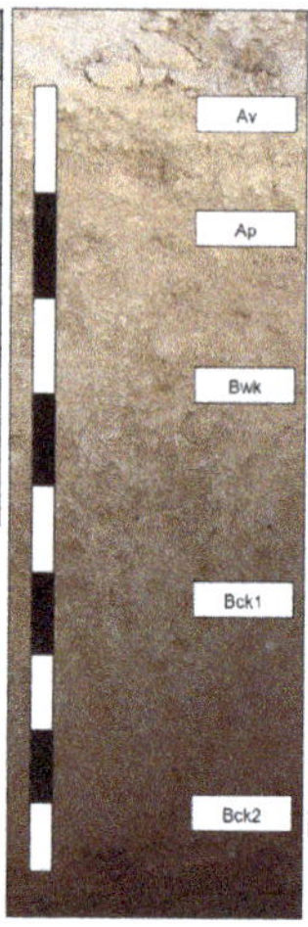

Depth [cm]	Description
0 - 0,5	very pale brown (10 YR 7/4); Tu3; silty clay loam (SiCL); vesicular crust; strongly calcareous; very few fine gravel
- 15	very pale brown (10 YR 7/4); Tu3; silty clay loam (SiCL); massive (coherent) / subangular blocky structure, extremely calcareous; very few fine gravel
- 40	light yellowish brown (10 YR 6/4); Tu2; silty clay (SiC); massive (coherent) / subangular blocky structure, extremely calcareous; very few fine gravel
- 70	light yellowish brown (10 YR 6/4); Lt3; silty clay (SiC); massive (coherent) / subangular blocky structure, extremely calcareous; very few fine gravel
- 110⁺	light yellowish brown (10 YR 6/4); Lt3; silty clay (SiC); massive (coherent) / subangular blocky structure, extremely calcareous; very few fine gravel

Figure 11. Two sections—to the **right** in the runoff area, to the **left** in the runin area on the tableland of Wadi Umm el-Ashdan—clearly show the varying soil depth of the natural soil and of the anthropogenically transformed soil. A. Nicolay.

3.5. Natural Vegetation

The endemic vegetation is made up of a plant spectrum typical of steppe environments, with the exception of a small strip of Mediterranean climate along the coast with a typical vegetation of smaller trees and shrubs [73,74] and ref. [75] (as comparison, however, the Marmarica is drier because of the Lee position to the Jebel Akhdar). On the Northern Tableland (down to 25 km south of the coast), plants have to be more drought resistant. The endemic species today are shrubs such as *Artemisia inculta*, *Atriplex halimus* or *Salsola*, a shrub belonging to the family of *Amaranthaceae*. Remains of these bushes were also found in an archaeological context [76]. *Asphodelus gymnocarpus*, *Thymelia hirsuta* can be found today, but were not found in the archaeobotanical remains [72,76,77] (Figure 12).

Figure 12. Vegetation on the Marmarica Plateau: Scattered shrubs on the Northern tableland (**left**). A.-K. Rieger; clockwise (**from top right**): *Asphodelus gymnocarpus*, *Salsola* (also found in Roman layers, see Figure 13), *Thymelaea hirsuta*, *Tamarix* (also found in Roman layers at Wadi Qasaba dated archaeologically from the 2nd to 4th century CE, and *Haloxyron salincornicum*. S. Valtin; O. Klammer; A. Heiss/VIAS Vienna.

Figure 13. Steppe and desert vegetation which is archaeobotanically attested includes Acacia found in a destruction horizon at Abar el-Kanayis dated archaeologically to the 5th/6th century CE and in a mortar from a Roman basin at Wadi Umm el-Ashdan A.-K. Rieger/V. Asensi; and *Salsola* sp., a shrub belonging to the family of Chenopodiaceae found, for example, at Wadi Umm el-Ashda in a levelling layer for the construction of a Roman building, at Abar el-Kanayis above the destruction layer of the 6th century CE, and at Zawiyet el-Agdab in mortar dated to the 6th century CE via C14 dating (for the location of Zawiyet el-Agdab see below Figure 21). T. Pokorny and A. Pokorna, https://cegu.ff.cuni.cz/cs/veda-a-vyzkum/projekty/terenni-projekty/zapadni-poust-2/zapadni-poust/electronic-herbarium/ accessed date 5 May 2023.

Their distribution is scattered and patchy due to the limited water availability and nutrient-rich soil. Wherever larger soil accumulations or depressions appear, the vegetation becomes denser. Grassland, however, does not exist.

Further south on the Premarmarican plain, vegetation of the above-mentioned kind is contracted in favourable places that provide enough water, such as depressions (see Figure 6 for the zoning). Desert conditions prevail from 90 km south of the coast to the Qattara Depression. Only *Acacia* trees grow in very favourable places, of which we have also archaeobotanical evidence (Figure 13, left) [18,73].

Even though we could not conduct an in-depth palaeoenvironmental analysis, the vegetation (and environmental conditions) do not seem to have changed much from the 2nd millennium BCE to today (they could, however, change in the near future because of global warming) [9]. Some archaeobotanical remains point to exactly this fact: it is

Chenopodiacaea and *Amaranthaceae*; endemic trees were *Tamarix*, *Acacia*, and *Phoenix* (date palm) as is the case still today (Figure 13).

3.6. Water and Soil Harvesting

As already mentioned above, people living in the Eastern Marmarica in the Bronze and Iron age, i.e., Graeco-Roman period, built various and far-reaching installations and structures from the abundant field stones to enhance water control and availability. These regional ancient water harvesting systems were developed and adapted over decades and centuries and reflect the experiential knowledge of their builders and users about the hydrological regime that they had acquired over a long period of time. This knowledge was gained through the special water and sediment dynamics typical of the region and its micro-units and zones.

The systems consist of dams across the wadis or tributaries, of lateral terraces along the wadi slopes, of diversion bunds on the plains on the Northern Tableland in which the valleys are cut (Figures 14 and 15) as well as of terraced fields on the tableland, made by embankments from earth with stabilising rows of stone inside. The latter ones—the tableland fields—are a unique feature in the Eastern Marmarica in comparison to other arid landscapes of the Mediterranean (such as the Negev, Tripolitania or Tunisia), since they are located in the runoff area and not in a classical runin area [63,78] (Figure 16). According to the ratio of catchment to cropping area (ccr factor)—the root-available water on the tableland field is sufficient for barley, in the valley for wheat and tree cultures (Figure 17). The oldest installations could be dated via OSL to the 2nd millennium BCE, continuing through the 1st millennium BCE to reach a peak with the most extensive installations during Roman times. We dated thirteen sediment layers in five sections via OSL. Samples were taken by night, analysis was done in Bayreuth by Markus Fuchs and is described in more detail in [63,71].

Another structure made for water harvesting (rather than for soil harvesting) and used as agricultural unit is stone mounds. They have a diameter of approx. 1.0 m to 2.0 m and rise between approx. 0.5 and 0.8 m from the surrounding surface. Filled with sediment, reaching down also to approx. 0.5 and 0.8 m is covered by cobbles and leads to the assumption that they were used as planting mounds by the inhabitants of the Graeco-Roman settlements on the tableland. The sediment could have stored the soil moisture, whereas the stone cover prevented it from evaporation. Even though their hydrological and pedological characteristics are not yet finally analysed because of the end of the project, their features, their location and distribution indicate their function as *teleilat el-einab*—for grape growing, comparable to a kind of mound in the Negev [79,80].

The most classical water harvesting installation is physically not so much changing [80] the landscape as the water and soil harvesting installations, however, it is extremely numerous: the cistern. On the Northern Tableland as well as on the Premarmarican Plain and the Marmarica Plateau, cisterns are dug into the limestone underground to collect the water from the winter rains. The cistern of Bir Helua on the Marmarica Plateau in Figure 18 is one of several such water supply points, which we surveyed in the Eastern Marmarica Survey. Many have been constructed only recently, as there was a huge development plan of the Egyptian government. However, many date back to pre-modern and even ancient, often Graeco-Roman, times, as the analysis of old cartographic material and the surveys show. The density of the cisterns corresponds to the rainfall gradient, decreasing from north to south. While several cisterns can be counted per square kilometre in the north, only two to four can be found 50 km south and less than one cistern per square kilometre can be found 150 km south of the coast (Figure 19).

Figure 14. Types of water and soil harvesting installations in the Eastern Marmarica: lateral terraces (Wadi Senab, **top**); a solidly constructed cross dam (Wadi Kharouba, **bottom**). Since the most intense settlement and production activities along these wadi can be dated to Roman times, we assume these installations to originate from the same period. In other wadi, OSL dating attests to a history of such structures from the 2nd millennium BCE [63,71]. A.-K. Rieger.

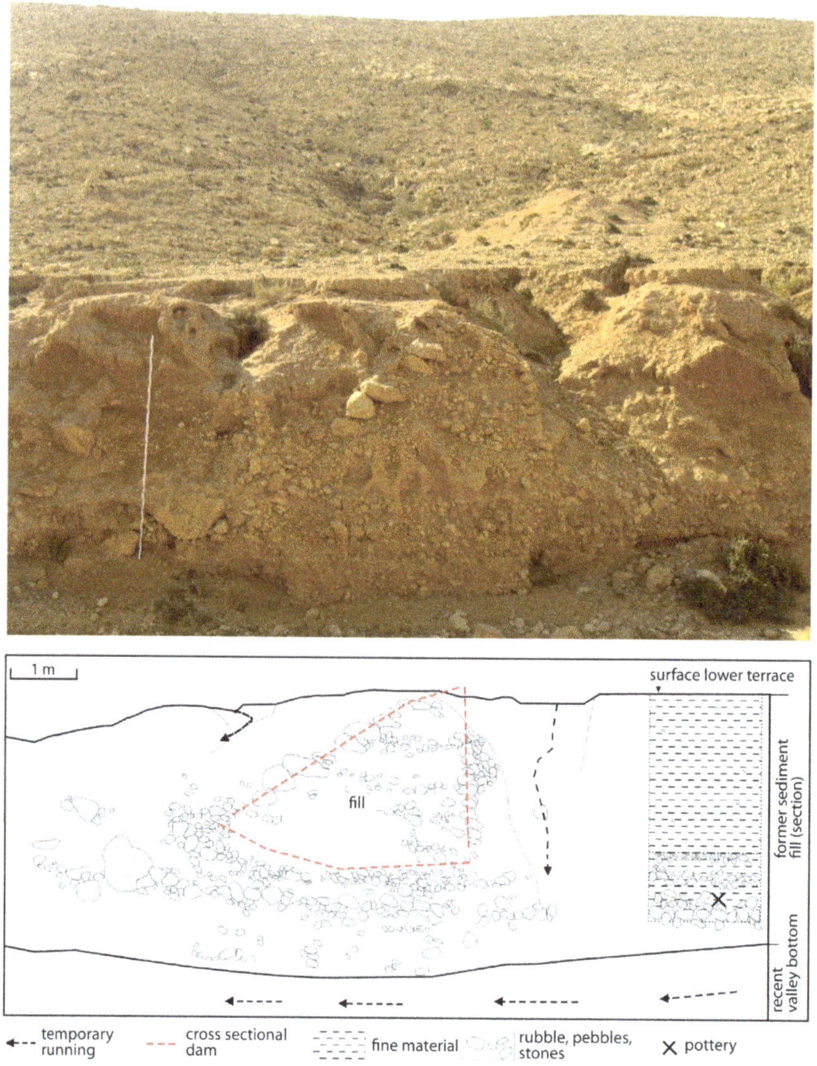

Figure 15. A loosely constructed cross dam (**top**) in a sediment fill in Wadi Umm el-Ashdan; the schematic drawing shows its construction (**bottom**) with the upstream (right) and a downstream (left) armour and the stepped rows of cobbles. Such structures were adapted to the continuously growing sediment fill. Dams of this kind were built in the Ptolemaic and early Roman period [78]. A. Nicolay.

Figure 16. Some water and soil harvesting installations are characteristic of the Eastern Marmarica because they are located in the runoff area on the tableland but still receive sufficient water without diminishing the runoff for the terraces along the wadi slopes and in the wadi too much (see Figure 17): embanked fields on the tableland of Wadi Umm el-Ashdan which date in their earliest phase to the 2nd millennium BCE via OSL dating (see [63] Table 4) and are still in use today (**left**); stone mounds of approx. 1.0 m to 2.0 m diameter with an enclosing ring made from field stones (**right**) on the tableland of Wadi Umm el-Ashdan close to the settlement from Graeco-Roman times. The fields are suitable for barley, whereas grapes could be grown on the mounds. A.-K. Rieger.

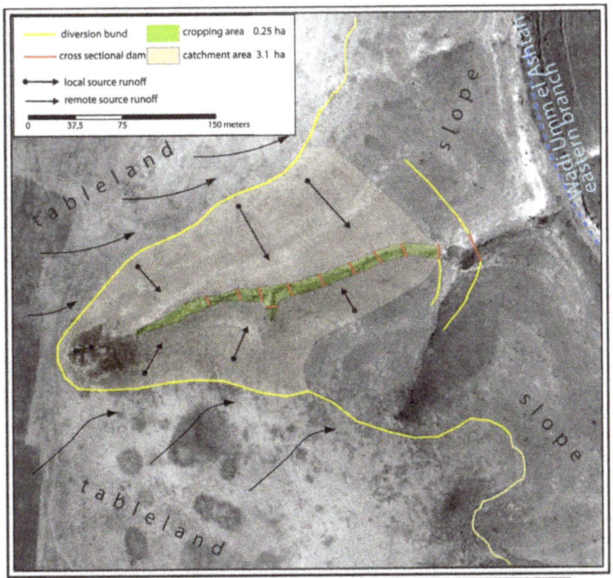

Precipitation / year	Water yield / root-available water with a CCR of 12.4 and three different runoff coefficients of arid landscapes 0.15	0.2	0.25
100 mm	186.0 mm	248.0 mm	310.0 mm
125 mm	232.5 mm	310.0 mm	387.5 mm
150 mm	279.0 mm	372.0 mm	465.0 mm

Figure 17. Schematic map of the water and soil harvesting system in a tributary to Wadi Umm el-Ashdan. A total of 8% of the entire valley is cultivable land, which receives water through the remote

runoff from the tableland and the local runoff from the slopes. The terraces in the valley of approx. 300 m reduce the natural longitudinal gradient of approx. 4.2% to 3.8% in the upper part and 2.5% in the lower part of the tributary. A. Nicolay.The table below the schematic map shows the various amounts of root-available water depending on the runoff-coefficient and the amount of precipitation. The CCR is the ratio (R) between the runoff/catchment (C) area (3.1 ha) and runin/cropping (C) area (0.25 ha), which is in this case 12.4. With the variated runoff coefficient from 0.15 to 0.25 the different water yields are calculated. A minimum of 150 mm of root-available water is sufficient for the cultivation of barley, 300 mm is necessary for tree cultivation as well as for wheat. A. Nicolay.

Figure 18. The cistern site of Bir Helua on the Marmarica Plateau (see cistern map below in Figure 20) is still in use today, but shows remains (pot shards) also from Graeco-Roman times. It is one of the several water supply points between the Mediterranean coast and the Qattara Depression with a long history of frequentation. In the foreground one sees the manhole to the right and a water trough to the **left**. The heap in the background are the remains from a building, most probably from Graeco-Roman times comparable to but smaller than the ones at Abar el-Kanayis (see cistern map below in Figure 20). A.-K. Rieger.

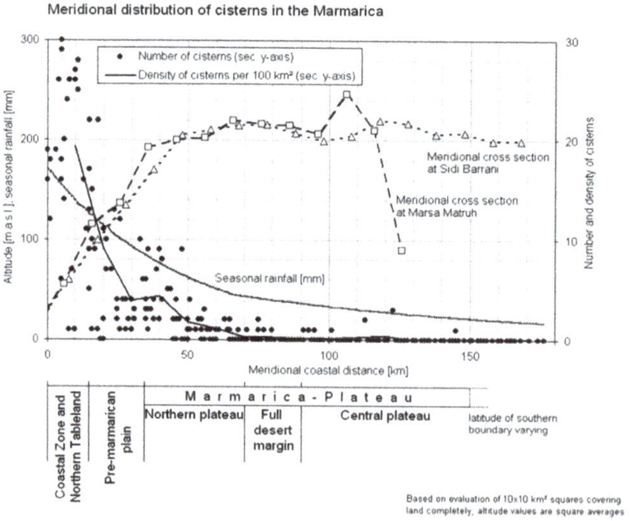

Figure 19. Distribution and number of cisterns according to rainfall and the ecological zoning in the Eastern Marmarica. Data are taken from the series P502 of the US Army Map Service, Corps of Engineers,

scale 1:250,000, using map sheets of the Middle East Forces (UK) 1941/42 based on information compiled by the British "Survey of Egypt". Since the rainfall pattern did not change very much from antiquity to today and all cisterns beyond the 20 km line south of the coast we surveyed date back to the Graeco-Roman or an even older period, the general ratio might have been similar in antiquity. Owing to denser settlement and government support, the closer to the coast (0–20 km) you look, the more modern constructions of cisterns can be found. Th. Vetter in [17] (after Figure 2).

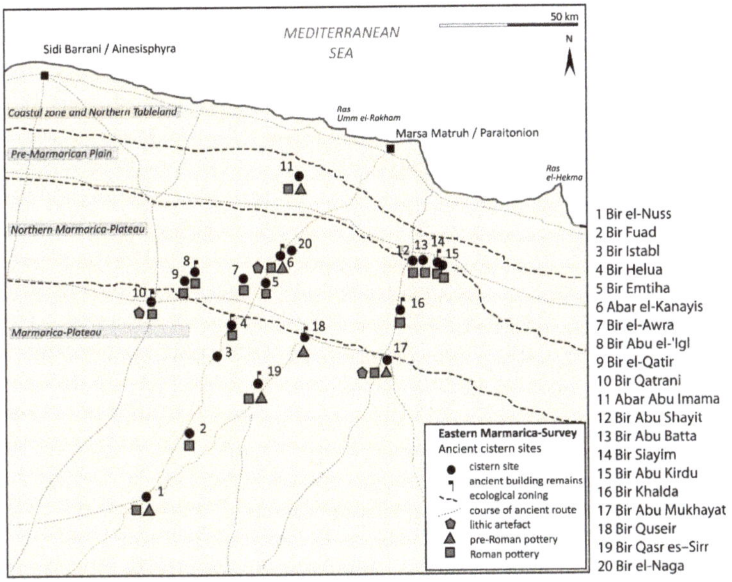

Figure 20. The mapped cistern sites surveyed in the Eastern Marmarica Survey all date from Graeco-Roman times at the latest and some of them were equipped with rest houses for the travellers and merchants on the routes between Siwa and Paraitonion/Marsa Matruh and the Mediterranean coast in antiquity. H. Möller/A.-K.Rieger [18] (Figure 2).

3.7. Living in Ancient Eastern Marmarica

The ancient runoff agriculture in the Eastern Marmarica and its potential is of particular interest in the context of resilience and water as a stressor, because it reveals the ability of people to establish forms of livelihoods and dwell in a niche in an ecological sense [39,51,81,82]. It could be discussed whether arid landscapes still form a niche for dwelling, as almost one-third (26.8%) of the Earth's land surface is arid regions with a population of 1.1 billion people (excluding populations in deserts and dry subhumid rangelands) [83].

Since the conditions did not change so much from the millennia BCE to today, as explained above [9,66], people in the Pharaonic (if we apply the Egyptian terminology) and Graeco-Roman times had to live in semi-arid to arid conditions in the Marmarica.

From the ecological marginality a political marginality can be deduced, since we have no evidence of political organisations based in the Nile Valley, in Siwa or coming from the Mediterranean in the region during the New Kingdom and in Persian times (Figure 21). Located in-between the better-known and economically powerful regions to the east and west, people in the ancient Marmarica seem to have been rather independent, except for a short-lived attempt at controlling the region under Ramses II, during whose reign the fortress of Zawiyet Umm el-Rakham close to the later Paraitonion was constructed, but without any long-lasting success [84]. Contemporaneous with the fortress, but because of the hydrological knowledge an autochthonous action, are the oldest traces of a water and soil harvesting system and agricultural activity in the region on the Tableland, dated to the late Bronze Age based on

OSL data [63]. Even though there are no habitations or other materials that can be attributed to the people using these early agricultural plots, the evidence correlates with the Late Bronze Age presence of Nile Valley Egyptians in the fortress of Zawiyet Umm el-Rakham [84] as well as with Mediterranean connections reflected in Cypriot, Aegean and Canaanite pottery found on Bates' Island in the lagoon of Paraitonion/Marsa Matruh [85,86].

As a marginal region with a position always between desert steppe and sea, between Nile Valley and Cyrenaica region, the Eastern Marmarica was never the centre of interest and remained rather independent throughout most of its history, even though officially being administratively under Ptolemaic rule from the 3rd century BCE. In the 1st century CE it became a province of the Roman Empire as part of the Crete and Cyrenaica [63].

Who the people inhabiting the Marmarica were is not easy to say, the Temehu, Tehenu, and Meschwesh are mentioned in Pharaonic sources, while various tribes were later listed and repeated by Graeco-Roman authors using ethnic stereotypes for desert dwellers [87]. However, during the later period focused on in our context (roughly the 5th century BCE to 5th century CE), the inhabitants extended and intensified the land-use and changed the patterns of economic exploitation, which is reflected in a higher find density: From the 5th century BCE onwards, the transfer of goods can be traced between Siwa and the coast, while an extension of the route system and its facilities as well as the intensification of its use is attested to from the 1st/2nd century CE onwards, and coming to an end—as far as the pottery indicates—with the 5th century CE [17] (476–480, pre-Roman finds, i.e., finds older than the mid-1st century BCE at Abar el-Kanayis, Bir Qatrani, Bir Abu Mukhayat, Bir Quseir) and ref. [20] (Figure 20). Accordingly, the described water and soil harvesting systems in the arid environment of the Eastern Marmarica reached their largest extent in the period between the 2nd century BCE and the 4th century CE. Barley as well as grapes seem to have been the most important crops, the latter also allowing for a surplus economy, which is inferred from the thousands of stone mounds and the output of the pottery workshops. Wheat played only a marginal role. Tree fruits such as figs as well as legumes could be traced archaeobotanically, but olives could not [63,88].

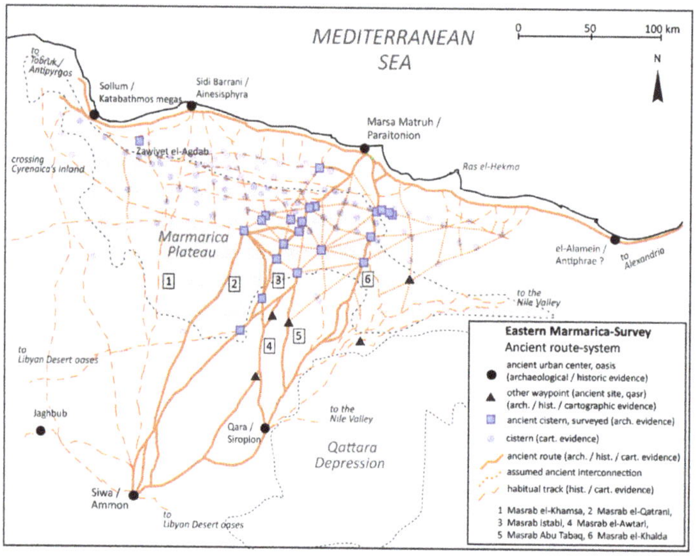

Figure 21. The route network attesting to trade mobility in Graeco-Roman times as well as to roaming as far south as pastures are available (approx. 60 to 80 km south of the coast). The frequentation of the routes in Graeco-Roman times is attested mainly by the findings of ceramic material. A.-K. Rieger/Th. Vetter [89] (Figure 16).

Centres of commerce with an even urban lifestyle included the harbour of Paraitonion (Marsa Matruh) and the smaller Katabathmos Megas (Sollum) on the Mediterranean coast. To the south, the Oasis of Siwa and Jarabub represent hubs for trade in the Libyan Desert (Figure 21).

4. Resilience of and in an Arid Social Landscape—Living with Uncertainties

The three concepts introduced above—event, practice, and knowledge—allow, in combination with the characteristics of the arid environment as can be found in the Eastern Marmarica (scarceness of resources; adaptive strategies for production; patchiness of utilised areas; space, communication, and mobility), to look at the various risks and stressors in the arid social landscape's system with co-occurring as well as subsystems and how they are interrelated, shifting and depending on each other. With the relational approach we can decipher even more precisely how the entities and subsystems are dynamically established and risks minimized [40].

Having explicated the details of the physical-geographical conditions and the human alterations and the living they could make from it, I will focus on water as a resource and agent, but also on soil when interpreting the social landscape through the three concepts [90,91]. It allows the assessment of biophysical as well as social evidence and material remains.

4.1. Socio-Natural Events—Uncertainties in an Arid Landscape Requiring Resilient Strategies

The concept of 'event' disentangles the dynamics active in and having an impact on a system [92] and ref. [93] (p. 9). The term 'event' describes a contingent event, which is capable of creating ruptures. The latter can trigger a reaction that strives for resilience [94].

In the context of analysing resilience in and of an arid social landscape, the perspective on water coming down as rainfall helps in the search for long-term continuities and short-term disruptions. Rainfall is of great variability and a very temporary phenomenon, but at the same time periodically appearing in the rain season in winter and thus a persisting phenomenon.

Water can therefore be conceptualised as an event in the Eastern Marmarica (as could be wind in the region): Precipitation is the primary event, overland flow and floods are secondary events in the arid environment of the Eastern Marmarica. They have an impact on the availability and how plants, animals and people are able to live. Water materialises in two ways: First, wherever the water slows down or comes to a halt, the sediment carried with it sinks down and accumulates. On the Marmarica Plateau and on the tableland, this occurs in depressions; in the valleys incised into the tableland and escarpment it is bends of the wadis where sediment packages accumulate. These are natural processes: through the recurrent events, sediment is swept into depressions and 'stratifies' there as an archive, so to speak, of the processes the water went through. The water not stored in the sediments drains into the Mediterranean Sea or disappears in the karstic underground forming water bearing stratum (which at some points leaks out at the coast.

The reaction by the inhabitants in the Eastern Marmarica to the eventful precipitations can be described as resilient since they knew how to live with the uncertainty of variable rainfalls and the overland flow generated by it. The varying magnitudes of rain events can lead to floods or droughts—the first being a shock, the latter being a stress, which can result in a crisis or at least a discontinuity of activities such as agriculture. The answer to water stress in the ancient Marmarica is an elaborated water and soil management. Heavy rainfall events can be accompanied by flash floods that do not only threaten animals, plants, and humans, but can have a long-term impact on the accumulated soils which erode and are swept away. Gullies show the destructive force of water events which people prevented by slowing down, deviating, and channelling the runoff of the rainwater. Constructing and maintaining such water and soil harvesting installations against the events that would change the setting is a resilient practice (Figure 9).

The event of rainfall, which is to a certain degree contingent and uncertain, had negative effects when it was too much or too little. People tried to minimize these effects in order to amplify the positive ones—water availability. They had (and have) influence though—not on the precipitation, but on its effects—through (wo)man-made measures. However, hope for rain or fear of heavy rains which is, in past as well as in contemporary societies, often framed in religious rituals, can be understood as a resilient strategy to cope with the risk of a lack of water.

In the sense of spatial theory, events make places [95]: The water in the Eastern Marmarica forms areas in natural processes as well as in (wo)man-induced processes in the form of sediment accumulations and the pertaining walls, dams, embankments and berms or the digging of a cistern, and the agricultural plots they can be used for.

Events appear in rhythms—in the case of rain it is the annual wet period. This influences the agricultural activities and practices, as for example when to plough, to sow, to harvest. These rhythms of actions can be disrupted by the lack of water in the arid climate of the Eastern Marmarica. Droughts are also events, which in sum could result in a crisis of livelihood for the inhabitants (see below Section 5). Yet, events can also be socio-economic events, such as putting together a trade caravan or making a deal, with effects "over the long term and their realisation of original structures of action and meaning" [96]. However, for such events the data sets from the ancient Marmarica are rather poor when we want to understand risk-minimising measures [88]. However, along the routes connecting the cistern sites the events of passing and stopping caravans are manifest in form of the rest houses from Roman times (3rd to 6th century CE), which offered shelter, protection, and water supply. Merchants had to invest and take high risks for transport and trade in the arid environment. The row of rest houses can be seen as an investment in infrastructure and a resilient strategy towards the adversity of harmful events, such as thirst, injuries, or loss of goods (and revenues) (Figure 22).

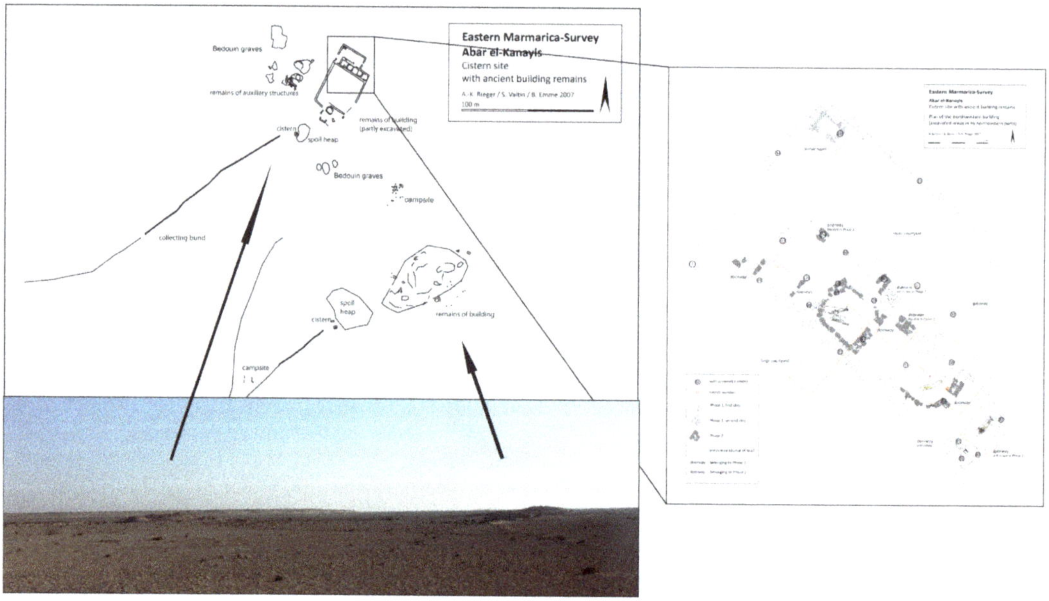

Figure 22. The cistern site of Abar el-Kanayis with two cisterns as well as two rubble heaps of destroyed buildings (**left**). The northern building, a courtyard with rows of rooms at each side, which can be called a rest house, was excavated and could be dated to the 3rd to the 6th century CE (**right**). Findings attest to foodstuffs and ceramic fine ware which was traded along the route passing through Abar el-Kanayis. In Roman times, such infrastructure for trade mobility was provided across the Marmarica Plateau (see Figures 20 and 21), [18] (Figures 4 and 9).

Another form of events could be conflicts and political changes, of which we know very little and cannot see any reflection in the archaeological material (*pólemos marmarikós*, a Marmarican war in the Augustan period, *Orientis Graeci Inscriptiones Selectae* 767). Only the disruption through the political and administrative change in the 7th century CE with the Arab rule in Egypt corresponds with a change in the settlement pattern and the economic strategies in the Marmarica (see below). However, one can assume that debates about the access to water, pasture, or cultivable land was a regular source of conflict between people. Such debates are events that in general are part of the negotiation of how many people, crops and herds can live on the available resources (land, pasture, and water) in an arid social landscape. However, how many people can make a living there depends on the kind of livelihood and economy they pursue—whether it is a subsistence or a surplus economy. How resilient their behaviour is depends on either the interest in revenues or just in survival.

4.2. Communities of Practices and Knowledge—The Formation of a Social Landscape and Its Resiliencies

We have already pointed out the practices employed when dealing with the events—rainfall, caravans, negotiations about water. Practices summarize that inhabitants and populations 'do things' in certain ways depending on the biophysical, natural landscape. The practices form the landscape but are at the same time embedded in it [97]. To think in practices regarding resilience allows for a focus on strategies of adaptation changing people's behaviour as well as influencing the landscape. A practice such as growing crops or making pottery is a knowledge-based routine of activity manifest in social and material relations [25,98,99]. Adaptations can be looked for in the different options for livelihood depending on ecological zones in the Eastern Marmarica, reflected, for example, in the kind of pottery produced and used.

Rainfall events in the arid environment of the Eastern Marmarica materialises in human practices. Remains of practices related to water can be seen in the wadi incisions and on the tableland as well as on the Plateau, where, in a reaction to a 'too much' or 'too little of rain', water and soil harvesting installations are constructed in the form of terraces walls, dams, embankments, berms, or the digging of a cistern.

The installations and structures help control and channel in the literal sense of the word the "eventness" of the water, which also entails its variability and unpredictability, even though one could (before modern-day climate change) count on the wetter season between December and February and a dry season for the rest of the year. As stated above, the accumulations of sediment carried with the overland flow serve as water storage and hence also created arable plots and allowed agriculture as practice of livelihood on the Northern tableland. In the south, sediment accumulates in depressions, where shrubs can grow, serving as pasture for the practice of herding and livestock breeding. Soil can be conceived of as part of "a convivial landscape" where it obtains an active role [58]. The practices of water and soil harvesting of the people living there control and make use of the events [49] (pp. 53–54). They show the wide-ranging knowledge of hydrological regime people in the ancient Eastern Marmarica had. Embankments of earthen material or terraces walls of stones allowing infiltration, the installation of channels and sluits, are the material evidence of the water and soil harvesting practices which can be dated from the 2nd millennium BCE to the first half of the 1st millennium CE. The plots, fields and terraces are signs of the type of agriculture people practiced, whereas the huge number of bones of small ruminants (sheep and goat) among the archaeozoological remains, mostly from contexts of the 2nd to the 3rd century CE, attest to livestock breeding as a second basis for people's livelihood [69].

The two practices—agriculture and livestock breeding—enhance the resilience of people living in the Eastern Marmarica, since they can react to variabilities in the water availability and cope with the effects of drier years with smaller crops, where the livestock breeding can be easier maintained. Having not only one basis of livelihood but growing crops and breeding small ruminants meant stability in a landscape of uncertainties.

People pursuing more the one or the other livelihood—livestock breeding and/or agriculture, depending on the temporary ecological conditions—formed communities of practice which is a suitable notion in order to describe the social organisation [99,100]. According to their practiced lifestyle, people came together, established a group, moved on, and re-grouped [101] (esp. ch. 5 on "Meso-networks: Communities of practice", pp. 98–123). Communities of practices formed by water and soil harvesting are either crops-, fruits- and/or grape-growers or herders—communities of production. They can also become the processors and vendors of the produced goods (wine, cereals, fruits, milk, cheese, meat, wool). Communities of economic practices such as packing or trading goods and establishing the infrastructure for this resulted from the productive practices, which are related, first of all, to packing goods, and secondly to a trade infrastructure such as the rest houses from Roman times mentioned above.

Two different ways of pottery production and use can be attributed to the two communities of practice in the Eastern Marmarica, which is another practice, embodied by those doing it regularly. In the northern zone of the study area, there are numerous production sites of wheel-made pottery (amphoras mainly), active between the 2nd century BCE and the 4th century CE, which required knowledge of technology and forms [20] (Figures 2, 23 and 24). The large quantities in which the wheel-made pottery appears is part of the economic practice of trade, since the form is apt for a Mediterranean-wide exchange of goods [102]. For sure the exchange reached down to Siwa, as the remains of the locally produced amphoras at a cistern site and rest house on the Marmarica Plateau show, where merchants travelled between the economic hubs Siwa and Paraitonion [18]. The second practice of making and using pottery is handmade pottery, the so-called Northern Libyan Desert Ware, which was made on the spot in small updraught kilns. The vessels found are cooking pots, not storage vessels such as the wheel-made amphoras. This kind of pottery was found in the areas which were used for pasture and at cisterns, also where rest houses existed. They reflect the other kind of mobility practiced in the region, roaming with the herds of sheep and goats or camels [19]: One is roaming, the other is headed to the market hubs in Siwa and Paraitonion. The adapted livelihood practices of people dwelling in the niche of the arid landscape, which requires various forms of mobility, is also reflected in the way of producing and using ceramic vessels.

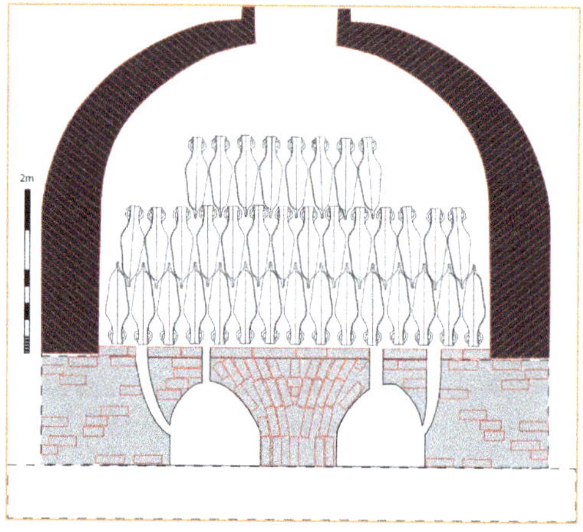

Figure 23. More than 45 pottery production sites were found in the coastal zone and on the Northern tableland in the Eastern Marmarica [20]. They produced wheel-made pottery, mainly amphoras, and were active between the 2nd century BCE and the 4th century CE. The reconstruction of the kiln shown here is based on the findings at Wadi Qasaba [68] (Figure 6).

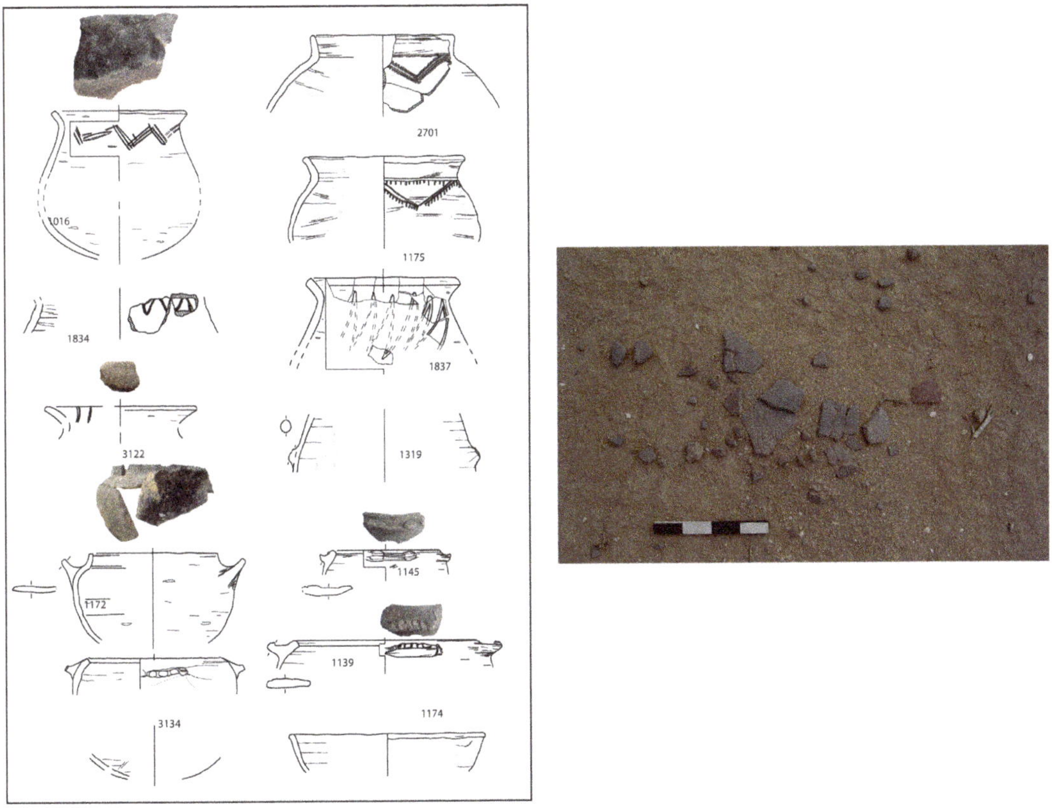

Figure 24. The Northern Libyan Desert Ware is a hand-made pottery, which could be documented in the Eastern Marmarica and dated stratigraphically to Roman times, but also in much younger contexts. The typology consists mainly of closed forms and cooking vessels. They were made on the spot in updraught kilns [19] (Figures 4 and 11).

Mobility is the practice most typical of the arid environment. Aridity requires that people utilise and inhabit larger spaces than in humid regions in order to have access to sufficient amounts of resources. Large spaces are covered by the patchy fields and wadi cultivation areas, which imply that fields and plots did not necessarily adjoin to each other, requiring to be mobile to a small extent. Second, the livestock breeding and grazing practices imply roaming and looking for pastures. Third, it is trade which the people from the area specialised in, because they knew the routes, water supply points and the topography. These three ways of being mobile materialise in the form of the patchy areas of agricultural land and rather small but numerous villages on the Northern Tableland for those taking care of the fields and trees; or in the form of camp sites and shelters and only temporarily or periodically used shelters in every zone of the study area where grazing areas and water were available; the facilities on the Marmarica Plateau, consisting of cisterns and rest houses for merchants and their caravans of animals and goods, but also for herders (Figures 3 and 22).

The various kinds of living and moving are materialised in the settlements in the north, in the cisterns and the route network across the Marmarica Plateau [17,18]. The construction of cisterns in the entire region, which has no surface groundwater, is a practice related to the practice of livestock breeding and trading. Cisterns were dug into the karstic underground of the limestone plateau in Roman times at the latest, but probably already much earlier. Many are still in use today and have a capacity of at least 130 m^3 (Figure 25). We know little about the practice of organising access to water. We can only draw on comparisons to today's practice of naming the cisterns after families or persons (Bir Abu Imama), if not according to location or particularities (Bir el-Nuss, Abar el-Kanayis) and the Bedouin water regulations: cisterns are the property of families who can give access to the water or deny it.

Figure 25. The cistern Abar Abu Imama close to Dardouma in the Premarmarican Plain (see Figure 20) dates to Graeco-Roman times according to the pot shards on surface at the site (3rd/2nd century BCE to 5th/6th century CE). The capacity of the cistern could be roughly measured and reaches up to 300 m^3. It is still in use today. A.-K. Rieger/Th. Vetter.

The rest houses at the cisterns reflect the practice of trading across the plateau, whereas the shelters and fireplaces around cisterns where also the rest houses stood mirror the roaming herders who had their herds but no goods to trade with them. The merchants had a number of stopovers with water supply, maybe also in need of new draft animals, whereas herders searched the pasture in depressions where cisterns also provided water.

According to the ecological zone (Figure 2), people were more mobile and lived a multi-sited life or a more stable life. If the ecological conditions changed, they could change their life-strategy and form communities of practice. These practices allow to easier absorb the repercussions of dry years and failed harvests.

Whatever the socio-economic and -cultural practices were, their precondition is the knowledge of what to do, when and where. Knowing what materials, resources, tools, and partners are necessary for pursuing a practice depends on environment and purpose. If passed on from generation to generation, knowledge guarantees a long-term functioning of socio-cultural practices. Knowledge is the umbrella under which strategies of behaving and acting can be developed and adjusted. They embrace the embodied and transmitted individual experiences made against the backdrop of the landscape in its biophysical appearance. Knowledge embedded into an arid social landscape goes far beyond 'mind maps', etc., but enables inhabitants to create a disposition for resilience over time and topography. In the ancient Eastern Marmarica, such knowledge can be summarised by where to find water, how to enhance its availability, where to send the herds, how to grow crops and how to make pots. The resilience of the system of this arid social landscape consists of its resiliencies in the subsystems, their shifting communities of practices. The approach via event, practices and knowledge differentiates the assessment of resilience of and in the landscape.

The water and soil harvesting systems are only manageable with a good understanding of the morphology and the fluvial-hydrological conditions in the individual wadi system. Hence, the people exploiting the northern coastal zones as well as the tableland had autochthonous experience and knowledge. The same is true for livestock breeding, where finding pasture for sustaining the herds was crucial. The hydrological regime was known to the people through observation and living there.

The more strategies inhabitants in the ancient Eastern Marmarica had, the easier the socio-ecological system could be in a functioning balance, absorbing 'outliers' of drier and hotter or more humid and cooler periods. Mobility is the best adapted life strategy for living in such variable circumstances the arid region is characterised by, and a precondition for survival. Through mobility and the mixed life strategies people created a prevention system and redundancies to be more resilient [103].

They had a dynamic knowledge, acquired through a landscape that was changing because of water events and human impact. Much knowledge and experience (habitual, experiential and tacit knowledge) was necessary to build and maintain walls, dams and terraces on the tableland and in the wadis, as well as the diversion bunds for the cisterns [104,105]. Knowledge about soil conditions, its humidity and fertility were basic for growing crops, vegetables, trees, and legumes, for subsistence or even an economic surplus. The knowledge which made people living in the ancient Eastern Marmarica resilient is an "enearthed and encultured" knowledge [4]. The social landscape develops from such knowledge and practices, passed on from generation to generation by observation, becoming habitual knowledge [89] enabling people to prevent or react to stressors and uncertainties.

Practices and knowledge create the disposition of resilience of people and the social landscape. If knowledge about how to cultivate, where to find water, how to make a vessel, how to settle a hassle with the neighbour is disrupted by crisis and shock, the resilience of the people decreases.

A disruption or better a series of disruptions happened in the ancient Eastern Marmarica in the 5th to 7th century CE, where we see a change in the economic system. The production of wheel-made pottery decreases, settlements stop to be used, cistern sites and rest houses along the routes across the Marmarica Plateau show a decline in traffic and trade (Figure 26). The reasons for this discontinuity can only be explained as multi-factored: climatic variations at a very local scale (rather than global climatic changes), a decrease in productivity, as well as shifts in the economic network of the Mediterranean and Eastern North Africa (rather than a disruption because of the "Arab conquest", which did not directly affect the Marmarica) may have led to the general change in the for centuries resilient social arid landscape of the Eastern Marmarica [106,107].

Figure 26. The settlements of the Ptolemaic (pink) and the Roman periods (green) are much more numerous than the five settlements which show material evidence from the 6th century CE. Either the way of life of the inhabitants of the Marmarica leaving material traces changed in such a way that archaeologically they are invisible, or the number of inhabitants in the region decreased considerably. A.-K. Rieger on base of Esri, Maxar, Earthstar Geographics.

5. Conclusion: Resilience and Niche Dwellings in the Eastern Marmarica

People in the past relied much more on a fine-tuned interaction with environmental regimes and conditions than today, as technological support was not as developed as it is today. This resulted in a greater exposure to risks and uncertainties. Resilience as an etic and non-normative term applied to a past social landscape can help us understand the strategies to cope with risks and stress. Looking at the arid social landscape as an assemblage of subsystems (climate, social groups, economy, etc.) allows to acknowledge the various subsystems or co-occurring systems that could react differently to shocks or stress, so that different resiliencies are provoked.

The case study of the arid social landscape of the ancient Eastern Marmarica showed how the adapted life strategies, sufficient for a subsistence or even a surplus economy, depended on the successful management of local resources and the adaptation to the uncertainties and variabilities in resource availability in the arid environment. Survival and life in the arid landscape of the Eastern Marmarica meant dwelling in a niche, which this socio-ecological system represents, and depended mainly on the dynamics of water and soil availability, emerging through the interaction of the components rainfall, morphology, as well as water and soil harvesting installations [44].

To make a living in a region which lies below the limit of rainfed agriculture, people applied specific strategies based on elaborate water and soil management and the combination of agriculture and mobile livestock breeding, in varying proportions depending on the variable climatic and micro-locational conditions of pastures and agricultural plots. This means they could compensate for crop shortages in drier years. The flexibility and thus resilience embedded in this mixed livelihood lies in the mobility that allows for the utilisation of large areas of the resource-poor environment—roaming with the herds, trading across the Marmarica Plateau or extending the arable plots on the Northern Tableland enlarged the redundancy in the system in order to cope with stressors (Figure 27).

As a consequence, this practice- and knowledge-based view on people allows to break with an explanation of fixed groups pursuing certain life strategies—'nomads' in the south on the Marmarica Plateau doing livestock breeding and 'sedentary' people in the north doing agriculture. We need to think more in terms of communities of practices grouped according to the more promising life strategy in a mixed system: depending on the variable precipitation, grazing grounds and cultivable areas, the inhabitants used more the one or the other source of livelihood. Groupings are more fluid and form and re-form as communities of practice. In an arid environment, such a monolithic livelihood is by far the most un-resilient. Related to these communities of practice of production were practices of pottery production that reflect the various ways of being mobile. People's ability to balance variabilities through the knowledge they gained by observation of events, by adapted practices, and experiential and habitual knowledge was the most resilient way of living.

The extension of the production to cash crops—mainly grapes—and a surplus production in Roman times led to the export of wine with the associated vast pottery production for transport vessels and the participation of people from the ancient Marmarica in an economic network. The risks of investing in such an economy in an arid environment with its variable climatic conditions was high.

With the surplus economy another risk enters the complex system of human-environment relations: possible shocks in the economic system of the Eastern Mediterranean to which the relations were closer than with that of the Western Mediterranean when it comes to imports. From this we infer a larger dependency on the export of goods to the Eastern Mediterranean. Because of the hiatus in the 6th/7th century CE we can surmise that a shift in economic interrelations and productivities in the Eastern Mediterranean had repercussions on the Eastern Marmarica during this period [106]. However, a multi-factorial crisis—we assume—could have caused the collapse of the surplus economy. With this the visibility in the archaeological record of people making use of the region decreases significantly. However, some people still lived there in the 7th century CE, even though the resilient strategies developed and practiced in the centuries before—from the 2nd century

BCE to the 6th century CE—did not suffice any longer. The niche they dwelled in had become smaller. It was not until the 19th century that the water and soil harvesting systems were reactivated [14,108].

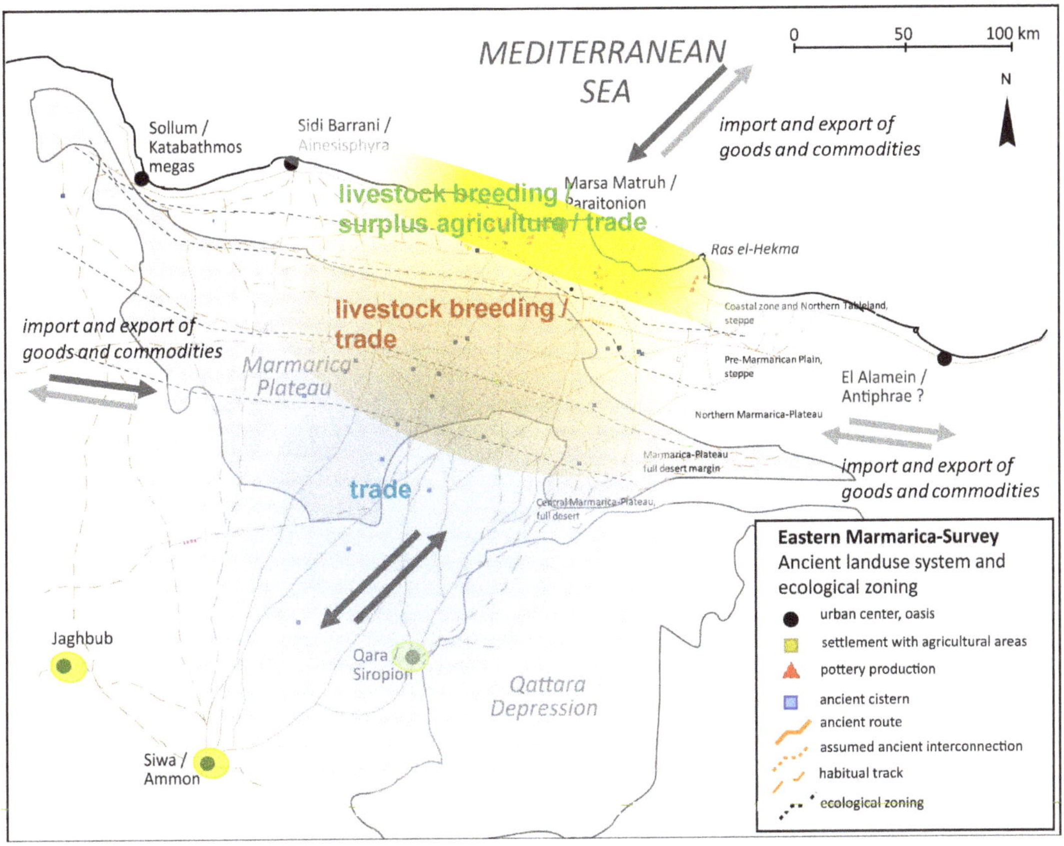

Figure 27. Scheme of the mixed life strategies and mobilities in the Eastern Marmarica in Graeco-Roman times, from the 5th century BCE to the 5th century CE, which enabled people to cope with the stressors of water scarcity and variability. These life strategies formed a resilient social landscape. From the 6th century CE onwards, surplus agriculture and trade came to an end, a change for which the reasons might be multi-factored (local climatic variations, shifts in the trade networks, administrational changes etc.) A.-K. Rieger.

Funding: The research in Egypt was funded by the Deutsche Forschungsgemeinschaft (DFG) in the frame of CRC 586, Universities Leipzig and Halle—Wittenberg, project A7 and the project "Eastern Marmarica Survey", grant no. Ri 1960/1-1 (2004–2011), whereas the research on the concept of resilience, also funded by the DFG, is part of grant no. Schlie 2255/1-1. The writing was supported by the Centre of the Research on Ancient Civilizations, University of Warsaw. Proofreading was financed by the Faculty of Arts and Humanities of the University of Graz; Open Access Funding was granted by the University of Graz.

Data Availability Statement: Data sharing is not applicable; all data can be found in the referenced papers (co-)authored by A.-K. Rieger.

Acknowledgments: I thank the editors to include me to this special issue, which was a welcome opportunity to combine two of my research interests: Landscape archaeology and the concept of resilience in Classics. For the support for the "Eastern Marmarica Survey" in NW-Egypt (see above "Funding") I thank the "Supreme Council of Antiquities" in Egypt as well as all team members. The network "Resources of Resilience in Antiquity: ecological, social and cultural systems between persistence and transformation" co-directed by Ch. Schliephake, A. Hartmann, and A.-K. Rieger forms the basis for the ideas on resilience as a tool for research in Classics (see above "Funding"). I am grateful to all its members and guests, especially A. Busch, M. Endreß, C. Grey, and Greg Woolf, for opening so nuanced perspectives on resilience in historical research on the one hand and in ancient societies on the other hand. During a fellowship granted by the Centre of the Research on Ancient Civilizations under the direction of R. Wisnewski, University of Warsaw, in 2021/2022 I could enjoy the freedom and intellectual atmosphere to work on this paper. At Warsaw I also had the chance to meet one of the co-editors of the special issue, thanks to A. Sarantis. Finally, I would like to express my gratitude to the spokesperson of the International Graduate School "Resonant Self-World Relations in Ancient and Modern Socio-Religious Practices" (FWF/DFG) at Graz, W. Spickermann, who is always supportive of my research activities. I owe thanks to mdpi, for their fast and professional processing of the paper. Most grateful I am to the anonymous reviewers for their clear and constructive comments, which helped improve the targument. All remaining flaws and errors are mine.

Conflicts of Interest: The author declares no conflict of interest. The funders had no role in the design of the study; in the collection, analyses, or interpretation of data; in the writing of the manuscript; or in the decision to publish the results.

References

1. Bernbeck, R. An Archaeology of Multi-Sited Communities. In *The Archaeology of Mobility: Old World and New World Nomadism*; Barnard, H., Wendrich, W., Eds.; Cotsen Institute of Archaeology Press at UCLA: Los Angeles, CA, USA, 2008; pp. 43–77.
2. *Environmental Humanities: A Rethinking of Landscape Archaeology?: Interdisciplinary Academic Research Related to Different Perspectives of Landscapes*; Kluiving, S.J.; Lidén, K.; Fredengren, C. (Eds.) Sidestone Press: Leiden, The Netherlands, 2021; ISBN 9789464270051.
3. Walker, B.; Gunderson, L.; Kinzig, A.; Folke, C.; Carpenter, S.; Schultz, L. A Handful of Heuristics and Some Propositions for Understanding Resilience in Social-Ecological Systems. *Ecol. Soc.* **2006**, *11*, 13. [CrossRef]
4. Schill, C.; Anderies, J.M.; Lindahl, T.; Folke, C.; Polasky, S.; Cárdenas, J.C.; Crépin, A.-S.; Janssen, M.A.; Norberg, J.; Schlüter, M. A more dynamic understanding of human behaviour for the Anthropocene. *Nat. Sustain.* **2019**, *2*, 1075–1082. [CrossRef]
5. Otok, S. Nature of Social Landscape. *Misc. Geogr.* **1988**, *3*, 239–244. [CrossRef]
6. Bollig, M. Visions of landscapes: An introduction. In *African Landscapes: Interdisciplinary Approaches*; Bubenzer, O., Bollig, M., Eds.; Springer: New York, NY, USA, 2009; pp. 1–38.
7. Carson, M.; Peterson, G. *An Arctic Resilience Assessment. Arctic Resilience Report*; Environment Institute and Stockholm Resilience Centre: Stockholm, Sweden, 2016; pp. 2–26. Available online: http://www.arctic-council.org/arr (accessed on 11 May 2023).
8. Horden, P.; Purcell, N. *The Corrupting Sea: A Study of Mediterranean History*; Oxford University Press: Oxford, UK, 2000.
9. Kuper, R.; Kröpelin, S. Climate-controlled Holocene occupation in the Sahara: Motor of Africa's evolution. *Science* **2006**, *313*, 803–807. [CrossRef] [PubMed]
10. Gosselain, O.P. To hell with ethnoarchaeology! *Archaeol. Dialogues* **2016**, *23*, 215–228. [CrossRef]
11. *Naturgeschichtliche Reisen durch Nord-Afrika und West-Asien in den Jahren 1820 bis 1825 von Dr. W. F. Hemprich und Dr. C. G. Ehrenberg*; Ehrenberg, C.G.; Hemprich, F.W. (Eds.) Mittler; Schade: Berlin, Germany, 1828.
12. Cole, D.P.; Altorki, S. *Bedouin, Settlers, and Holiday-Makers: Egypt's Changing Northwest Coast*; American University in Cairo Press: Cairo, Egypt, 1998; ISBN 9774244842.
13. Hüsken, T.; Roenpage, O. *Jenseits von Traditionalismus und Stagnation: Analyse einer beduinischen Ökonomie in der Westlichen Wüste Ägyptens*; Lit: Münster, Germany, 1998; ISBN 3825837629.
14. Müller-Mahn, H.-D. *Die Aulad Ali zwischen Stamm und Staat: Entwicklung und sozialer Wandel bei den Beduinen im Nordwestlichen Ägypten*; Reimer Verlag: Berlin, Germany, 1989.
15. Lancelotti, C.; Biagetti, S.; Zerboni, A.; Usai, D.; Madella, M. The archaeology and ethnoarchaeology of rain-fed cultivation in arid and hyper-arid North Africa. *Antiquity* **2019**, *93*, 1026–1039. [CrossRef]
16. Tucker, D. Tracking Mobility in the Syrian Desert. Potential of Simple Features for Mapping Landscapes of Mobile Pastoralists. In Proceedings of the 37th International Conference, Making History Interactive: Computer Applications and Quantitative Methods in Archaeology (CAA), Williamsburg, VA, USA, 22–26 March 2009; Frischer, B., Crawford, J.W., Koller, D., Eds.; Archaeopress: Oxford, UK, 2010; pp. 1–8.
17. Vetter, T.; Rieger, A.-K.; Möller, H. Water, Routes and Rangelands: Ancient Traffic and Grazing Infrastructure in the Eastern Marmarica (Northwestern Egypt) (with H. Möller, Th. Vetter). In *Desert Road Archaeology in the Eastern Sahara and Beyond, Africa Praehistorica 27*; Riemer, H., Förster, F., Eds.; Heinrich-Barth-Institut: Köln, Germany, 2013; pp. 455–484.

18. Rieger, A.-K.; Möller, H.; Valtin, S.; Vetter, T. On the route to Siwa: The Roadhouse of Abar al Kanayis on the Marmarica-Plateau. *Mitt. des Dtsch. Archäologischen Inst. Abt. Kairo* **2012**, *68*, 135–174.
19. Rieger, A.-K.; Möller, H. Northern Libyan Desert Ware: New thoughts on 'Shell-tempered Ware' and other handmade pottery from the Eastern Marmarica (north-west Egypt). *Libyan Stud.* **2012**, *43*, 11–31. [CrossRef]
20. Rieger, A.-K.; Möller, H. Kilns, commodities and consumers: Greco-roman pottery production in Eastern Marmarica (Northwestern Egypt). *Archäologischer Anz.* **2011**, *1*, 141–170.
21. Holling, C.S. Resilience and Stability of Ecological Systems. *Annu. Rev. Ecol. Syst.* **1973**, *4*, 1–23. [CrossRef]
22. Haraway, D.J. *When Species Meet*; University of Minnesota Press: Minneapolis, MN, USA, 2007.
23. Donges, J.F.; Lucht, W.; Cornell, S.; Heitzig, J.; Barfuss, W.; Lade, S.J.; Schlüter, M. Taxonomies for Structuring Models for World–Earth Systems Analysis of the Anthropocene: Subsystems, their Interactions and Social-Ecological Feedback Loops. *Earth Syst. Dyn.* **2021**, *12*, 1115–1137. [CrossRef]
24. Aldred, O.; Lucas, G. Events, temporalities and landscapes in Iceland (2010). In *Eventful Archaeologies: New Approaches to Social Transformation in the Archaeological Record*; Bolender, D.J., Ed.; State University of New York Press: Albany, NY, USA, 2010; pp. 189–198.
25. Reckwitz, A. Toward a Theory of Social Practices. *Eur. J. Soc. Theory* **2002**, *5*, 243–263. [CrossRef]
26. Werlen, B. Action, Knowledge, and Social Relations of Space. In *Knowledge and Action*; Meusburger, P., Werlen, B., Kaeding-Suarsana, L., Eds.; Springer Open: Cham, Switzerland, 2017; pp. 31–56.
27. Walker, B.; Holling, C.S.; Carpenter, S.R.; Kinzig, A.P. Resilience, Adaptability and Transformability in Social-ecological Systems. *Ecol. Soc.* **2004**, *9*, 1–9. [CrossRef]
28. *Actor Network Theory and after*; Law, J.; Hassard, J. (Eds.) Blackwell: Oxford, UK, 1999.
29. *Panarchy: Understanding Transformations in Human and Natural Systems*; Gunderson, L.H.; Holling, C.S. (Eds.) Island Press: Washington, DC, USA, 2002.
30. Rampp, B. Strategies, Dispositions and Resources in Multi-Level Resilience Processes. In *Strategies, Dispositions and Resources of Social Resilience*; Endress, M., Clemens, L., Rampp, B., Eds.; Springer Fachmedien Wiesbaden: Wiesbaden, Germany, 2020; pp. 59–79.
31. Endreß, M. Resilienz im Diskurs—Für ein reflexives Resilienzverständnis. Eine Einleitung. In *Resilienz als Prozess*; Beiträge zu einer Soziologie von Resilienz; Endreß, M.; Rampp, B., Eds.; Springer: Wiesbaden, Germany, 2022; pp. 1–16.
32. Brown, K. Global environmental change I: A social turn for resilience? *Prog. Hum. Geogr.* **2014**, *38*, 107–117. [CrossRef]
33. Cote, M.; Nightingale, A.J. Resilience thinking meets social theory. *Prog. Hum. Geogr.* **2012**, *36*, 475–489. [CrossRef]
34. Busch, A.W.; Lieb, K. Resilienzfaktoren in diachroner und interkultureller Perspektive—Was macht den Menschen widerstandsfähig? Available online: https://web.rgzm.de/no_cache/forschung/schwerpunkte-und-projekte/details-forschungsprojekte/article/resilienzfaktoren-in-diachroner-und-interkultureller-perspektive-was-macht-den-menschen-widerstand/ (accessed on 15 February 2023).
35. *Resilienz als Prozess: Beiträge zu einer Soziologie von Resilienz*; Endreß, M.; Rampp, B. (Eds.) Springer VS: Wiesbaden, Germany, 2022.
36. Redman, C.L.; Kinzig, A.P. Resilience of Past Landscapes: Resilience Theory, Society, and the Longue Durée. *Conserv. Ecol.* **2003**, *7*, 14. [CrossRef]
37. Turner, M.D. Political ecology I. *Prog. Hum. Geogr.* **2014**, *38*, 616–623. [CrossRef]
38. Schilling, T. Resilienz und Transitionen in komplexen adaptiven Systemen: Gemeinsamkeiten zweier systemdynamischer Perspektiven. In *Resilienz als Prozess*; Beiträge zu einer Soziologie von Resilienz; Endreß, M.; Rampp, B., Eds.; Springer VS: Wiesbaden, Germany, 2022; pp. 19–54.
39. Løvschal, M. Retranslating Resilience Theory in Archaeology. *Annu. Rev. Anthropol.* **2022**, *51*, 195–211. [CrossRef]
40. Schreiber, S. Resilienz als Resonanzfähigkeit. Ein affekttheoretischer Blick auf ein psycho-sozio-archäologisches Forschungsfeld. In *Resilienz als Prozess: Beiträge zu einer Soziologie von Resilienz*; Endreß, M., Rampp, B., Eds.; Springer VS: Wiesbaden, Germany, 2022; pp. 81–122. ISBN 978-3-658-38269-8.
41. Martin-Breen, P.; Anderies, J.M. Resilience: A Literature Review. In *Bellagio Initiative Background Paper*; Institute of Development Studies: Brighton, UK, 2011; Available online: https://opendocs.ids.ac.uk/opendocs/bitstream/handle/20.500.12413/3692/Bellagio-Rockefeller%20bp.pdf?sequence=1&isAllowed=y (accessed on 15 February 2023).
42. Walker, B.H.; Anderies, J.M.; Kinzig, A.P.; Ryan, P. Exploring Resilience in Social-Ecological Systems through Comparative Studies and Theory Development: Introduction to the Special Issue. *Ecol. Soc.* **2006**, *11*, 12. [CrossRef]
43. Anderies, J.M.; Barfuss, W.; Donges, J.F.; Fetzer, I.; Heitzig, J.; Rockström, J. Conceptualizing World-Earth System Resilience: Exploring Transformation Pathways Towards a Safe and Just Operating Space for Humanity. 2022. Available online: https://arxiv.org/pdf/2204.04471 (accessed on 11 May 2023).
44. Flaherty, E. *Complexity and Resilience in the Social and Ecological Sciences*; Palgrave Macmillan: London, UK, 2019.
45. Ungar, M. Systemic resilience: Principles and processes for a science of change in contexts of adversity. *Ecol. Soc.* **2018**, *23*, 34. [CrossRef]
46. Izdebski, A.; Mordechai, L.; White, S. The Social Burden of Resilience: A Historical Perspective. *Hum. Ecol. Interdiscip. J.* **2018**, *46*, 291–303. [CrossRef]

47. Haraway, D. Situated knowledges: The science question in feminism and the privilege of partial perspectives. *Fem. Stud.* **1988**, *14*, 575–599. [CrossRef]
48. Rampp, B. Von sozialen Prozessen, Relationen und Konstruktionen. Analytische Potenziale des Resilienzansatzes aus soziologischer Perspektive. In *Resilienz als Prozess: Beiträge zu einer Soziologie von Resilienz*; Endreß, M., Rampp, B., Eds.; Springer VS: Wiesbaden, Germany, 2022; pp. 123–158.
49. Endreß, M. Strategies, Dispositions and Resources as Socio-Historical Constructions. In *Strategies, Dispositions and Resources of Social Resilience*; Endress, M., Clemens, L., Rampp, B., Eds.; Springer Fachmedien Wiesbaden: Wiesbaden, Germany, 2020; pp. 43–57.
50. Cannon, T.; Müller-Mahn, D. Vulnerability, resilience and development discourses in context of climate change. *Nat. Hazards* **2010**, *55*, 621–635. [CrossRef]
51. Heras-Escribano, M.; de Pinedo-García, M. Affordances and Landscapes: Overcoming the Nature-Culture Dichotomy through Niche Construction Theory. *Front. Psychol.* **2017**, *8*, 2294. [CrossRef] [PubMed]
52. Rieger, A.-K. Landscape as artefact and as agent: The potentials of Landscape Archaeology in arid landscapes for research in Classical Archaeology. In *Proceedings of the 5th Landscape Archaeology Conference 2018, Classical Archaeology and Landscapes: Session, Newcastle/Durham, UK, 19 September 2018*; Kennedy, W.M., Eser, R.A., Eds.; Brepols: Turnhout, Belgium, in print.
53. *Landscape and Agency: Critical Essays*; Wall, E.; Waterman, T. (Eds.) Routledge Taylor & Francis Group: London, UK; New York, NY, USA, 2018.
54. Labuhn, I.; Finné, M.; Izdebski, A.; Roberts, N.; Woodbridge, J. Climatic Changes and their Impacts in the Mediterranean during the First Millennium AD. *Late Antiq. Archaeol.* **2016**, *12*, 65–88. [CrossRef]
55. Pennos, C.; Pechlivanidou, S.; Aidona, E.; Bourliva, A.; Lauritzen, S.-E.; Scholger, R.; Kantiranis, N. Decoding short-term climatic variations from cave sediments over the Mid-Holocene: Implications for human occupation in the Katarraktes Cave System, Northern Greece. *Zeitschr. f. Geomorphol.* **2021**, *63*, 67–80. [CrossRef]
56. Manning, S.W.; Kocik, C.; Lorentzen, B.; Sparks, J.P. Severe multi-year drought coincident with Hittite collapse around 1198-1196 BC. *Nature* **2023**, *614*, 719–724. [CrossRef]
57. Balbo, A.L.; Gmez-Baggethun, E.; Salpeteur, M.; Puy, A.; Biagetti, S.; Scheffran, J. Resilience of small-scale societies: A view from drylands. *Ecol. Soc.* **2016**, *21*, 53. [CrossRef]
58. Given, M. Conviviality and the Life of Soil. *Camb. Archaeol. J.* **2018**, *28*, 127–143. [CrossRef]
59. Antczak, K.A.; Beaudry, M.C. Assemblages of practice. A conceptual framework for exploring human–thing relations in archaeology. *Archaeol. Dialogues* **2019**, *26*, 87–110. [CrossRef]
60. Anderson, B.; McFarlane, C. Assemblage and geography. *Area* **2011**, *43*, 124–127. [CrossRef]
61. Tozzi, A. An approach to pluralizing socionatural resilience through assemblages. *Prog. Hum. Geogr.* **2021**, *45*, 1083–1104. [CrossRef]
62. Deleuze, G.; Guattari, F. *Capitalisme et Schizophrénie: Mille Plateaux*; Édition de Minuit: Paris, France, 1980.
63. Vetter, T.; Rieger, A.-K. Water harvesting as a key for understanding adapted ancient livelihoods in an arid environment—Approaches to dryland archaeology in the Eastern Marmarica (NW-Egypt). *J. Arid Environ.* **2019**, *171*, 103940. [CrossRef]
64. Vetter, T. Die rezente äolische Dynamik an der ägyptischen Nordwestküste. *Hallesches Jahrb. Geowiss. Reihe A* **2001**, *23*, 79–90.
65. Vetter, T.; Rieger, A.-K. Ancient water harvesting in the Old World Dry Belt—Synopsis and outlook. *J. Arid Environ.* **2019**, *169*, 42–53. [CrossRef]
66. Kröpelin, S.; Verschuren, D.; Lézine, A.-M.; Eggermont, H.; Cocquyt, C.; Francus, P.; Cazet, J.-P.; Fagot, M.; Rumes, B.; Russell, J.M.; et al. Climate-driven ecosystem succession in the Sahara: The past 6000 years. *Science* **2008**, *320*, 765–768. [CrossRef]
67. Vetter, T.; Rieger, A.-K.; Nicolay, A. Disconnected runoff contributing areas: Evidence provided by ancient watershed management systems in arid north-eastern Marmarica (NW-Egypt). *Geomorphology* **2014**, *212*, 41–57. [CrossRef]
68. Möller, H.; Rieger, A.-K. Necessity is the Mother of Invention: The Fuel of Graeco-Roman Pottery Kilns in the Semi-Arid Eastern Marmarica. In *Fuel and Fire in the Ancient Roman World: Towards an Integrated Economic Understanding*; McDonald Institute for Archaeological Research; Veal, R., Leitch, V., Eds.; University of Cambridge Press: Cambridge, UK, 2019; pp. 71–86. ISBN 978-1-902937-91-5.
69. Pöllath, N.; Rieger, A.-K. Insights in Diet and Economy of the Eastern Marmarica—Faunal remains from Greco-Roman sites in northwestern Egypt (Abar el-Kanayis, Wadi Umm el-Ashdan and Wadi Qasaba). *Mitt. Dtsch. Archäologischen Inst. Kairo* **2011**, *67*, 163–180.
70. Gueraud, O. Signature d'un Platrier. *Bull. Soc. d'Archeol. Alex.* **1936**, *30*, 31–33.
71. Vetter, T.; Rieger, A.-K.; Klammer, O.; Fuchs, M.; Nicolay, A. Spätholozäne Abfluss- und Sedimentdydnamik im semiariden Nordwest-Ägypten und ihre anthropogene Modifikation. *Zentralbl. Geol. Paläontologie Teil 1* **2009**, *Heft 1/2*, 171–187.
72. Nicolay, A. *Zusammenhänge Zwischen Rainwater Harvesting Anlagen und Kolluvien in den Wadi der Nordwestküstenregion Ägyptens. Diplomarbeit*; Martin-Luther-Universität: Halle, Germany, 2009.
73. Kehl, H.; Bornkamm, R. Landscape Ecology and Vegetation Units of the Western Desert of Egypt. In *Geopotential and Ecology: Analysis of a Desert Region*; Meissner, B., Wycisk, P., Eds.; Catena-Verl.: Cremlingen-Destedt, Germany, 1993; pp. 155–178. ISBN 3923381352.
74. Bornkamm, R.; Kehl, H. The plant communities of the Western Desert of Egypt. *Phytocoenologia* **1990**, *19*, 149–231. [CrossRef]

75. Furnari, F.; Brullo, S. Phytogeographical considerations on the coastal vegetation of Cyrenaica. *Anales Jard. Bot. Madrid* **1981**, *37*, 765–772.
76. Le Houérou, H.N. Plant diversity in Egyptian Marmarica: Catalogue of the vascular species with their bioclimatic distribution patterns, frequency, habitat and main ecological attributes. *Candollea* **2004**, *59*, 259–308.
77. Kehl, H.; Gauer, J.; Stahr, K. Soil vegetation relationship of a small catchment area on the Libyan Plateau in NW-Egypt. *Berl. Geowiss. Abh. (A)* **1984**, *50*, 303–324.
78. Vetter, T.; Rieger, A.-K.; Nicolay, A. Ancient rainwater harvesting systems in the north-eastern Marmarica (north-western Egypt). *Libyan Stud.* **2009**, *40*, 9–23. [CrossRef]
79. Mayerson, P. Ancient agricultural remains in the central Negev. The Teleilât el-Anab. *Bull. Amer. Sch. Orient. Res.* **1959**, *153*, 19–32. [CrossRef]
80. Evenari, M.; Shanan, L.; Tadmor, N. *The Negev: The Challenge of a Desert*; Harvard University Press: Cambridge, MA, USA, 1971.
81. Spengler, R.N. Niche Dwelling vs. Niche Construction: Landscape Modification in the Bronze and Iron Ages of Central Asia. *Hum. Ecol.* **2014**, *42*, 813–821. [CrossRef]
82. Laland, K.; Matthews, B.; Feldman, M.W. An introduction to niche construction theory. *Evol. Ecol.* **2016**, *30*, 191–202. [CrossRef]
83. United Nations. *United Nations Decade on Deserts and the Fight against Desertification: (UNDDD)*; United Nations: New York, NY, USA, 2011.
84. Snape, S.R.; Wilson, P. *Zawiyet Umm el-Rakham. The Temple and Chapels*; Rutherford Press: Bolton, ON, Canada, 2007.
85. White, D. *Marsa Matruh I: The Excavation*; INSTAP Academic Press: Havertown, PA, USA, 2002.
86. White, D. *Marsa Matruh II: The Objects*; INSTAP Academic Press: Havertown, PA, USA, 2002.
87. Rieger, A.-K.; Vetter, T.; Möller, H. The desert dwellers of the Marmarica (Western Desert) as a case study for the Eastern Desert (?). In *The History of the Peoples of the Eastern Desert*; Barnard, H., Wendrich, W., Eds.; Cotsen Institute of Archaeology Press: Los Angeles, CA, USA, 2012; pp. 156–173.
88. Rieger, A.-K. Text and Landscape: The Complementarity of the Papiro Vaticano Greco 11 R (PMarm) to Landscape-Archaeological Results from the Arid Marmarica (NW-Egypt/NE-Libya). *Topoi* **2017**, *21*, 105–146. [CrossRef]
89. Rieger, A.-K. The Various Ways of Being Mobile: Habitual Knowledge, Life-Strategies and the Ancient Route Networks on the Eastern Marmarica- Plateau (Northern Libyan Desert). *Open Archaeol.* **2017**, *3*, 49–68. [CrossRef]
90. Krause, F.; Strang, V. Thinking Relationships Through Water. *Soc. Nat. Resour.* **2016**, *29*, 633–638. [CrossRef]
91. Strang, V. The Social Construction of Water. In *Handbook of Landscape Archaeology*; David, B., Thomas, J., Eds.; Routledge: London, UK, 2008; pp. 123–130.
92. Sanchez, A.X.; Osmond, P.; van der Heijden, J. Are Some Forms of Resilience More Sustainable than Others? *Procedia Eng.* **2017**, *180*, 881–889. [CrossRef]
93. Klein, R.J.T.; Nicholls, R.J.; Thomalla, F. Resilience to natural hazards: How useful is this concept? *Environ. Hazards* **2003**, *5*, 35–45. [CrossRef]
94. Beck, R.A., Jr.; Bolender, D.J.; Brown, J.A.; Earle, T.K. Eventful Archaeology. *Curr. Anthropol.* **2007**, *48*, 833–860. [CrossRef]
95. Casey, E.S. How to Get from Space to Place in a Fairly Short Stretch of Time: Phenomenological Prolegomena. In *Senses of Place*; Feld, S., Basso, K.H., Eds.; School of American Research Press: Santa Fe, NM, USA, 1996; pp. 13–52.
96. Kapferer, B. Introduction: In the Event—Toward an Anthropology of Generic Moments. In *In the Event: Toward an Anthropology of Generic Moments*; Kapferer, B., Meinert, L., Eds.; Berghahn Books: New York, NY, USA; Oxford, UK, 2015; pp. 1–28. ISBN 9781782388906.
97. de Certeau, M. *L'invention du Quotidien, I: Arts de Faire*; Gallimard: Paris, France, 1990; ISBN 9782070325764.
98. Bourdieu, P. *Outline of a Theory of Practice*; Cambridge University Press: Cambridge, UK, 2010.
99. *The Practice Turn in Contemporary Theory*; Knorr Cetina, K.; Schatzki, T.R.; von Savigny, E. (Eds.) Routledge: London, UK, 2001.
100. *Archaeology and Apprenticeship: Body Knowledge, Identity, and Communities of Practice*; Wendrich, W. (Ed.) University of Arizona Press: Tucson, AZ, USA, 2012.
101. Knappett, C. *An Archaeology of Interaction: Network Perspectives on Material Culture and Society*; Oxford University Press: Oxford, UK, 2011; pp. 98–123. [CrossRef]
102. *Transport Amphorae and Trade in the Eastern Mediterranean: Acts of the International Colloquium at the Danish Institute at Athens, September 26–29, 2002*; Eiring, J.; Lund, J. (Eds.) Aarhus University Press: Aarhus, Denmark, 2004.
103. Biggs, C.R.; Yeager, L.A.; Bolser, D.G.; Bonsell, C.; Dichiera, A.M.; Hou, Z.; Keyser, S.R.; Khursigara, A.J.; Lu, K.; Muth, A.F.; et al. Does functional redundancy affect ecological stability and resilience? A review and meta-analysis. *Ecosphere* **2020**, *11*, e03184. [CrossRef]
104. Gorman, M.E. Types of Knowledge and their Roles in Technology Transfer. *J. Technol. Transf.* **2002**, *27*, 219–231. [CrossRef]
105. Kaldewey, D.; Schatzki, T.R. Questions to Theodore R. Schatzki. In *Revealing Tacit Knowledge: Embodiment and Explication*; Adloff, F., Gerund, K., Kaldewey, D., Eds.; Transcript Verlag: Bielefeld, Germany, 2015; pp. 113–118.
106. Rieger, A.-K.; Möller, H. Keramik spätrömischer Zeit in der östlichen Marmarica (NW-Ägypten): Eine marginale Wirtschaftsregion und ihre überregionalen Handelsbezüge (Mareotis, Kyrenaika, Siwa). In *Himmelwärts und Erdverbunden? Religiöse und Wirtschaftliche Aspekte Spätantiker Lebensrealität*; Haensch, R., von Rummel, P., Eds.; Verlag Marie Leidorf: Rahden/Westfalen, Germany, 2021; pp. 303–322.

107. Kuhlmann, K.P. Roman and Byzantine Siwa. Developing a latent picture. In *Life on the Fringe, Living in the Southern Egyptian Deserts during the Roman and Early-Byzantine Periods, Proceedings of a Colloquium Cairo, Egypt, 9–12 December 1996*; Kaper, O., Ed.; CNWS Research School, Leiden University: Leiden, The Netherlands, 1998; pp. 159–180. ISBN 978-90-5789-015-4.
108. Ball, J. Problems of the Libyan Desert. *Geogr. J.* **1927**, *70*, 21–38. [CrossRef]

land

MDPI

Review

The Anthropocene in the Aspiring UNESCO Global Geopark Schelde Delta Area: Geological History, Human Resilience and Future Landscape Management

Sjoerd Kluiving [1,*] and Ronald Waterman [2]

1 Faculty of Humanities, Department of Art & Culture, History, and Antiquity/Faculty of Science, Institute for Environmental Studies, Vrije Universiteit Amsterdam, NL-1081 HV Amsterdam, The Netherlands
2 Faculty of Civil Engineering and Geosciences, Delft University of Technology, NL-2628 CN-1 Delft, The Netherlands; buildingwithnature@gmail.com
* Correspondence: s.j.kluiving@vu.nl

Abstract: In north-western Europe, the Rhine, Meuse and Scheldt rivers have created a large river delta over the past 3 million years. Geological phenomena in the Scheldt region in north-western Belgium and in the southern Netherlands testify from a highly dynamic landscape, showing a range from very old (50 MY) to very young (recent) geological processes. The great diversity in geological processes and resulting landscapes is unprecedented on a global scale and has had its impact on the region's cultural and economic history, shaping today's reality in the global polycrisis. However, the area is usually observed by people as a flat and featureless type of terrain, although sometimes, unexpected elevation differences and sharp contrasts in landscapes occur alternating with omnipresent waterways. Therefore, here, the seven most conspicuous landforms are reviewed and presented in conjunction with the geological history of the area, including the typical lowland theme of the human battle against water. This study aims to (a) reconstruct the Tertiary and Quaternary to Holocene Dutch–Flemish Schelde Delta history, (b) review a cultural history that evolves into the present of the Anthropocene, and (c) project the desired future for sustainable landscapes in the aspiring UNESCO Global Geopark Schelde Delta between plural landscape management scenarios of Revitalised Land- and Waterscape and Improved Biodiversity.

Keywords: Anthropocene; geology; archaeology; land- and waterscape; sustainable landscape management scenarios; AquaPuncture; Nature Futures Framework; UNESCO Geopark

1. Introduction

The aspiring UNESCO Global Geopark Flemish-Dutch Schelde Delta (aUGGP ScheldeDelta) covers an area of 5500 km^2, with 63 municipalities in total and a population of 1,500,000 divided over two countries (Belgium and The Netherlands) [1]. It comprises three Flemish provinces (Antwerpen, Oost-Vlaanderen and West-Vlaanderen), the Dutch provinces Zeeland and the western part of Noord-Brabant. This international and interregional area comprises, beyond urban and agricultural areas, geologic morphological sites, sites with nature and landscape values, archaeological and cultural historic sites, touristic education sites, and military sites and fortifications.

Importantly, the aUGGP ScheldeDelta depicts not only terrestrial sites (75%), but also complementary (often neglected) aquatic phenomena (25%) in the form of estuaries, rivers, lagoons, lakes, canals, aquaculture, harbour basins, and small-scale canals and brooks [1]. Within the latter, a complete salt (40%) and freshwater (60%) infrastructure is present, which has been primarily formed by nature and has, in most cases, been subsequently adapted and newly shaped by the action of humans.

Although the candidacy for a UNESCO Geopark is based on the unquestionable exceptional natural (and cultural) values of the area [1], the growing impact of humans on our environment has also been called the Anthropocene, alluding to the unsustainable use of the landscape and its resources [2–4]. For the very first time in geological history, humankind has become a significant factor through lifestyle and numbers and may have caused a new geological epoch/era [5,6]. Many scholars agree that currently we live in the Anthropocene, even if they are still debating the dates from which this epoch of humans can be said to have started [2,7–9]. In the Anthropocene, the forceful actions of mankind have so far caused a serious loss of biodiversity, notable climate change and severe ocean acidification in the natural world, next to a social and humanitarian world in different phases of a planet in transition [4–6,10].

Therefore, in this paper, sustainable landscape scenarios for the aUGGP ScheldeDelta will be discussed, targeted with an emphasis on the proposed landscape management of the aquatic nature of the area. Our partial focus on the water landscape (waterscape, next to the terrestrial landscape) of the aUGGP ScheldeDelta is also given in the present and future threats to our water quality, salinity % and all kinds of pollution and water quantity (shortage or excess) [11,12].

The aim of this paper is to address future landscape management strategies for the aUGGP ScheldeDelta in light of the geological, archaeological and historical developments from the Tertiary and Quaternary periods to the Holocene era. While the holistic desirable view on future landscape management in the Nature Futures Framework is captured in [13], we combine the approach here with AquaPuncture [14] and the future scenarios of the aUGGP ScheldeDelta [1]. This paper will focus on the land–water interface that is so apparent throughout the history of the area and analyse the potential for future developments in these marginal landscapes.

The bid book for the aUGGP ScheldeDelta, describing human action as a geological force, poses important 21st-century questions: (1) How do we (re)shape our (intrinsic) relationship with nature? (2) How do we inspire people to change their behaviour in a sustainable way [1]? The proposed area stimulates cross-border international cooperation and broad coalitions with a prominent role for educational and knowledge institutions, governments, businesses and end users (residents and visitors). Experience and imagination are the ultimate tools for engaging people (young and old) and stimulating awareness [1].

The urgency of landscape change from a continental via a regional to a local scale [10,15–17] addresses the importance of stakeholder participatory design of sustainable landscape management strategies that have been adequately indicated and advocated, including the potential of abandoned landscapes [18–20]. In addition, a continental approach to biodiversity loss is found in variable landscape management practices, such as extensive agriculture, and nature restoration tied to rewilding [21,22]. Hence, from global to EU perspectives, a sustainable agricultural landscape with improved biodiversity is desired that coops with climate adaptation and healthier soils, sustainable agricultural use and preservation and, wherever possible, increasing biodiversity values [17,23].

AquaPuncture is defined as an instrument for the optimal use, adaptation and management of inland waterways, their waterfronts and their adjacent terrain. It is an instrument for the benefit of safety against flooding, water storage, water level regulation, navigability, water quality, economy, employment, spatial quality and environmental values, water management, sewage, all forms of (renewable) energy, sustainable fisheries, seabed mining, eco-friendly construction, tourism, and governance [24]. In this paper, we review AquaPuncture and the Nature Futures Framework in its overall completeness, striving for sustainability on land and water, and the parts of the approaches that might or might not be applicable in the area will be discussed.

2. Background

Inland waterways always linked urban and rural areas. Globally, these waterways were always a focal point for settlement and economic activities along the slow waterways system through cities and lakes. However, these waterways were used for everything, from drainage and irrigation, drinking water supply, beer production, fishing, and transport of persons and goods, to open sewers [25]. Furthermore, many industrial activities along these waterways were present and resulted in emissions of chemicals in the water. The inland waterway system became, to a certain extent, obsolete, and its main transport function was taken over by the faster rail and road systems. The spatial relation between the waterway and urban development became neglected. Currently, the deteriorating state of nature and biodiversity is calling for transformative action [13,17]; therefore, we are once again fully aware of the significance of this unique relationship between the inland waterways and the adjacent urban and rural habitats [14,24].

Even the oldest exposed sediments at Rupel or at Nieuw-Namen testify from water processes and seas, such as rising sea levels, storm surges and tidal currents, which built these deposits and landforms [26]. The clear presence of water in the northern lower part of the proposed Geopark is dominated by the Grevelingen Lake and the Eastern and Western Schelde. The river Scheldt in the Flemish Valley and the Rupel River are dominating the southern higher part [27]. Secondary waterways originate from peat reclamation, industrialisation (channels) or natural causes, such as demarcating old islands that are now embedded in a polder structure, drained from the seawater. Nature reserve areas, such as the tidal flats of Land van Saeftinghe, with multiple tidal gullies, comprise and are adjacent to the remains of lost villages and polder structures [1].

Geoheritage in the aUGGP ScheldeDelta with 40 geosites in total is explored, developed and celebrated, linked to all other aspects of the natural, cultural and intangible heritage of the region [1]. In this paper, the deep history of the area will be demonstrated along the seven most conspicuous landmarks. The main geological highlights in the proposed Geopark Schelde Delta (Figure 1), divided over coastal dunes, the marine clay region with polders, the Campine coversand region, the sand region of the Flemish Valley, and the cuestas [1], are (1) the coastal dune complex of Schouwen, (2) the tidal delta of the Land of Saeftinghe, (3) the escarpment of the Brabantse Wal, (4) the Stekene-Maldegem coversand ridge, (5) the river dunes in the lower Scheldt valley, (6) the Pliocene beach at Nieuw-Namen, and (7) the cuestas and escarpments of Waasland, Temse (Oligocene), Zomergem-Oedelem and near Wetteren and Wichelen and Burghout (Figure 1). The cuestas and escarpments are all higher-situated areas along the southern rim of the Geopark. These elevations are part of Tertiary sand and clay layers that give significance to the deep history that dips (slope) to the north in the whole region of Ghent, Antwerp and Brussels [1,28]. Therefore, everywhere in this part of Flanders, this cuesta-like morphology in the oldest rocks appears because the cohesive (Tertiary) clay layers are more resistant to erosion than the loose-sand layers of the Quaternary processes [28]. A younger outcrop is The Netherlands' oldest beach, a Pliocene uplift near Nieuw-Namen beautifully exposed with evidence of sea, frost and soil formation [26]. The escarpment of the Brabantse Wal shows a Quaternary fluvial cliff face erosion in exposed Lower Pleistocene estuarine deposits [29]. In the Flemish Valley, a Pleistocene large valley during the last ice age created a polar desert in the area, while a forceful circum-polar wind created parabolic dunes in the river valley (and also on the Brabant escarpment). The same periglacial environment produced an extensive and elongated Stekene-Maldegem ridge that has effectively blocked an older Scheldt river course that was originally flowing northwards centrally into the Schelde delta. Through the formation of the cover sand ridge 11,000 years ago, the Scheldt river since then flows into the erosional valley of Hoboken via Antwerpen into the Westerschelde [27].

We demonstrate the significance of the unique relationship between the geoheritage, the inland waterways and the adjacent urban and rural habitats. Because the water quality has been improved considerably through various measures and regulations, we want to rediscover and revitalise the waterway network and its waterfronts through AquaPuncture for the benefit of its users. Waterways are a vital backbone in the urban and rural landscapes. AquaPuncture is an already successfully proposed development level elsewhere [14,24]. Important for the AquaPuncture development is taking into account the cultural and natural history of the region and its biodiversity. The Schelde Delta region is by its aqueous nature fit for revitalisation development and water transport. A new and sustainable vision for Geopark tourism needs to be defined alongside the application of AquaPuncture. Transport through the UNESCO Geopark by foot, bicycle and electric boats will add to the experience of the area.

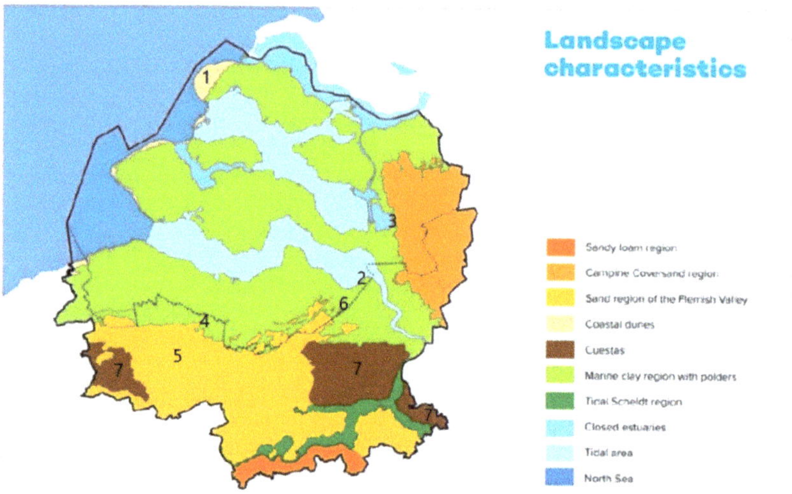

Figure 1. Overview map with landscape characteristics in SW Netherlands and W Flanders region with annotated the seven most conspicuous landmarks in the aUGGP Schelde Delta: (1) Coastal dune complex Schouwen, (2) Tidal delta of Land van Saeftinghe, (3) Escarpment of Brabantse Wal, (4) Cover sand ridge Stekene-Maldegem, (5) River dunes in the lower Schelde valley, (6) Pliocene beach at Nieuw-Namen, (7) Cuestas and escarpments of Waasland and Temse, Zomergem-Oedelem and near Wetteren and Wichelen (after and copyright from [1]).

Due to the interaction between geology and humans at the transitions of land and water, high and low, fresh and salt, and sand and clay, an enormous wealth of geomorphology, landscapes, cultural history (material and intangible) and nature has been created [1]. We assume that inhabitants and visitors of this area, when confronted with a deep history of its geology, can take a strong identity, belonging and a sense of pride and place value to this region [30].

3. Review: Aspiring UNESCO Global Geopark Schelde Delta Area

3.1. Geological and Geomorphological Landscape Reconstruction and Delta History

The Waasland cuesta, the Rupel area (Oligocene high of Temse), as well as the Nieuw-Namen sites, share in common the Tertiary history of the landscape in the Geopark. A diachronic history can be drawn from south to north with 50 MY old Eocene deposits at the Oedelgem-Zomergem cuesta, Oligocene deposits exposed at Temse and the Waasland cuesta front in the south to the 3 MY old Late-Pliocene beach deposits at Nieuw-Namen [1,26,31]. A distinct landscape feature within Brabant and Flanders is the Brabantse Wal, with a suggested tectonic origin and being formed by Pleistocene and Holocene processes and in particular the flow of the river Schelde [29]. In the southern part from the lower to the upper Scheldt river valley, Late-Glacial large-scale meanders with associated river dunes testify of the river and aeolian processes during the Late-Glacial period [27].

On geological time scales, the Dutch and Belgium river deltas all are a result of the interplay between fluvial and tectonic processes, as they represent a structural part of the European Cenozoic Rift System. Tectonic processes in this region are driven by the rift system of the Lower Rhine Graben that causes faults, uplift and downwarping of terrain in NW Europe, including within the aUGGP Schelde Delta area [28,32–34]. The tectonic setting of and resulting processes in the region are exemplified by the general tilt of the terrain towards the north, as well as the escarpments and cuesta fronts throughout the area [1,28]. Due to the continental tilting towards the North Sea basin, all major rivers (Rhine, Meuse, Schelde and Weser) are discharging in this marine basin. In the aUGGP Schelde Delta area, we observe the overall terrain sloping north, as well as the Schelde discharging northwards.

The Waasland cuesta in the southern part of the Geopark area is a so-called clay cuesta. The maximum height of the cuesta near Waasmunster is about 30 m above sea level. This cuesta, together with its continuation in the Rupel area (cuesta of Boom), is the geomorphologically clearest example of a clay cuesta in Flanders. Other clay cuestas with similar orientation, but less distinct than the Waasland-Boom cuesta, comprise the cuesta of Oedelem-Zomergem [1] and the cuesta of the Campine Clay towards the east [35] (7 in Figure 1).

The Waasland cuesta is a landscape feature that originates from the fact that NNE-sloping geological layers have a differential erosional resistance [28,29]. In the Waasland area, the Oligocene clay layer offered more resistance to erosion than under- and overlying sands, which resulted in a sharply pronounced cuesta front that extends from Waasmunster over Temse to Rupelmonde, with a gentle northward slope to Vrasene (on the line Beveren-Sint-Gillis-Waas) that has been covered with a (Weichselian) coversand sand layer with a thickness of multiple metres [27,29]. Although the Rupel clay layer extends in the eastern direction in the area around Boom, the cuesta is bounded at Rupelmonde by the erosional valley of Hoboken (a 'water gap'), which has been used by the river Schelde for at least the past 20,000 years [1,27].

The geomorphology of the Flemish part of the region is dominated by an extended coversand ridge, stretching from Stekene to Maldegem in a west–east orientation [27]. The Flemish Valley is positioned south of this ridge, in which several rivers converged, along with the river Scheldt as the most important one. In the Late-Pleistocene period, the coversand ridge blocked the water of the river Scheldt from discharging directly into the North Sea and forced it in an eastward direction, crossing the ridge near Antwerp [27].

During the last ice age (Weichselian) in the Flemish sand landscapes, Middle Palaeolithic habitation has been found [36]. More intensive habitation occurred during the transition to the Holocene (Late-Glacial period). Numerous prehistoric sites are present with a discontinuous spread of sites dating from the Final Palaeolithic to the Neolithic [37,38]. During the Holocene, as higher places in a wetland environment, the sand ridges such as the Stekene-Maldegem were also preferred dwelling spots for (Late-)Mesolithic hunter–gatherers, whose presence has been amply documented by archaeological investigations [39]. When rapid sea-level rise started to slow down between 6000 and 5000 BP, peat marshes were formed, covering the valleys in between the sand ridges and, increasingly, the sand ridges as well. Many Mesolithic archaeological spots were covered by peat (and later clay or sand deposits), ensuring their preservation [39].

The delta processes that have shaped the entire province of Zeeland and adjacent areas started at the beginning of the Holocene epoch when the sea level was rising with the Scheldt river continuously discharging [40]. The coastal dune complex of Schouwen is a conspicuous landform through its multiple parabolic shapes of land dunes that are formed within one giant parabolic shape [1]. The particular shape and formative processes of these coastal dunes have been anthropogenically influenced. The adoption of agriculture with associated deforestation caused widespread aeolian erosion and deposition in parabolic land dunes. Finally, the Land of Saeftinghe shows a unique modern-day marine tidal landscape around a major inner bend of the Scheldt river, with gullies and tidal flats, biodiversity plants and fauna [40].

3.2. The Lowland Theme of Human Entanglement with Water

A remarkable feature of the landscape of the Scheldt Delta Geopark is the omnipresent theme of living with the water, formerly known as the human battle against water in this region. A specific relict of these human water entanglements is the widely occurring drowned landscapes and villages in an extended area of Zeeland and along the fringes of Noord-Brabant and Flanders [41–43]. These landscapes and villages contribute significantly to the historical, current and future themes of the interaction of humans with water, coping with extreme conditions, as well as the vulnerability and resilience to a changing wetland environment and increasing flood risk [43]. In this region, the drowned landscape is a predominantly covered landscape; different archaeological layers testify of different periods of occupation and settlement, and each layer—whether Mesolithic or Late Medieval—has been sealed by subsequent deposits of peat, sand and clay [41–44].

Specific historical features of the vanished and drowned landscapes in the western part of the Dutch province of Noord-Brabant previously have been used as key features of future landscape redevelopment [42,43]. In the highest part of the area of the Brabant Wal Pleistocene, sandy soils are dominant on a surface that is gently sloping down towards the north and more steeply to the west. Here, mainly clayey soils and remnants of peat occur, lying slightly above sea level (barely) and covering human activity [42].

Around 5000–6000 years ago, the rate of sea-level rise slowed down, which formed the start of major peat growth behind sand barriers. The relative rise of sea level in the last part of the Holocene was caused by subsidence and erosion of the peat layer [40]. The major floods that occurred during the first part of the 2nd millennium AD were the cause of more than 100 drowned villages in the Zeeland, Brabant and Flanders regions combined [41–44]. Islands of the current province Zeeland were formed as islands in the major river delta that the Scheldt river formed over the past millennia.

In medieval times, long-term flooding affected the lowlands of the landscape tremendously. Large parts of the flooded landscape were eventually covered with a thick layer of clay that also covered most medieval settlements and large parts of the old infrastructure, freezing it in time [42]. In other parts, flooding was quite destructive, because new channels were formed that completely washed away all archaeological remnants. Then, a second settlement period began by gradually re-embanking the flooded lands, which was completed by the end of the 19th century [43]. The Holocene history of Zeeland is a major record of Scheldt fluvial discharge in the North Sea, while the sea level was rapidly rising in the first half of the Holocene [40].

Additionally, in Flanders, there was renewed flooding over the pre-existing peat marshes in the (late-)medieval period. In some periods, an alternation of floods and reclamations occurred, resulting in deposits of sand and clay, which covered medieval field systems dating back to the 12th and 13th centuries [45]. Only in the higher sandy areas in the south of the region (such as the 'Waasland cuesta'), settlements from these periods have been preserved. Large-scale peat extraction was an important goal of the reclamation, as peat was the main fuel for the booming Flemish Cities in the 13th and 14th centuries [45]. Floods of 1530 and 1570 AD created the drowned landscapes of Zuid-Beveland and Saeftinghe. By the 16th century, most of the late medieval floods had been undone, and the entire area was reclaimed. Close to the river Scheldt, the polder Doel (from c. 1567) was one of the last major embankment projects before the Eighty Years' War (1568–1648) [45].

Between the 17th and the early 20th centuries, the drowned medieval landscape was again reclaimed in what can be called an 'extreme make-over'; what remained of the medieval landscape was often deliberately erased by new, much more rational and geometric landscapes, which symbolised both the cost-efficiency of the Polder Capitalism that flourished in this area, and the appetite for landscape design by its wealthy merchant and noble creators [46]. Just as in neighbouring Zeeland, the Flemish part of the Geopark still contains several of these 'grandiose' 17th-, 18th- and 19th-century Polders, with their highly recognisable visual and historical landscape qualities [1].

4. Review: Future Landscape Management Scenarios

Based on the cultural development within the polder landscape and from the start of the Industrial Revolution, the Flemish part of the aUGGP ScheldeDelta had an earlier start of the Industrial Revolution than The Netherlands [47]. Whereas Flanders developed industrialisation after the 1830 independence of The Netherlands as the most industrialised country of Europe, The Netherlands followed only half a century later, and the prevalence of agriculture played a role. After the Second World War, through the Great Acceleration [48], the entire Scheldt Delta area has developed in an Anthropocene landscape, where most, if not all, features and surfaces of the landscape are influenced or made by humans.

This influence is omnipresent in the harbours of Antwerp, Gent-Terneuzen and Sluis-Sas van Gent and, to a lesser extent, in the smaller harbours of Vlissingen and Bergen op Zoom. The harbour of Antwerp and its associated industries, petroleum refineries, petrochemical industries, chemical plants and the automotive industry, restructured large scale as one of the most important petrochemical centres globally. The harbour of Antwerp, together with the nuclear energy plant at Doel, seen from the north, forms a fascinating (and threatening) anthropogenic façade. The area has from the past centuries until today also experienced a north–south transport route for industry, business travel and tourists [42]. The great accessibility also attracted wars; from multiple periods, battlefields and signs of battles and defence lines are prominently seen in the landscape [42].

In the Zeeland area, all land is continuously anthropogenically influenced as a consequence of its long history of flooding and disasters associated with low elevation. The name Zeeland, 'land of the sea' in Dutch, implies that the name is derived from islands and peninsulas (former islands) [41].

The preceding part of this paper has highlighted the dynamic and deep history of the entire Scheldt Delta area. In the past 150 years, the influence of industry has increased in this area with an acceleration in the past 70 years. The results of these human-induced influences are more modern and convenient landscapes suitable for 20th- and 21st-century lifestyles, but also delivered contaminations due to the chemical industry (e.g., PFAS from 3M). The impact of all these contaminating influences is still continuing to be received and recorded in the land- and waterscape of today and calls urgently for a global societal and political response [16].

Based on these considerations and taking into account the (deep) history of the landscape and adjoining waters, we review two scenarios for the future landscape management of the Scheldt Delta area. The first scenario is the Revitalised Landscape and Waterscape and introduces and discusses the concept of AquaPuncture in the Scheldt Delta area (4.1) [14,24,25]. The second scenario for future landscape management is Improved Biodiversity, discussing the introduction of the Nature Futures Framework in the area to revitalise the landscape (4.2) [13]. The outcome is important in addressing the future of the transnational aUGGP Schelde Delta [1] and therefore inspires other global Geoparks and beyond these in other areas.

4.1. Scenario: Revitalised Land- and Waterscape

The scenario describes AcquaPuncture to revitalise the landscape and waterscape. AquaPuncture is meant to revitalise the inland waterways and their waterfronts. In this scenario, the UcGP vision of sustainable tourism and economy is expanded to come to a more societal and economic approach [1,14,24]. Water was always omnipresent and has determined the fate of many inhabitants first in creating the delta and dry land in a period of sea-level rise. Later, the area lost property or houses or experienced damages due to flooding. Here, we propose AquaPuncture as a potential future 'land- and waterscape' management based on a thorough analysis of the urban and rural characteristics of the inland waterways and their waterfronts. First of all, the characteristics, history, typology and classification of the waterway itself (river, canal, lake) must be studied, followed by a study of the waterfronts and adjacent territories. Good plans have their roots in the past and are pointing towards the future [24]. To implement AquaPuncture successfully on these rural and urban characteristics, within the region, the following order of elements and pathways need to be taken into account [14]: (1) inland waterways and their waterfronts, (2) user groups, (3) physical adaptations (interventions), and (4) organisation for waterway and waterfront development. Within physical adaptations (interventions), preventing further contamination and improving water quality are also incorporated (see below 12. in Waste water purification.

AquaPuncture is defined as the optimal use, adaptation, experience and management of inland waterways and their waterfronts for safety against flooding, water storage, water level regulation, water quality, navigability, economy, employment, environment and natural landscape. For the sustainable future of inland waterways, the first item to consider when applying AquaPuncture is the urban and rural characteristics of the inland waterways and their waterfronts (Table 1). Within the aUGGP Schelde Delta report, a significant portion of these characteristics is covered [1]. Landscape and nature have been thoroughly described in the bid book, which is summarised above in 'Review: Geological and Geomorphological Landscape Reconstruction'. Therefore, in this paper, we describe the Revitalised Land- and Waterscape scenario through the available waterways, the user groups, the physical adaptations or interventions needed and the organisation for the waterway and waterfront development (Table 2).

Table 1. Presence and characteristics of areas with regard to the waterfront in the Revitalised Landscape and Waterscape scenario of AquaPuncture.

Presence and Characteristics of Waterfront Areas
Residential areas
Leisure parks
Tourist and recreational facilities
Museums and monuments
Commercial and industrial zoning
Infrastructure
Agricultural areas
Landscape and nature

4.1.1. Water (Ways) in the Schelde Delta Area

To describe the potential of applying AquaPuncture in this region, here, a large part of available waterways in the area of the Scheldt Delta has been listed. The Schelde Delta area has in the north three major waterways: Westerschelde, Oosterschelde (salt water) and the Grevelingen (brackish water), which are, in the east, connected via the Oesterdam and Markiezaatskade to the Schelde-Rijn canal and, in the west, by the North Sea. In this aqueous area of large waterways, some local water bodies are present, such as Krabbenkreek between Anna Jacoba polder and Tholen, Markiezaatsmeer near Bergen op Zoom, and Veerse Meer between Noord-Beveland and Zuid-Beveland. On a smaller scale, the Kanaal door Walcheren is present, with an equal waterway through Zuid-Beveland.

The southern part of the Scheldt Delta area is dominated by the Scheldt river and the canal Gent-Terneuzen; in the far west, the area is bounded by the North Sea (Het Zwin). Additionally, the so-called 'double canal' Schipdonk Kanaal dominates the landscape in western Flanders. The Scheldt river has a paleo-meander south of Temse and westward a tributary river course, the Durme, including a paleo-meander. A major waterway in the southern part is the harbour of Antwerp, with its many canals and tributaries. Harbour basins exist as well as rivers and tributaries, such as Rupel, Dender 'grote' Nete and 'kleine' Nete. In addition, smaller harbours are present, such as Gent, Terneuzen, Vlissingen, Bergen op Zoom, with significant waterways such as the Damse Vaart/Kanaal Brugge Sluis and Zoom lake, as well as multiple marinas along all waterways north and south.

Table 2. Physical adaptations and interventions (n = 24) in and along the waterways.

Potential Physical Adaptations and Interventions
1. Height of bridges above water level
2. Dredging depth via environment-friendly dredging
3. Expanding sluice capacity and bridge and sluice servicing
4. Dike/levee adaptation, river/canal widening and room for the river
5. Aqueducts and boat conveyors
6. Water level regulation via sluices, pumping stations and weirs;
7. Facilities for drainage/irrigation
8. Pier/jetty/quay wall/moorings and berths with facilities; realisation of facilities:
9. Loading/unloading platforms
10. Yachting harbours and inland container terminals
11. Introduction of environment-friendly banks/shores
12. Waste water purification
13. Implementation of water framework directive for physical, chemical and biological quality
14. Introduction of hotels, restaurants, cafés/pubs, museums and companies along the waterway
15. Linking inland waterways
16. Urban development with connecting waterways
17. Infrastructure, including bicycle lanes and footpaths and parking spaces along the waterway
18. Enhancing blue-green spatial qualities of urban and rural areas
19. Restoring and purposeful using cultural heritage values in and along the waterway;
20. Introduction of environment-friendly powered vessels
21. Nautical safety
22. Enhancing spatial quality
23. Environmental measures
24. Mitigating measures with regard to climate change: water use for agriculture, aquaculture, drinking water, cooling water and energy.

Old remnants of tidal gullies and streams are present in most terrains in the marine clay region, e.g., Zwaakse Weel, de Braakmankreek and Isabellakanaal, northeast of Sint-Laurens multiple waterways in natural and anthropogenically shapes. Another example is shown in Stierskreek in Zeeuwsch Vlaanderen. On the micro-scale, a multitude of wielen (results of dike breakthroughs) are present in Southern Beveland and to a lesser extent on other Zeeuwsche islands. Everywhere, a plethora of creeks and ditches are present. Scattered through the landscape are also the waters present that are clearly anthropogenic remnants of dredging and excavation activities resulting from mineral (clay, sand) extraction for construction purposes.

4.1.2. User Groups

For the optimal use of AquaPuncture and adaptation and management of the waterways and their waterfronts, six actual and potential user groups in and along the waterways should be considered [14] (Figure 2). All these user groups can be clearly identified in the Schelde Delta region, where all waterways and water infrastructures are related to them. In terms of environmental sustainability, the right and just balance between the five anthropogenic user groups and the sixth natural one, Acquatic/Terrestrial Flora and Fauna, need to be kept in place. Therefore, stakeholder interaction with the Nature Futures scenario is needed to safeguard and promote nature values (Section 4.2).

4.1.3. Physical Adaptations (Interventions)

To achieve a revitalised land- and waterscape for these user groups, physical adaptations (interventions) and organising measures are necessary [14] (Figure 3). Examples of these interventions and measures are physical adaptations from sluices to pumping stations, realisation of facilities from water depth to marinas, nautical safety, enhancing spatial qualities, and environmental and climate adaptation measures. For the aUGGP Schelde Delta, these physical adaptations and interventions in and along the waterways are listed from 24 potential choices to holistically manage the land- and waterscape (see full list in Table 2).

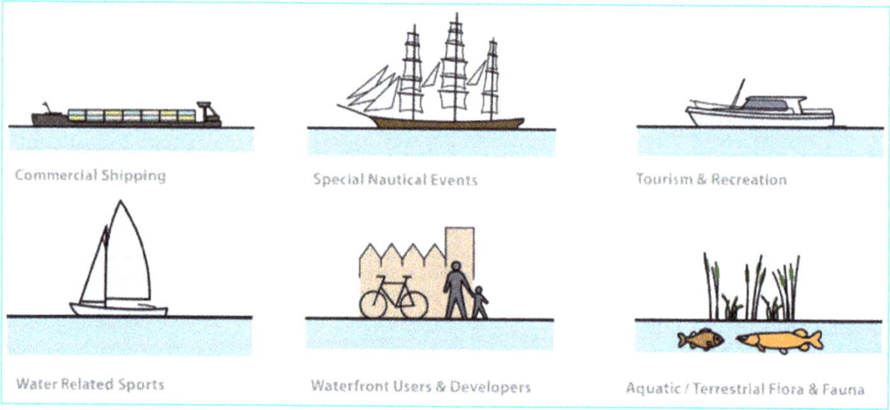

Figure 2. Actual and potential user groups in and along the waterways: Commercial shipping for persons and goods; Tourism and recreation; Special nautical events ("Sail", floating flower shows, regattas of heritage ships, dragon boat races, concerts on water, special cruises); Water related sports (sailing, surfing, rafting, rowing, canoeing, fishing, swimming); Waterfront users and developers; Aquatic/terrestrial flora and fauna and micro-organisms (copyright from [14]).

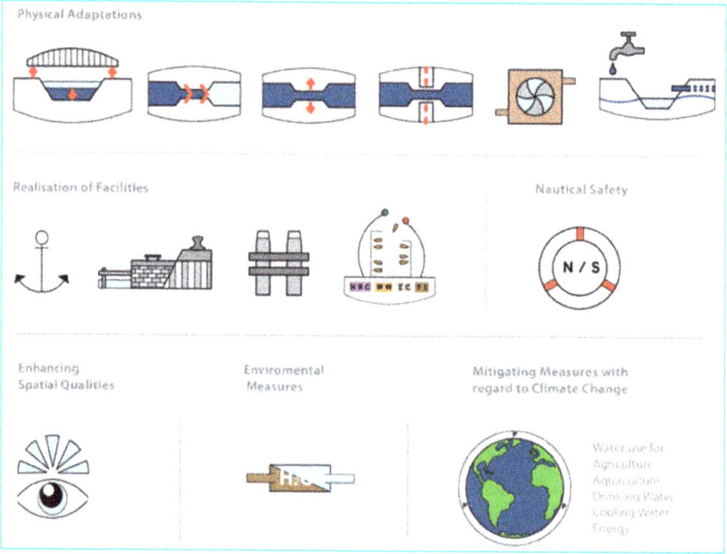

Figure 3. Physical adaptations and interventions managing the land- and aquascape (copyright from [14]).

4.1.4. Organisation for Waterway and Waterfront Development

To acquire adequate support for revitalised land- and waterscape, stakeholder analysis and participation are important, as well as representatives of the public–private partnership, in order to come to a societal cost–benefit analysis [14] (Figure 4). Different municipalities will take variable degree measures, e.g., water quantity (values) and drinking water supply (objectives). In these latter activities, a humane economy approach should be apparent that brings in balance the economical, societal and environmental interests [4]. Within the aUGGP Schelde Delta, the five levels of Government in NL and Belgium are international, national, provincial, communal and local. To organise these interactions, the network of the aUGGP ScheldeDelta is important in tying the private and societal partners to local governmental networks, as well as education, promotion, general outreach and communication tools [1].

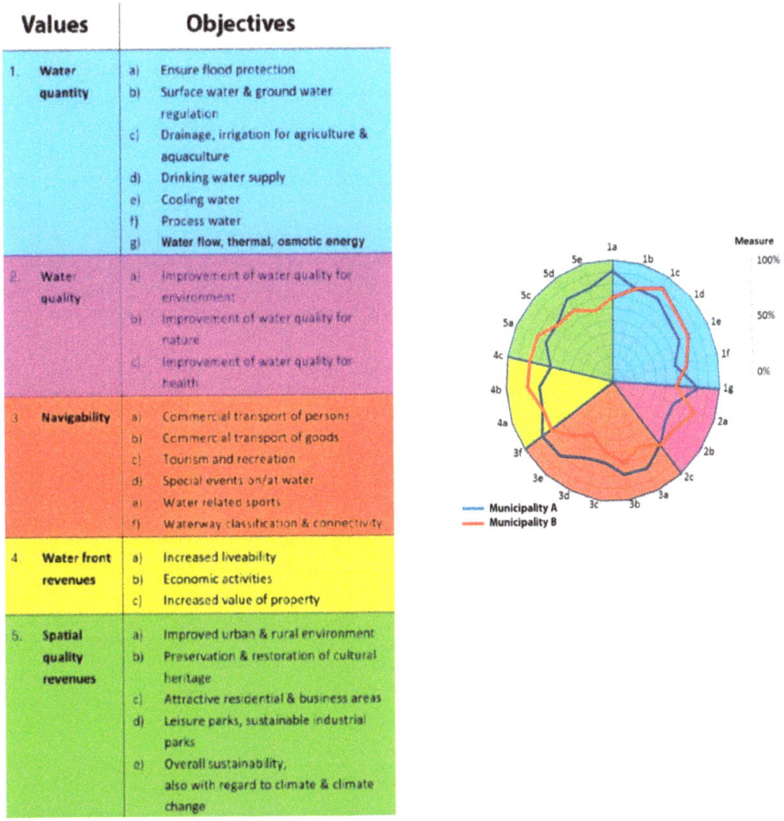

(A) (B)

Figure 4. AquaPuncture: societal costs and benefits measurement model for municipalities along the connecting water ways. (**A**) (left). The values and corresponding objectives in the measurement model. (**B**) (right). The model can be used for benchmarking here as example for municipalities A and B (from [14]).

The adoption of AquaPuncture in the aUGGPScheldeDelta area will evolve differently in different regions, as will the opportunities for development vary as well [14,24]. The most chances to start with a development of revitalised land- and waterscape will be the Leisure domain, which is central in the management of the aUGGPScheldeDelta: 'The leisure domain offers opportunities as a trusted experimental space where behavioural change can be brought about' [1].

4.2. Scenario: Improved Biodiversity

The scenario Improved Biodiversity describes the emergence of the Anthropocene landscape of the aUGGP Schelde Delta and the Nature Futures Framework (NFF) [13] to revitalise the landscape.

The growing urgency of the biodiversity crisis [17], as well as the ongoing climate crisis [49], is demanding holistic approaches to future landscape management scenarios. The future plan of the aUGGP Schelde Delta calls also upon important questions such as how we do (re)shape our (intrinsic) relation with nature, and how do we inspire people to change their behaviour in a sustainable way [1,16]?

The essence of the NFF approach is to populate the NFF triangle with examples of how nature values are represented in different regions of the aUGGP Schelde Delta, across local scales, and how they could change in the future (Figure 5) [13]. For the implementation of new nature values, it is important to engage with society as stakeholders in diverse appearances within the region [19–22]. In addition, the already-established aUGGP Schelde Delta network can be elaborated and extended with the NFF approach (Figure 6). At first, we need to mobilise the scientific community (scientific chamber) and encourage local research groups to align ongoing work with the activities of IPPC, IPBES, as well as the (Dutch) national environmental agency [17,50,51]. Second, the aUGGP Schelde Delta indicates the leisure domain to engage with the general public, addressing major social issues such as those formulated in the SDGs [1]. Potential user groups are geotourists, local ambassadors, excursion guides and park visitors in general. Educational activities are aimed at (a) initiating 100 % Geopark product offer, mainly but not exclusively for the audience of the geotourist; (b) enriching existing touristic offers with Geopark content; and (c) creating tools the partners can use in their own products and communication. These roads of communication can be informed and enriched with the NFF approach in an iterative way, inspiring park users to (re)connect with nature in a co-design approach [13] (Figure 6).

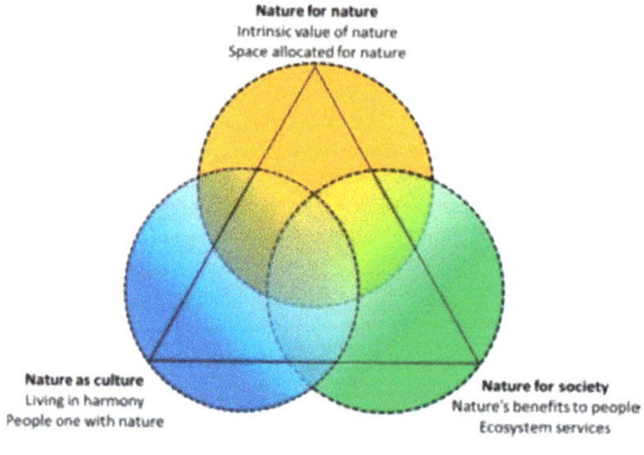

Figure 5. The Nature Futures Framework showing Nature for nature, Nature as culture and Nature for Society (copyright from [13]).

Within the aUGGP Schelde Delta, sustainable tourism is stimulated, as well as the distribution of visitors over the whole aspiring Geopark. Tourism entrepreneurs are stimulated to make their businesses more circular and with a lower carbon footprint [1]. Hiking and biking are promoted via routes and networks to motivate visitors circular in a climate-friendly way. The routes and networks are complemented by visitor centres with an important spill function. The aUGGp Schelde Delta adheres to the vision that leisure has to be in balance with society as a whole and contribute to flourishing communities [1].

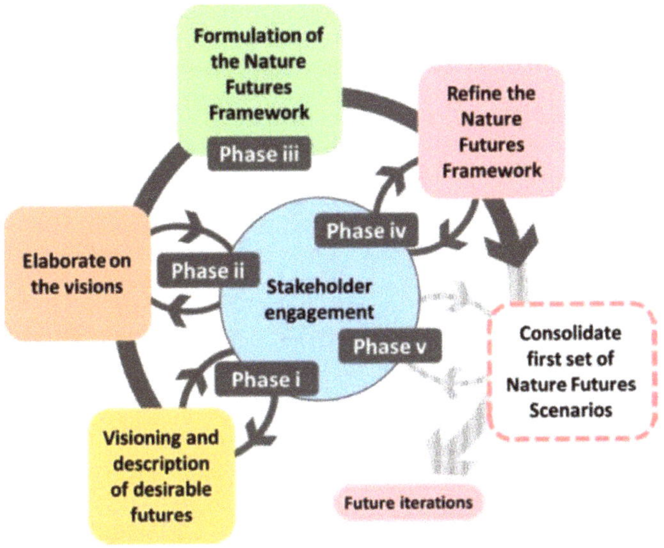

Figure 6. Five main methodological phases used for the development and co-design of Nature Future scenarios, whose process illustrates how the Nature Futures Framework evolved (copyright permitted from [13]).

The start of applying the NFF approach to the aUGGP Schelde Delta applies to the iterative steps on the 2nd Horizon of the stakeholder interaction. Via IPBES (Intergovernmental Science-Policy Platform on the Biodiversity and Ecosystem Services) [17] and The Netherlands Environmental Assessment Agency [50], local cycles, multiple-level schools and community groups and buildings need to be informed and inspired to reach more incremental steps in the horizon framework (Figure 7). Within these local cycles, the process of transition is clearly highly dependent on the stakeholder interaction, resulting in incremental steps, as well as progress over a long-term period [22]. Stakeholder interactions should be part of day-to-day or week-to-week returning frequency to build on incremental adjustments, e.g., to formulate diverse value perspectives for nature in the future [13].

Discussion and Synthesis

The deep-time history of the Scheldt Delta area with the UNESCO Geopark candidacy [1] bolsters regional cohesion in landscape history and between five provinces in two countries. The sometimes-sharp (Escarpment of Brabantse Wal, Oligocene cuesta of Temse, coastal dunes of Schouwen), often-modest (Waasland cuesta, outcrop Nieuw-Namen, coversand ridge Stekene Maldegem) to no-elevation differences (Land of Saeftinghe and many polder landscapes) are shaped by processes such as river (fluvial deposition and erosion [44]), sea (marine deposition), frost (periglacial action), wind (eolian dunes), and faults and uplift (tectonics).

Due to the dynamic interaction between geology and humans at the transitions of land and water, high and low, fresh and salt, and sand and clay, an enormous wealth of geomorphology, landscapes, cultural history (material and intangible) and nature has been created [1]. From 50 million years, from the Eocene to the Oligocene Boom clay until the ongoing sedimentation in the Wadden landscape in the Land van Saeftinghe today, different landforms may give a unique identity to the wider inhabitants of the region, stimulating the acceptance of landscape as heritage, suggesting a sense of pride that may facilitate a sustainable transition in the Anthropocene [30]. The views on Revitalised Land- and Waterscapes and Nature Futures adopted in this paper allow a new perspective on landscape management, including watercourses, big to small, in the Scheldt Delta area

(Figures 8 and 9). The overall nature of the geological processes relates to the aqueous nature of the area, expressed in major waterways, rivers, canals, tidal gullies, streams, pools caused by dyke breaches, creeks, ditches and many anthropogenic water bodies, such as from mineral extraction. Even the extended polder landscape of the northern part of the proposed Scheldt Delta area in Zeeland has been drained multiple times and with disastrous forces from water in the past 500 years. Good plans point to the future, rooted in the past, and humans play a crucial role but need to be steered in favourable directions, e.g., as demonstrated with revitalising land- and waterscapes in Singapore from sewer to stream [14] (Figure 8). Additionally, the iterative process of Nature Futures has the attractive value that it invites visitors and users of the landscape and the water to reflect on the type of nature values and the choice or selection of a sustainable society for itself (Figure 9).

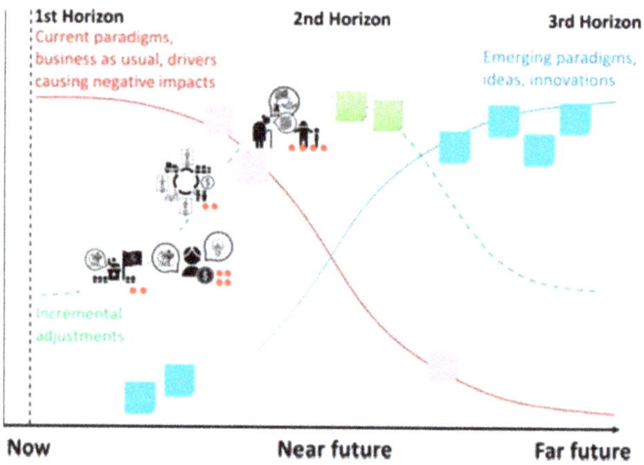

Figure 7. First horizon represents current paradigms and business as usual, while third horizon displays the emerging paradigms, ideas and innovations. The intermediate, second horizon represents the incremental adjustments needed for the transition between the first and the third horizon. The horizontal axis shows the time dimension, showing that all horizons can occur at the same time. The vertical axis represents the intensity and dominancy of the stakeholder interaction process (copyright permitted from [13]).

Humans have become a geological factor for the first time in the earth's history. Therefore, now, it is about letting human activity carry out natural processes. Humans, in their lifestyles and numbers, are acting as a geological factor, with the result that human action is present in natural processes as much as possible. What is needed to safeguard the environment for the complex and chaotic future in a sustainable landscape design? How can a deep and recent environmental history of an area contribute to acknowledging a desirable change? The importance of making people conscious of the landscape around them and making them aware of possible futures has never been more important than today. For two decades, a scientific and societal discussion has been taking place around the concept and potential starting date of the Anthropocene [2–4,6–9,51]. Therefore, popular awareness of the long-term formation of the landscape, as well as informing identity, belonging and a sense of pride and place to citizens and policy makers, in combination with the AquaPuncture approach, will also encourage a positive approach to creating a sustainable future for the planet.

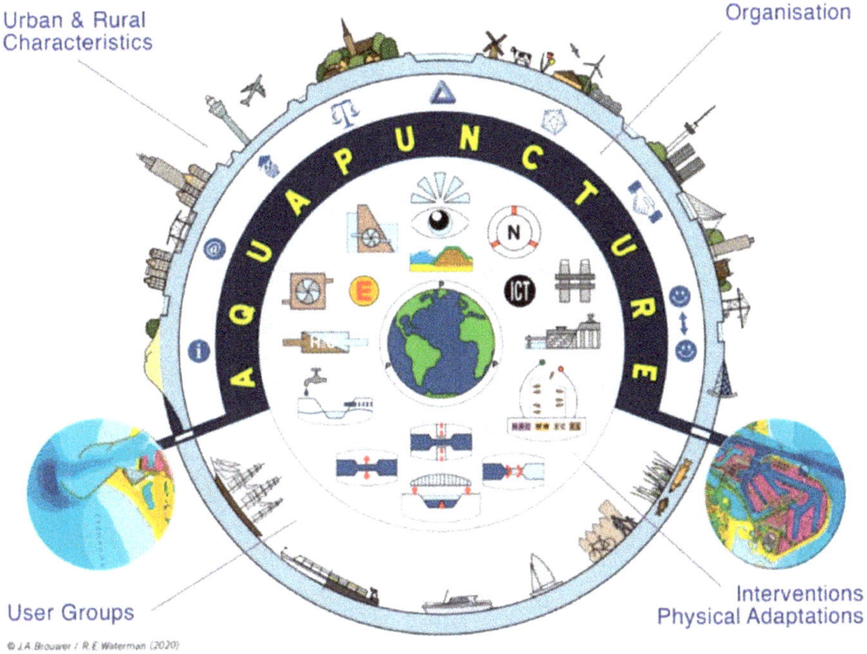

Figure 8. Relationship between user groups, urban and rural characteristics, organisation, and interventions and physical adaptations in a Dutch setting; note also the signs for E = environmental sustainability and ICT = digitisation of processes (copyright permitted from [14,24]).

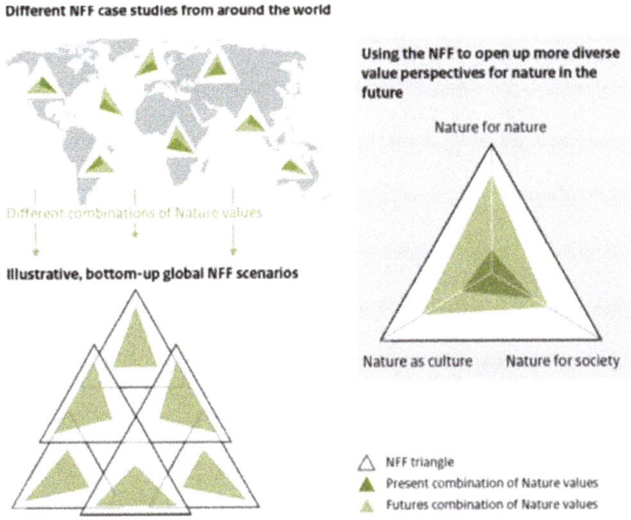

Figure 9. Perspective of Nature Futures Framework illustrating the bottom-up process on a global scale and to use the NFF to open up more diverse value perspectives for nature in the future (copyright permitted from: [13]).

Inhabitants and visitors of this area when confronted with a deep history of the geology can take a strong identity, belonging and a sense of pride and place value to this region [30]. To put emphasis on AquaPuncture and Nature Futures to be adopted by the proposed Geopark follows logically out of the deep history with ample focus on pure nature and water. The revaluation and revitalisation of waterways offer many potential avenues for sustainable development in the 21st century (Figure 8), e.g., Geopark visitors, tourists and citizens can enjoy electrical boats (whisper boats) to explore the natural and cultural landscape of the Scheldt area. These same visitors can marvel at rewilding initiatives and improved biodiversity habitats that will surround and be embedded within cultural centres such as villages and cities.

Apart from the dramatic history and drowned landscapes from the Middle Ages into the New and Modern Ages (1953), the entire area of the Scheldt Delta Geopark has witnessed two major transitions that have profoundly shaped the landscape. At first, the Industrial Revolution introduced mechanical power instead of horsepower and introduced energy fuels such as coal, oil and gas. In the centuries preceding this fossil revolution, the area has also witnessed peat (and salt) extraction as a precursor to the later more widely available fossil energy sources [16].

Second, our society, including the Scheldt Delta area, experienced a Great Acceleration after the Second World War [7,48], accelerating the use of resources, including the growth of industry and consumption, to currently unsustainable levels [4]. Observing the landscape in the Schelde Delta area is traditionally performed on a land-based approach, i.e., all features described are terrigenous and comprise mineral or rock materials. Rarely, the aqueous nature of the area has been described or witnessed. From historical records, a great example of a painting observed out of boats floating in the Medieval Scheldt depict villages and settlements at the Brabant scarp. In modern times, a dominance of land-based forms, landscapes and attributes is based on an anthropocentric behaviour that steers people's observations always on a terrigenous spectre e.g., [42]. This also is based on the notion that humans in the path of their long time evolution have always progressed and moved over land.

Whereas acupuncture is applied to revitalise the nervous system and the human organs, AquaPuncture is applied to revitalise the inland waterways and their waterfronts [24]. The unique geography of the proposed Schelde Delta UNESCO Geopark application [1] is that due to its dynamic landscape history of flooding and river change, most locations and areas are located at waterways. The Geopark has as its main status, showing the beauty of the earth and environmental phenomena. To do that sustainably, the use of EV boats, potentially circular in energy, can be demonstrated to visit multiple Geosite locations throughout the area. In addition, the Nature Futures Framework will be targeted on the land part of the Geopark (75% of area [1]).

Since we are currently facing multiple crises (war, energy, and refugees) and are in a period of transitions (e.g., food, energy and climate), it is therefore important to address sustainability in transport systems, water quality, economic development, energy use and social welfare through more healthy environments and use the concept of AquaPuncture to revitalise the inland waterways and their waterfronts. However, water management, sewerage, all forms of energy, sustainable fisheries, see bed mining, eco-friendly building, tourism, a sustainable socio-sphere and governance can also be much more concrete than adopting 17 SDGs as a way of greenwashing organisations [24]. It needs to be further openly discussed how the 22 main functions (and derived) of nature, landscape and water [2] relate to the 17 SDGs and how we implement that in the Nature Future Framework. We also may need these multiple frameworks to couple and diversify functions to action and to alleviate class differences.

Adaptation of the inland waterways through AquaPuncture will continue to play a significant role in the total concept of achieving a smaller global footprint [24]. Because good plans have their roots in the past and are pointing to the future, the adoption of AquaPuncture and the Nature Future Framework also points to sustainable landscape management [15], as has been moved forwards in current discussions and political debates in the world around the nitrogen crisis [52]. Many political processes currently appear to be accelerating, caused by the pandemic, energy crisis and war, as well as climate and biodiversity targets for 2030 [17,46]. We propose that an approach of AquaPuncture and Nature Futures Framework applied to the area of the aUGGP Schelde Delta holds a promise for a sustainable future, especially when this novel approach will be coupled with circularity, waste control and biodiversity improvement in respectful and efficient stakeholder interaction [13,18,22,53].

5. Conclusions

- The Anthropocene landscape and waterscape of the aUGGP Schelde Delta will inspire and motivate inhabitants and visitors with geological history, identity, belonging and a sense of pride and place value.
- A deeper insight is achieved into the geological history of both aquatic and terrestrial surface and underground landscapes, including the interplay of rivers with the specific tectonic framework in the landscape.
- AquaPuncture's Revitalised Land- and Waterscape is an already elsewhere successfully implemented development while Nature Futures has also promising and inspiring values for the Geopark Scheldt Delta.
- Important for the AquaPuncture and Nature Futures Framework development is taking into account the special 'aqueous' cultural and natural history of the region and its biodiversity. It strives for stakeholder interaction with incremental steps in integrated, multifunctional, sustainable, development of co-designing landscapes and waterscapes.
- The aUGGP Schelde Delta region is by its aqueous nature fit for water transport and further (sustainable) developments. The surrounding area in all its aspects can be a source of inspiration for nature values in the NFF.
- A new and sustainable vision for Geopark tourism (leisure and economy) is defined and is prone to be followed up by AquaPuncture and Nature Futures as a response from bottom-up stakeholder interaction in society and to be supported and implemented by policy makers.

Author Contributions: All authors contributed equally 50% (S.K.) and 50% (R.W.). All authors have read and agreed to the published version of the manuscript.

Funding: The research presented in this paper by Kluiving has been undertaken within the project TERRANOVA: the European Landscape Learning Initiative, funded by the European Union's Horizon 2020 research and innovation programme, and Marie Skłodowska-Curie Actions–Innovative Training Networks MSCA-ITN, under Grant Agreement No. 813904.

Data Availability Statement: All data are available in the cited research.

Acknowledgments: Kees Kasse (VU-NL), Alexandra Brown (AUC-NL) and Kay Koster (TNO-NL) are acknowledged for reviews of earlier drafts of the paper. We thank the two anonymous reviewers who have contributed constructively with critical remarks on the final version of the paper.

Conflicts of Interest: The authors declare no conflict of interest.

References

1. Westrik, C.; Bernaerts, G.; Kiden, P.; Jonkers, W. (Eds.) *aUGGp Schelde Delta UNESCO Global Geopark Candidate's Application*; Aspiring UNESCO Global Geopark Schelde Delta: Middelburg, The Netherlands, 2022; pp. 1–50.
2. Waters, C.N.; Zalasiewicz, J.; Summerhayes, C.; Barnosky, A.D.; Poirier, C.; Gałuszka, A.; Cearreta, A.; Edgeworth, M.; Ellis, E.; Ellis, M.; et al. The Anthropocene Is Functionally and Stratigraphically Distinct from the Holocene. *Science* **2016**, *351*, aad2622. [CrossRef] [PubMed]
3. Lewis, S.; Maslin, M. Defining the Anthropocene. *Nature* **2015**, *519*, 171–180. [CrossRef] [PubMed]
4. Pattberg, P.; Kluiving, S.J. Introduction to the Special Issue: Accelerating and Scaling Transformative Change in Anthropocene: A Multi-Disciplinary Perspective. *Int. J. Environ. Sustain.* **2022**, *18*, I–VII. [CrossRef]
5. Crutzen, P. Geology of mankind. *Nature* **2002**, *415*, 23. [CrossRef] [PubMed]
6. Crutzen, P.J.; Stoermer, E.F. The "Anthropocene". *Glob. Change Lett.* **2000**, *41*, 17.

7. Waters, C.N.; Williams, M.; Zalasiewicz, J.; Turner, S.D.; Barnosky, A.D.; Head, M.J.; Wing, S.L.; Wagreich, M.; Steffen, W.; Summerhayes, C.P.; et al. Epochs, events and episodes: Marking the geological impact of humans. *Earth-Sci. Rev.* **2022**, *234*, 104171. [CrossRef]
8. Gibbard, P.L.; Bauer, A.M.; Edgeworth, M.; Ruddiman, W.F.; Gill, J.L.; Merritts, D.J.; Finney, S.; Edwards, L.; Walker, M.; Ellis, E.; et al. A practical solution: The Anthropocene is a geological event, not a formal epoch. *Epis. J. Int. Geosci.* **2021**, *45*, 349–357. [CrossRef]
9. Merritts, D.; Edwards, L.E.; Ellis, E.; Walker, M.; Finney, S.; Gibbard, P.; Gill, J.; Maslin, M.; Bauer, A.; Edgeworth, M.; et al. The Anthropocene is complex. Defining it is not. *Earth-Sci. Rev.* **2023**, *238*, 104340. [CrossRef]
10. Rockström, J.; Steffen, W.; Noone, K.; Persson, Å.; Chapin, F.S.; Lambin, E.F.; Lenton, T.M.; Scheffer, M.; Folke, C.; Schellnhuber, H.J.; et al. A safe operating space for humanity. *Nature* **2009**, *461*, 472–475. [CrossRef]
11. Sustainable water solutions. *Nat. Sustain.* **2020**, *3*, 73. [CrossRef]
12. Greve, P.; Kahil, T.; Mochizuki, J.; Schinko, T.; Satoh, Y.; Burek, P.; Fischer, G.; Tramberend, S.; Burtscher, R.; Langan, S. Global assessment of water challenges under uncertainty in water scarcity projections. *Nat. Sustain.* **2018**, *1*, 486–494. [CrossRef]
13. Pereira, L.M.; Davies, K.K.; den Belder, E.; Ferrier, S.; Karlsson-Vinkhuyzen, S.; Kim, H.; Kuiper, J.J.; Okayasu, S.; Palomo, M.G.; Pereira, H.M.; et al. Developing multiscale and integrative nature-people scenarios using the Nature Futures Framework. *People Nat.* **2020**, *2*, 1172–1195. [CrossRef]
14. Waterman, R.E. *Integrated Coastal Policy with Building with Nature*; Drukkerij Banda: Heerenveen, The Neterlands, 2010; 71p, ISBN 978-90-80522-3-7.
15. Lindholm, K.-J.; Fernández, N.; Svenning, J.-C.; Pereira, H.; Kluiving, S. Policy recommendations for sustainable landscape management strategies. *TERRA NOVA* **2020**, *2022*, 1–20. [CrossRef]
16. Snow, K.; Kluiving, S. Reading the Anthropocene through Landscape Archaeology: Historical and Contemporary Landscapes of Accumulation in the Netherlands and Lebanon. *Int. J. Environ. Sustain.* **2022**, *18*, 3–25. [CrossRef]
17. IPBES. *Summary for Policymakers of the Thematic Assessment of the Sustainable Use of Wild Species of the Intergovernmental Science-Policy Platform on Biodiversity and Ecosystem Services*; Fromentin, J.-M., Emery, M.R., Donaldson, J., Danner, M.-C., Hallosserie, A., Kieling, D., Balachander, G., Barron, E.S., Chaudhary, R.P., Gasalla, M., Eds.; IPBES Secretariat: Bonn, Germany, 2022. [CrossRef]
18. Fayet, C.M.J.; Reilly, K.H.; Van Ham, C.; Verburg, P.H. The potential of European abandoned agricultural lands to contribute to the Green Deal objectives: Policy perspectives. *Environ. Sci. Policy* **2022**, *133*, 43–44. [CrossRef]
19. Segar, J.; Pereira, H.M.; Filgueiras, R.; Karamanlidis, A.A.; Saavedra, D.; Fernández, N. Expert-based assessment of rewilding indicates progress at site-level, yet challenges for upscaling. *Ecography* **2022**, *2022*, 1–10. [CrossRef]
20. Diogo, V.; Helfenstein, J.; Mohr, F.; Varghese, V.; Debonne, N.; Levers, C.; Swart, R.; Sonderegger, G.; Nemecek, T.; Schader, C.; et al. Developing context-specific frameworks for integrated sustainability assessment of agricultural intensity change: An application for Europe. *Environ. Sci. Policy* **2022**, *137*, 128–142. [CrossRef]
21. Pereira, H.M. (Ed.) *Rewilding European Landscapes*; Springer International Publishing: New York, NY, USA, 2015.
22. Quintero-Uribe, L.C.; Navarro, L.M.; Pereira, H.M.; Fernández, N. Participatory scenarios for restoring European landscapes show a plurality of nature values. *Ecography* **2022**, *2022*, e06292. [CrossRef]
23. Bianchi, G.; Pisiotis, U.; Cabrera Giraldez, M. *GreenComp the European Sustainability Competence Framework*; Punie, Y., Bacigalupo, M., Eds.; Publications Office of the European Union: Luxembourg, 2022.
24. Waterman, R.E.; Brouwer, J. Aquapuncture for Sustainable Waterways. *Terra Et Aqua Nr 121* **2015**, 5–18.
25. Kranendonk, P.; Kluiving, S.; Troelstra, S. Chrono- and archaeostratigraphy and development of the River Amstel: Results of the North/South underground line excavations, Amsterdam, the Netherlands. *Neth. J. Geosci.* **2015**, *94*, 333–352. [CrossRef]
26. Kluiving, S.J.; Troelstra, S.R.; Kasse, C.; Lelivelt, R.A. Het oudste strand van Zeeland: Een hernieuwde kennismaking met de "Meester van der Heijden groeve" (de Kauter), Nieuw-Namen. *Grondboor Hamer* **2012**, *66*, 431–437.
27. Kiden, P. De evolutie van de Beneden-Schelde in België en Zuidwest-Nederland na de laatste ijstijd. *Belgeo. Rev. Belg. De Géogr.* **2006**, *30*, 279–294. [CrossRef]
28. Vandenberghe, N. Tectonic and climatic signals in the Oligocene sediments of the Southern North-Sea Basin (Ernest Van den Broeck medallist lecture 2016). *Geol. Belg.* **2017**, *20*, 105–123. [CrossRef]
29. Kasse, C. De Brabantse Wal, op de grens van hoog en laag, oud en jong. *Grondboor Hamer* **2009**, *63*, 173–178.
30. Kluiving, S.J. Heritage and Landscapes. In *The Encyclopedia of Archaeological Sciences*; López Varela, S.L., Ed.; Wiley Online Library: Hoboken, NJ, USA, 2018. [CrossRef]
31. Vandenberghe, N. *Sedimentology of the Boom Clay (Rupelian) in Belgium*; Paleis der Academiën: Brussel, Belgium, 1978; 137p.
32. Vanneste, K.; Meghraoui, M.; Camelbeeck, T. Late Quaternary earthquake-related soft-sediment deformation along the Belgian portion of the Feldbiss Fault, Lower Rhine Graben system. *Tectonophysics* **1999**, *309*, 57–79. [CrossRef]
33. Kübler, S.; Friedrich, A.M.; Gold, R.D.; Strecker, M.R. Historical coseismic surface deformation of fluvial gravel deposits, Schafberg fault, Lower Rhine Graben, Germany. *Int. J. Earth Sci.* **2018**, *107*, 571–585. [CrossRef]
34. Grützner, C.; Fischer, P.; Reicherter, K. Holocene surface ruptures of the Rurrand Fault, Germany—Insights from palaeoseismology, remote sensing and shallow geophysics. *Geophys. J. Int.* **2016**, *204*, 1662–1677. [CrossRef]
35. Kasse, C. *Early-Pleistocene Tidal and Fluviatile Environments in the Southern Netherlands and Northern Belgium*; Free University Press: Amsterdam, The Netherlands, 1988; 190p.

36. Crombé, P.; Van Der Haegen, G. *Het Midden-Paleolithicum in Noordwestelijk België*; Archeologische Inventaris Vlaanderen: Ghent, Belgium, 1994; Volume 3.
37. Crombé, P.; Van Strydonck, M.; Boudin, M.; Van den Brande, T.; Derese, C.; Vandenberghe, D.; Zwertvaegher, A. Absolute Dating (14C and OSL) of the Formation of Coversand Ridges Occupied by Prehistoric Hunter-Gatherers in NW Belgium. *Radiocarbon* **2012**, *54*, 715–726. [CrossRef]
38. Derese, C.; Vandenberghe, D.; Gils, M.; Mees, F.; Paulissen, E. Final Palaeolithic settlements of the Campine region (NE Belgium) in their environmental context: Optical age constraints. *Quat. Int.* **2012**, *251*, 7–21. [CrossRef]
39. Crombé, P. (Ed.) The last hunter-gatherer-fishermen in Sandy Flanders (NW Belgium). In *The Verrebroek and Doel Excavation Projects. Volume 1: Palaeo-Environment, Chronology and Features. Archaeological Reports Ghent University 3*; Academia Press: Ghent, Belgium, 2005; p. 334.
40. Vos., P.C.; van Heeringen, R.M. Holocene geology and occupation history of the province of Zeeland. *Meded. Ned. Inst. Voor Toegep. Geowetenschappen TNO* **1997**, *59*, 109.
41. Kuiper, J.J.B.; van Dierendonck, R.M. (Eds.) *Sluimerend in Slik*; Den Boer: Middelburg, The Netherlands, 2004; 120p, ISBN 9789074576505.
42. Kluiving, S.J.; Brand, N.; Borger, G.J. *De West-Brabantse Delta: Een Verdronken Landschap Vormgeven. Geo- and Bioarchaeological Studies*; Instituut voor Geo- en Bioarcheologie: Amsterdam, The Netherlands, 2007; Volume 7.
43. de Kraker, A.M.J.; Kluiving, S.J. Verdronken landschap voor het voetlicht. Onderzoek naar de locatie en specifieke kenmerken van verdronken nederzettingen en structuren van West-Brabant voor 1600. In *Noordbrabants Historisch Jaarboek*; Stichting Matrijs: Utrecht, The Netherlands, 2007; pp. 59–70.
44. Missiaen, T.; Jongepier, I.; Heirman, K.; Soens, T.; Gelorini, V.; Verniers, J.; Crombé, P. Holocene landscape evolution of an estuarine wetland in relation to its human occupation and exploitation: Waasland Scheldt polders, northern Belgium. *Neth. J. Geosci.* **2017**, *96*, 35–62. [CrossRef]
45. Soens, T.; Tys, D.; Thoen, E. Landscape transformation and social change in the North Sea polders, the example of Flanders (1000-1800 AD). *Siedlungsforschung. Archäologie-Gesch.-Geogr.* **2014**, *31*, S.133–S.160.
46. Soens, T.; De Graef, P. Polder mania or marsh fever? Risk and risk management in early modern drainage projects: The case of Kallopolder, Flanders, 1649 to 1662. *Agric. Hist. Rev.* **2014**, *62*, 231–255.
47. Anzion, F.K.M. Hoe Jan Salie op stoom kwam: Nederland en de Industriële Revolutie, Themanummer. Spiegel Historiael: Zeist, the Netherlands, 2003; p. 38.
48. Steffen, W.; Richardson, K.; Rockström, J.; Cornell, S.E.; Fetzer, I.; Bennett, E.M.; Biggs, R.; Carpenter, S.; Vries, W.; de Wit, C.; et al. Planetary Boundaries: Guiding Human Development on a Changing Planet. *Science* **2015**, *10*, 1259855. [CrossRef]
49. IPCC. *Climate Change 2022: Impacts, Adaptation, and Vulnerability. Contribution of Working Group II to the Sixth Assessment Report of the Intergovernmental Panel on Climate Change*; Pörtner, H.-O., Roberts, D.C., Tignor, M., Poloczanska, E.S., Mintenbeck, K., Alegría, A., Craig, M., Langsdorf, S., Löschke, S., Möller, V., et al., Eds.; Cambridge University Press: Cambridge, UK; New York, NY, USA, 2022; p. 3056. [CrossRef]
50. Schoolenberg, M.; den Belder, E.; Okayasu, S.; Alkemade, R.; Lundquist, C.; Pereira, H.; Chettri, N.; Cheung, W.; Ferrier, S.; Hauck, J.; et al. Report on the Workshop 'Next Steps in Developing Nature Futures', PBL Netherlands Environmental Assessment Agency, the Hague, the Netherlands, nr 3411. Available online: https://www.pbl.nl/sites/default/files/downloads/3314_Report_on_The_Hague_workshop_June_2018_V9.pdf (accessed on 18 March 2023).
51. Martinez, A.; Kluiving, S.J.; Muñoz Rojas, J.; Borja Barrera, C.; Fraile Jurado, P. From hunter-gatherer subsistence strategies to the Agricultural Revolution: Disentangling Energy Regimes as a complement to cultural phases in Northern Spain. *Holocene* **2022**, *32*, 884–896. [CrossRef]
52. Gu, B.; Zhang, X.; Lam, S.K.; Yu, Y.; van Grinsven, H.J.; Zhang, S.; Wang, X.; Bodirsky, B.L.; Wang, S.; Duan, J. Cost-effective mitigation of nitrogen pollution from global croplands. *Nature* **2023**, *613*, 77–84. [CrossRef] [PubMed]
53. Fayet, C.M.; Leen, F.J.; Quintero Uribe, L.C.; Rigo, R.; Houet, T.; Lindholm, K.J.; Kluiving, S.J. TERRANOVA White Paper 2. An explorative opinion paper: 'Why do we need stakeholders' engagement in knowledge production: TERRANOVA's vision on landscape transformation. *Zenodo* **2021**, 1–11. [CrossRef]

land

MDPI

Article

Identifying the Authenticity of Plantscapes through Classics: A Case Study of Beijing Suburbs in the Qing Dynasty

Dong Xu [1], Junda Zhu [2], Zhiyu Chen [1], Nan Hu [1], Peiyan Wang [3,*] and Yunyuan Li [1]

1 School of Landscape Architecture, Beijing Forestry University, 35-Qinghua East Road, Beijing 100083, China
2 Zhong Jiao Highway Planning and Design Institute, 296-Chaoyangmen Inner Street, Beijing 100010, China
3 School of Art and Design, Beijing Forestry University, 35-Qinghua East Road, Beijing 100083, China
* Correspondence: imasder@bjfu.edu.cn

Abstract: The plantscapes surrounding historical gardens hold significant value, reflecting the natural pristine state as well as demonstrating the cultural attributes of the landscape. This study aims to develop a method for identifying the characteristics of historic plantscapes and to recognize the authenticity of historic landscapes from the perspective of plant elements. Our method combines textual and geospatial data analysis to examine the plant species, their relationships and combinations, and spatial distribution. The case study focuses on the Beijing suburbs during the Qing Dynasty, as documented in *A Collection of Past Events in Beijing*. We identified 658 plants recorded, encompassing 44 families and 58 genera. These plants were categorized into 7 groups based on the growth type and morphological characteristics, leading to 54 plant relationship outcomes, 107 plant combination scenarios, 5 plant combination categories, and 7 representative plant combinations. Additionally, we mapped the spatial distribution of plants, forming 16 plantscape groups and depicting the spatial kernel density distribution of important plants. We also determined the characteristics of plantscapes in different directions in the suburb. Our findings advocate for respecting the historical development of the plantscape and understanding its evolution, particularly emphasizing the use of high-quality native plants and plant combinations.

Keywords: historic gardens; vegetation history; landscape history; urban greening; text-mining analysis

1. Introduction

The design and formation of historic gardens not only reflect the profound respect ancient peoples had for and adaptation to the natural environment but also reflect the results of a long-term dynamic interaction between humans and nature [1–3]. In this process, plant elements played a pivotal role, serving as a link between nature and artistry and as vital living elements that bridge the internal and external spaces of historic gardens [4]. Plants have played a crucial role in the formation and evolution of garden landscapes, which encompasses not only the specific species planted but also the proportions, color schemes, spacing, and respective heights of the vegetation [5], demonstrating adaptability to climatic and environmental changes, as well as to human aesthetic and cultural preferences [4,6]. By meticulously arranging and positioning plants, the ancients were able to effectively respond to environmental shifts and shape the garden landscape in a manner that aligns with cultural and aesthetic standards while also fulfilling functional requirements [7,8]. This adaptability is reflected not only in the growth and evolution of plants but also in the long-term nature of garden design and the dynamic changes in the landscape [9], resulting in a historical garden system that is in harmony with the natural environment.

The Florence Charter [5] explicitly stated that in any work of maintenance, conservation, restoration, or reconstruction of a historic garden, or of any part of it, all its constituent features must be dealt with simultaneously. Isolating the various operations would damage the unity of the whole. However, previous cultural heritage landscape conservation theories and approaches predominantly concentrated on specific historic gardens or buildings [10–12], with a notable absence of consideration for the natural relationships and contexts. It is therefore imperative to consider historic gardens in a more holistic, sustainable manner, within the context of the landscape [13,14]. This entails focusing on the ecological and cultural landscapes of the areas in which they are situated. For decades, the ecological environment and landscape characteristics of vegetation in historic gardens have been altered by the process of urbanization [15,16], and the changes in vegetation planning practices have led to the homogenization of urban vegetation landscape [17]. In this context, scientists and natural resource conservationists have endeavored to identify multiple strategies for the protection of urban vegetation, biodiversity, and cultural values [15,18,19]. However, the majority of studies on historic gardens or cultural heritage vegetation tend to concentrate on modern perceptions of biodiversity and cultural diversity [18–20]. It is, therefore, necessary to reassess the authenticity of plantscapes from a historical perspective. This can facilitate a more profound comprehension of the intricate dynamics between humans and their environment [21], thereby paving the way for the long-term resilience of historic landscapes.

Chinese historic gardens are an important part of the world's gardens and cultural heritage [22]. Beijing, as the capital of China and one of the first batch of declared historical and cultural cities, has many historic gardens and cultural heritage [23], which has had a profound influence on Chinese gardens and even oriental gardens [24]. During the Qing Dynasty (1644–1911 AD), the creation of Beijing's historic gardens reached its peak, forming the "Three Hills and Five Gardens" historic garden clusters, and the urban plantscapes surrounding the garden clusters also paid more attention to the beauty and symbolic meanings of plant configuration [25]. Since the beginning of the 21st century, a substantial number of scholars have engaged in research on Beijing's historical gardens [26–28], encompassing a range of topics including garden art and rationale [29], cultural thought [30], heritage protection and utilization [31], and other related areas of inquiry. The scope of the research encompasses a variety of garden-related subjects, with a particular focus on the Beijing Mountains and the Three Hills and Five Gardens Historic Garden Cluster region, as well as the Beihai areas [31–33]. These studies provide a comprehensive and valuable reference for introducing unique and individual information about Beijing's historic gardens. Nevertheless, there is a paucity of studies that concentrate on plants within the broader public domain during the historical period. From our perspective, this is largely constrained by the historical material and the methodologies employed to examine it. It is, therefore, necessary to construct a comprehensive research framework in order to facilitate a more nuanced understanding of the relationship between historical documents and planted landscapes.

The historical material, especially works from specific periods, provides insight into the political and economic conditions of the region at the time [34]. However, these historical materials contain a substantial quantity of textual information and encompass comprehensive and intricate content, rendering it challenging to conduct a comprehensive and detailed review of a single issue. In recent years, interdisciplinary methods have emerged that can be applied between science and the humanities. These methods have the potential to be applied to the study of historical plantscapes. Text mining, as an interdisciplinary method drawing upon information retrieval, statistics, computational linguistics, and especially data mining, can disclose concealed information and can deal with a large number of words and structures in natural language on the one hand and ambiguity, uncertainty, and vagueness on the other hand [35,36]. As a result, it has gained wide application and attention in many research fields [37–39]. Furthermore, the Geographic Information System (GIS) spatial analysis is particularly important for studying the spatial distribution of plants, which can improve the methodology of historical plantscape studies and enhance the planning and management of plantscapes [40,41]. The combination of text mining and GIS is more likely to improve the disadvantage of the lack of objectivity in the study of historical evidence. Therefore, this study takes Beijing, China as an example to explore the possibility of text and geospatial data analysis in identifying the authenticity of plantscapes.

This study employs an authoritative official book of the Qing Dynasty, *A Collection of Past Events in Beijing* (hereafter, the *Collection*) [42], to examine the general pattern of distribution of regional native plants in Beijing during the historical period. It considers the application of plants with regional characteristics and cultural connotations, beginning in the suburban natural environment. In addition, it offers a novel perspective on the conservation of historic gardens, public spaces, and contemporary parks. In the past, Beijing's urban growth, population increase, and land use patterns have changed considerably in the modern period (18th–19th century) [43]. Since urbanization, the number of urban vegetation has been drastically reduced, and a large number of domesticated plants have been introduced, leading to a drastic change in the ecological environment and landscape characteristics of Beijing's urban greening [44,45], and the emergence of the problem of homogenization of the plantscape [27]. Therefore, how to balance the ecological development and the cultural heritage of historic plantscapes will be a research topic, as this will contribute significantly to the resilience of historical landscapes.

This study aims to add knowledge pertinent to the conservation of historic gardens, the assessment of historic plantscapes, and the cultural management of urban plantscapes. Furthermore, we examined an important historical document, the *Collection*, identified urban plantscapes outside of Beijing's historic garden clusters during the Qing Dynasty, and quantified three main factors that may have influenced urban green landscapes, including plant species, plant configurations, and spatial distribution of plants.

2. Materials and Methods

2.1. Study Materials

The data material for this study is collected from an official book of the Qing Dynasty named *A Collection of Past Events in Beijing*. In 1773, the emperor ordered the publication of the book during the 50th to 52nd years of the reign of Emperor Qianlong (1785–1787 AD) [42]. As an official local journal of the Qing Dynasty, the book was compiled by nearly 1000 individuals with reference to more than 2000 ancient texts, combined with fieldwork to ensure its high degree of authenticity and objectivity. In contrast to other historical materials, this text provides a comprehensive account of the historical geography of Beijing. It encompasses a detailed topography, historical development, a delineation of notable sites, and their cultural backgrounds, as well as an examination of the products and customs of each region [46,47]. These meticulous records offer invaluable insights into the study of history landscapes, while also serving as a crucial reference for modern landscape planning and conservation.

The book is composed of 18 distinct volumes. One of the 18 volumes is entitled Suburbs. It provides a comprehensive account of the distribution of plants across the diverse geographic spaces of the suburbs of Beijing, encompassing the city's eastern, western, southern, and northern suburbs. The volume is composed of over 190,000 Chinese characters and offers a detailed overview of the plants in these diverse urban environments. This volume contains 658 plant-related descriptive phrases (Figure 1) and includes records of plant cultivation, encompassing not only the natural growth of native plants and large areas of open field cultivation but also the ornamental plant cultivation activities and the floriculture activities. Therefore, the *Collection* is of paramount importance and significant influence in the study of historic plant landscapes.

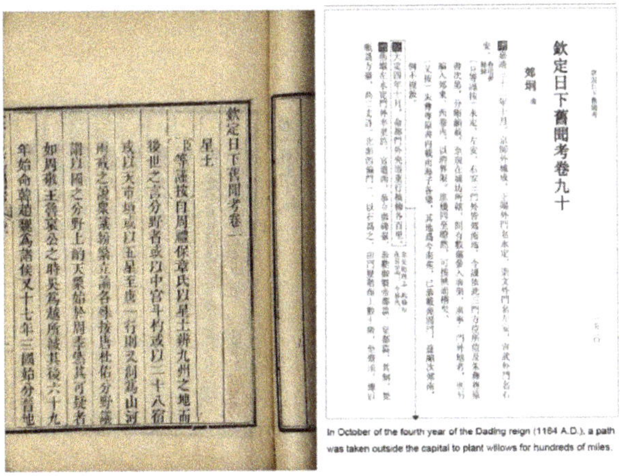

Figure 1. The 1788 edition (**left**) and the 1985 edition (**right**) of the *Collection*.

2.2. Study Area

The study was conducted in the capital city of China, Beijing (115°24′–117°3′ E, 39°28′–41°05′ N). Beijing City is located in the north of the North China Plain, with a sub-humid continental monsoon climate. The elevation of the terrain in Beijing City is characterized by a high elevation in the northwest and a low elevation in the southeast. The mountainous area accounts for approximately 62% of the total area, while the plain area is characterized by a flat and open topography, accounting for approximately 38% of the total area [44,45].

However, the Suburbs volume of the *Collection* did not specify the exact extent of the suburbs. Furthermore, due to the political and historical reasons in Beijing during the Qing Dynasty, the boundaries of the suburbs that were part of the capital city at that time were unclear. Therefore, it is necessary to establish the study area. Our research area was defined based on the following resources (Figure 2):

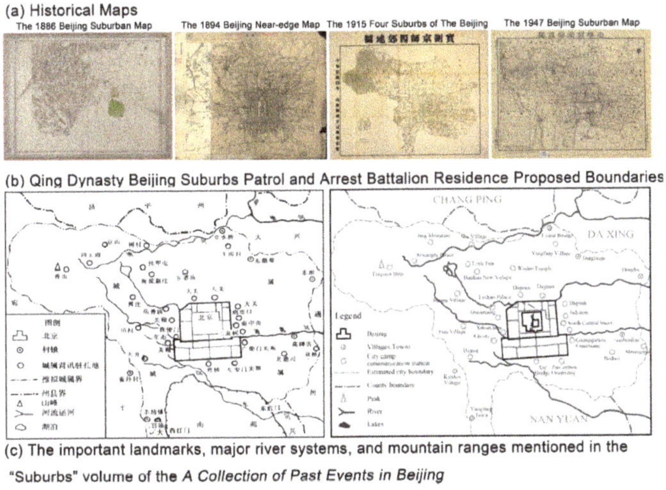

Figure 2. Methodology for the identification of the study area. (**a**) Historical maps containing the suburbs of Beijing during the Qing Dynasty. (**b**) The Qing government sent officials to survey the borders of certain provinces and regions [48]. (**c**) The important landmarks, major river systems, and mountain ranges mentioned in the Suburbs volume of the *Collection*.

Firstly, the 4 historical maps were superimposed based on the Qing Dynasty Beijing Suburbs Patrol and Arrest Battalion Residence Proposed Boundaries to obtain the core area and key geographical points. Secondly, the important landmarks, major river systems, and mountain ranges mentioned in the Suburbs volume of the *Collection* were marked out on the map. Finally, after careful consideration of the aforementioned content and expert consultation, the study area was determined (Figure 3a).

The geomorphology of the study area has remained static to date, with numerous World Heritage sites and historic gardens still extant in the central Beijing region and its surrounding suburbs. These include the Beijing Ancient City Historic Garden Cluster and the Three Hills and Five Gardens Historic Garden Cluster (Figure 3b). In addition to these historic gardens, the Beijing Municipal Government initiated the "Urban Park Loop" project in 2000 [49] (Figure 3c). The objective of the project is to mitigate the expansion of urban areas. As of May 2024, a total of 153 urban parks have been constructed, encompassing an area of approximately 203.2 square kilometers. The "Urban Park Loop" serves as a vital location for public recreation and fitness activities. However, the homogenization of these urban parks has become evident, underscoring the significance of this study in guiding plant landscape planning in contemporary urban environments.

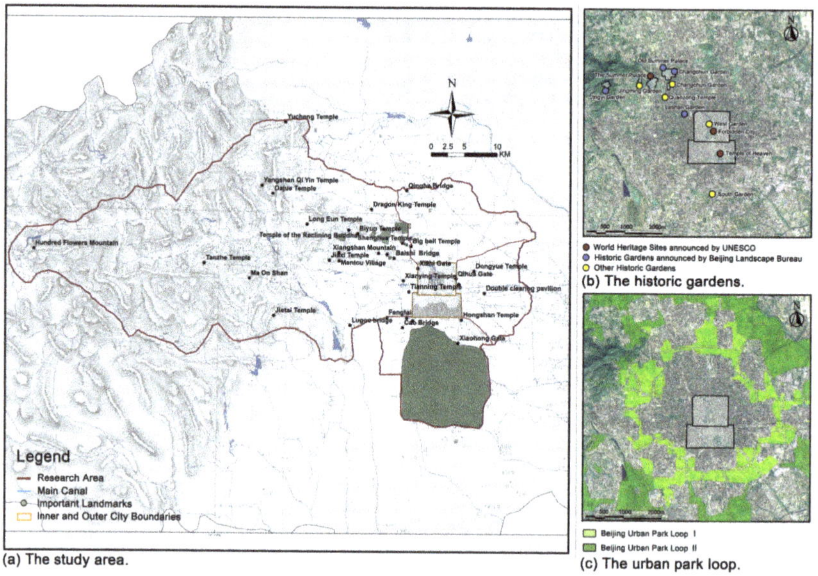

Figure 3. The location of the study area. (**a**) Study area after overlay of historical information. (**b**) Major historic gardens within the study area. (**c**) Modern urban park ring within the study area.

2.3. Data Collection

The textual data originate from the 1985 Beijing Publishing House publication, the *Collection*. The historical map of Beijing was sourced from the Historical Map Network "http://www.txlzp.com/" (accessed on 1 March 2024). Information on historical gardens was obtained from the Beijing Parks Management Centre "https://gygl.beijing.gov.cn/" (accessed on 1 March 2024). Geospatial databases from multiple sources were utilized to identify the locations of historic sites. The first source was the National Catalogue Service for Geographic Information "https://www.webmap.cn" (accessed on 8 March 2024), which provides data on city boundaries, lakes, and rivers extracted in October 2023. The second data source was the Baidu Map Open Platform "https://lbs.baidu.com/" (accessed on 15 March 2024), which contains points of interest (POIs) of historical gardens in Beijing.

2.4. Data Preparation

The major steps of this study are shown in Figure 4.

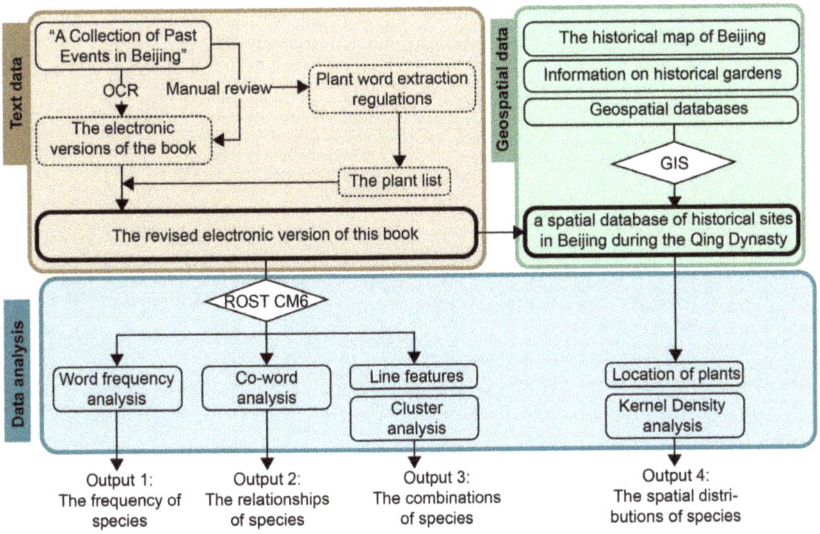

Figure 4. Workflow of research.

The *Collection* is presented vertically in the traditional Chinese script and contains press notes in different font sizes (Figure 1); thus, it is difficult for direct text mining. Therefore, OCR text recognition software was used for text recognition, and the contents of the book were then subjected to a manual review and proofreading process in order to obtain accurate electronic versions of the book.

Concurrently, it is imperative to establish plant word extraction regulations during the manual inspection process, because certain plant names are utilized in the names of individuals, locations, and landscapes in the book. There are 6 specific scenarios:

1. Plant names refer to surnames, for instance, Li (Prunus salicina), Yang (Populus), and Liu (Salix) were often used as surnames, so we removed personal names and surnames from the text.
2. Plant names referred to place names, for instance, some Mulan (Yulania) and Shuangliushu (Salix), respectively, represented the names of the Mulan paddock and the scenic spots inside the South Garden, so we preserved the plant words inside the scene name when it was first mentioned.
3. Plant names referred to verbs, for instance, Man (Vines) and Zao (Algae) were often used to express diffusion or modification, so we selected and excluded some words based on their context.
4. Plant names referred to other meanings, for instance, some Lian (Nelumbo) represented supplies of Buddhism, so we selected and excluded some words based on their context.
5. Plant homonymy, for instance, some Furong (Hibiscus) represented Hehua (Nelumbo), so we replaced plant words by combining relevant textual research [25,50].
6. Plant synonyms, for instance, Ou (Lotus root), Jihe (Lotus leaf), Fuqu, Furong, and Hehua were another name for Lian (Nelumbo), so we unified synonymous words into standard words.

Subsequently, a plant list was compiled for the Suburbs volume of the *Collection* based on the data from the electronic version book and the plant word extraction regulations (Table 1). Due to the fact that the ancient people recorded plants in a manner that differed from the contemporary recording method, the original plant words were collated with reference to the Chinese Flora. (The generic terms algae, moss, bamboo, melon, fruit, and vegetable, which are often used in articles, have been retained.) Although we identified all the plant species, and most of them presented a one-to-one correspondence, in cases of under-differentiation, we used the botanical genus as the standard plant words [51].

Table 1. A plant list for the Suburbs volume of the *Collection*.

Plant Types	Original Plant Words	Standard Words of Plants	The Genera of Plants	Plant Types	Original Plant Words	Standard Words of Plants	The Genera of Plants
Tree	song, gua, huangshansong, heisong, baisong, guosong, qiaosong, yansong	Pine	*Pinus, Larix*	Herbaceous flowers and ground cover	ju, ganju, huangju, yeju, jinzhan	Chrysanthemum	*Chrysanthemum*
	bai, kuai	Cypress	*Juniperus, Sabina, Platycladus*		lan, lanshao	Orchid	*Cymbidium*
	gui	Fragrans	*Osmanthus*		shaoyao	Chinese Herbaceous Peony	*Paeonia lactiflora*
	shan	China fir	*Picea, Abies*		jiguan	Cockscomb	*Celosia cristata*
	putishu	Linden	*Tilia*		fengxian	Impatiens balsamina	*Impatiens*
	zibai	Taxus	*Taxus*		kui	Sunflower	*Helianthus*
	liu, chuiliu, chuiyang	Willow	*Salix, Tamarix*		shuixiang, zelan	Lycopus Lucidus	*Lycopus*
	huai, panhuai	Pagoda tree	*Styphnolobium*		sou, qian	Radix rubiae	*Rubia*
	yang, baiyang	Poplar	*Populus*		song	Artemisia	*Artemisia*
	yu	Elm	*Ulmus*		luo, luoshi	China StarJasmine	*Trachelospermum*
	wutong, wu,tong	Phoenix Tree	*Firmiana*		zhi, baizhi	Angelica dahurica	*Angelica dahurica*
	zhe	Cudrania	*Maclura*		ba, jiao	Plantain	*Musa*
	suoluo	Horse Chestnut	*Aesculus*	Aquatic plant	he, jihe, furong, fuqu, ou, lian	Lotus	*Nelumbo*
	sang	Mulberry	*Morus*		wei, lu, luhua	Reed	*Phragmites*
	chun	Chinese Toon	*Toona*		puwei, pu	Pampas Grass	*Cortaderia*
	pingpo	Apple	*Malus pumila*		gu, jiao	Zizania Latifolia	*Zizania*
	jiu	Chinese Tallow	*Triadica*		jianjia, di	Silvergrass	*Miscanthus sacchariflorus*
	chu	Ailanthus Altissima	*Ailanthus*		xing	Nymphoides Peltata	*Nymphoides*
	feng, danfeng, qi	Maple	*Acer*		ping	Duckweed	*Lemna*
	yanshu	Cotinus coggygria	*Cotinus*		qian	Euryale	*Euryale*
	yinxing	Ginkgo	*Ginkgo biloba*		ling	Trapa	*Trapa*
	qiu, jia	Catalpa Bungei	*Catalpa*		liao	Polygonum	*Persicaria*
	shi, wubi	Persimmon	*Diospyros*		suo	Sedge	*Cyperus*
	lizi	Chestnut	*Castanea*		zao, jue	Algae	——
	zao	Jujube	*Ziziphus*		meitai, taimei, tai, qingtai	Moss	——
	hetao	Walnut	*Juglans*	Lianas	songluo, nvluo, biluo, luo	Cuscuta	*Cuscuta*
	wuhuaguo	Fig	*Ficus carica*		teng, ziteng, tenghua	Wisteria	*Wisteria*
	tao, shantao, bitao	Peach	*Prunus persica*		rendong	Honeysuckle	*Lonicera*
	xing, wenxing, shanxing	Apricot	*Prunus armeniaca*		lingxiao	Campsis Grandiflora	*Campsis*
	mei, hanmei, lamei, huangmei	Wintersweet	*Chimonanthus*	Crop	mai	Wheat	*Triticum*
	tang, haitang, xifuhaitang	Crabapple	*Malus spectabilis*		he, shu	Sorghum	*Sorghum*
	li	Pear	*Pyrus*		dao, daotian, daowa, daoting	Rice	*Oryza*
	li, lishu	Plum	*Prunus salicina*		gua, paogua, huanggua	Melon and Fruit	——
	yulan, mulan	Yulan Mangnolia	*Yulania*		cai, caiyuan, caipu, shupu. shuwa, yuanwa	Vegetable	——
	naizi	Crab Apple	*Malus asiatica*	Shrub	mudan	Peony	*Paeonia × suffruticosa*
	zhuying	Cherry Tree	*Prunus pseudocerasus*		jiazhutao	Oleander	*Nerium*
	ziwei	Crape myrtle	*Lagerstroemia*		yueji	Rosa Chinensis	*Rosa chinensis*
Bamboo	zhu, huang, jun, maozhu, zizhu	Bamboo	——		meigui	Rosa Rugosa	*Rosa rugosa*
					haitong	Pittosporum Tobira	*Pittosporum*

Finally, the original text was revised, and the electronic version of the Suburbs volume of the *Collection* in which the plants were corrected was obtained.

To analyze the results from a geographic perspective, the research data were obtained using the geographic coordinate picker in Google Earth to calibrate the geospatial coordinates of study areas. Subsequently, the historical maps and relevant data were imported into ArcGIS 10.8. The spatial coordinates in ArcGIS 10.8 were then matched and projected to create a spatial database of historical sites in Beijing during the Qing Dynasty and to visualize the location of plants on a map.

2.5. Data Analysis

To comprehensively examine and quantify the plant characteristics of the suburbs in which the historic gardens of Qing Dynasty Beijing were situated, this study adopted well-established methods and models that are widely accepted in textual analyses and geospatial studies. This study employs word frequency analysis and co-word analysis of ROST CM6, as well as clustering analysis of Origin 2022. The spatial distribution density of plants was estimated using kernel density analysis in ArcGIS 10.8.

2.5.1. Word Frequency Analysis

Word frequency analysis is employed to ascertain the frequency with which standard plant words are recorded in a text, thereby enabling the quantification of the ancient people's attention and application of different plants [52]. The word frequency analysis function of ROST CM6 is employed to ascertain the frequency of each plant species. Origin 2022 is then used to sort the frequency of plants and calculate the outliers, mean, and median of plants [53,54].

2.5.2. Co-Word Analysis

Co-word analysis can be employed to describe the strength of connections between different plants, thereby enabling the inference of the relationships between different plants [55,56]. The social and semantic network analysis functions of ROST CM6 were employed to obtain the semantic networks of plants, which includes the co-word analysis. The formula for the co-word analysis is as follows:

$$S_{ij} = \frac{C_{ij}}{min(C_i, C_j)} \tag{1}$$

where C_{ij} represents the number of $Plant_i$ to $Plant_j$ in the text; C_i is the frequency of $Plant_i$ in the text; C_j is the frequency of $Plant_j$ in the text; min (C_i,C_j) represents the minimum value of C_i and C_j frequencies.

2.5.3. Cluster Analysis

Cluster analysis can be employed to demonstrate the similarity between different plants, thereby inferring combinations formed by multiple plants [57]. The social and semantic network analysis functions of ROST CM6 were employed to obtain line features of text, which may be regarded as scenarios in which the plants are cultivated. The line features were transformed into a matrix of plant combination scenarios. If a specific plant was applied in a given scenario, this was noted as a value of 1, while the remaining plants were noted as a value of 0 (Table 2). The plant combination scenario matrix was subjected to a systematic cluster analysis using Origin 2022 in order to generate clustering results for plant combination.

Table 2. Two examples of the correspondence between the original text, line features, and plant combination scene matrix.

Scenario	Volume	Original Text	Line Features	The Matrix of Plant Combination Scenarios			
				Sorghum	Pampas Grass	Reed	Moss
Scenario 1	Volume 90	Moss covers the water. In the setting sun, passing pedestrians stop their horses, the air is filled with loss and despondency, and sorghum is sparse.	Sorghum, Moss	1	0	0	1
Scenario 2	Volume 98	On both sides of the waterfront, the wind brings the scent of sorghum, the water rises and then falls a few feet, and the roots of the pampas grass still carry mud. The sudden rain on the night of the 21st of last month caused the water of Kunming Lake to vent, and you can still see the water marks on the mud of the reed roots.	Sorghum, Pampas Grass, Reed	1	1	1	0

2.5.4. Kernel Density Analysis

Kernel density analysis is a nonparametric density estimation method, which represents the pattern of spatial points by analyzing the spatial density of the location of plants [58]. It can intuitively reflect the aggregation state of plants. The location of plant information was transformed into discrete point data with attributes using ArcGIS 10.8. Kernel density analysis was performed with a search radius of 5 km, and the kernel density results were classified into 10 classes using the natural discontinuity grading method. The formula for the kernel density analysis is as follows:

$$f_n(x) = \frac{1}{nh}\sum_{i=1}^{n} k\left(\frac{x - X_i}{h}\right) \tag{2}$$

where n is the number of the plants; $k((x - X_i)/h)$ denotes the kernel function; $x - X_i$ denotes the distance from the valuation point x to the X_i; and h denotes the bandwidth.

3. Results

3.1. The Composition and Frequency of Species

A total of 658 plants were recorded in the suburbs of Beijing during the Qing Dynasty, as documented in the Suburbs volume of the *Collection*. These plants can be classified into 77 standard plant words (Table 1), including 44 families and 58 genera. The families with a variety of plants are Rosaceae (10 species), Poaceae (7 species), Apocynaceae (2 species), Paeoniaceae (2 species), Lythraceae (2 species), Fabaceae (2 species), and Bignoniaceae (2 species).

The standard plant words were divided into 7 categories based on the type of growth and morphological characteristics of the plant. The most frequently recorded plant type was trees, with 391 records and 37 species. This was followed by aquatic plants, with 83 records and 13 species. The next most frequently recorded category was that of crops, with 65 records and 5 species. Bamboo was recorded on 45 occasions, herbaceous ground covers on 41 occasions, shrubs on 19 occasions, and lianas on 9 occasions (Figure 5a).

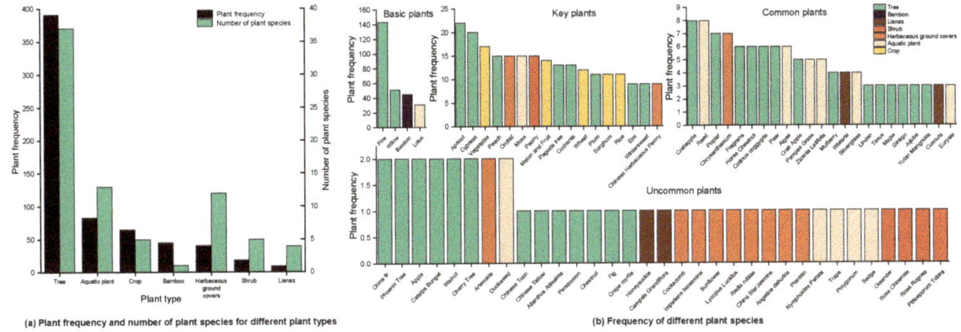

Figure 5. The frequency of plant types and species: (**a**) plant frequency and number of plant species for different plant types and (**b**) frequency of different plant species.

The standard plant words were divided into four categories: basic, key, common and uncommon plants, based on the outliers (17.57), average value (8), and median value (3) of plant-recording frequency (Figure 5b). The basic plants with outliers were pine (143), willow (51), bamboo (45), and lotus (31), accounting for 41.03% of all plants. The key plants recorded more frequently than average contained 8 trees, 5 crops, 2 herbaceous ground covers, and 2 aquatic plants, accounting for 35.10% of all plants. The common plants with recording frequencies between the average and the median contained 14 trees, 6 aquatic plants, 2 lianas, and 1 herbaceous ground cover, accounting for 16.87% of all plants. The uncommon plants with recorded less frequently than median contained 13 trees, 9 herbaceous ground covers, 5 aquatic plants, 4 shrubs, and 2 lianas, accounting for 43.42% of the plant species.

3.2. The Composition and Frequency of Species

The relationships between plant species were generated from an analysis of co-words in the Suburbs volume of the *Collection*; a total of 54 plants are related to other plants, of which 25 have a plant combination of greater than or equal to two species (Figure 6a). The plants that formed the greatest number of combinations with other plants were pine (20 species), willow (14 species), bamboo (8 species), lotus (8 species), and peach (6 species). The most frequent plant combinations were pine-bamboo (15), pine-willow (8), pine-cypress (7), pine-pagoda tree (7), and willow-elm (5).

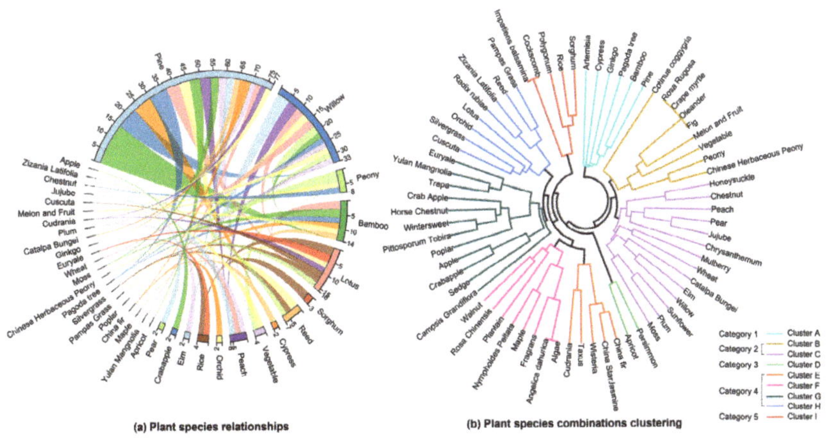

Figure 6. The proportional relationships and combinations clustering of plant species: (**a**) chordal graph of plant species relationships and (**b**) dendrogram of plant species combinations clustering.

The character lines of the Suburbs volume of the *Collection* reflect 107 plant combination scenarios (Table 2). The plants present within the scenarios were recorded and analyzed by clustering, resulting in the production of a dendrogram of plant combinations. This dendrogram shows five cluster categories and seven representative plant combinations (Figure 6b). These plant assemblages consisted of an average of eight plant species and, with the exception of Cluster D, were mixed with a variety of plant types that could form a tree–shrub–grass or tree–grass plant community.

3.3. *The Spatial Distributions of Species*

Figure 7 illustrates that the distribution of plants in the suburbs of Beijing during the Qing Dynasty was more concentrated, forming several plant groups that were scattered across various areas and closely related to historical sites. It can be demonstrated that the plant groups in Suburb West are mainly concentrated in the vicinity of the Three Hills and Five Gardens historical cluster, historical temples, and water systems by delineating a 5×5 km grid.

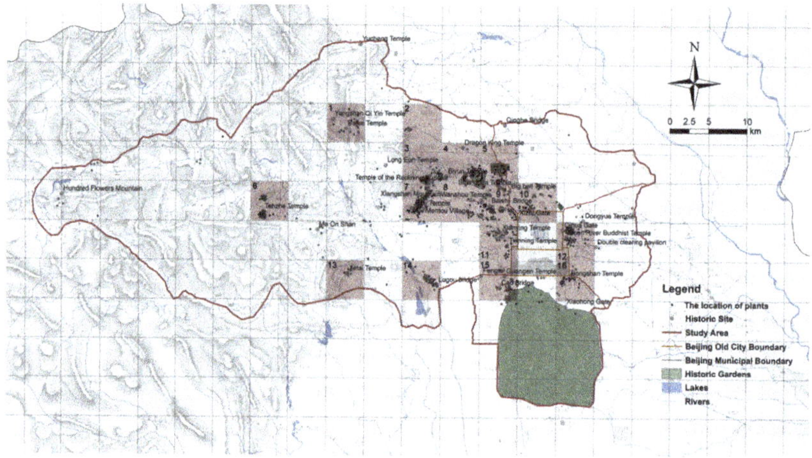

Figure 7. Recorded plant locations and history sites. Grid areas where plant locations are more clustered are highlighted.

Among the clusters, the number of plant records in the Three Hills and the Five Gardens cluster is the largest (Grid 3–5, 7–9), with the majority concentrated on the edges of the historical gardens and among the Fragrant Hills. The historical temples with plant records include Dajue Temple (Grid 1), Tanzhe Temple (Grid 6), and Jietai Temple (Grid 13). The principal waterways with documented plant records include the Nanjian River (now it is called Jingmi Water Diversion Canal, Grid 2) and the Lugou Bridge (Grid 14). In contrast, the plant groups in Suburban North, Suburban East, and Suburban South are situated in close proximity to the old city of Beijing and are primarily concentrated around city gates and bridges, including Xizhi Gate, Qihua Gate (Grid 10 and 12), and the area around Cao Bridge (Grid 15).

As illustrated in Figure 8, an examination of the kernel densities of the basic and key plants reveals a noteworthy pattern. Firstly, from the perspective of the suburbs' diverse geomorphic environmental characteristics, the western suburbs are predominantly mountainous, with a transition zone between the mountains and the plains. This complexity and variability in geomorphology provide a diverse range of habitats for plants, resulting in a rich variety of plant species and forms within the region. The region's transition zone boasts superior natural conditions, proximity to the historic city center, many historical gardens, and a proclivity for developing external public spaces that are meticulously landscaped. The suburban western mountainous area is home to a multitude of temples, many of which feature ornamental flowers such as lotus and orchids. These flowers, with their strong religious and cultural connotations, serve to enhance the ornamental and cultural nature of the temple environments.

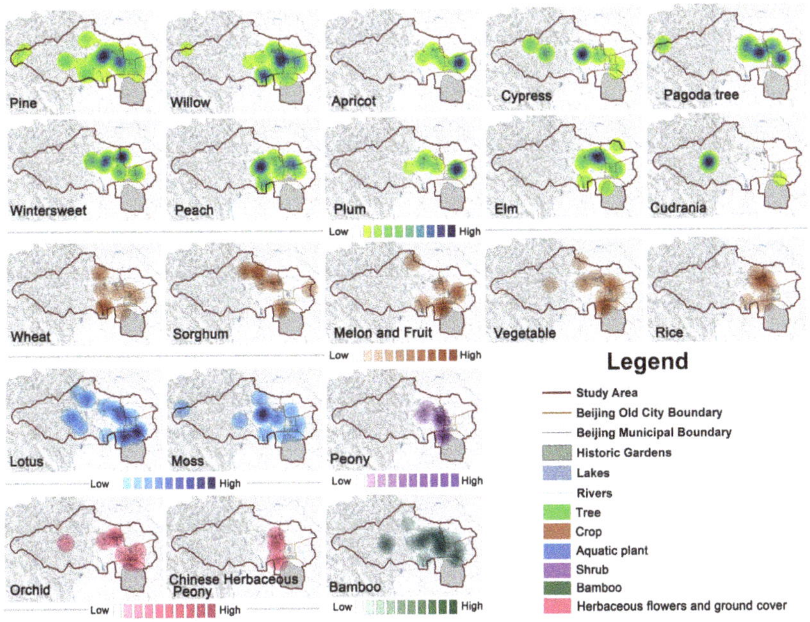

Figure 8. Kernel density of basic and key plants.

The Suburb North is situated in a transition zone. However, in contrast to the Suburb West, it exhibits a paucity of ornamental plants, including apricot trees, orchids, peonies, and crops such as rice, wheat, and sorghum. The northern suburb is less affected by human activities and presents an unaltered natural landscape. Suburb East and Suburb South are situated in a vast plain region. In comparison to Suburb West and Suburb North, the soil and water conditions are relatively concentrated and stable. This is evidenced by the prevalence of plants such as lotus, vegetables, and melons in the vicinity of water, which reflects the interdependence between plants and their growing environment in the plain region.

Secondly, from the perspective of the plant types, the spatial distribution of trees is the most extensive. Evergreen plants, such as pines and cypresses, are widely distributed in mountainous and transition zones. In contrast, willows, pagoda trees, and various types of ornamental deciduous trees are distributed in the plains. Aquatic plants are also widely distributed, though the locations of points with high nuclear density are not uniform. For example, lotus is distributed along the water bodies in the southern suburbs, while moss is distributed in the dense forests of the western mountains. With regard to crops, grain crops such as rice, wheat, and sorghum are primarily situated in the northwestern suburbs, whereas vegetables, fruits, and melons are distributed in proximity to water bodies in the southern suburbs. The Three Hills and Five Gardens area and the vicinity of the old city are the main locations for the growth of flowers and shrubs.

Finally, our findings revealed that certain areas exhibited a higher concentration of plants with elevated nucleation densities. These areas demonstrated a degree of specificity, with notable examples including the prevalence of wintersweet trees in the northern suburb and the clustering of plum and apricot trees in the eastern part of the suburb. Cudrania trees were planted in the vicinity of Tanzhe Temple (the name of the temple is derived from the large tracts of Cudrania trees in this area), sorghum was mainly planted along the drainage canal, and rice was concentrated in the area of the Three Hills and Five Gardens Historic Garden Cluster. This is due to two factors: the presence of microclimates conducive to the growth of specific plants and the influence of cultural and political factors on the distribution of plants.

4. Discussion

4.1. Authenticity of Plant Species

A total of 77 standard plant words were identified in the Suburbs volume of the *Collection*. The majority of these plants are native to Beijing, having originated from the city's natural forests and cultivated land. They are also largely consistent with the types of plants that are currently common in the area [59], because the identification of native plants is based on their natural geographical distribution [60]. Beijing is situated at the intersection of the Yanshan and Taihang Mountains, where the climate is characterized by high temperatures and rainy summers, cold and dry winters, and low temperatures and precipitation in the winter. This climatic variation results in a relatively limited number of native plant species in Beijing [61].

However, the natural environment of Beijing has been subjected to human influence for a considerable period of time, during which the natural forests and cultivated land have undergone the process of plant introduction and replacement [62]. Some of the plants listed in Table 1 are not native to Beijing, particularly ornamental plants such as peonies, Chinese herbaceous peonies, fragrans, and ginkgoes, which were introduced to Beijing for artificial cultivation and breeding around the Liao Dynasty (907–1234 AD). Plum trees, the earliest recorded instance of which was planted in Beijing during the Yuan Dynasty (1271–1368 AD), are also included [63]. Other plants such as magnolias, hepaticas, and begonias were also introduced and planted in different historical periods [64]. The plant composition of the suburbs during the Qing Dynasty can be explained by the inherent natural properties of the environment and selective human interference based on ornamental value.

In addition, the number of plant species documented is considerably lower than the results of modern surveys of Beijing's plant species [65]. This is partly because the ancients recorded plants in a different way and did not classify them as meticulously as we do today, and partly because a large number of exotic plants and new ornamental varieties were introduced to Beijing after the modern era [63,64]. For example, Zhao et al. found that the composition of plant species in Beijing's urban areas differed significantly from that of natural habitats, with exotic species accounting for 53% [61]. Therefore, the plant species documented in ancient books from the Qing Dynasty can, to a certain extent, be used to reconstruct the authenticity of the plantscapes during the historical period.

4.2. Preferences of Plant Application

The application of plants is subject to intricate interrelationships with their social and environmental contexts [66]. The application of plants reflects the people's understanding of plants from three aspects: culture, nature, and economy. The frequency of plant species demonstrated that the ancients were most concerned with trees, followed by aquatic plants, crops, and bamboos, and less concerned with groundcovers, shrubs, and vines (Figure 5). This may be attributed to the fact that the types of plants under consideration frequently possess cultural and productive value. Previous studies have shown that culture of plants is seen as powerful forces that play a positive role in people's lives [67], and even in modern times, people are more concerned about plants that are useful to humans and plants with cultural symbols [68].

For instance, pines are perceived to possess the quality of longevity due to their resilience to drought and cold. The Chinese pronunciation of willow is similar to the phrase "stay behind", which is often used to convey a sense of reluctance to part ways [69]. Conversely, the relatively limited number of documented shrubs and grasses may also be attributed to the absence of cultural and plant imagery [18]. Furthermore, the classification of basic plants, key plants, common plants, and uncommon plants is also related to the cultural value of plants.

We found in the relationships of species (Figure 6a) that only 15 species have more than two phyletic relationships. This result demonstrates the importance of ancient preferences for specific plants and the importance of these plant relationships. It is proposed that species relationships may be influenced by a combination of natural perceptions and cultural preferences [70]. Plants with two or more plant relationships are usually the dominant plants in coniferous and mixed coniferous-broad forests in the mountainous and shallow montane zones of Beijing and also grow in the vicinity of water systems in the plains [65]. The utilization of land in proximity to shallow knolls and watercourses for landscaping has the potential to exert a considerable influence on the environment, where plants are selectively retained and play a pivotal role in local natural and cultural systems.

Other relationships of species, in part derived from plant idioms, proverbs, and poems in the Chinese language and culture, reflect human-dominated cultural preferences. For example, the phrase "peaches and plums" represents a metaphor for the students cultivated by the teacher. This cultural preference for the use of plants was favored by the literati in ancient times. Previous studies have demonstrated that plant relationships can convey cultural values, enabling individuals to construct specific images or scenarios in their minds [51,71]. There is also a part of the plant relationship derived from the plant's inter-organism relationship and the plant's dependence on the environment [72], which reflects the ancient people's observation of the environment, for example, the Cuscuta parasitized on pine trees, and the rice's dependence on the water system makes the rice and the lotus also have a close connection.

4.3. *Shaping of Plant Landscape*

Plantscapes involve the integration of natural and cultural processes and are interdependent with vegetation management issues in human landscapes [73]. The results of the cluster analysis of plant combinations demonstrate the value proposition for shaping the plantscape (Figure 6b).

Category 1 (Cluster A) depicts a fundamental plant combination comprising evergreens and colorful foliage trees within a shallow suburban landscape. Category 2 (Cluster B and C) comprises vegetable, floral, medicinal, and fruit trees, which illustrate the economic and ornamental aspects of the plant combinations. Category 3 (Cluster D) is composed exclusively of apricot and persimmon trees. This is due to the traditional belief that planting an apricot tree and a persimmon tree in a farmyard is considered auspicious and has some feng shui value. Category 4 (Cluster E to H) encompasses plant combinations from disparate geographical locations. These combinations are observed in mountainous areas, courtyards, water, and the edges of water. In contrast, Category 5 (Cluster I) represents a plant combination with a distinctive, vibrant character.

These five plant combinations can be employed to inform contemporary plantscape design, while also reflecting the ancient peoples' focus on respecting the conditions of the natural environment and balancing ornamental and economic values in their vegetation management. This is evidenced by the fact that even the royal palace plantations had to have a self-sufficient business model [74].

The spatial distribution of plants has shaped the wider botanical landscape. Our study reveals a phenomenon from a holistic perspective, demonstrating that the plantscapes in the suburbs of Beijing during the Qing Dynasty exhibited notable variations in different directions of the city. The primary reason for this outcome is that plant growth must be adapted to the natural environment. The Beijing region encompasses a diverse range of topographical features, including mountainous, moderately elevated, and low-lying areas [44,45]. The northwest region is situated upstream of the river, which is characterized by abundant water resources and a relatively favorable natural environment [31]. Accordingly, from the perspective of plant species diversity, the northwest suburbs exhibit the greatest richness, although melons, fruits, vegetables, and certain flowers are more conducive to cultivation in the plains. Furthermore, specific microclimates are present in select areas. For instance, some caves in the western suburbs experience winter temperatures that are sufficiently high to permit the overwintering of ornamental flowers. These areas are primarily utilized for the breeding of flowers.

The urban planning of Beijing in the Qing Dynasty was also influenced by the natural environment, which had an impact on the distribution of plants. In particular, the suburbs on the northwest side of the city had abundant water resources because they were situated at the upper water level of the water conservancy system [33]. Consequently, the number of royal gardens was much higher, and the external public environment was landscaped to a greater extent. Additionally, a significant quantity of rice and other crops were cultivated in the vicinity of the royal gardens on the west side of the suburb. This reflected the emperor's political inclination to prioritize agriculture. However, the suburbs in the southeast of the city are situated at the lower water level. A considerable number of gardens in the upper reaches of the lower reaches have limited access to water, which presents a challenge in terms of water for gardening and living, water transport, and irrigation. This has led to difficulties in implementing large-scale, complex landscape transformations of the suburban environment [29]. Consequently, the species of plants are relatively homogeneous, often emphasizing a specific type of plant, and the plant landscape is more focused on economic value.

In addition, within the boundaries of Qing Dynasty Beijing, the closer one gets to the suburbs, the larger the area of the natural environment, the higher the percentage of the area occupied by plants, and even the more natural pristine landscapes are presented [33]. We compared the plants of the suburbs in the Qing Dynasty with scholars' studies on the plants of historical gardens based on historical images and surviving remains [29,32,75] and found that there is a consistency between the plant species of the suburbs and the basic plant species within the historical gardens, which reflects the fusion of the public environment of the suburbs and the historical gardens. However, we also found that the plants applied in many characteristic nodes in the historic gardens were not consistent with the suburban public environment, mainly because more kinds of plants in the historic gardens were imported here from all over the country, which required fine management and care, and these plants were not suitable for growing on land in Beijing. Therefore, the research in this paper is more for the public environment construction when choosing tone tree species to provide reference.

At the same time, we also identified a relationship between the spatial distribution of plants in Beijing during the Qing Dynasty and the modern landscape layout of Beijing. This relationship reflects the continuity of the cultural lineage. For example, even in the modern era, the southern suburbs of Beijing remain the epicenter of the flower industry [76], boasting flower-themed parks such as the Flower Fragrance Park and the Beijing World Flower Grand View Garden. The eastern suburbs continue to house a significant number of nursery industries. The western suburbs remain mountainous, situated at the junction of mountains and plains. The natural substrate has undergone minimal change, yet the majority of the plants that were characteristic of the historical period have been supplanted by urbanized garden plants. The northern suburbs have been significantly impacted by the urbanization process, resulting in the replacement of the original natural landscape with residential areas and other construction sites.

Overall, it can be observed that, with the exception of the flower and nursery industries in the southern and eastern suburbs and the mountainous forested environment in the western suburbs, which have retained their historical characteristics, the remaining green areas that overlap with the study area have been replaced by ordinary green areas following urbanization. The plant species, combinations, and overall plantscape appearance of these areas have lost their regional characteristics and now exhibit a homogeneous appearance, which represents a significant difference in the existing green area structure in Beijing [61]. This phenomenon is particularly evident in the existing green space structure of Beijing, especially in the urban park loop.

5. Conclusions and Recommendations for Sustainable Conservation

The planting of native plants with authenticity contributes significantly to the sustainability of urban ecosystems and cultural systems in urban greening [19]. With the ongoing process of global migration and cultural integration, it is likely that the homogenization of plants across cities will become more pronounced [77]. This has the potential to negatively impact the ecological and cultural resilience of the landscape, particularly in cities with historic gardens.

As a culturally unified country, China's similar cultural background has led to the emergence of greater similarities in the flora of various regions [70]. As the capital of China, Beijing is home to numerous historical gardens. Any significant alterations to the flora within these gardens, whether through the introduction of new species or the removal of existing ones, could have a detrimental impact on the gardens' originality and cultural value. Accordingly, this study aims to reconstruct the botanical landscape of Beijing during the Qing Dynasty, employing a historical approach that integrates textual and geospatial data analysis. The main conclusions are as follows:

A total of 658 plants, including 44 families and 58 genera, were recorded in the suburbs of Beijing during the Qing Dynasty. These can be classified into seven categories according to the growth type and morphological characteristics of plants, forming 54 plant relationship results, 107 plant combination scenarios, 5 plant combination cluster categories, and 7 representative plant combinations. This illustrates the spatial distribution of the plants and the characteristics of the plantscape in different directions. Additionally, a graphical illustration of the spatial distribution of plants was created, resulting in the identification of 16 plantscape aggregation points. Furthermore, the spatial kernel density distribution of important plants was determined to gain insight into the characteristics of plantscapes in different aspects of the suburb.

We suggest that the use of high-quality native plants can be enhanced while emphasizing landscaping and biodiversity to avoid homogenization of urban greenery and that some native plants can be given appropriate cultural values based on historical information. At the same time, we recognize the introduction of new species into historic gardens and their external environments, and that over time introduced species can be adapted to local traditions through cultural adaptation [19]. Consequently, the ratio of native to introduced plants should be subject to reasonable control. Furthermore, our findings offer insights into the urban plantscapes of Beijing during the Qing Dynasty. They provide recommendations for the relationships, combinations, and spatial distribution of plants, which may inform the practices of those involved in urban greening.

In conclusion, in urban areas with historical gardens, it is imperative to consider not only the interior of these gardens but also the natural environment surrounding them. This approach is fundamental to an understanding of the authenticity of urban vegetation and may also provide a foundation for the resilience of historic landscapes. This approach also allows for a more balanced approach to the ecological development and the cultural heritage of historical plantscapes, which is also a crucial issue that must be taken into account when assessing the necessity of enhancing management in urban green spaces.

The present study is subject to certain constraints. One such limitation is that ancient people recorded plants in a manner that differs from the methods used today. This discrepancy may have introduced bias in the identification of plant species and the analysis of results. This study, meanwhile, focuses on exploring the characteristics of plant landscapes at the urban scale. This approach, however, does not provide direct guidance for the analysis of specific historical gardens or parks. Consequently, future studies will expand the sample size of the literature, enhance the scientific rigor and precision of the excavation of ancient texts, and incorporate specific historical gardens and modern parks into the study. This will facilitate the provision of references for the study of plant landscapes in traditional gardens and the planning of modern urban plant landscapes.

Author Contributions: Conceptualization, D.X. and N.H.; methodology, D.X.; software, J.Z. and Z.C.; validation, D.X., J.Z. and Z.C.; formal analysis, D.X., J.Z. and Z.C.; investigation, N.H.; resources, P.W.; data curation, P.W.; writing—original draft preparation, D.X.; writing—review and editing, D.X. and N.H.; visualization, J.Z. and Z.C.; supervision, P.W.; project administration, Y.L.; funding acquisition, N.H. All authors have read and agreed to the published version of the manuscript.

Funding: This research was funded by the BEIJING PHILOSOPHY AND SOCIAL SCIENCE PLANNING PROJECT, grant number 22YTC026.

Data Availability Statement: The original contributions presented in the study are included in the article material; further inquiries can be directed to the corresponding author/s.

Acknowledgments: We would like to thank Ph.D. student Tianze Li for proofreading.

Conflicts of Interest: The authors declare no conflicts of interest.

References

1. Fekete, A.; Kollányi, L. Research-Based Design Approaches in Historic Garden Renovation. *Land* **2019**, *8*, 192. [CrossRef]
2. de Oliveira Paiva, P.D.; de Brito Sousa, R.; da Silva Cavalcante, S.F.N. Patchwork quilt: A methodology proposed for the study of historic gardens. *Urban For. Urban Green.* **2021**, *62*, 127169. [CrossRef]
3. Gao, L.; Dietze-Schirdewahn, A. Garden culture as heritage: A pilot study of garden culture conservation based on Norwegian examples. *Urban For. Urban Green.* **2018**, *30*, 239–246. [CrossRef]
4. Carrari, E.; Aglietti, C.; Bellandi, A.; Dibari, C.; Ferrini, F.; Fineschi, S.; Galeotti, P.; Giuntoli, A.; Del Fa, R.M.; Moriondo, M.; et al. The management of plants and their impact on monuments in historic gardens: Current threats and solutions. *Urban For. Urban Green.* **2022**, *76*, 127727. [CrossRef]
5. ICOMOS. Florence Charter on Historic Gardens. Available online: https://www.icomos.org/en/newsletters-archives/179-articles-en-francais/ressources/charters-and-standards/158-the-florence-charter (accessed on 18 April 2024).
6. Hodor, K.; Przybylak, Ł.; Kuśmierski, J.; Wilkosz-Mamcarczyk, M. Identification and Analysis of Problems in Selected European Historic Gardens during the COVID-19 Pandemic. *Sustainability* **2021**, *13*, 1332. [CrossRef]
7. Athanasiadou, E. Historic Gardens and Parks Worldwide and in Greece: Principles of Acknowledgement, Conservation, Restoration and Management. *Heritage* **2019**, *2*, 2678–2690. [CrossRef]
8. Prigioniero, A.; Sciarrillo, R.; Spuria, L.; Zuzolo, D.; Marziano, M.; Scarano, P.; Tartaglia, M.; Guarino, C. Role of historic gardens in biodiversity-conservation strategy: The example of the Giardino Inglese of Reggia di Caserta (UNESCO) (Italy). *Plant Biosyst.—Int. J. Deal. All Asp. Plant Biol.* **2021**, *155*, 983–993. [CrossRef]
9. Bourke, M. How will my garden grow? A philosophy for the restoration of historic gardens. *J. Gard. Hist.* **1983**, *3*, 49–54. [CrossRef]
10. Millward, A.A.; Sabir, S. Benefits of a forested urban park: What is the value of Allan Gardens to the city of Toronto, Canada? *Landsc. Urban Plan.* **2010**, *100*, 177–188. [CrossRef]
11. Muhamad, K. Land use-based stakeholders mapping for natural heritage preservation—Case study: Hampstead Heath Ponds, London. *Urban For. Urban Green.* **2023**, *80*, 127821.
12. Scazzosi, L. Reading and assessing the landscape as cultural and historical heritage. *Landsc. Res.* **2004**, *29*, 335–355. [CrossRef]
13. Ginzarly, M.; Houbart, C.; Teller, J. The Historic Urban Landscape approach to urban management: A systematic review. *Int. J. Herit. Stud.* **2019**, *25*, 999–1019. [CrossRef]
14. Antrop, M. Sustainable landscapes: Contradiction, fiction or utopia? *Landsc. Urban Plan.* **2005**, *75*, 187–197. [CrossRef]
15. Gong, C.; Chen, J.; Yu, S. Biotic homogenization and differentiation of the flora in artificial and near-natural habitats across urban green spaces. *Landsc. Urban Plan.* **2013**, *120*, 158–169. [CrossRef]
16. Altman, A.; Shennan, S.; Odling-Smee, J. Ornamental plant domestication by aesthetics-driven human cultural niche construction. *Trends Plant Sci.* **2022**, *27*, 124–138. [CrossRef] [PubMed]
17. Dunn, C.P.; Heneghan, L. Composition and Diversity of Urban Vegetation. In *Urban Ecology*; Niemelä, J., Breuste, J.H., Elmqvist, T., Guntenspergen, G., James, P., McIntyre, N.E., Eds.; Oxford University Press: Oxford, UK, 2011; pp. 103–115.
18. Jin, C.; Zheng, M.; Huang, L.; Qian, S.; Jim, C.Y.; Lin, D.; Zhao, L.; Minor, J.; Coggins, C.; Chen, B.; et al. Co-existence between humans and nature: Heritage trees in China's yangtze River region. *Urban For. Urban Green.* **2020**, *54*, 126748. [CrossRef]
19. Zhang, M.-J.; Chen, B.; Xu, C. Cultural tree preference and its influence on tree biodiversity in urban public spaces in Nanjing city, China. *Urban For. Urban Green.* **2020**, *48*, 126568. [CrossRef]
20. Uchida, K.; Karakida, K.; Iwachido, Y.; Mori, T.; Okuro, T. The designation of a historical site to maintain plant diversity in the Tokyo metropolitan region. *Urban For. Urban Green.* **2023**, *84*, 127919. [CrossRef]
21. van den Brink, M.; van den Brink, A.; Bruns, D. Boundary thinking in landscape architecture and boundary-spanning roles of landscape architects. *Landsc. Res.* **2022**, *47*, 1087–1099. [CrossRef]
22. Dong, L. Plants and their cultural meaning in Chinese historical gardens. *Acta Hortic.* **2010**, *881*, 885–889. [CrossRef]
23. Wu, L. Planning Structure, Old Town Renewal and Urban Design of Historic and Cultural Cities. *City Plan. Rev.* **1983**, *6*, 2–12.
24. Song, F. The art of Chinese gardens and its influence on Japan. *J. Zhejiang Agric. Univ.* **1992**, *3*, 110–116.

25. Zeng, X.; Yin, H. Landscape Feature and Evolution of Plants Place-names in Beijing Area. *Landsc. Archit.* **2017**, *11*, 99–104. [CrossRef]
26. Whiteman, S.H. From Upper Camp to Mountain Estate: Recovering historical narratives in Qing imperial landscapes. *Stud. Hist. Gard. Des. Landsc.* **2013**, *33*, 249–279. [CrossRef]
27. Barmé, G.R. Beijing, a garden of violence. *Inter-Asia Cult. Stud.* **2008**, *9*, 612–639. [CrossRef]
28. Jia, J. An analysis of the phenomenon of writing imitation in the royal gardens of the Qing Dynasty. *Zhuangshi* **2010**, *2*, 16–21. [CrossRef]
29. Xiao, Y.; Zhu, Q.; Zhuo, K. Study on the Plant Landscape and Garden Management System of the Royal Garden in Beijing during the Qing Dynasty. *Landsc. Archit.* **2018**, *25*, 110–115. [CrossRef]
30. Zhang, X.; Li, X. Research on Health Thought in Chinese Classical Gardens: Taking the Qing Dynasty Royal Gardens as an Example. *China Landsc. Archit.* **2019**, *35*, 28–33. [CrossRef]
31. Liu, J.; Hu, L.; Li, S. An Analysis on the Definition of the Protection and Control Scope of the Historical and Cultural Protection Zone of the Qing Dynasty Royal Gardens in the Western Suburbs of Beijing. *Chin. Landsc. Archit.* **2009**, *25*, 15–18.
32. Tang, S.; Zhou, X. Study on the Plant Landscaping of Field Scenic Spots in the Three Mountains and Five Gardens of the Qing Dynasty. *China Urban For.* **2019**, *17*, 68–72.
33. Zhang, C.; Wang, M.; Wang, Q. A Study of Urban Landscaping in Beijing during the Qing Dynasty. *Chin. Landsc. Archit.* **2022**, *38*, 14–19. [CrossRef]
34. Noble, P. The importance of Old French chronicles as historical sources of the Fourth Crusade and the early Latin Empire of Constantinople. *J. Mediev. Hist.* **2012**, *27*, 399–416. [CrossRef]
35. Hotho, A.; Nürnberger, A.; Paass, G.J.L.F. A Brief Survey of Text Mining. *J. Lang. Technol. Comput. Linguist.* **2005**, *20*, 19–62. [CrossRef]
36. Hassani, H.; Beneki, C.; Unger, S.; Mazinani, M.T.; Yeganegi, M.R. Text Mining in Big Data Analytics. *Big Data Cogn. Comput.* **2020**, *4*, 1. [CrossRef]
37. Hu, X.; Downie, J.S.; Ehmann, A.F. Lyric text mining in music mood classification. In Proceedings of the International Society for Music Information Retrieval Conference, Kobe, Japan, 26–30 October 2009.
38. Ceusters, W. Medical Natural Language Understanding as a Supporting Technology for Data Mining in Healthcare. In *Medical Data Mining and Knowledge Discovery*; Physica-Verlag: Heidelberg, Germany, 2001.
39. Segall, R.S.; Zhang, Q.; Cao, M. Web-Based Text Mining of Hotel Customer Comments Using SAS Text Miner and Megaputer Polyanalyst. In Proceedings of the SWDSI 2009, New Orleans, LA, USA, 14–17 November 2009.
40. Nelson, M.R.; Orum, T.V.; Jaime-Garcia, R.; Nadeem, A. Applications of Geographic Information Systems and Geostatistics in Plant Disease Epidemiology and Management. *Plant Dis.* **1999**, *83*, 308–319. [CrossRef]
41. Samat, N. Applications of Geographic Information Systems in Urban Land Use Planning in Malaysia. In Proceedings of the 4th Taipei International Conference on Digital Earth, Taipei, Taiwan, 25–26 May 2006.
42. Yu, M. *A Collection of Past Events in Beijing*; Beijing Ancient Books Press: Beijing, China, 1985; Volume 1.
43. Xie, Y.; Fang, C.; Lin, G.C.S.; Gong, H.; Qiao, B. Tempo-Spatial Patterns of Land Use Changes and Urban Development in Globalizing China: A Study of Beijing. *Sensors* **2007**, *7*, 2881–2906. [CrossRef]
44. Du, J.F.; Thill, J.C.; Peiser, R.B.; Feng, C.C. Urban land market and land-use changes in post-reform China: A case study of Beijing. *Landsc. Urban Plan.* **2014**, *124*, 118–128. [CrossRef]
45. Li, H.L.; Peng, J.; Liu, Y.X.; Hu, Y.N. Urbanization impact on landscape patterns in Beijing City, China: A spatial heterogeneity perspective. *Ecol. Indic.* **2017**, *82*, 50–60. [CrossRef]
46. Wu, Y. The Compilation Time and Historical Value of the Book "A Collection of Past Events in Beijing". *Documentation* **1992**, *3*, 199–206.
47. Miao, R. A Study on the Compilation of "A Collection of Past Events in Beijing"—Also on the Newly Discovered Four Library Manuscripts. *J. Chin. Lit. Hist.* **2015**, *4*, 237–265.
48. Han, G. Exploring the Administrative Boundary of Beijing's Suburbs in the Qing Dynasty. *Acta Geogr. Sin.* **1999**, *2*, 56–63.
49. Wang, Y.; Chang, Q.; Li, X. Promoting sustainable carbon sequestration of plants in urban greenspace by planting design: A case study in parks of Beijing. *Urban For. Urban Green.* **2021**, *64*, 127291. [CrossRef]
50. Guo, C.; Zhu, Q.; Yin, H.; Jiang, J.; Tang, Y. Study on the Plants and the Plant Landscape of Changchun Garden. *Chin. Landsc. Archit.* **2020**, *36*, 123–128. [CrossRef]
51. Tardío, J.; Pardo-de-Santayana, M. Cultural Importance Indices: A Comparative Analysis Based on the Useful Wild Plants of Southern Cantabria (Northern Spain). *Econ. Bot.* **2008**, *62*, 24–39. [CrossRef]
52. Dicle, M.F.; Dicle, B. Content analysis: Frequency distribution of words. *Stata J.* **2018**, *18*, 379–386. [CrossRef]
53. Leys, C.; Ley, C.; Klein, O.; Bernard, P.; Licata, L. Detecting outliers: Do not use standard deviation around the mean, use absolute deviation around the median. *J. Exp. Soc. Psychol.* **2013**, *49*, 764–766. [CrossRef]
54. Lydersen, S. Mean and standard deviation or median and quartiles? *Tidsskr Nor Laegeforen* **2020**, *140*, 9. [CrossRef]
55. Feng, J.; Zhang, Y.Q.; Zhang, H. Improving the co-word analysis method based on semantic distance. *Scientometrics* **2017**, *111*, 1521–1531. [CrossRef]
56. Braam, R.R.; Moed, H.F.; Van Raan, A.F. Mapping of science by combined co-citation and word analysis, i. Structural aspects. *J. Am. Soc. Inf. Sci.* **1991**, *42*, 233–251. [CrossRef]

57. Bridges, C.C. Hierarchical Cluster Analysis. *Psychol. Rep.* **2016**, *18*, 851–854. [CrossRef]
58. Okabe, A.; Satoh, T.; Sugihara, K. A kernel density estimation method for networks, its computational method and a GIS-based tool. *Int. J. Geogr. Inf. Sci.* **2009**, *23*, 7–32. [CrossRef]
59. Zheng, X.; Zhang, Q. Status and Prospects of Urban Landscape Plants' Application in Bejing. *Chin. Landsc. Archit.* **2011**, *27*, 81–85.
60. Huang, L.; Tian, L.J.; Zhou, L.H.; Jin, C.; Qian, S.H.; Jim, C.Y.; Lin, D.M.; Zhao, L.; Minor, J.; Coggins, C.; et al. Local cultural beliefs and practices promote conservation of large old trees in an ethnic minority region in southwestern China. *Urban For. Urban Green.* **2020**, *49*, 126584. [CrossRef]
61. Zhao, J.J.; Ouyang, Z.Y.; Zheng, H.; Zhou, W.Q.; Wang, X.K.; Xu, W.H.; Ni, Y.M. Plant species composition in green spaces within the built-up areas of Beijing, China. *Plant Ecol.* **2010**, *209*, 189–204. [CrossRef]
62. Wang, G.M.; Jiang, G.M.; Zhou, Y.L.; Liu, Q.R.; Ji, Y.S.; Wang, S.X.; Chen, S.B.; Liu, H. Biodiversity conservation in a fast-growing metropolitan area in China: A case study of plant diversity in Beijing. *Biodivers. Conserv.* **2007**, *16*, 4025–4038. [CrossRef]
63. Xu, L. Introduction and cultural appreciation of Prunus mume in Beijing. *J. Beijing For. Univ.* **2012**, *34*, 158–163. [CrossRef]
64. Zhang, B.; Yang, H.; Cheng, F. Landscape Plants in Bejjing: Introduction and Planting History. *J. Agric.* **2017**, *7*, 74–79.
65. Meng, X.S.; Ouyang, Z.Y.; Cui, G.F.; Li, W.F.; Zheng, H. Composition of plant species and their distribution patterns in Beijing urban ecosystem. *Acta Ecol. Sin.* **2004**, *24*, 2200–2206. [CrossRef]
66. Carvajal-Contreras, D.R. Ancient Plants and People: Contemporary Trends in Archaeobotany. *Ethnobiol. Lett.* **2015**, *6*, 203–207. [CrossRef]
67. Taylor, B. Plants as persons: Perceptions of the natural world in the North European Mesolithic. *Time Mind-J. Archaeol. Conscious. Cult.* **2020**, *13*, 307–330. [CrossRef]
68. Vardi, R.; Mittermeier, J.C.; Roll, U. Combining culturomic sources to uncover trends in popularity and seasonal interest in plants. *Conserv. Biol.* **2021**, *35*, 460–471. [CrossRef] [PubMed]
69. Yang, Y.; Zhang, A. A Comparative Study of the Connotations of Plant Words in English and Chinese Languages. *Lang. Teach. Linguist. Stud.* **2002**, *4*, 74–80.
70. Kendal, D.; Williams, K.J.H.; Williams, N.S.G. Plant traits link people's plant preferences to the composition of their gardens. *Landsc. Urban Plan.* **2012**, *105*, 34–42. [CrossRef]
71. Wertz, A.E. How Plants Shape the Mind. *Trends Cogn. Sci.* **2019**, *23*, 528–531. [CrossRef] [PubMed]
72. Brooker, R.W. Plant-plant interactions and environmental change. *New Phytol.* **2006**, *171*, 271–284. [CrossRef]
73. De Blois, S.; Domon, G.; Bouchard, A. Landscape issues in plant ecology. *Ecography* **2002**, *25*, 244–256. [CrossRef]
74. Xiao, Y.; Liu, L. Royal Land Use and Management in Beijing in the Qing Dynasty. *Land* **2021**, *10*, 1093. [CrossRef]
75. Hu, N.; Wang, P.; Li, X. Research on Plant Species Identification and Plant Configuration in the Inner Court Garden of Beijing Royal Garden. *J. Beijing For. Univ.* **2022**, *44*, 100–114.
76. Wu, W.-T. The Historical Situation and Environment of the Rise of Flower Culture in the South Area Beside Dadu City Since Yuan Dynasty. *Soc. Sci. Beijing* **2010**, *2*, 92–97. [CrossRef]
77. Trigger, D.S.; Head, L. Restored Nature, Familiar Culture: Contesting Visions for Preferred Environments in Australian Cities. *Nat. Cult.* **2010**, *5*, 231–250. [CrossRef]

land

MDPI

Article

Resilience of Terraced Landscapes to Human and Natural Impacts: A GIS-Based Reconstruction of Land Use Evolution in a Mediterranean Mountain Valley

Titouan Le Vot [1,2], Marianne Cohen [2,3,*], Maciej Nowak [2,4], Paul Passy [5] and Franck Sumera [6]

1 UFR de Geographie, Université Panthéon-Sorbonne 1, 75005 Paris, France
2 Médiations, Science des Lieux Science des Liens, Sorbonne Université, 75006 Paris, France; maciej.nowak@amu.edu.pl
3 UFR de Géographie et Aménagement, Sorbonne Université, 75005 Paris, France
4 Faculty of Biology, Adam Mickiewicz University, 67-712 Poznań, Poland
5 Prodig UMR 8586, GeoTeCa, Université Paris Cité, 75013 Paris, France; paul.passy@univ-paris.fr
6 Centre Camille Julian UMR 7299, Ministry of Culture, Aix-Marseille Université, 13097 Aix-en-Provence, France; franck.sumera@culture.gouv.fr
* Correspondence: marianne.cohen@sorbonne-universite.fr

Highlights:

- Terraced lands are a historical landscape under risk in Mediterranean mountains;
- Their surface and land-use trajectories are analyzed across five centuries in a GIS;
- Although mostly abandoned, they persist in the ground landscape;
- Their environmental and productive functions are part of their future resilience.

Abstract: Terraced historical landscapes have multiple functions in mountain land, limiting erosion, enabling agricultural production and constituting cultural heritage. Currently, they are largely abandoned in Mediterranean regions and facing the ongoing impacts of climate change. Our aim is to reconstruct the evolution of land use on the terraces in order to test the hypothesis of the resilience of these landscapes and their age in recent history (17th–21st century). To achieve this, we used various current and archive spatial datasets and GIS knowledge to detect and map terraces and the changes in land use. We tested this hypothesis in a territory impacted by a recent extreme event, facing the challenge of its reconstruction. Our main outcome showed that the optimal use of the terraces corresponded to the demographic optimum of the mid-19th century, and they were gradually abandoned after the Second World War, with significant differences between Mediterranean and mountain lands. Despite this evolution, the terraces persisted and withstood an extreme event, validating our resilience hypothesis and opening avenues for the revitalization of this territory based on this heritage. These findings are drawing perspectives for the future of terraced landscapes in Mediterranean mountains in the context of climate change.

Keywords: lidar; historic maps; past and future resilience; climate change

1. Introduction

"The landscape is part of the land, as perceived by local people or visitors, which evolves through time as a result of being acted upon by natural forces and human beings" (https://www.coe.int/en/web/landscape/the-european-landscape-convention, accessed on 1 April 2023). From this perspective, the resilience of historical landscape has to be understood as the result of interactions between natural and anthropogenic drivers over time. Terraced landscapes are a good example of resilience of historical landscapes, because they have been maintained until the present time despite important environmental and socio-economic changes. These landscapes have received increasing attention in recent years as cultural landscapes have been threatened by agrarian abandonment [1–3], as have their functions of hydro-climatic regulation, food production, and preservation of natural and cultural heritage, both tangible and intangible [4]. This issue has been renewed by the notion of resilience, applied to socio-ecological landscapes [5] and traditional agro-systems [6], and the ability to overcome disturbances, assessed over the long term and more recently in the context of the acceleration of history. These disturbances are materialized by the occurrence of extreme meteorological events (droughts, floods) attributed to climatic irregularity, particularly in the Mediterranean region. In the past, the Little Ice Age and irregular rainfall were combined with land clearance and overgrazing during the demographic maximum ("torrential crisis") [7–12]. More recently, these extreme events have been attributed to global warming [13], and their frequency is expected to increase in the future [14,15].

Agricultural terraces should be considered to be "nature-based solutions" in response to climate change [4] [https://www.iucn.org/our-work/nature-based-solutions, accessed on 1 April 2023]. Indeed, they may regulate natural processes, even though they rely on human labor for the good quality of their construction, appropriate drainage and maintenance. They contribute to carbon storage [16] and the reduction in erosion when the soil is covered with vegetation and not saturated [17–19]. Agricultural abandonment can lead to degradation and the reactivation of erosive processes [10,17–19], even if the degradation of dry-stone walls is not always related to the stage of the vegetal succession [20]. This abandonment mainly concerns the terraces furthest from the villages [21]. Their rehabilitation is costly [22], but it helps to restore agro-biodiversity and the spontaneous biodiversity of the orchard's herbaceous stratum [20,23]. The resilience of this historical heritage is, therefore, a major challenge in mountainous regions facing the impact of climate change, even though they are relatively depopulated [24–27].

To address this issue, it is important to detect terraced areas and trace their historical trajectories using appropriate methods. The analysis of changes in land use, in particular in terms of historical forms of land development such as terrace-based development, requires the provision of spatially accurate field information [28,29]. Unfortunately, topographic maps rarely contain information about terraces and their land use. It is often presented in a conventional way, which does not reflect the real estate, enabling, e.g., the calculation of the surface and length of terraces. Moreover, the use of satellite and aerial images limits the detection of terraces to areas not covered by shrub and tree vegetation [29] and periods after the 1950s [25,26]. Old maps and land registers provide information on land use prior to the 1950s, without necessarily identifying terraced areas [3]. Several authors have retraced the evolution of forested areas since the 18th or 19th century, using properly georeferenced and cautiously interpreted old maps [30–32]. Other studies focused on the evolution of agricultural surfaces based on comparison between old cadasters and aerial photography (1950s to 2000s) [23,32,33]. These studies evidence significant differences in the evolution of land use in Mediterranean mountain regions, according to their market integration and demographic changes.

To detect the topographic signatures of anthropogenic structures such as terraces, remotely sensed high-resolution LiDAR data (Light Detection and Ranging) based on laser scanning technology provide detailed information on terrain surfaces [19,34,35], including terraces overgrown by trees or covered by the soil surface [36]. The spread of LiDAR technology has enabled significant advances in archaeology, particularly in forested areas, revealing the roles of ancient societies in shaping and developing their territories long considered "wild" or "virgin" [37,38]. In Mediterranean Europe, the mapping of terraces has been exhaustive in Italy, with Lidar data making it possible to complete pre-existing mapping work based on other sources to calculate the surface area by region and estimate the lengths of the terraces (Ferrarese et al., in [2]). In France, qualitative terrace mapping based on aerial photographs [38,39] can now rely on the progressive provision of LiDAR data by the French National Institute of Geographic and Forest Information (IGN).

By mapping the terraced plots and their historical trajectories since the 17th century, we aim to verify the hypothesis of the resilience of these historical landscapes and draw lessons for their future. We assume that despite human and natural impacts across 5 centuries, terraces maintained and are bringing ecosystem services in a high-risk environment. The hypotheses have been applied in a valley presenting a diversity of landscapes and facing the challenge of rebuilding the territory in a post-disaster context, drawing potential lessons for other Mediterranean valleys under risk due to climate change. In addition to scientific knowledge, our study aims to contribute to the decision-making process regarding the rehabilitation of terraces by residents' associations and professional and hobby farmers [22]. For this aim, Lidar data are associated with cartographic, photographic and historical sources in a multi-source and multi-date approach, thanks to the use of GIS and statistical processing.

2. Materials and Methods

2.1. Study Site

Our study site, the Roya valley, is a cross-border catchment located in the Maritime Alps (France, 636 km^2) and Imperia Province (Italy, 69 km^2), flowing into the Mediterranean Sea (Figure 1). It has very steep slopes along a bioclimatic gradient, from the Mediterranean Sea to alpine domains (maximum altitude 2755 m). These mountains, forming an orographic barrier, are a source of intense rainfall events, particularly in the context of climate change. This valley is a representative case of Mediterranean-montane terroir, raising the question of resilience, and it is, therefore, particularly rich in lessons.

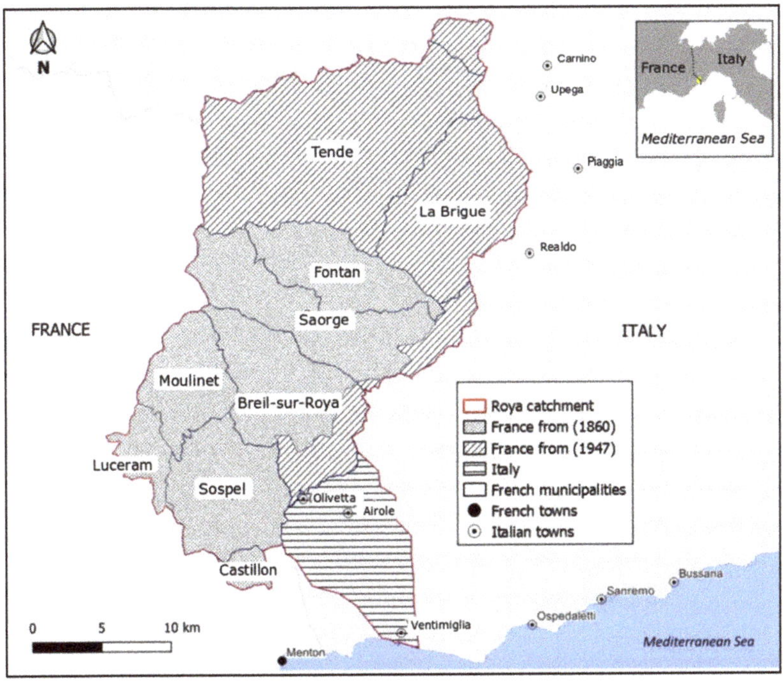

Figure 1. Location map. The Italian towns Carnino, Upega, Piaggia and Realdo were part of La Brigue before 1947. Realization by T. Le Vot.

The Roya valley was occupied by man from proto-historic times to the Middle Ages at the subalpine level and from Antiquity at the montane level [40]. Tangible remains of ancient breeding activity niversité Paris Cité.1983.2107; are still visible in the landscape (https://www.vermenagna-roya.eu/fr/patrimoine/enclos-de-la-causega/, accessed on 1 April 2023). In this work, we extend the study area to the entire French watershed and reduce the time scale to the last five centuries, based on map sources. During this period, the Roya valley was part of the County of Nice that alternated between its attachment to France and the kingdoms, duchies or principalities that made up part of Italy before 1860–1870 [41,42].

In 1860, the middle part of the Roya valley became French, while the upstream part and mouth remained in the Kingdom of Sardinia. There was a significant increase in population (+69%) between 1793 and 1861 (Cassini.ehess.fr, accessed on 1 April 2023, Table S1), a period that corresponds to the demographic peak observed in Mediterranean rural territories [27], a trend amplified here by the easing of military conflict. Although varied, agricultural production was not considered self-sufficient at the time of the demographic peak [43,44], particularly in northern municipalities highly specialized in breeding (Tables S2 and S3).

After the 19th century, the rural economy diversified from agriculture and forestry to other activities (tourism, energy, mine, hay marketing [43,45]; Table S4). The smallest and most remote municipalities underwent strong depopulation, particularly after the Second World War (the population was divided by four), while it was maintained in the most important urban nuclei that are well connected with Italy and the Riviera (−45% on average). In 1947, the eastern and upstream part became part of France, while the mouth remained Italian (Figure 1). Since the 2000s, there has been a slight demographic revival in all the municipalities, but most of the population did not participate in an agricultural activity in 2020 [https://www.insee.fr/fr/statistiques/2011101?geo=DEP-06, accessed on 1 April 2023]. Agriculture w specialized in mixed farming or poly-livestock farming (https://vizagreste.agriculture.gouv.fr/#/indicateur/06/2020, accessed on 1 April 2023).

Throughout this history, several violent climatic events occurred, especially during the Little Ice Age. Periods of intense hydrologic activity in the Rhône valley (1551–1600, 1651–1700, and particularly 1701–1710) alternated with periods of calm (1601–1650) [46]. This chronology corresponds partly with the 500–1000-year-old deposits found in the Roya valley [47,48]. In Southern France, erosion and torrential activity stages remained intense until the end of the 19th century, linked to the demographic maximum [9]. They decreased with the development of dams and reforestation from 1860 onward, after two important floods in 1840 and 1856 in the Rhône Valley (Table S4) [8,9,45]. During the 20th century, only two intense events occurred (1926 and 1952, a maximum of 200 mm/day) in the Roya Valley (http://pluiesextremes.meteo.fr/, accessed on 1 April 2023). This period corresponds to a decrease in the width of the river channel and forest reclamation.

In October 2020, Storm Alex, amplified by global warming and the moisture reserves of the Mediterranean Sea, hit the Roya Valley (663 mm in one day) and caused extensive damage, particularly to the road network and infrastructure concentrated on the valley floor [49]. This was exacerbated by the uprooting of trees that formed ice jams [50]. The valley was profoundly transformed, becoming unrecognizable to local residents, with a wide active strip filled with coarse sediments, tree trunks, vehicles and ruins. Farmers were affected by the storm: loss of land, displacement of herds, communication difficulties for many months and even the death of a shepherd trying to save his sheep occurred. Following 2020, after three years of drought, Storm Aline occurred in October 2023, fortunately being less intense than Storm Alex (200 mm/day). It is now the local people's turn to get involved and find original ways to rebuild the area and the valley's economic fabric, following a major investment by the French government to restore networks. The event has strengthened local solidarity, supported by several associations and groups of residents. In this context, the agricultural terraces represent the area's cultural and natural potential. They are an innovative nature-based solution for thinking and acting in a resilient territory [4].

2.2. Materials

2.2.1. Terrace Mapping

Three databases of the French National Institute of Geographic and Forest Information (IGN) were used and analyzed in a GIS for terrace mapping in the French Roya catchment via reasoned and multi-criteria photo-interpretation (Table 1).

Table 1. Materials used for terrace and land-use change mapping. Georf: georeferenced; Topo: topography.

Date	Name	Author [Source]	Scale	Georf	Topo	Land Use	Forest	Gaps
1672–1675	Carta generale de stati di sua Altezza Reale	Cartographer GT. Borgonio, [www.gallica.fr, accessed on 1 April 2023]	1/90,000	no	rough	Intercrop, tree-crop	No	Eastern slopes
17th	Engravings (Saorge and Tende)	GT. Borgonio [www.archives06.fr, accessed on 1 April 2023]	ND	no	rough	Detailed	No	no
1749–1754	Carte géométrique du comté de Nice	Engeneer P.J. Bourcet [https://www.servicehistorique.sga.defense.gouv.fr/le-shd-en-france/vincennes-centre-historique-des-archives, accessed on 1 April 2023]	1/14,400	no	rough	Intercrop, crop, tree-crop	No	Western slopes, northern limit
1850	Carta topografica degli Stati in terraferma di S. M. Il Re di Sardegna	Royal staff corps [https://www.defense.gouv.fr/academ/membres-lacadem/bibliotheque-lecole-militaire, accessed on 1 April 2023]	1/50,000 Sheet no. NLXXXI	no	regular	Crop, pasture, wood, olive, vine	Yes	no
1864–1866	Napoleonic cadaster	French State, [www.archives06.fr, accessed on 1 April 2023, 40]	2500	yes	no	Crop, pasture, wood, olive, vine	Yes	no
1947–1951	Aerial photography		1/40,000	yes	no	Visual interpretation of the proportion of the plot covered with scrubland and forest		no
2023	BD TOPO, Zone de Végétation	IGN [www.geoservices.ign.fr, accessed on 1 April 2022]	1/25,000	yes	no			Crops, pasture, orchards in 2020
2023	Plots of land		1/25,000	yes	no	No		no
2020	Lidar HD post-Storm Alex		DEM 1 m resolution	yes	good	No		no
2023	Lidar HD 06			yes	good	No		no

- LIDAR HD

LiDAR data were used to detect the layout of terraces in the French Roya catchment. LiDAR HD point clouds came from aerial LiDAR acquisition with densities of at least 10 pulses per m^2 and 5 pulses per m^2 above a 3200 m altitude. Using class no. 2 of the classification attribute of LiDAR data (laser beams reflected from the ground), a Digital Elevation Model (DEM) with a spatial resolution of 1 m was created by the authors. The DEM enabled the generation of a shaded relief model, which made it possible to visualize terrain microforms in terms of the terraces (parallel stripes), disregarding the trees above ground level. Lidar may be generated with a different sun inclination to improve micro-topography detection. All output data were saved in *asc format. The above task was conducted using SAGA GIS© 3.6.2 software.

- Plots of land layer

To avoid confusion between terraces and natural shapes, the elongated and narrow shapes of plots of land layers were a first clue, because terraces were built perpendicularly to the slope within this frame, which corresponds to the property of the landowner who invested labor, materials and expertise in the construction and maintenance of the terraces in order to obtain significant harvest. A small square plot, corresponding to a permanent or seasonal habitat (small house or hut), is often associated with clusters of terraces, testifying to the need to have a place to rest or store equipment during agricultural work on plots far from the village, the low distance to habitat being observed as a favorable factor in the literature [21,38].

- Aerial photography 1947–1951

These images give an image of traditional land use at the beginning of the rural depopulation that followed the Second World War, the abandonment of agriculture and the development of tourism [2,3,22,23,25,45]. Terraces are recognizable via aerial photography when they are delineated by white dry stone walls, which generate a white-striped texture (Figure S1) or a dark grey-striped texture as a function of slope illumination. Aerial photography was largely used to check our interpretation of terraces based on a couple of Lidar-plots of land (Table 1).

- Data analysis in a GIS

The Lidar layer was overlaid in a GIS, using Qgis 3.6.2© software, with the plots of land layer, which gave the frame of land use. Plots of land carved into terraces were selected with a multi-selection tool and identified automatically in the attribute table of the plots of land layer with the raster calculator. When terraces were occupying only part of the plot of land, a new polygon was created, following the design of the terraces. The other part of the plot was generally plain and in very few cases covered with rock blocks (Figure S1) or shaped by a ravine. This methodology is time-demanding, although very cautious and without error. It is similar to those used to map terraced land in Italy [2]. These criteria should be integrated into an automatic mapping process, based on algorithms, and compared with the results obtained with reasoned photo-interpretation. It is the same for the calculation of the total length of the terraces, which requires an accurate and automatic method (in progress).

2.2.2. Land-Use Change Mapping

- Building a multi-date geographical database

Our study is based on the conception of a multi-date geographical database during the period 17th–21st century (Le Vot, unpublished master thesis), with different images giving information of variable precision about past and current agricultural land use (Table 1). Except for the Napoleonic cadaster, which provides information on the land-use history of each parcel of property, the other sources provide this information in the form of areas with figurative colors or letters. Land use in the oldest maps is not described in the legend and has been interpreted (Figure S1).

- Old maps analysis

Ortho-rectification of the oldest maps proved impossible due to the projection errors made (Table 1). For this reason, the boundaries of cultivated areas were identified based on the geographical information available in our database (topographical map, the DEM, slope values, hydrographic network, roads and habitat visible on aerial photographs of 1947–1951). To compensate for the spatial imprecision of the oldest maps, we selected only terraced plots present in the area designated as "cultivated" because they were most probably cultivated. The information about land use is more or less detailed (Table 1, Figure S1). We considered that terraces described as "wood" correspond mainly to chestnut groves [40]). Timber harvesting was carried out on large plots of land every 50 years [43], whereas terracing is better suited to annual fruit picking.

- Napoleonic cadaster

The land use was described in 1864–1866 in the mid and southern catchment by the Napoleonic cadaster, realized after the region's attachment to France in 1860 (Figure 1, Table 1). This map was georeferenced following Bonne's projection with the reference coordinate system ESRI 53024, adapted to old maps [51], with Arcgis Pro 3.2© software.

The current shape of terraced plots was edited to fit with the one of the Napoleonic cadaster. This work was carried out plot by plot, for each land-use type, and further merged in a synthetic layer. Following previous studies [40], we considered that "pastures" referred to a large and plain plot, without terraces, freely grazed by the herd during the summer and generally distant from the human habitat. Terraced plots that are referred to "pasture" in the Napoleonic cadaster were probably meadows used for hay production or grazed during spring and autumn [43].

- Aerial photography 1947–1951

Land use during the 20th century is documented by aerial photography from 1947–1951 (Table 1) because this period is representative of the beginning of rural abandonment in Mediterranean mountains [2,3,29]. The abandonment of terraces was mapped plot-by-plot via multiple selection, considering two forest reclamation stages, the early (tree cover 10–49%) and mature stages (tree cover >50%). Additionally, we considered the spatial arrangement of trees to identify the first reclamation stage (colonization front, irregular arrangement of trees) and scrubland cover and to avoid including in this category plots with regular arrangement of trees, corresponding to intercropped system.

- BD TOPO 2023

During the 21st century, the vegetation layer of the BD TOPO 2023 (IGN) gives the following information about land cover: the different types and densities of forest and scrublands and vine and olive groves. Because of gaps (Table 1), the automatic method of intersecting plot centroids with this layer created errors. The plot-by-plot method was applied to extract the information about the dominant land use in the terraced plots. Despite being time-demanding, this method uses the same criteria as the analysis of the photography of 1947–1951.

- Data criticism

Comparisons of the surfaces of terraced plots at different periods of time must be considered with caution, because each cartographer followed their own rules (Table 1). To minimize the gaps existing in the Borgonio and Bourcet maps, we created a merged attribute, considering cultivated terraced plots mapped either by one or by both authors. The Napoleonic cadaster was used to calculate the property taxes, and for this reason, it may underestimate the forest cover and the frequency of intercrop. The topographic map was established by the Royal Staff Corps to describe the entire Kingdom of Sardinia. The cover of groves has not been updated in the BD TOPO 2023 since the 2000s (pers.com, Laurence Jolivet, IGN). This map is less accurate than the high-resolution photo-interpretation at the land-plot level in the municipality of Sospel [23].

2.3. Data Analysis

The information about the land use of terraced plots is available in the terraced plots layer for five periods: 1672–1675 (Borgonio), 1749–1754 (Bourcet), 1850 (topographic map) or 1864–1866 (Napoleonic cadaster), 1947–1951 and 2000–2023. Mapping with Qgis 3.6.2© software was followed by data analysis with Excel 2016© software. The data analysis had two objectives. The first objective was to compute the repartition of the terraced plots and land-use change in the French Roya valley. For this aim, pivot tables were realized to compute the surfaces of terraced plots, detailed by municipality, land use and period.

The second objective was to map and analyze the trajectories of land-use change for terraced plots. For this aim, we created a new attribute summarizing the change in or persistence of land use for each terraced plot. The land use described during the 19th century was the most precise, and it was, therefore, used as a reference. The abandonment of terraced chestnut woods was only considered if resinous or mixed forests were present in 2023, following field observations. A new attribute, named 17–21, with five letters corresponding to the five dates, was generated with the Concat function of Qgis 3.6.2© software. The 332 categories were further merged in 24 categories, with three letters, the first corresponding to the seniority of the terraces (17th and/or 18th century), the second to land use during the 19th century, and the third to evolution during the 20th and 21st centuries. Swinging trajectories (conversion from intercrops or vine in the 19th century to olive groves in the 21st century, or rehabilitation after the 1950s) were described separately. Intercrops were simplified to reduce the number of categories. The same simplification was applied for the elaboration of the legend of the map (Figure 2).

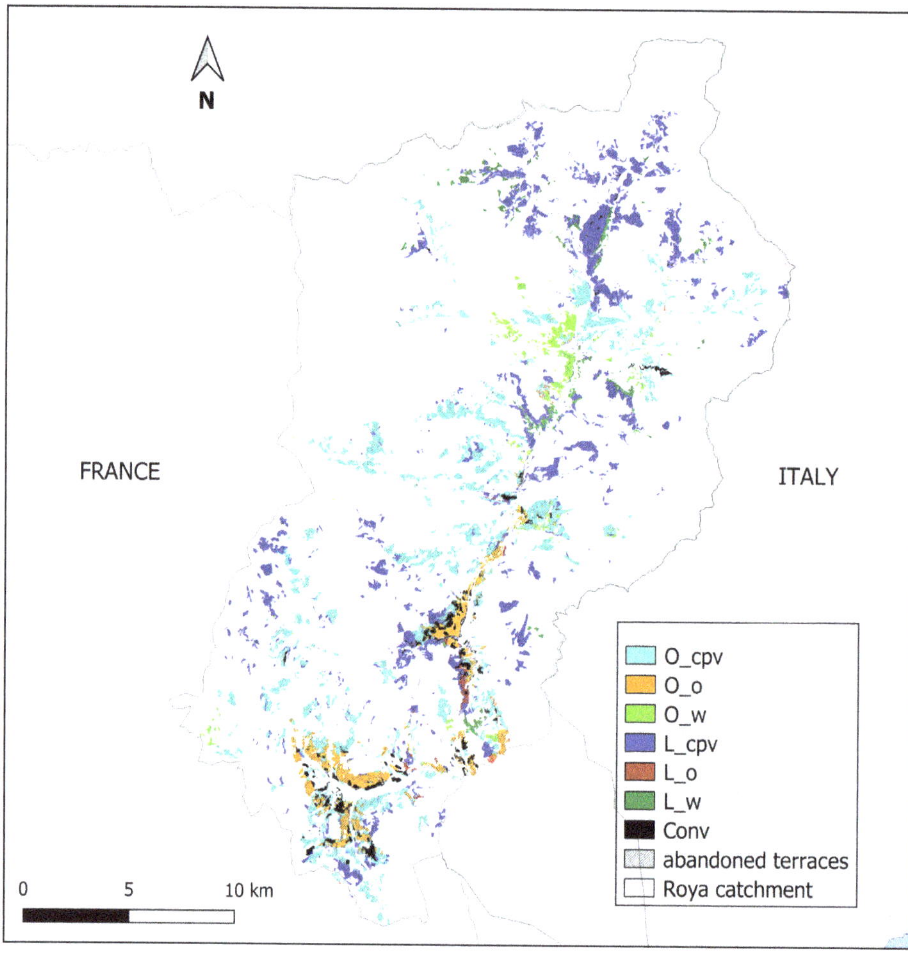

Figure 2. Land-use trajectories of terraced plots during the period 1680–2023 in the French Roya catchment. 0: old (17–18th century); L: late (19th century); cpv: crops, pasture or vine; o: olive; w: chestnut wood. Sources: Borgonio map, Bourcet map, Topographic military map, Napoleonic cadaster, aerial photography 1947–51, plots of land and BD Topo 2023 (IGN). Realization: T. Le Vot.

Different colors were given to the main land uses during the 19th century, depending on their seniority. Stripes were added for abandoned plots. Despite being time-demanding, this method is better adapted to the variable size of the terraced plots, partially covering the surface of the catchment, compared with cellular automata, based on a fixed size of continuous pixels [52].

3. Results

3.1. Terraces and Land Use

In total, 29,445 terraced plots have been identified in the French Roya catchment, covering a surface area of 90.8 km^2, corresponding to 14.26% of the whole catchment and 14.96% of the seven main municipalities. The total length of the terraces has been estimated to be 9000 km, following the literature [2], representing an important potential legacy. The average size of the terraced plots varies in a range from 0.23 ha (Breil-sur-Roya) to 0.41 ha (Moulinet). Sospel, the southernmost municipality, has the highest proportion of surface in terraces (28.3%), followed by Breil-sur-Roya (20.3%), Fontan (16.9%) and Moulinet (14.7%). The northernmost and largest municipalities, covering 60% of the catchment, have a lesser surface proportion of terraced plots (La Brigue, Saorge and Tende, around 11%). This suggests a gradient of increasing surface of terraces from the mountainous northern municipalities to the Mediterranean southern municipalities.

The comparison between the areas occupied by cropped terraces in the French Roya during the period suggests a dramatic increase from the 17th to the mid-19th century, doubling every century, followed by a decrease beginning in the 1950s (−33% in a century) and accentuating dramatically in 2023 (−73% in half a century) (Table 2). Considering the merged attribute (1680–1750), the total surface that was cropped in 1672-75 and/or 1749-54 was 57.70 km^2. Only one sixth of the cultivated areas during the 19th century persist in the 21st century (−81%), the abandonment being more notable in La Brigue, Fontan, Saorge and Moulinet and less dramatic in Breil-sur-Roya and Sospel (one third abandoned).

Table 2. Surface areas of cultivated terraced plots and total surface area of the municipality (in km^2). Period: 1672–2023.

Municipality	1672–1675	1749–1754	1850–1864	1947–1951	2020–2023	Total Surface Area
Tende	3.09	6.96	20.28	15	4.69	177.47
La Brigue	3.63	4.89	10.66	6.9	0.73	91.77
Fontan	3.70	3.86	8.45	5	1.35	49.61
Saorge	4.27	7.60	8.77	5.7	1.46	86.78
Breil-sur-Roya	2.76	7.41	16.43	11.5	3.85	81.31
Moulinet	3.34	1.70	6.08	2.8	0.30	41.07
Sospel	5.80	14.13	17.72	11.8	4.26	62.48
Total	26.62	48.79	90.8	61.1	16.86	636.46

During the 19th century, the terraces were cultivated with cereals, forage and groves (olive, vine, chestnut, fruit trees) organized along the bioclimatic gradient (Table 3).

Table 3. Surface of different land uses on terraced land in the Roya Valley in the 19th century (in km^2).

Municipality	Crops	Pasture	Chesnut	Vine	Olive	Intercrop
Tende	6.6	2.0	2.6	0.1	0	0.6
La Brigue	9.1	0.4	1.3	0.9	0	0.1
Fontan	13.0	1.6	1.3	1.4	0.1	0.2
Saorge	6.5	0.4	0.1	2.5	0.5	0.2
Moulinet	10.5	2.7	0	1.0	0	0.1
Breil-sur-Roya	10.8	0.9	0.9	2.4	4.8	1.8
Sospel	11.9	2.2	0	4.1	8.6	1.2

This was due to the decreasing heat needs of olive trees, vines and chestnuts, which are growing, respectively, in Mediterranean, temperate and mountain climates, with these differences also having a strong effect on the natural flora [53]. Despite the specialization in olive-growing and breeding according to the increasing latitude, land use was diversified on terraced plots.

3.2. Trajectories of Land Use

As expected, the abandonment trend is the most frequent case (85% of the total surface, Table 4). Other trends are detectable depending on the seniority or land use of the plots. Terraces mapped in the 17th or 18th century are still more frequently cropped than the additional terraces mapped in the 19th century. This suggests that the terraces built during the demographic maximum have been more widely abandoned (91%) than the oldest ones (81%) and are now almost completely covered with open or closed forest. Parts of the oldest terraces maintained or even increased across the period, particularly with half of olive grove being maintained (50%), one fifth (21%) thanks to either conversion, intercrop abandonment or renaissance. Half of chestnut woods (46%) still maintained. Conversely, cereal crops have decreased drastically (93% abandoned), even on old terraces, though their previous extension gives an idea of the potential surfaces available to develop local food production after clearing the ligneous species. Vineyards almost disappeared (99.8%), particularly in La Brigue, though this has been documented since the 17th century. It is the same for pasture located in Tende, as observed since the 19th century (95% abandoned).

Table 4. Land-use trajectories of terraced plots in the French Roya catchment (in ha). Period: 1672–2023.

Trajectory		Land Use	Breil s/Roya	Fontan	La Brigue	Moulinet	Saorge	Sospel	Tende	Total
since 17–18 th centuries	abandoned	chestnut	21	1	31		30	1	134	217
		crops	334	334	380	243	485	543	273	2591
		olive	122	4			50	317	0	493
		pasture	29	44	1	56	34	104	34	302
		vine	123	25	78	38	160	167	21	613
	cropped	chestnut	4	12	2	1	7		176	202
		crops	21	15	7	4	9	26	38	119
		olive	143	0			4	166		313
		pasture	1	4				5	4	15
		vine	0		1			0	0	1
		conversion	127	12	27	5	11	142	16	341
since 19th century	abandoned	chestnut	24		43		0		117	184
		crops	457	238	403	122		125	778	2123
		olive	34	3			5	15		58
		pasture	48	30	37	55	3	27	294	496
		vine	25	35			9	16	1	86
	cropped	chestnut	22	53	37	1	1		37	152
		crops	7	12	3		1	3	48	74
		olive	13				0	2		15
		pasture	0		0		0		19	20
		conversion	26	3	1	1	0	2	10	43

The map evidences the site-specific historical trajectories (Figure 2). The most remote parts of the catchment were cultivated later (striped dark blue) than those near the Roya and Bevera rivers. Old olive groves are persisting on the slope facing the villages of Sospel and Breil-sur-Roya. Parts of them have been abandoned, with others rehabilitated, in different positions on the slopes in these two villages. Old chestnut woods have been maintained on almost the whole slope overlooking Saint-Dalmas-de-Tende. On the contrary, the slope facing the village of La Brigue that had been covered with vineyards since the 17th century is abandoned, according to the BD TOPO 2023.

4. Discussion

This discussion is organized in five sections. First, we compare our results with the literature on terraced land and land use in the Roya Valley, in surface, time and space; second, we discuss how far this spatiotemporal organization validates our hypothesis of the resilience of historical landscapes. Finally, we recall the uncertainties in our results and suggest ways of reducing them.

4.1. *How Many Terraces?*

The surface area occupied by the terraced plots is supposed to correspond to the following cropped areas: vine, olive, cereals, fruit trees and meadows, and for this reason, we compare our findings to the estimations of cropped surfaces during the 19th century. The proportion of the surface of terraced plots that we found, almost 15%, is three times higher than the estimations of Fodéré (1801, [43]) and twice as high as estimates dating from 1874 [43] in the French "County of Nice". It is nearer to the inventory of Durante made at the time of the Kingdom of Sardinia in 1846, even if it probably overestimates the proportions of crops (14%) and meadows (11%) [cited by 43]. The highest proportions of terraced plots in Sospel and Breil-sur-Roya are consistent with the literature [44,53,54] and their Mediterranean-type agriculture.

Our innovative methodology, based on a multi-source photo-interpretation, including Lidar data, brings more precision than the use of outdated aerial photography [2,38, 39]. Our findings are inferior to the cover estimated via photo-interpretation (30 to 50%) in the Maritime Alps and the Var [38], and the highest cover—28.6%—is found in the southernmost municipality of Sospel. They are superior to the cover of 10% considered representative of the whole Roya valley ([1], Blanc in [2]) and 8% in the Italian region of Liguria [2]. These lower proportions are nearer to those found by our mapping process in the northernmost municipalities specializing in breeding activity (La Brigue, Saorge and Tende, around 11%). These differences are explained by the non-exhaustive nature of previous studies and the lack of Lidar data. In conclusion, the legacy of terraced plots is important, considering its surface and length in the Roya Valley, and representative of the heterogeneity of this territory.

4.2. *Time Variation*

- 1672–1749

The increase in terraced land during the period 1672–1749 may be due to two different causes. First, we assume that the strong hydrological activity [9], combined with the increase in cultivated surfaces between the Borgonio and Bourcet maps, may explain the braided form of the confluence of the Roya–Bieugne–Lavensa rivers mapped by Bourcet in 1749–54 (Figure 3), despite the mitigation effect of the terraces on geomorphic processes. The presence of terraces, as well as their functions, were reported in 1736 ("To take advantage of the mountains on the left where there is a little land, they formed amphitheatre steps to support them and prevent the waters from carrying them away") [55]. The increase in terraced surfaces occurred during the military conflict between the Kingdom of France and the Duchy of Savoy (Kingdom of Sardinia after 1720), with severe consequences in Sospel and Breil-sur-Roya (plague, smallpox), as well as Tende and La Brigue (seasonal migration) [54]. Despite a decrease in the population noticeable in the first French Census (1793, https://www.departement06.fr/de-la-population-et-des-causes-de-sa-diminution-37200.html, accessed on 1 April 2023), the anthropogenic pressure was mostly due to the obligation of producing locally food for the population [42,54,56].

Second, the increase in the surface of terraced land is partly due to an underestimation of the cultivated areas by Borgonio and Bourcet, both in the shaded slopes and at the limits of the catchment. The intercropped landscapes mapped by Borgonio and Bourcet do not give an exhaustive idea, and the landscapes that retained their attention probably have a specific meaning that we have ignored. The large scale (Borgonio) and the context of a military campaign (Bourcet) are possible explanations for the gaps existing between these maps. The merged attribute is giving a more realistic assessment of formerly terraced plots.

- 1749–1850

Between 1749 and 1850–1866, the increasing trend in terraced land cover (+86%, +57% by considering the merged attribute) is partly linked to the demographic increase (+69%) that occurred between 1797 and 1861, as is true in rural French territories [27]. According to the literature [57], ten square-kilometers of terraces can be built in half a century if the demographic and economic conditions are favorable. This estimation is lower if compared with the extension of terraces between Borgonio's map and the mid-19th century maps (64.18 km^2). When considering the merged attribute 1672–1749, the increase, 36.1 km^2, in approximately 150 years is more consistent with this rate [57]. Most works agree with this chronology [2], while another author considers that terraced landscapes were already built in the 18th century or even before in the Roya valley [58].

Figure 3. Braided river at the confluence between the Roya, Bieugne and Levensa rivers, Saint-Dalmas-de-Tende. Source: Bourcet map, 1749–1754. Green lines: river; red: town and roads; green dots with orange lines: intercrop; light brown: uncultivated.

- 1850–1947

This period has been more frequently studied in the literature [2,3,25,32] thanks to the availability of historical sources. As in other Mediterranean mountains [3,25], the abandonment process began in the 1950s, or even before in the Maritime Alps [45]. The forest reclamation trend (33%) is probably due to the demographic decrease (−45%) during the past century and the Second World War (the war was violent, with destruction and military campaigns). The development of trade between France and Italy, following the new division of the territory (1860) [43] and the development of transport infrastructures (Col de Tende Tunnel in 1882; railway lines in 1928), with a resultant decline in self-sufficiency, are other explanations for the decrease in cultivated terraces, as observed in other studies [2,3,23,25,32].

- 1947–2023

Many authors have underlined the importance of the decline of traditional agriculture and the incentive towards modernization that produced the abandonment of terraced land in mountainous regions [2] and, more generally, of agricultural activity except in regions where agricultural production has intensified with market integration [3,6,25,26]. During that period, the population of the Roya Valley maintained at a low level until the 1980s and then began to recover, increasing slightly from the 2000s onward. Yet, the number of farmers probably followed the downward trend (-88%) observed in the Maritime Alps since the 1970s (https://vizagreste.agriculture.gouv.fr/#/indicateur/06/2020, accessed on 1 April 2023). For those reasons, three quarters of terraced plots were abandoned (72%) during this period, a value among the highest observed in the Mediterranean and mountainous European regions [2,3,6,25,26].

4.3. Spatial Analysis

The northern part of the Roya Valley is less terraced than the southern one due to the importance of high-altitude pasture and woods [54]. This difference is also due to the density of population during the 19th century, which is correlated with the percentage of terraced plots ($r = 0.86$), because highly populated municipalities need more food and have a larger labor force to build and maintain terraces. Most of the northern population were shepherds and migrated to the Riviera during the winter to feed their herd [41,54,55]. The orientation of the economy interacting with the density of population brings about another example of the importance of these socio-economic factors in the spatial organization of a territory contrary to a deterministic vision [40].

The diversity of land use during the 19th century suggests an economy partly based on auto-consumption (vine, cereals) and partly based on formal or informal markets (olive, breeding, hay) [43]. Intercrops, which characterize Mediterranean agro-systems [23], are noticeable in southern municipalities. Compared with the description of Mediterranean and mountain economies [56], land use was more diversified in the southern municipalities than in the Mediterranean territories, including pasture for breeding activity, and the northern municipalities were even more specialized in breeding activity [42–44]. While in southern municipalities, the economy of the terraced plots was based on olive oil, cereals, etc. in the northern municipalities, terraced plots were partly occupied by pasture, contributing to hay production for the livestock [43,44]. During the 21st century, abandoned and cropped terraced land does not depend on altitude or distance to the valley and villages, contrary to previous studies [21], including those in southern villages, where there is a significant persistence or renaissance of olive-growing, as observed at a regional scale, due to professional or hobby farmers [22,59]. Access by car is now decisive.

4.4. Resilience of Historical Landscapes

Our hypothesis of the resilience of historical landscapes is validated by our study, following "contemporary definitions considering resilience to consist of (1) the amount of disturbance that a system can absorb while still remaining within the same state or domain of attraction; (2) the degree to which the system is capable of self-organization (versus lack of organization or organization forced by external factors); and (3) the degree to which the system can build and increase its capacity for learning and adaptation" [5]. Meanwhile, almost all the processes that characterize the resilience of traditional landscapes are represented in our study [3,6]: abandonment, renaissance, continuity, etc.

First, a high number of terraces persist (continuity), despite the amount of disturbance since the 19th century (abandonment, demographic loss, Storm Alex) [5]. We observe hardly any degradation of dry-stone walls in abandoned terraces with the Lidar data, contrariwise to other studies [2,10,17–19]; this finding should be verified by field work ([20], Story team, Tterrcat project, https://projetstory.wordpress.com/actualites/, accessed on 30 March 2024).

Second, co-variations between the surfaces of cropped terraces and the population (more recently the number of farmers) suggest the self-organization of the socio-ecological system [5]. The legacy of terraces has been built in a tough historical context, characterized by military conflicts [41,42], fluctuating demographics [cassini.ehess.fr], extreme events until the end of the 19th century [8,9,43,44], and the changing economy since the end of the Second World War [25,26] or even before (after the First World War in the Maritime Alps, [45]). Terraced plots covered with cereals and pasture that no longer correspond to the needs of food supply and modern breeding activity almost disappeared, similar to in other regions [2,3,6,25,26,32].

Third, the continuity of cropped terraces since the 17th–18th century suggests a "persistence despite the amount of disturbance" [5]. This legacy is maintained particularly in some ancient terraces that are still cropped with olive and chestnut trees, as observed in other studies [3,6,45]. This is due to the longevity of these trees, the insignificant damages suffered by olive groves in Sospel and Breil during the freeze event in 1956 and the small labor force required to obtain a harvest, whether sold or self-consumed, fitting with the requirements of hobby and professional farmers [22,59].

Fourth, the renaissance of olive growing observed in the Roya Valley, as in other regions [3,6,22], suggests a "capacity for learning and adaptation" [5]. It is less pronounced than the doubling of the number of trees observed between 1956 and 2011 in France, but part of this difference is due to the lack of frost damage in the Roya Valley in 1956 [23]. This renaissance is still ongoing after the extreme events that hit the Roya Valley based on the strong participation of local communities [2,60] (olive growing: Cultures en terrasses, https://www.terrassesroya.eu/; vineyards in La Brigue (https://miimosa.com/fr/projects/la-ciappea-renaissance-de-la-viticulture-en-haute-roya; chestnut woods in Tende (https://www.chataigne-roya.fr/).

Finally, the resilience of this historical landscape relies on the adaptation capacity of the local population itself. First, despite fluctuations, the demography recovered after the military conflicts between France and Italy. What could have led to the total collapse and depopulation of this valley has forged a cross-border culture. This cross-border culture is superimposed with differences between Piedmont's culture to the north and Liguria's culture to the south. Second, Storm Alex in October 2020, which could have led to the collapse of the territory, triggered a strong social response and solidarity. The legacy of terraces is mostly invisible though important culturally and ecologically for the resilience of this valley.

4.5. Limits and Prospects of Our Study

It is important to bear in mind that the accuracy of the results obtained depends on the reliability of the materials and methods (Table 1). Differences in interpretation, variations in mapping methods and other variables may lead to uncertainties in our results. Consequently, the results of this analysis are specific to the protocol adopted and may not correspond perfectly to the reality on the ground. These uncertainties could be reduced by using other maps and cadasters drawn up during the 18th century (Turin archives, in progress). A dating campaign is required [33], using innovative methods for the terraces [61] and the remains of the huts built next to the terraces. Orchard surfaces mapped by the BD TOPO IGN should be verified with recent aerial photography and field surveys. The acceleration of the increase in forested surfaces between 1990 and 2018 [52] suggests that more frequent dates should be considered, similar to previous studies [23,25]. This sensitivity to data must be kept in mind when considering the results and to recognize the inherent limitations of using historical data in landscape studies.

As the Roya Valley is a cross-border territory, it is important to widen this study to the whole catchment, including the mouth of the river in Italy. In the Italian, southern part of the catchment (municipalities of Olivetta, Airole and Ventimiglia), terrace mapping is still in progress. Partial information is available on forests (Period 1897–2023), scrubland, vine, olive and crops (Period 1950–2023). If we consider both Italian and French parts of the valley, the forest cover multiplied by two between the 19th century and the 1950s, and once more since then, similar to other Mediterranean mountain territories [26]. In total, forest cover has been multiplied by four (Le Vot, unpublished master thesis), a higher increase than that observed in the whole Provence–Côte d'Azur region (excluding the County of Nice) during a longer period (18th–21st century, *3.5) [30].

Scrubland was not mentioned during the 19th century and was probably lacking or merged with pasture [45]. According to the digitized SCAN Historique 1950 (https://geoservices.ign.fr/scanhisto, accessed on 1 April 2023), scrubland covered 6.2% of the surface of the French part of the catchment and less than 0.1% of the Italian part. Agricultural activity was maintained better in the Italian municipalities, for a diversity of crops, depending on the market, including flowers, market gardening; 52% of olive groves were still cropped in 2006 [https://srvcarto.regione.liguria.it/geoviewer2/pages/apps/geoportale/index.html?id=680, accessed on 1 April 2023].

5. Conclusions

This study, based on multi-source data, has mapped the extension of terraced land and reconstructed their land-use trajectories in a Mediterranean mountain valley. By comparing the land use of the terraces since the 17th century with demographic data, the literature and local community testimonies, the resilience of terraced landscapes has been based on the ability of local populations to overcome human and natural hazards through the past 5 centuries. Today, these historical landscapes have largely been swallowed up by the reclaiming forest, but they persist in the ground and represent potential for the future in a post-disaster context.

Our methodology and results could be applied in other Mediterranean mountain valleys facing the same problems and able to rely on the heritage of historical landscapes and the potential participation of local communities. Valleys similar to Roya, arranged all around the Mediterranean Basin and flowing into the sea, are subject to land abandonment and increasingly frequent and violent extreme events as a result of climate change. This configuration leads to high vulnerability to climate change, with coastal erosion problems combining with those linked to devastating floods coming from upstream. The resilience of historical landscapes constitutes, in this context, an asset, providing a nature-based solution to face the challenges and uncertainties of the present and the future. This kind of study should also give useful information to local communities to assess the capacities of terraced plots and, more generally, historical landscapes to be rehabilitated for the revitalization of the territory (Story team, Tterrcat project, https://projetstory.wordpress.com/actualites/, accessed on 30 March 2024).

The transposition of our methodology will be tested in the future at the mouth of the Roya watershed, located in Italy. The cross-border context, a major difficulty in the Roya valley, does not arise in all Mediterranean mountain valleys, which should facilitate the transfer of this method. In European countries, historical sources, dating from at least since the 19th century, are available on-line. Lidar data are currently available, and automatized methods would allow us to extend this kind of study, as is already the case in the whole of Italy. The results obtained in the Roya catchment make it a "sentinel territory" for socio-ecological and historical landscape resilience and an example of the kind of resilience that can be achieved in the future. The citation of Marc Antrop [28], "Landscapes of the past are important for the future", is increasingly true in a changing world.

Supplementary Materials: The following supporting information can be downloaded via this link: https://www.mdpi.com/article/10.3390/land13050592/s1, Table S1: Populations of the main municipalities of the Roya Valley 1793–2017, Table S2: Economy of the County of Nice and the municipalities of Tende and La Brigue in the 19th century [43], Table S3: Economy of the Roya Valley in the 19th century [43,44,54], Table S4: Land-use change in the Maritime Alps in 1896–1929 [45], Figure S1: Detection of terraces and land use with different sources.

Author Contributions: T.L.V. and M.C.: conceptualization, terrace mapping and analysis, writing, editing; M.C.: supervision; M.N.: Lidar analysis, writing, editing; F.S.: Napoleonic cadaster, supervision; P.P.: supervision, funding acquisition. All authors have read and agreed to the published version of the manuscript.

Funding: Funding of the master thesis by Geoteca, Université Paris Cité. No funding for APC (invited manuscript).

Data Availability Statement: The datasets presented in this article are not readily available because the data are part of an ongoing study. Raw materials are fully available on line and in the libraries of regional and military archives.

Acknowledgments: We are grateful for the support given by the story team (Sorbonne University), GeoTeCa (Université Paris Cité), archives services (scan supply, authorized photographs), M. Le Vot (voluntary revision of English language) and the fruitful discussions with local associations and researchers involved in the Cultures en terrasses project. We also warmly thank the three reviewers and editors for their suggestions for improving the manuscript.

Conflicts of Interest: The authors declare no conflicts of interest.

References

1. De Reparaz, A. La culture en terrasses, expression de la petite paysannerie méditerranéenne traditionnelle. *Méditerranée* **1990**, *71*, 23–29. Available online: https://www.persee.fr/doc/medit_0025-8296_1990_num_71_3_2679 (accessed on 1 April 2023). [CrossRef]
2. Varotto, M.; Bonardi, L.; Tarolli, P. (Eds.) *World Terraced Landscapes: History, Environment, Quality of Life*; Environmental History book series (Envhis) Volume 9; Springer: Cham, Switzerland, 2019. [CrossRef]
3. Wolpert, F.; Quintas-Soriano, C.; Plieninger, T. Exploring land-use histories of tree-crop landscapes: A cross-site comparison in the Mediterranean Basin. *Sustain. Sci.* **2020**, *15*, 1267–1283. [CrossRef]
4. FAO. Globally Agricultural Heritage Systems. Available online: https://www.fao.org/giahs/fr/ (accessed on 1 April 2023).
5. Cumming, G.S. Spatial resilience: Integrating landscape ecology, resilience, and sustainability. *Landsc. Ecol.* **2021**, *26*, 899–909. [CrossRef]
6. Vallés-Planells, M.; Galiana, F.; Torrijos, I.D. Agricultural abandonment and resilience in a Mediterranean periurban traditional agroecosystem: A landscape approach. *Ecol. Soc.* **2020**, *25*, 5. [CrossRef]
7. Surell, A. *Étude sur les Torrents des Hautes-Alpes*; Carilian-Gœury et Victor Dalmont: Paris, France, 1841.
8. Bravard, J.P. Le traitement des versants dans le Département de la Drome. Des inondations de 1840 à la loi du 27 juillet 1860. *Ann. Des Ponts Et Chaussées* **2002**, *103*, 37–43. [CrossRef]
9. Descroix, L.; Gautier, E. Water erosion in the southern French Alps: Climatic and human mechanisms. *Catena* **2002**, *50*, 53–85. [CrossRef]
10. García-Ruiz, J.M.; Lana-Renault, N. Hydrological and erosive consequences of farmland abandonment in Europe, with special reference to the Mediterranean region—A review. *Agric. Ecosyst. Environ.* **2011**, *140*, 317–338. [CrossRef]
11. Benito, G.; Macklin, M.G.; Zielhofer, C.; Jones, A.F.; Machado, M.J. Holocene flooding and climate change in the Mediterranean. *Catena* **2015**, *130*, 13–33. [CrossRef]
12. Wilhelm, B.; Rapuc, W.; Amann, B.; Anselmetti, F.S.; Arnaud, F.; Blanchet, J.; Brauer, A.; Czymzik, M.; Giguet-Covex, C.; Gilli, A.; et al. Impact of warmer climate periods on flood hazard in the European Alps. *Nat. Geosci.* **2022**, *15*, 118–123. [CrossRef]
13. Ginesta, M.; Yiou, P.; Messori, G.; Faranda, D. A methodology for attributing severe extratropical cyclones to climate change based on reanalysis data: The case study of storm Alex 2020. *Clim. Dyn.* **2023**, *61*, 229–253. [CrossRef]
14. Tramblay, Y.; Somot, S. Future evolution of extreme precipitation in the Mediterranean. *Clim. Chang.* **2018**, *151*, 289–302. [CrossRef]
15. Kreibich, H.; Van Loon, A.F.; Schröter, K.; Ward, P.J.; Mazzoleni, M.; Sairam, N.; Abeshu, G.W.; Agafonova, S.; AghaKouchak, A.; Aksoy, H.; et al. The challenge of unprecedented floods and droughts in risk management. *Nature* **2022**, *608*, 80–86. [CrossRef] [PubMed]
16. Brown, A.G.; Fallu, D.; Walsh, K.; Cucchiaro, S.; Tarolli, P.; Zhao, P.; Pears, B.R.; van Oost, K.; Snape, L.; Lang, A.; et al. Ending the Cinderella status of terraces and lynchets in Europe: The geomorphology of agricultural terraces and implications for ecosystem services and climate adaptation. *Geomorphology* **2021**, *379*, 107579. [CrossRef]
17. Stanchi, S.; Freppaz, M.; Agnelli, A.; Reinsch, T.; Zanini, E. Properties, best management practices and conservation of terraced soils in Southern Europe (from Mediterranean areas to the Alps): A review. *Quat. Inter.* **2012**, *265*, 90–100. [CrossRef]
18. Dorren, L.; Rey, F. A review of the effect of terracing on erosion. In *Briefing Papers of the 2nd SCAPE Workshop*; Boix-Fayons, C., Imeson, A., Eds.; 2004; pp. 97–108. Available online: https://www.ecorisq.org/docs/Dorren_Rey.pdf (accessed on 1 April 2023).
19. Arnáez, J.; Lana-Renault, N.; Lasanta, T.; Ruiz-Flaño, P.; Castroviejo, J. Effects of farming terraces on hydrological and geomorphological processes. A review. *Catena* **2015**, *128*, 122–134. [CrossRef]
20. Boillat, S.; Burga, C.A.; Gigon, A.; Backhaus, N. La succession végétale sur les cultures en terrasses de la Vallée de la Roya (Alpes-Maritimes, France) et sa perception par la population locale. *Geogr. Helv.* **2004**, *59*, 154–167. [CrossRef]
21. Slámová, M.; Krčmářová, J.; Hronček, P.; Kaštierová, M. Environmental factors influencing the distribution of agricultural terraces: Case study of Horný Tisovník, Slovakia. *Morav. Geogr. Rep.* **2017**, *25*, 34–45. [CrossRef]
22. Angles, S.; Grésillon, E.; Minvielle, P. L'oléiculture en France: Quand le paysage participe au renouveau d'une activité agricole. *Proj. Paysage* **2017**, *17*, 4157. [CrossRef]
23. Cohen, M.; Godron, M.; Cretin-Pablo, R.; Pujos, R. Plant biodiversity in Mediterranean orchards is related to historical land use: Perspectives for biodiversity-friendly olive production. *Reg. Environ. Chang.* **2023**, *23*, 70. [CrossRef]
24. Antrop, M. Why landscapes of the past are important for the future. *Landsc. Urban Plan.* **2005**, *70*, 21–34. [CrossRef]
25. Lasanta, T.; Arnáez, J.; Pascual, N.; Ruiz-Flaño, P.; Errea, M.P.; Lana-Renault, N. Space–time process and drivers of land abandonment in Europe. *Catena* **2017**, *149*, 810–823. [CrossRef]

26. Luginbuhl, Y.; Howard, P.; Terrasson, D. (Eds.) *Landscape and Sustainable Development, The French Perspective*; Routledge: London, UK, 2016.
27. Duby, G. *Histoire de la France Rurale*; Tomes 1-2-3-4; Editions du Seuil: Paris, France, 1992.
28. Diarte-Blasco, P.; Castro-Priego, M. Is Historical Cartography a Useful Tool for Landscape Analysis? A Perspective from Inland Spain (Zorita de los Canes, Guadalajara) from the Middle Ages to the Present. *Land* **2023**, *12*, 1627. [CrossRef]
29. Diaz-Varela, R.A.; Zarco-Tejada, P.J.; Angileri, V.; Loudjani, P. Automatic identification of agricultural terraces through object-oriented analysis of very high resolution DSMs and multispectral imagery obtained from an unmanned aerial vehicle. *J. Environ. Manag.* **2014**, *134*, 117–126. [CrossRef] [PubMed]
30. Vallauri, D.; Grel, A.; Granier, E.; Dupouey, J.L. *Les forêts de Cassini. Analyse Quantitative et Comparaison Avec les Forêts Actuelles*; [Technical report]; Rapport WWF/INRA, Marseille; 2012; Available online: https://hal.science/hal-01267936 (accessed on 1 April 2023).
31. Herrault, P.A.; Sheeren, D.; Fauvel, M.; Monteil, C.; Paegelow, M. A comparative study of geometric transformation models for the historical "Map of France" registration. *Geogr. Tech.* **2013**, *1*, 34–46.
32. Cohen, M.; Alexandre, F.; Mathieu, N. Modelling interactions between biophysical and social systems: The example of bush progression in Causse Méjan. *CyberGeo* **1998**, *39*, 1551. [CrossRef]
33. Font, P.G. Sobre la datación y límites de las terrazas de cultivo. Algunas observaciones de campo. In *Paisaje, Cultura Territorial y Vivencia de la Geografía: Libro Homenaje al Profesor Alfredo Morales Gil*; Instituto Interuniversitario de Geografía: Alicante Universidad, Spain, 2016; pp. 111–124.
34. Lieskovský, J.; Lieskovský, T.; Hladíková, K.; Štefunková, D.; Hurajtová, N. Potential of airborne LiDAR data in detecting cultural landscape features in Slovakia. *Landsc. Res.* **2022**, *47*, 539–554. [CrossRef]
35. Inomata, T.; Triadan, D.; Pinzón, F.; Burham, M.; Ranchos, J.L.; Aoyama, K.; Haraguchi, T. Archaeological application of airborne LiDAR to examine social changes in the Ceibal region of the Maya lowlands. *PLoS ONE* **2018**, *13*, e0191619. [CrossRef] [PubMed]
36. Matyukira, C.; Mhangara, P. Advancement in the Application of Geospatial Technology in Archaeology and Cultural Heritage in South Africa: A Scientometric Review. *Remote Sens.* **2023**, *15*, 4781. [CrossRef]
37. Trier, Ø.D.; Reksten, J.H.; Løseth, K. Automated mapping of cultural heritage in Norway from airborne lidar data using faster R-CNN. *Int. J. Appl. Earth Obs. Geoinf.* **2021**, *95*, 102241. [CrossRef]
38. Castex, J.M. L'aménagement des pentes et des sols dans les Alpes-Maritimes et le Var. *Méditerranée* **1983**, *47*, 3–15. [CrossRef]
39. Benjeddou, M.; Dagorne, A.; Castex, J.-M.; Davtian, G. Maps and comments concerning some terraced sites of the Alpes-Maritimes (06-F): Granile (Tende), Roquebilière, Auribeau. In *Terraced Landscapes of the Alpine Arc, Atlas/Interpretations, Project Alpter*; Marsilio Editore: Milano, Italy, 2008; pp. 76–79.
40. Sumera, F. Signature des Occupations Protohistoriques et Antiques dans L'évolution des Paysages et Dans la Construction de la Géographie Humaine du Massif du Mercantour (Alpes-Maritimes). Ph.D. Thesis, University Aix-Marseille, Marseille, France, 2015.
41. Bourrier, M.; Bourrier-Reynaud, C. Le Comté de Nice: Enjeu et Théâtre de Cinq Siècles de Guerres Pour Devenir Français (1388–1860). In *Fortifications et Frontières Entre Guerre et Paix. Actes du 136e Congrès National des Sociétés Historiques et Scientifiques, «Faire la Guerre, Faire la Paix», Perpignan, 2011*; Editions du CTHS: Paris, France, 2015; pp. 21–33. Available online: https://www.persee.fr/doc/acths_1764-7355_2015_act_136_12_2749#:~:text=Le%2024%20mars%201860,%20en,la%20Savoie%20contre%20la%20Lombardie (accessed on 1 April 2023).
42. Costamagna, H. Les guerres et leurs conséquences dans le Comté de Nice au XVIIIe siècle. *Cah. Méditerranée* **1973**, *6*, 77–93. Available online: https://www.persee.fr/doc/camed_0395-9317_1973_num_6_1_1377 (accessed on 1 April 2023). [CrossRef]
43. Guiot, L. *Les Forêts et les Pâturages du Comte de Nice, Mémoire de la Société Générale d'Agriculture de France*; Imprimerie Bouchard-Huzard: Paris, France, 1875. Available online: https://gallica.bnf.fr/ark:/12148/bpt6k6540493n.texteImage (accessed on 1 April 2023).
44. Casalis, G. Dizionario Geografico Storico Statistico Commerciale Compilato per Cura del Professore e Dottore di Belle Lettere, Cavaliere Dell'ordine de SS Maurizio e Lazzaro Opera Molto Utile Agli Impiegati nei Pubblici e Private Uffizi a Tutte le Persone Applicate al Foro Alla Milizia al Commercio e Singolarmente Agli Amatori Delle Cose Patrie. 1855. Available online: http://www.Archeo-alpi-maritimi.com/ethno_hautes_vallees.php (accessed on 1 April 2023).
45. Casimir, J. Statistique Agricole de la France. Monographie Agricole du Département des Alpes Maritimes, Vol. MCMXXXVII, 1929. Available online: https://gallica.bnf.fr/ (accessed on 1 April 2023).
46. Pichard, G.; Provansal, M.; Sabatier, F. Les embouchures du Rhône. L'apport de la cartographie à l'étude de leur évolution géomorphologique au cours du Petit âge glaciaire (PAG). *Méditerranée* **2014**, *122*, 43–59. [CrossRef]
47. Fouache, E.; Desruelles, S.; Gorini, C.; Bianchi, N.; Marchiel, A. Evaluation of the return time of flash floods generated in the Roya Valley (Alpes Maritime, France) on October 2, 2020: What challenges does this type of event pose for risk management? In Proceedings of the Copernicus Meetings No. ICG2022-7, Coimbra, Portugal, 12–16 September 2022. [CrossRef]
48. Kerverdo, R.; Marchiel, A.; Gorini, C.; Lafuerza, S.; Fouache, E. Extreme debris flows events recorded on Roya river tributaries: What challenges does this type of event pose for risk management? In Proceedings of the 6th World Landslides Forum, Florence, Italy, 14–17 November 2023.

49. Carrega, P.; Michelot, N. Une catastrophe hors norme d'origine météorologique le 2 octobre 2020 dans les montagnes des Alpes-Maritimes. An over-standing disaster of meteorological origin on October 2, 2020, in the mountains of the Alpes-Maritimes. *Physiogéo* **2021**, *16*, 1–70. [CrossRef]
50. Piton, G.; Cohen, M.; Flipo, M.; Nowak, M.; Chapuis, M.; Melun, G.; Robert, Y.; Andréis, N.; Liebault, F. Large in-stream wood yield during an extreme flood (Storm Alex, October 2020, Roya Valley, France): Estimating the supply, transport, and deposition using GIS. *Geomorphology* **2023**, *446*, 108981. [CrossRef]
51. Balletti, C. Georeference in the analysis of the geometric content of early maps. *e-Perimetron* **2006**, *1*, 32–39. Available online: https://www.e-perimetron.org/Vol_1_1/Balletti/Balletti.pdf (accessed on 1 April 2023).
52. Vignal, M. L'impact du Changement Global sur la Flore du Sud-Est de la France: Modélisation Multiscalaire de la Répartition de 25 Espèces à L'horizon 2100 par la Dynamique de Population. Ph.D. Thesis, University of Nice, Nice, France, 2020. Available online: https://theses.hal.science/tel-03105744 (accessed on 1 April 2023).
53. Andrieu, J.; Lombard, F.; Vignal, M.; Godron, M.; Alexandre, F. A phyto-climatic transect in the Alpes Maritimes used to characterize the northern limit of the Mediterranean biogeographical area. *Biodivers. J.* **2020**, *11*, 679–688. [CrossRef]
54. Wagner, E. Mémoire sur la Reconnaissance des Hauteurs Entre la Vésubie et la Roya Occupées par les Armées Française et Austro-Sarde en 1742, 1793,1794, Collection du Ministere de la Défense, SHD, Département de L'armée de Terre, 1 VD 34art 4 sect 1 parag 5, 4° C1 N° 54. 1865. Available online: http://www.archeo-alpi-maritimi.com/ethno_hautes_vallees.php (accessed on 1 April 2023).
55. SHD. Routes d'Antibes à Coni et de Coni à Asti, et la Traversée d'Antibes à Barcelonnette. 1 VD 34 art 4 sect 1 parag 5, 1736, 4° C1, N°13, Collection du Ministère de la Défense, SHD, Département de L'armée de Terre. 1736. Available online: http://www.archeo-alpi-maritimi.com (accessed on 1 April 2023).
56. Costamagna, H. Le Comté de Nice Au XVIIIe Siècle. Introduction à la Thèse De 3e Cycle Recherches sur les Institutions Communales Dans le Comté de Nice au XVIIIe Siècle (1699–1792), Université de Nice. 1971. Available online: https://www.departement06.fr/documents/Import/decouvrir-les-am/rr46-1973-n2-02.pdf (accessed on 1 April 2023).
57. Parcs Naturels régionaux de France. Terrasses agricoles. Gestion des espaces naturels, agricoles et forestiers. In *Collection Expérimenter Pour Agir*; Fédération des parcs régionaux: Paris, France, 2000.
58. Rebours, F. Versants aménages et déprise rurale dans l'est des Alpes-Maritimes. *Méditerranée* **1990**, *71*, 31–36. Available online: https://www.persee.fr/doc/medit_0025-8296_1990_num_71_3_2680 (accessed on 1 April 2023). [CrossRef]
59. Gennai-Schott, S.; Sabbatini, T.; Rizzo, D.; Marraccini, E. Who remains when professional farmers give up? Some insights on hobby farming in an olive groves-oriented terraced Mediterranean area. *Land* **2020**, *9*, 168. [CrossRef]
60. Desbordes, E. Les Paysages de Terrasses Agricoles Dans le Bassin Méditerranéen Septentrional. Etude Comparée des Alpes Maritimes et de la Ligurie. Ph.D. Thesis, Université de Limoges, Limoges, France, 1999. Available online: https://theses.fr/1999LIMO2002 (accessed on 1 April 2023).
61. Kinnaird, T.; Bolos, J.; Turner, A.; Turner, S. Optically-stimulated luminescence profiling and dating of historic agricultural terraces in Catalonia (Spain). *J. Archaeol. Sci.* **2017**, *78*, 66–77. [CrossRef]

land

MDPI

Article

Aerial Remote Sensing Archaeology—A Short Review and Applications

Dimitris Kaimaris

School of Spatial Planning and Development, Aristotle University of Thessaloniki, 54124 Thessaloniki, Greece; kaimaris@auth.gr

Abstract: Aerial and remote sensing archaeology are tools for identifying marks on images of archaeological remains covered by soil. In other words, they are archaeological prospection tools that fall into the category of non-destructive research methods. In this paper, a short review of these valuable research tools is carried out, presenting the way marks appear, and also the categories of aerial and remote sensing archaeology, depending on the medium of storage of images and the type of platform of transfer of sensors. The timing of the emergence of each category is determined and examples of relevant surveys are presented. Focusing on the use of an Unmanned Aircraft System (UAS) with a multispectral sensor (aerial remote sensing archaeology), their partial utilization is revealed for collecting images in areas outside of the visible spectrum, aiming at the identification of covered archaeological remains. To this end, examples of the use of UAS with different sensors are presented and indexes that have been used so far in respective applications are gathered. Aerial remote sensing archaeology took place in two areas of particular interest in Northern Greece. UAS WingtraOne GEN II was used to collect multispectral images. In both study areas, soil and vegetation or crop marks were detected, possibly covered archaeological remains, initially in RGB orthophotomosaics and then more clearly in the index maps, such as the normalized difference vegetation index, simple ratio, brightness index, second brightness index, and anthocyanin reflectance index 2b.

Keywords: aerial remote sensing; UAS; multispectral sensor; index; marks; archaeology; covered remains

1. Introduction

1.1. Aerial and Remote Sensing Archaeology

The discovery and documentation of unknown covered archaeological remains allows the protection and preservation of our cultural heritage, which is threatened by uncontrolled agricultural exploitation and increasing urban development.

Towards this direction, aerial and remote sensing archaeology aims at the measurement and thematic documentation of existing above-ground archaeological remains, as well as the identification of unknown possible archaeological remains under soil or vegetation/crop. In other words, they are archaeological prospection tools that fall into the category of non-destructive research methods. This ability to locate new sites of archaeological interest is due to the effect that the structures have on the soils that cover them. In particular, in these soils, there is a lack of moisture (presence of a solid structure under the ground, e.g., building foundations) or excess moisture (presence of open construction under the ground, e.g., defensive ditch) in relation to the adjacent soils (i.e., below soils where there are no structures). As a result, there are differences in height, quality, and temperature of the vegetation or crop above the covered structures in relation to the adjacent vegetation or crop, as well as differences in surface temperature of the bare ground above the covered structures in relation to the adjacent bare ground. These differences are depicted as marks in images [1–3] (Figure 1).

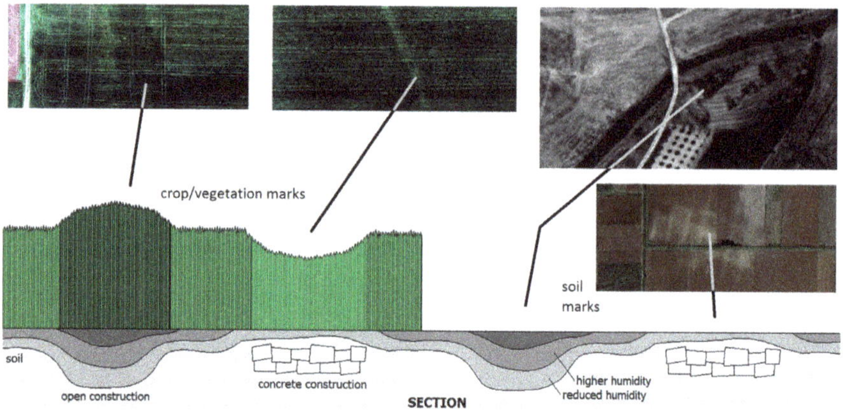

Figure 1. Anatomy of the marks' appearance. Positive crop/vegetation or soil marks (dark) and negative crop/vegetation or soil marks (light). Possible presence of open and compact structure, respectively, under the soil.

Aerial and remote sensing archaeology can be categorized into aerial photography archaeology (e.g., use of film on a manned aircraft [4], use of digital single lens reflex camera on a manned aircraft [5], or on an Unmanned Aircraft System (UAS) [6]), aerial remote sensing archaeology (e.g., use of a Multispectral (MS) sensor on a manned aircraft [7] or on a UAS [8]), space photography archaeology (e.g., use of film on a spaceborne platform [9]), and space remote sensing archaeology (e.g., use of an MS sensor on a spaceborne platform [10]). In reviewing the timing of their appearance, it should be mentioned that the founder of aerial archaeology is O.G.S. Crawford (archaeologist, 1886–1957), who made the first mark identifications in aerial photographs in the early 1920s. He concluded that human interventions of the past in the landscape create surface anomalies that are visible only when viewed from above [11]. While new archaeological discoveries were initiated from that time onwards, 46 years after the first marks were observed in aerial photographs (aerial photography archaeology), linear marks of prehistoric human intervention in the Phoenix, Arizona, area were identified (in 1969) in infrared photographs from the S065 multiband experiment flown on Apollo 11 (space photography archaeology) [12]. Prehistoric sites and roads in Chaco Canyon (New Mexico) were identified 8 years later (in 1977) in the MS images of the Landsat-1 satellite (space remote sensing archaeology). Landsat-1 (1972–1978) was the first satellite system for systematic observation of the Earth's surface and its objects and was capable of generating MS data in digital form [13]. In the 1980s, Airborne Imaging Spectrometer (AIS), Airborne Visible Infrared Imaging Spectrometer (AVIRIS), Airborne Thematic Mapper (ATM), and Compact Airborne Spectrographic Imager (CASI) sensors were created, the digital images of which were exploited in the following years in a variety of archaeological applications [14,15] (aerial remote sensing archaeology).

1.2. UAS and Remote Sensing in Locating Covered Archaeological Remains

To date, there have been countless discoveries of covered archaeological remains with the help of aerial and remote sensing archaeology, utilizing panchromatic (PAN) [2,7,10,16–37], RGB [5,29,31,38–43], multispectral (MS) [7,10,14–19,22–28,32–34,37,39,40,44,45], thermal [2,7,46–49], lidar [22,37,50–57], and radar [58–63] images. Sensor dependency platforms can be manned aircrafts [2,5,7,14,15,20–22,29,31,32,37–57] and satellites [10,16–19,23–28,30–37,58–64]. Unmanned Aircraft Systems (UAS) are widely used in archaeology [50,65–74], but little use is made of them in collecting images in areas outside of the visible spectrum, aiming to locate covered archaeological remains [75,76]. Among these (minimal) surveys [4,8,31,52,76–96], González and Hernández [91] identified marks of a covered ancient Roman road (Via XXXIV on the Antonine Itinerary) that crossed the province of Álava

(Northern Spain), but also of side-covered monuments, utilizing UAS with an MS sensor. NDVI (normalized difference vegetation index) orthomosaics allowed the observation of crop marks. At Ravenshall, Fife (Scotland), Moriarty et al. [85] identified marks of covered archaeological remains using RGB and MS sensors in UAS. The analysis revealed that several vegetation index maps, particularly the simple ratio (SR) and the NDVI, proved instrumental in identifying marks in the multispectral results. Fiz et al. [86] used UAS to identify marks at Vilanera (L'Escala, Catalonia, Spain) in RGB and MS images and produced vegetation index maps. Brooke and Clutterbuck [80] used RGB, MS, and thermal images from UAS to locate and interpret marks at two archaeological sites in Nottinghamshire, UK (Camp Hill, Hexgreave Park, and Shelford Priory). At Camp Hill, thermal imagery, in particular, yielded useful results, revealing many features that appear to be anthropogenic and which relate to the site's former use as a defensive hillfort and enclosure. In addition, the mark of a later Roman road is very clearly delineated on thermal images. In contrast, at Shelford Priory, mainly MS images coupled with Random Forest classification gave useful results, revealing apparent buried archaeology to varying degrees. Ronchi et al. [84] conducted systematic surveys of the archaeological remains of the imperial Domitian villa in Sabaudia (Italy), which is covered by dense pine forest. MS and lidar sensors mounted on UAS were used. Compared to vegetation indexes (from MS images), lidar's product (Digital Terrain Model (DTM)) excels in this survey, as it reveals in the best way the forest-covered remains of buildings, paths, and walls. The study by Casana et al. [95] utilized lidar in UAS at three distinct archaeological sites: tropical forests at Kealakekua Bay State Historic Park (Hawai'i), piñon–juniper forests on Mesa Verde's North Escarpment (Colorado), and mixed deciduous–evergreen forests at Enfield Shaker Village (New Hampshire). The results demonstrated the existence of a multitude of archaeological features at the three study sites, thereby illustrating the potential of UAS-based lidar as a tool in archaeological prospection. The Greater Khingan Range (Northeastern China) is the birthplace of numerous ancient nomadic peoples, including the Xianbei, Khitan, and Mongols. However, the region has only yielded a limited number of archaeological discoveries due to the dense forest cover. In a pioneering study, Li [96] employed UAS-based lidar for the first time to detect and map ancient sites concealed beneath the forest canopy in Shenshan Mountain (southeast of the Greater Khingan Range). The UAS missions achieved remarkable clarity in revealing the extent and complexity of the archaeological structures under the forest, which may date back to the Liao period (907–1125). In a publication, Cao et al. [90] introduced a new UAS-based magnetic system that proved effective in an archaeological area with a complex topography in Huizhou, China. The results demonstrated a high degree of correlation between the target anomaly marked by the UAS aeromagnetic survey and the location of a possible archaeological site. This study demonstrates that the UAS aeromagnetic system may have potential for future archaeological surveys, as it is in the early stages of its development and there are no other relevant bibliographic references.

1.3. Indexes from UAS Sensors for Locating Covered Archaeological Remains

Vegetation/crop and soil indexes number in the hundreds [97–99]. They were developed by (and for) the use of bands of satellite system images. Due to the limited availability of bands from sensors used in UAS, many of these indexes cannot be exploited. Table 1 presents indexes used in applications for detecting covered structures with UAS and remote sensing sensors (aerial remote sensing archaeology). Each vegetation/crop index is designed to analyze a specific parameter of the plant, e.g., vigor (which closely relates to soil conditions and water content), biomass, chlorophyll (which is associated with photosynthesis), anthocyanin content in leaves (which are pigments that shield leaves from excessive light), etc. There are fewer soil indexes and they are correlated with the humidity, the presence of carbonates or sulfates in the soils, etc. (obviously, reference is made to these characteristics of vegetation/crop and soil, as the appearance of marks is affected by the size of the covered structure, its depth, the type of soil, the water content, and the type of vegetation/crop).

Table 1. List of indexes.

Index	Abbreviation	Formula	Comment
NDVI	Normalized Difference Vegetation Index	$\frac{NIR-Red}{NIR+Red}$	Normalized ratio ranging for the measurement of green vegetation [100,101].
GNDVI	Green Normalized Difference Vegetation Index	$\frac{NIR-Green}{NIR+Green}$	Modification to NDVI, increased sensitivity to chlorophyll concentration [84,100,102].
RENDVI or NDRE	Red-Edge NDVI or Normalized Difference Red-Edge Index	$\frac{NIR-Red\ Edge}{NIR+Red\ Edge}$	Monitoring chlorophyll content is possible through the sensitivity and efficacy of the index [78,84,100,103].
NDVIRE or NDRER	Red-Edge Normalized Difference Vegetation Index or Normalized Difference Red-Edge/Red	$\frac{Red\ Edge-Red}{Red\ Edge+Red}$	For modification to NDVI, instead of Near Infrared (NIR) reflectance, red-edge information was used. Measurement of the amount of chlorophyll in the plants [78,100,104].
SAVI	Soil-Adjusted Vegetation Index	$1.5*\frac{NIR-Red}{NIR+Red+0.5}$	SAVI and NDVI are almost alike in their formulation, with the only difference being that an adjustment factor is used by SAVI to minimize the effects of soil on the spectral signal [105].
GSAVI	Green Soil-Adjusted Vegetation Index	$1.5*\frac{NIR-Green}{NIR+Green+0.5}$	The green band instead of the red is used in the calculation of GSAVI, which, at rest, follows the SAVI formula structure [106].
OSAVI	Optimized Soil-Adjusted Vegetation Index	$1.5*\frac{NIR-Red}{NIR+Red+0.16}$	It uses a standard value of 0.16 for the canopy background adjustment factor. This value provides greater soil variation than SAVI for low vegetation cover, while demonstrating increased sensitivity to vegetation cover greater than 50% [78].
SR	Simple Ratio	$\frac{NIR}{Red}$	Simple vegetation distinction is conducted by the ratio of NIR scattering to chlorophyll absorption [85].
MSR	Modified Simple Ratio	$\frac{(\frac{NIR}{Red})-1}{\sqrt{(\frac{NIR}{Red})-1}}$	Sensitivity to vegetation characteristics is improved by the combination of renormalized NDVI and SR [85].
GCI	Green Chlorophyll Index	$(\frac{NIR}{Green})-1$	The content of leaf chlorophyll in various species of crops is estimated through the index [106].
GRVI	Green Ratio Vegetation Index	$\frac{NIR}{Green}$	Simple ratio modification, sensitive to rates of photosynthesis [85].
NGRDI	Normalized Green Red Difference Index	$\frac{Green-Red}{Green+Red}$	Biomass showed a positive correlation to the index [100,107,108].
NLI	Non-Linear Vegetation Index	$\frac{NIR*NIR-Red}{NIR*NIR+Red}$	Linear relationships with vegetation parameters were emphasized by the modification to NDVI [85].
GEMI	Global Environment Monitoring Index	$\eta*(1-0.25*\eta)-\frac{Red-0.125}{1-Red}$ where $\eta=\frac{2*(NIR^2-Red^2)+1.5*NIR+0.5*Red}{NIR+Red+0.5}$	Alters the light seen by the instrument and can cause variations in the calculated values of vegetation indices [76].
TVI	Triangular Vegetation Index	$0.5*(120*(NIR-Green)-200*(Red-Green))$	Total area of the triangle (with vertices at green, red, and NIR wavelengths) increases with higher chlorophyll absorption [76].
MTVI	Modified Triangular Vegetation Index	$1.2*(1.2*(NIR-Green)-2.5*(Red-Green))$	Lower values indicate higher chlorophyll content [76].
TCARI	Transformed Chlorophyll Absorption Ratio Index	$3*((Red\ Edge-Red)-0.2*(Red\ Edge-Green)*(Red\ Edge-Red))$	Indicates the relative abundance of chlorophyll. It is affected by the underlying soil reflectance [76].
MCARI	Modified Chlorophyll Absorption in Reflectance Index	$((Red\ Edge-Red)-0.2*(Red\ Edge-Green))*\frac{Red\ Edge}{Red}$	Gives a measure of the depth of chlorophyll absorption and is very sensitive to variations in chlorophyll concentrations [76].

Table 1. *Cont.*

Index	Abbreviation	Formula	Comment
BI	Brightness Index	$\sqrt{\frac{Red^2+Green^2}{2}}$	Uses the albedo of the terrestrial surface to differentiate between the vegetal covers and the ground [78].
BI2	Second Brightness Index	$\sqrt{\frac{Red^2+Green^2+NIR^2}{3}}$	Sensitive to the brightness of soils, which is highly correlated with the humidity and the presence of salts in the surface [78].
CI	Color Index	$\frac{Red-Green}{Red+Green}$	Related to the concentration of carbonates or sulfates in the soils [78].
PSRI	Plant Senescence Reflectance Index	$\frac{Red-Green}{NIR}$	Vegetation health monitoring, plant physiological stress detection, crop production, and yield analysis. The value of this index ranges from −1 to 1, but the common range for green vegetation is −0.1 to 0.2 [78].
RI	Redness Index	$\frac{Red^2}{Green^2}$	Identifies soil color variations [78].
ARI1	Anthocyanin Reflectance Index 1	$\frac{1}{Green} - \frac{1}{Red\ Edge}$	Changes in green absorption relative to red indicate leaf anthocyanins [78].
ARI1B	Anthocyanin Reflectance Index 1B	$\frac{1}{Green} - \frac{1}{Red}$	Uses the red band in the original formula [78].
ARI2	Anthocyanin Reflectance Index 2	$NIR * \left(\frac{1}{Green} - \frac{1}{Red\ Edge}\right)$	A variant of the ARI1, which is sensitive to changes in green absorption relative to red, indicating leaf anthocyanins [78].
ARI2B	Anthocyanin Reflectance Index 2B	$NIR * \left(\frac{1}{Green} - \frac{1}{Red}\right)$	Uses the red band in the original formula [78].
LCI	Leaf Chlorophyll Index	$\frac{NIR-Red\ Edge}{NIR+Red}$	Chlorophyll content in areas with full leaf coverage [84].
RGRI	Red Green Ratio Index	$\frac{(Red+Red\ Edge)/2}{Green}$	Used to indicate leaf production and stress [85].

2. Study Areas to Identify Marks of Possible Covered Archaeological Remains

2.1. Archaeological Site of Europos, Northern Greece

The archaeological site of Europos (40°53′15.5″ N 22°33′20.0″ E, Figures 2 and 3) is located approximately 55 km northwest of Thessaloniki (Greece).

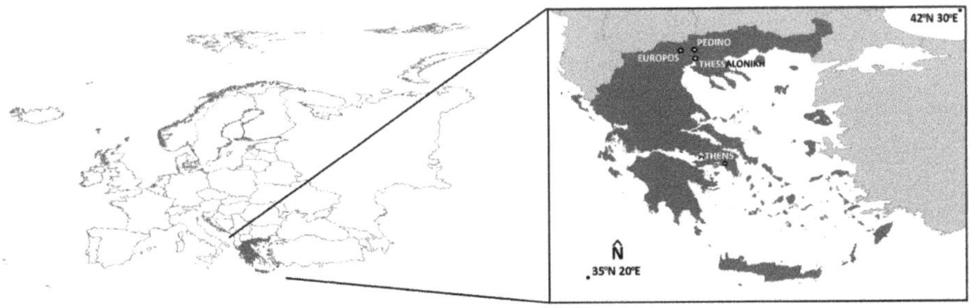

Figure 2. Greece in the European continent and Europos and Pedino locations in Greece.

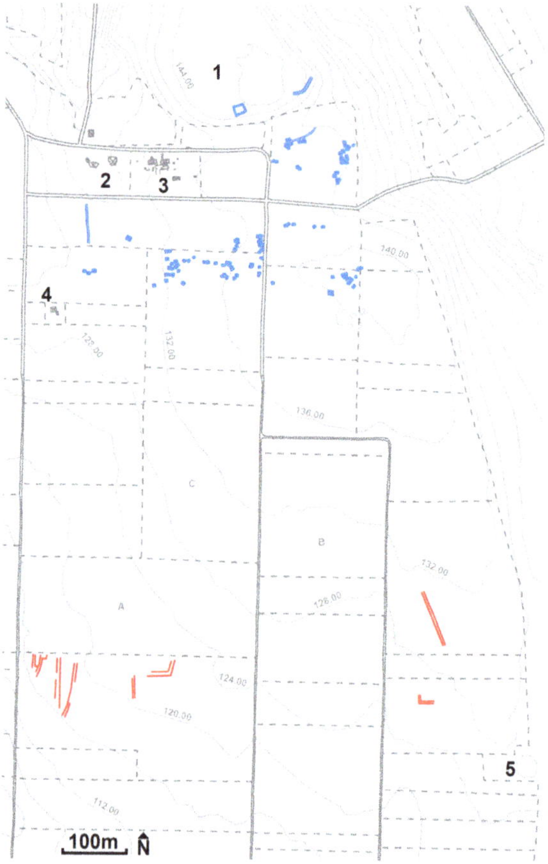

Figure 3. Location 1, the archaic acropolis of Europos. Location 2, a space for visitors to access information about the archaeological site. Location 3, an open-air archaeological area of a cemetery from the Roman and early Christian times, that is either sheltered or not (marked by black solid squares representing existing tombs that are open or closed). Location 4, three tombs from ancient times that are sheltered (Macedonian tomb, cist, and burial vault). Location 5, a burial mound dating to the late-4th and early-3rd century BC. The boundaries of the parcels of land are delineated by a dotted line, while the contour lines are shown in grey with a dimension of 4 m. Only the road west of the acropolis to Location 2 is paved; the rest are rural dirt roads. In blue, the rendering of the marks of previous research, and in red, the rendering of new marks. Center of figure: 40°52′56.30″ N 22°33′23.51″ E.

The archaeological findings of the wider area are of the Classical and Hellenistic era. Seleucus Nicator came from Europos and was the founder of important cities of the Hellenistic period and one of the most important generals of Alexander the Great. During the early Roman Empire, Europos continued to thrive, while the prosperity of the acropolis (Figure 3, Location 1) continued in the 1st to the 3rd century AD. In the 4th century AD, it gradually declined, while in the 5th to 6th century AD, the acropolis was abandoned [109]. Geophysical surveys allowed the mapping of the subsoil of the acropolis [110]. Targeted exploratory excavations uncovered portions of the urban infrastructure of the acropolis, including funerary monuments, among other findings [111]. The area of the Roman and early Christian necropolis (Figure 3, Location 3) situated to the southwest of the acropolis has been transformed into an accessible tourist destination and is situated within the boundaries of the designated archaeological site. Additionally, a three-tomb cluster

(Figure 3, Location 4), comprising a Macedonian tomb, a cist, and a burial vault, has been preserved in the area southwest of the acropolis [109]. In the area south bounded by this cluster of tombs and north by the organized archaeological site, aerial remote sensing archaeology research in 2022 [1] using an MS sensor in UAS led to the identification, among others, of a large number of marks of possible covered ancient tombs (Figure 3, rendering of marks in blue). The search for new marks continued south of the above area until the burial mound (late-4th, early-3rd century BC, Figure 3, Location 5), and its results are presented in this paper (Figure 3, rendering of marks in red).

2.2. *Archaeological Site of Pedino Hill, Northern Greece*

On Pedino Hill (40°55′1.97″ N 22°53′17.33″ E, Figures 2 and 4), approximately 40 km north of the city of Thessaloniki (Greece), a walled ancient city of the (at least) early-4th century BC up to Roman times (Figure 4, Location 1) was identified, initially by surface survey and then by systematic archaeological excavation. The site has been protected by archaeological law since the 1980s. The name of the city is unknown; it had a dense urban fabric with a wall at least 4 m high and 5.5 m wide. The area of the ancient city within the walls is approximately 22,000 m^2. Today, the south monumental central gate of the wall and part, approximately 60 m, of the fortification of the ancient city have been discovered (Figure 4, Location 2).

On the hill and in the area south of it, an aerial remote sensing archaeology survey was conducted using an MS sensor in a UAS, which led to the detection of marks of possible covered ancient remains (Figure 4).

Figure 4. Location 1, the ancient walled city of Pedino Hill. Location 2, with a gray line marking the positions of the excavations of the southern gate and part of the wall. The boundaries of each parcel are delineated by a dashed line and the contour lines are depicted in gray with a 4 m contour. The rural dirt roads are drawn with a continuous gray line. In blue, the drawing of marks outside the hill; in red, the drawing of marks on the hill. Center of figure: 40°54′47.42″ N 22°53′15.86″ E.

3. Equipment

In the above surveys, UAS WingtraOne GEN II was used (Figure 5).

Figure 5. WingtraOne GEN II at the study area. In the background, the hill of Pedino.

The UAS in question is a fixed-wing vertical take-off and landing (VTOL). In order to calculate the center coordinates of the images received, it utilizes an integrated multi-frequency PPK GNSS antenna (GPS: L1, L2; GLONASS: L1, L2; Galileo: L1; BeiDou: L1). All flight parameters and the plan are modelled via the WingtraPilot© (version 2.12.2) software. It features two propellers and an MS sensor (Table 2) [112].

Table 2. Technical specifications of the multispectral sensor [112].

Camera	Technical Specifications
MicaSense RedEdge-MX	Multispectral sensor Focal length: 5.5 mm 1.2 Mpixel (resolution 1280 × 960) Weight: 231.9 g (includes DLS 2 and cables) Five spectral cameras: Blue (465–485 nm); Green (550–570 nm); Red (662–673 nm); Red Edge (712–722 nm); Near Infrared (NIR) (820–860 nm) Ground Sample Distance: 8.2 cm/pixel at 120 m Field of View (FOV): 47.2° Horizontal FOV; 36.2° Vertical FOV

Additionally, the GNSS Topcon HiPer SR was employed (yielding horizontal and vertical positioning accuracy of approximately 10 mm and 15 mm, respectively; GPS: L1, L2, L2C; GLONASS: L1, L2, 2C; SBAS-QZAA: L1, L2C) for the acquisition of the requisite data that enabled the calculation of the coordinates of the acquisition centers pertaining to each image in the Greek Geodetic Reference System 87 (GGRS87).

4. Methods, Processing, and Products

4.1. Flight Plans and Image Collection

The two flights with the MS sensor south of the archaeological site of Europos took place on 16 April 2024 from 11:30 a.m. to 13:00 p.m. The wind speed was 2.1 m/s (2 Beaufort) and the air temperature was 20 °C. On the hill and south of the archaeological site of Pedino, the two flights with the MS sensor took place on 14 May 2024 from 12:10 p.m. to 13:45 p.m. The wind speed was 6.3 m/s (4 Beaufort) and the air temperature was 24 °C. It should first be noted that all flights were created with the use of 70% side and front overlap images (Figures 6–9). A total of 35 strips were implemented for Europos, with a 100 m flight height (the minimum permitted by the UAS software, WingtraPilot© version 2.12.2, for the MS sensor) on both flights. The anticipated (and ultimately achieved) spatial resolution of the 6.8 cm images, the complete flight duration of 41 min and 14 s, and the total number of 1362 MS images were successfully collected. For Pedino, a total of 58 strips were implemented, with a flight height of 100 m in both instances, while the projected (and ultimately achieved) spatial resolution of the 6.8 cm images, the total flight time of 51 min and 3 s, and a total of 1580 MS images were obtained.

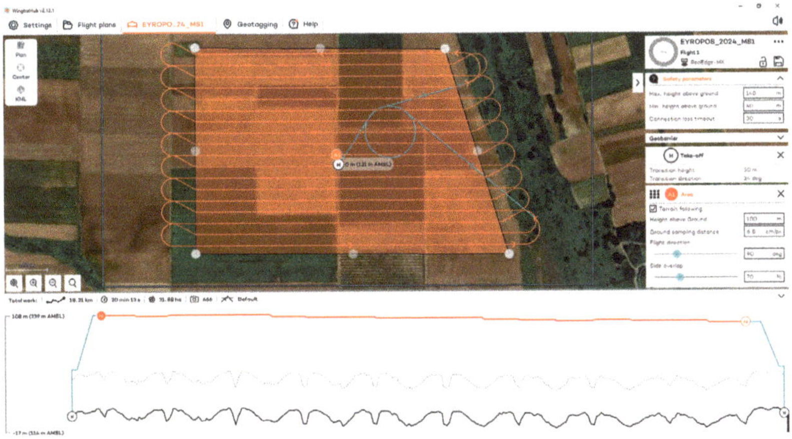

Figure 6. The first flight plan with the MS sensor south of the archaeological site of Europos.

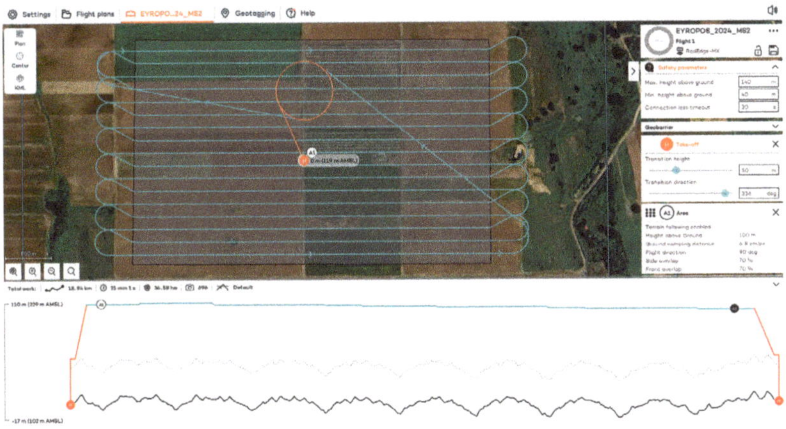

Figure 7. The second flight plan with the MS sensor, south of the flight plan of Figure 6.

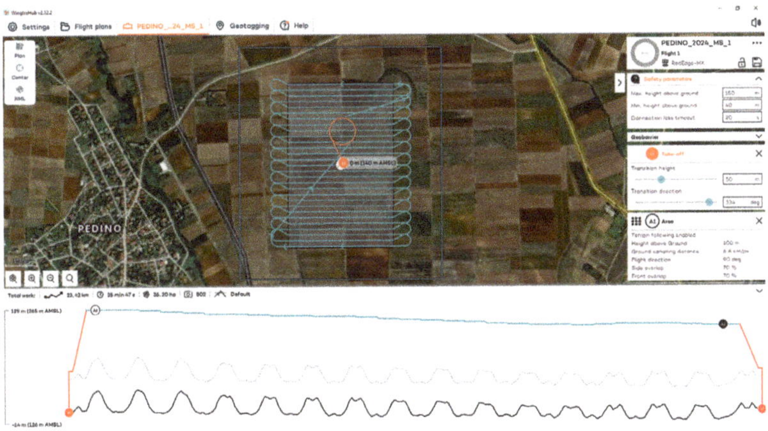

Figure 8. The first flight plan with the MS sensor on the hill and southwest of the hill at the archaeological site of Pedino.

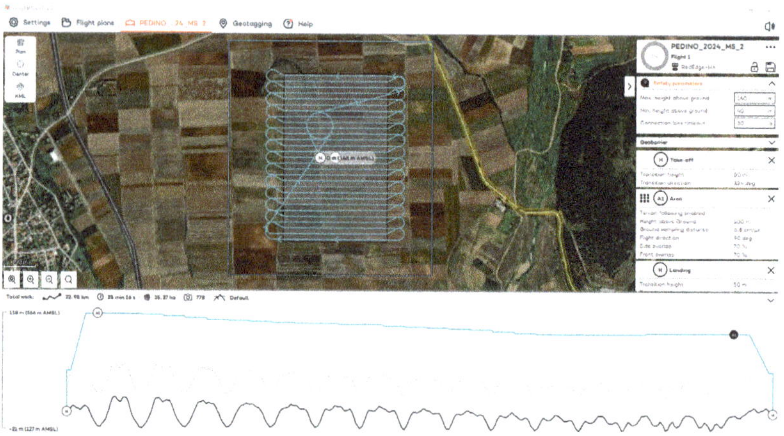

Figure 9. The second flight plan with the MS sensor, east of the flight plan of Figure 8.

4.2. Collection of Terrestrial Data, Image Processing, and Production of Digital Surface Models and Orthophotomosaics

The procedure that follows took place twice: once for the archaeological site of Europos and once for the archaeological site of Pedino Hill. The GNSS Topcon HiPer SR was initially utilized to identify the X, Y, and Z coordinates of a randomly selected point (which served as a reference point for subsequent measurements) at a relatively short distance from the home position of the UAS. The GNSS was linked to the national network of multiple fixed stations available to Topcon, and the position (X, Y, and Z) of the point was determined with an accuracy of a few millimeters using the real-time kinematic (RTK) method. Subsequently, continuous measurements of the position were made using the Static GPS surveying method with the same GNSS for the same point over a period of 30 min before the commencement of the first flight, during the two flights of each study area, and for 30 min after the completion of the second flight. In the office, the coordinates of the reception centers of each MS image were calculated using the high accuracy coordinates of the above point, its static measurements, and the PPK GNSS measurements of the UAS. This process yielded a 3D accuracy of the centers of the images ranging from 2 to 3 cm.

Subsequently, the calibration of spectral information with spectral targets is necessary using the Agisoft Metashape© (version 2.0.0) software. Prior to and following the flight, the suitable calibration target of the Micasense RedEdge-MX was mapped [100,113–116]. The target was automatically detected by the Agisoft Metashape© (version 2.0.0) software and the reflectance values of all spectral bands were calculated. This was followed by the resolution of the block of all MS images (without Ground Control Points, as the coordinates of the reception centers of each MS image have been calculated), and for the archaeological site of Europos, the RMSE was 1.4 cm, the produced digital surface model spatial resolution was 12.6 cm, and the orthophotomosaic spatial resolution was 7 cm (Figure 10). For the archaeological site of Pedino Hill, the RMSE was 1.5 cm, and the produced digital surface model had a spatial resolution of 13.7 cm and orthophotomosaic spatial resolution of 7 cm (Figure 11).

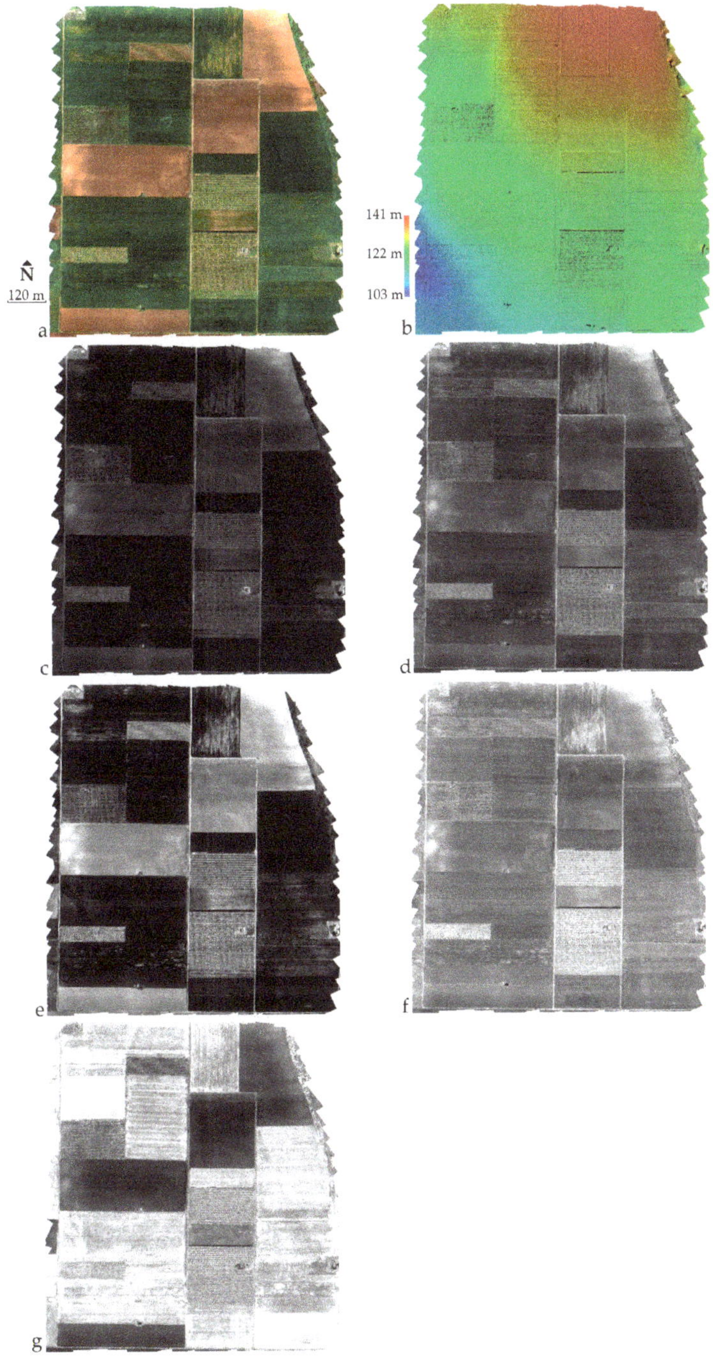

Figure 10. (**a**) The orthophotomosaic (RGB) of the archaeological site of Europos; (**b**) The digital surface model; The orthophotomosaics of MS images: (**c**) Blue; (**d**) Green; (**e**) Red; (**f**) Red Edge; (**g**) Near InfraRed. Center of images: 40°52′49.63″ N 22°33′25.50″ E.

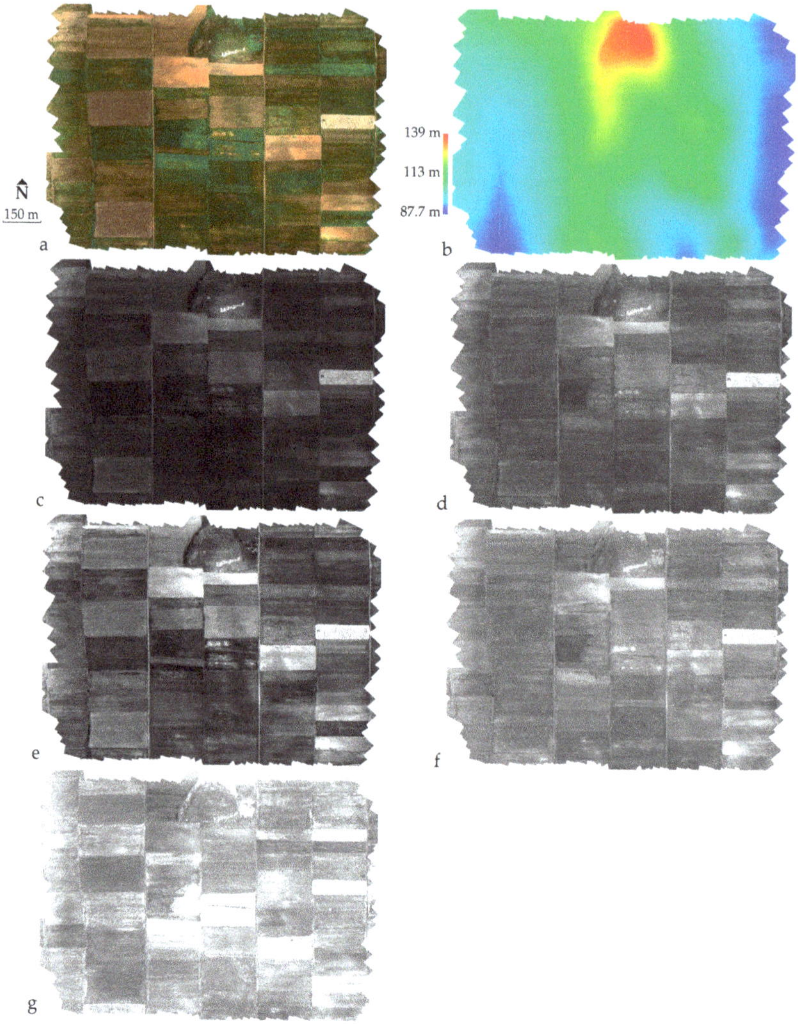

Figure 11. (**a**) The orthophotomosaic (RGB) of the archaeological site of the hill and south of the archaeological site of Pedino; (**b**) The digital surface model; The orthophotomosaics of MS images: (**c**) Blue; (**d**) Green; (**e**) Red; (**f**) Red Edge; (**g**) Near InfraRed. Center of figure: 40°54′47.42″ N 22°53′15.86″ E.

4.3. Production of Index Maps

The index formulas presented in Table 1 were created in the Agisoft Metashape© (version 2.0.0) software [1] and the corresponding index maps were produced (Figure 12).

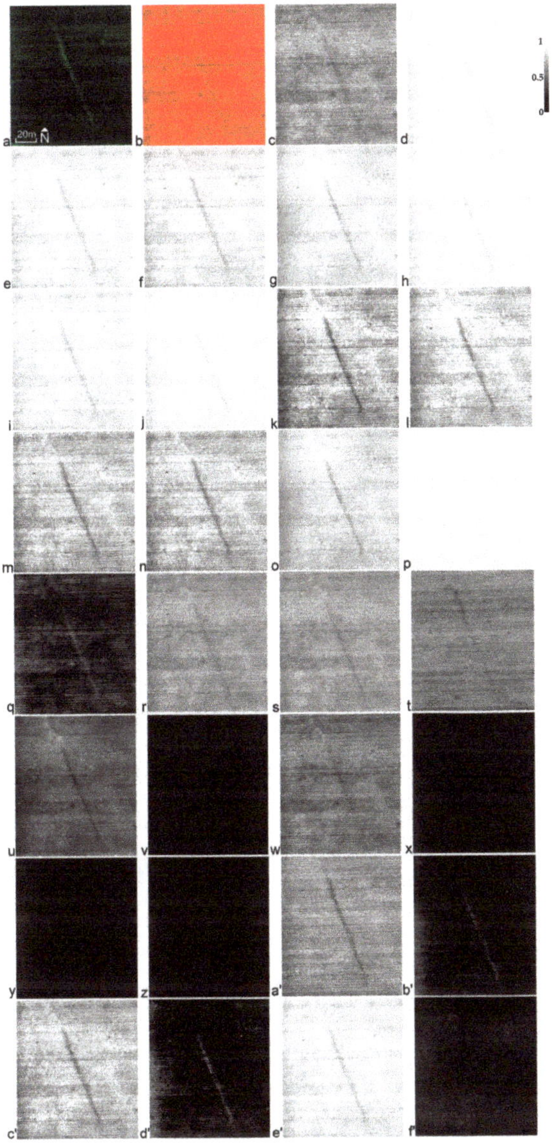

Figure 12. An observation site of negative crop marks 3.8 m wide south–southeast of the acropolis of Europos (Figure 3, north of Location 5). Excerpts from orthophotomosaics and indexes (**a**) bands: B-G-R (RGB); (**b**) bands: B-G-NIR; (**c**) NIR; (**d**) NDVI; (**e**) GNDVI; (**f**) RENDVI or NDRE; (**g**) NDVIRE or NDRER; (**h**) SAVI; (**i**) GSAVI; (**j**) OSAVI; (**k**) SR; (**l**) MSR; (**m**) GCI; (**n**) GRVI; (**o**) NGRDI; (**p**) NLI; (**q**) GEMI; (**r**) TVI; (**s**) MTVI; (**t**) TCARI; (**u**) MCARI; (**v**) BI; (**w**) BI2; (**x**) CI; (**y**) PSRI; (**z**) RI; (**a′**) ARI1; (**b′**) ARI1B; (**c′**) ARI2; (**d′**) ARI2B; (**e′**) LCI; (**f′**) RGRI. The grayscale of the images ranges from 0 (black) to 1 (white), with 0 representing the least favorable and 1 representing the most favorable result of the index. For instance, in the case of NDVI, a value of 0 corresponds to pixels without crop, a value of 0.5 to pixels with poor growth or poor crop health, and a value of 1 to pixels with good growth or healthy crop.

5. Discussion

In this survey, as in [1,117], the indexes were not sufficiently indicative of the presence of possible archaeological remains. It is important to note that the term "possible" should be used in such cases, as only systematic archaeological research and excavation can confirm the existence of anomalies in the soil or vegetation and determine whether they are due to the presence of archaeological remains underneath the soil. In this study, the RGB or near-infrared (NIR) bands of the orthophotomosaics were able to detect these possible archaeological remains. In the majority of instances, however, they markedly enhanced the visual observation of the marks, with the indexes that permitted this being SR and ARI2B (see, for example, Figure 12).

The set of new marks identified and their spatial distribution in the study area south of the acropolis of Europos are presented in Figure 3 (drawing of the marks in red). Northwest of Location 5 (Figure 3), there is a linear negative (possible presence of a solid structure under the ground) crop mark of approximately 112 m in length and 3.8 m in width, as well as a smaller length zigzagged negative crop mark, south of the previous one (Figure 13).

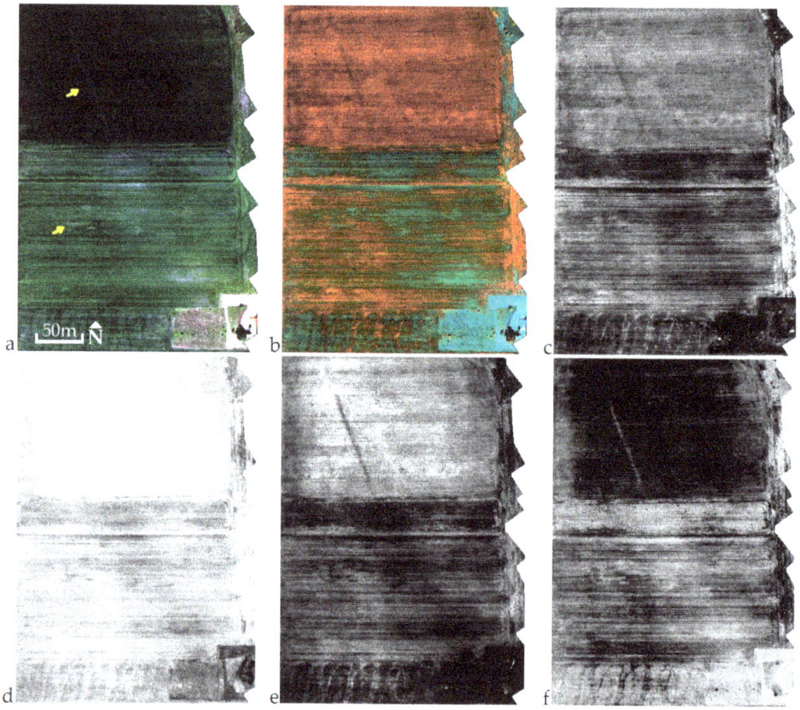

Figure 13. Two negative crop marks are located southeast of the acropolis of Europos, a short distance northwest of the burial mound (Figure 3, Location 5). The burial mound is in the lower right corner of the above images. Yellow arrows indicate the positions of the marks. Extract of orthophotomosaic and indexes: (**a**) bands: B-G-R; (**b**) bands: B-G-NIR; (**c**) NIR; (**d**) NDVI; (**e**) SR; (**f**) ARI2B.

Of all the indexes, those that improved the visual observation of marks were SR and ARI2B. The width of the linear mark does not refer to a covered main (big) road, but perhaps to a secondary covered road that may have been used exclusively for the communication of the acropolis with the burial mound of Location 5 (Figure 3). Interestingly, this linear mark and the linear mark of the previous study [1] (the mark between Locations 2 and 4 of Figure 3, with a north–south direction) have the same width and are spatially contained within the 128.00 m to 132.00 m contour lines. One could assume that these two marks are related and may possibly belong to the same covered road that provided communication between the acropolis and the three tombs of site 4 and the burial mound of site 5 (Figure 3). As already mentioned, the above are hypotheses that only archaeological research and excavation can prove.

Southwest of the acropolis of Europos, new positive (possible presence of an open structure under the ground) marks were identified (Figure 3 in red and Figure 14).

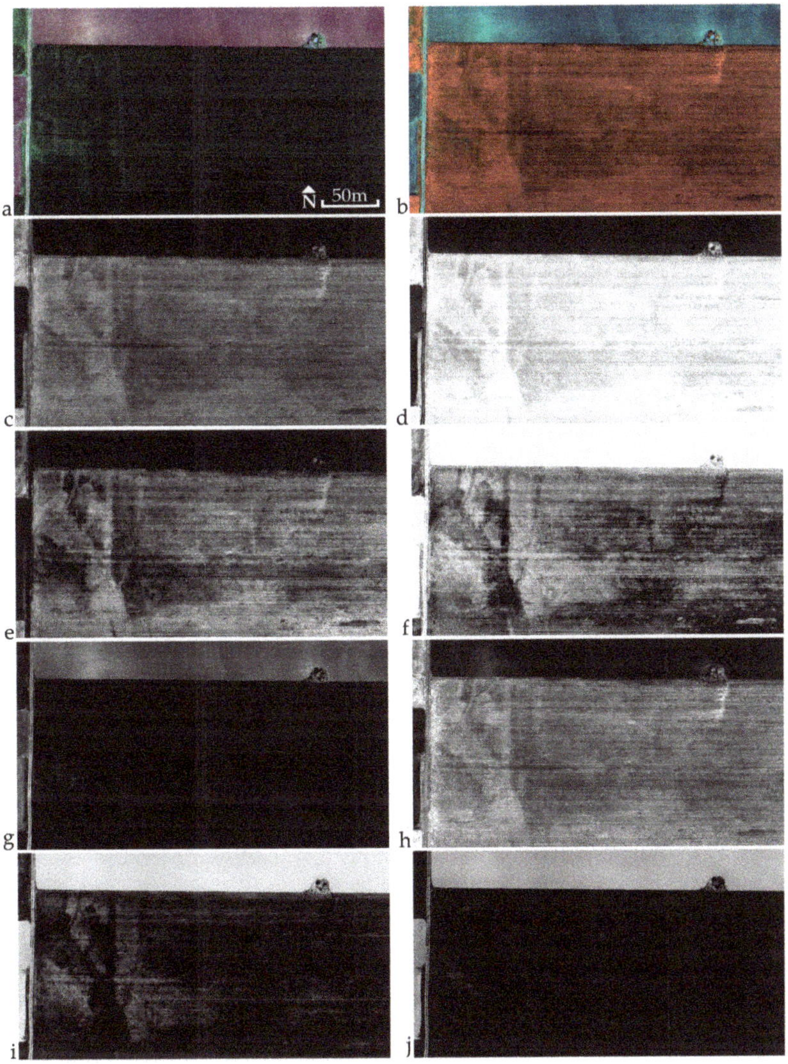

Figure 14. Linear positive crop marks are located southwest of the archaeological site of Europos (in red in Figure 3). Excerpt of orthophotomosaic and indexes: (**a**) bands: B-G-R; (**b**) bands: B-G-NIR; (**c**) NIR; (**d**) NDVI; (**e**) SR; (**f**) ARI2B; (**g**) BI; (**h**) BI2; (**i**) CI; (**j**) RI.

Of all the indexes, those that improved visual observation of the marks were SR and ARI2B. Marks do not refer to some typical possible structures under the ground that retain moisture, but their study is recommended in the future. North of the central parcel of land of Figure 14 (where marks are also located), there is a parcel of land without crops. For this reason, images from indexes (BI, BI2, CI, and RI) related to the soil were added to Figure 14 to check if the marks extend further north. Unfortunately, however, the above parcel of land without crops (Location A in Figures 3 and 15), as well as the parcel of land of Location B (Figures 3 and 15), had been irrigated with tons of water the day before the image collection (according to the in situ testimonies of their owners). A parcel of land of Location C (Figures 3 and 15) was irrigated from the dawn of the same day of image collection (as shown by the irrigation system in Figure 15, below the letter C). The above did not allow marks to appear on UAS images. It is unknown how many other parcels of land of the study area had been subjected to intense irrigation, which is abnormal for the season due to prolonged drought.

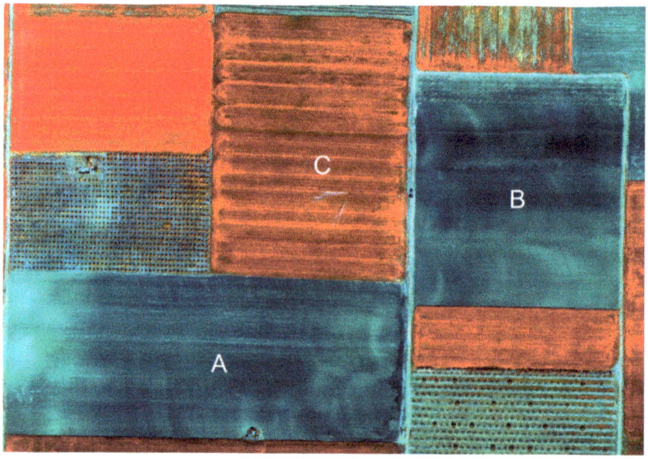

Figure 15. Parcels of land A, B, and C with high surface moisture. Bands: B-G-NIR.

The marks found in the study area of the archaeological site of Pedino Hill are shown in Figure 4.

Negative vegetation marks (Figure 4, Location 1, drawing of the marks in red) of possible internal structures (e.g., streets and/or buildings) and the wall of the ancient city (drawing of the marks in yellow in Figure 16) were identified on the hill. The same marks were found in the geophysical survey carried out on the hill years ago by Professor Grigoris Tsokas (Director of the Laboratory of Exploration Geophysics, School of Geology, Aristotle University of Thessaloniki, Greece) but have not yet been published. In the parcel of land southwest of the hill (in contact with the hill), linear negative soil marks (Figure 4, drawing of the marks in blue) were found that refer to possible solid structures under the ground. This parcel of land has no crops and, for this reason, images from indexes (BI, BI2, CI, and RI) related to the soil were added to Figure 16 (drawing of the marks in blue). Of all the indexes, those that improved the visual observation of the marks were NDVI, SR, BI, and BI2.

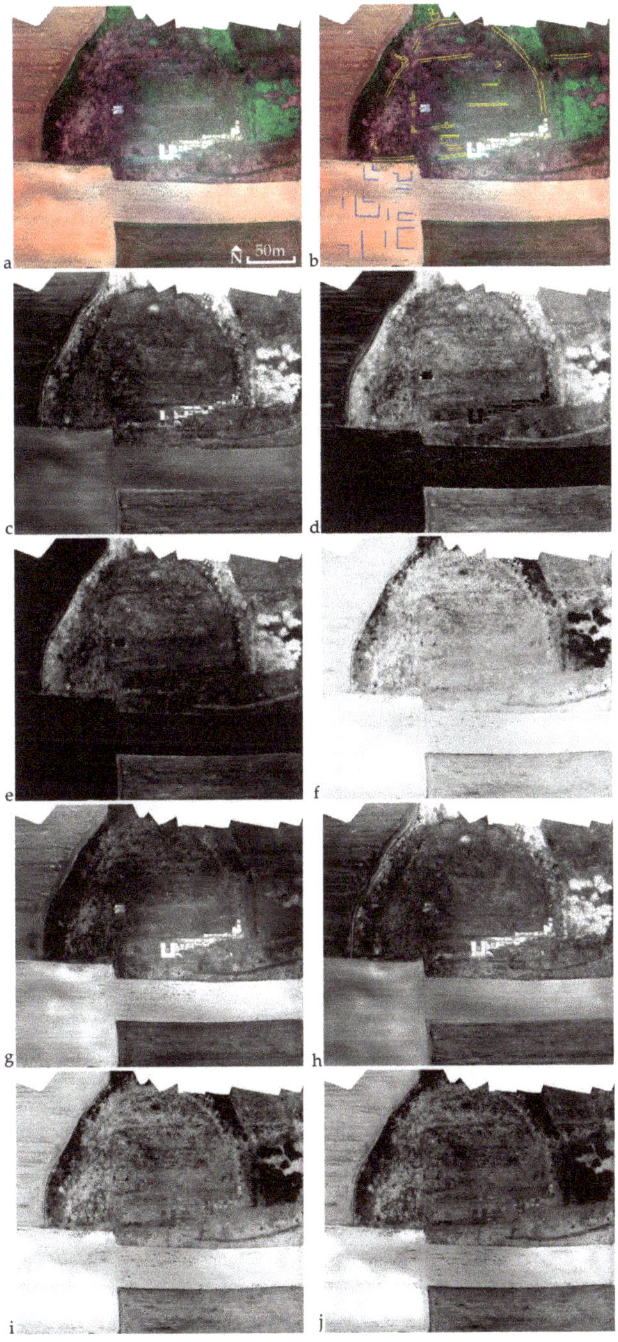

Figure 16. The archaeological site of Pedino Hill. Negative vegetation marks were found inside the walls (drawing of the marks in yellow) and soil marks (drawing of the marks in blue) outside the walls. Excerpt from orthophotomosaic and indexes: (**a**) B-G-R; (**b**) drawing of the marks; (**c**) NIR; (**d**) NDVI; (**e**) SR; (**f**) ARI2B; (**g**) BI; (**h**) BI2; (**i**) CI; (**j**) RI.

In the center of the study area (Figure 4), four circular negative crop marks ranging from 14 to 22 m in diameter were identified, along with one linear negative vegetation mark (Figure 17, drawing of the marks in yellow), with a width of approximately 4 m and a length of approximately 80 m. The contour lines (Figure 4) do not justify the presence of circular marks (i.e., they are not small hills). Of all the indexes, those that improved the visual observation of the marks were NDVI and SR.

Figure 17. The negative crop/vegetation marks in the center of the study area of the archaeological site of Pedino Hill. Excerpt from orthophotomosaic and indexes: (**a**) B-G-R; (**b**) drawing of the marks; (**c**) B-G-NIR; (**d**) NIR; (**e**) NDVI; (**f**) SR.

East of the study area of the archaeological site of Pedino Hill (Figure 4), square positive crop marks of dimensions from 2.5 × 6.5 m to 8.5 × 8.5 m were identified, as well as irregularly shaped positive crop marks (at the top of the images of Figure 18). Linear negative soil marks were also detected (at the bottom of the images of Figure 18). Of particular interest are the positive crop marks, as they are located in a hill of the study area (Figure 4). Of all the indexes, those that improved the visual observation of the marks were NDVI, SR, BI, and BI2.

Figure 18. Positive crop marks at the top and negative soil marks at the bottom of the images, east of the study area of the archaeological site of Pedino Hill. Excerpt from orthophotomosaic and indexes: (**a**) B-G-R; (**b**) drawing of the marks; (**c**) NIR; (**d**) NDVI; (**e**) SR; (**f**) ARI2B; (**g**) BI; (**h**) BI2; (**i**) CI; (**j**) RI.

6. Conclusions

Aerial and remote sensing archaeology are tools of archaeological prospection. They are exploited by interdisciplinary teams (surveyors, archaeologists, historians, etc.) on the one hand for the discovery of new covered archaeological remains, and on the other hand for the efficient management and allocation of human resources in fieldwork, carrying out surface surveys and/or excavation in specific geospatial locations. This reduces the time and cost of surveys (avoiding extensive surface surveys and/or excavations). Through the marks, the "resilience" of structures and historic complexes beneath the soil is revealed. While not visible when walking in the field, they have the potential and ability to cause "alterations" to the soil or vegetation/crop, which allows them to be observed in aerial and satellite images.

On the other hand, the resilience of historic landscapes refers to their potential to retain their cultural, natural, and historic values despite environmental and anthropogenic pressures. Aerial and remote sensing archaeology reveals historic landscapes that were unable to survive and maintain these values when the aforementioned pressures were triggered. For example, extreme weather events, earthquakes, and other natural disasters led to the levelling of ancient structures, which over time were abandoned and covered by soil. The transformation of large areas of land in the early-20th century into fertile farmland led to the covering of ancient roads and other structures.

Currently, there is a partial use of UAS with an MS sensor for collecting images in areas outside of the visible spectrum, aiming at the detection of covered archaeological remains. Almost all UAS can, at a relatively small cost of EUR 4000–6000, be equipped with MS sensors as well. The benefits will be manifold since the marks either appear for the first time in their images or are more clearly visible than in RGB images. Of particular interest is the use of a UAS aeromagnetic system for the detection of covered archaeological remains, a technique that will be of particular interest to researchers in the near future, as it is currently in the early stages of development.

Of the indexes utilized in aerial remote sensing archaeology, it seems that it is no coincidence that many researchers use NDVI and SR to locate marks, as both indexes gave good results in this study. It is also evident that there is room for the creation of new indexes after experimentation and systematic study of the spectral information of the sensor bands used in UAS.

Climate change adds an unpredictable cause, among those (size of the covered structure, its depth, type of soil, water content, and type of vegetation/crop) responsible for the appearance and intensity of visual observation of the marks. According to the phenological cycle of the dominant crop (wheat) in both study areas, the ideal period for detecting marks is theoretically from mid-April to mid-May. In Europos, while the timing of the image collection would have been ideal under normal climatic and meteorological conditions, the prolonged drought and the forced heavy irrigation by farmers did not allow the appearance of a large number and visually bold marks.

Both study areas identified soil and crop or vegetation marks, which could be attributed to archaeological remains beneath soil. This, however, is a hypothesis that can only be substantiated after systematic archaeological research and excavations.

Funding: This research received no external funding.

Data Availability Statement: The figures in this paper have a print resolution similar to or inferior to the images, e.g., of Google Maps or Google Earth. No original images or raw data will be made available on the locations, as they concern archaeological sites.

Acknowledgments: Thanks to Georgia Stratouli (Head of the Ephorate of Antiquities of Kilkis, Greece) and Nektario Poulakaki (Head of the Department of Prehistoric and Classical Antiquities and Museums of the Ephorate of Antiquities of Kilkis, Greece) for permission to collect data.

Conflicts of Interest: The author declares no conflicts of interest.

References

1. Kaimaris, D.; Tsokas, D. Application of UAS with Remote Sensing Sensors for the Location of Marks in the Archaeological Site of the Europos, Greece. *Remote Sens.* **2023**, *15*, 3843. [CrossRef]
2. Ben-Dor, E.; Kochavi, M.; Vinizki, L.; Shionim, M.; Portugali, J. Detection of buried ancient walls using airborne thermal video radiometry. *Int. J. Remote Sens.* **2001**, *22*, 3689–3702. [CrossRef]
3. Wilson, D.R. *Air Photo Interpretation for Archaeologists*, 2nd ed.; Tempus: Stroud, UK, 2000.
4. McLeester, M.; Casana, J.; Schurr, M.R.; Hill, A.C.; Wheeler, J.H. Detecting prehistoric landscape features using thermal, multi-spectral, and historical imagery analysis at Midewin National Tallgrass Prairie, Illinois. *J. Archaeol. Sci. Rep.* **2018**, *21*, 450–459.
5. Monterroso-Checa, A.; Gasparini, M. Aerial archaeology and photogrammetric surveys along the roman way from corduba to emerita. Digitalizing the ager cordubensis and the ager mellariensis. *SCIRES-IT SCIentific Res. Inf. Technol.* **2016**, *6*, 175–188.
6. Eisenbeiss, H.; Zhang, L. Comparison of DSMs generated from mini UAV imagery and terrestrial laser scanner in a cultural heritage application. *Int. Arch. Photogramm. Remote Sens. Spat. Inf. Sci.* **2006**, *36*, 90–96.
7. Challis, K.; Kincey, M.; Howard, A. Airborne remote sensing of valley floor geoarchaeology using Daedalus ATM and CASI. *Archaeol. Prospect.* **2009**, *16*, 17–33. [CrossRef]
8. Agudo, P.U.; Pajas, J.A.; Pérez-Cabello, F.; Redón, J.V.; Lebrón, B.E. The Potential of Drones and Sensors to Enhance Detection of Archaeological Cropmarks: A Comparative Study Between Multi-Spectral and Thermal Imagery. *Drones* **2018**, *2*, 29. [CrossRef]
9. Ur, J. CORONA satellite photography and ancient road networks: A northern Mesopotamian case study. *Antiquity* **2003**, *77*, 102–115. [CrossRef]
10. Traviglia, A.; Cottica, D. Remote sensing applications and archaeological research in the Northern Lagoon of Venice: The case of the lost settlement of Constanciacus. *J. Archaeol. Sci.* **2011**, *38*, 2040–2050. [CrossRef]
11. Luo, L.; Wang, X.; Guo, H.; Lasaponara, R.; Zong, X.; Masini, N.; Wang, G.; Shi, P.; Khatteli, H.; Chen, F.; et al. Airborne and spaceborne remote sensing for archaeological and cultural heritage applications: A review of the century (1907–2017). *Remote Sens. Environ.* **2019**, *232*, 111280. [CrossRef]
12. Giardino, J.M. A history of NASA remote sensing contributions to archaeology. *J. Archaeol. Sci.* **2011**, *38*, 2003–2009. [CrossRef]
13. Ware, A.J.; Gumerman, J.G. Remote sensing methodology and the Chaco Canyon prehistoric road system. In *Aerial Remote Sensing Techniques in Archeology*; Lyons, T., Hitchcock, R., Eds.; Reports of the Chaco Center, No. 2; Cultural Resource Management Division: Washington, DC, USA, 1977; pp. 135–168.
14. Aqdus, A.S.; Drummond, J.; Hanson, S.W. Discovering archaeological cropmarks: A hyperspectral approach. *Int. Arch. Photogramm. Remote Sens. Spat. Inf. Sci.* **2008**, *37*, 361–365.
15. Pascucci, S.; Cavalli, M.R.; Palombo, A.; Pignatti, S. Suitability of CASI and ATM airborne remote sensing data for archaeological subsurface structure detection under different land cover: The Arpi case study (Italy). *J. Geophys. Eng.* **2010**, *7*, 183–189. [CrossRef]
16. Kalayci, T.; Lasaponara, R.; Wainwright, J.; Masini, N. Multispectral contrast of archaeological features: A quantitative evaluation. *Remote Sens.* **2019**, *11*, 913. [CrossRef]
17. Lasaponara, R.; Masini, N. Detection of archaeological crop marks by using satellite QuickBird multispectral imagery. *J. Archaeol. Sci.* **2007**, *34*, 214–221. [CrossRef]
18. Trier, Ø.D.; Larsen, S.Ø.; Solberg, R. Automatic detection of circular structures in high-resolution satellite images of agricultural land. *Archaeol. Prospect.* **2009**, *16*, 1–15. [CrossRef]
19. Calleja, F.J.; Pagés, R.O.; Díaz-Álvarez, N.; Peón, J.; Gutiérrez, N.; Martín-Hernández, E.; Relea, C.A.; Melendi, R.D.; Álvarez, F.P. Detection of buried archaeological remains with the combined use of satellite multispectral data and UAV data. *Int. J. Appl. Earth Obs. Geoinf.* **2018**, *73*, 555–573. [CrossRef]
20. Ciminale, M.; Becker, H.; Gallo, D. Integrated technologies for archaeological investigation; the Celone Valley project. *Archaeol. Prospect.* **2007**, *14*, 167–181. [CrossRef]
21. Campana, S.; Dabas, M.; Marasco, L.; Piro, S.; Zamuner, D. Integration of remote sensing, geophysical surveys and archaeological excavation for the study of a medieval mound (Tuscany, Italy). *Archaeol. Prospect.* **2009**, *16*, 167–176. [CrossRef]
22. Rowlands, A.; Sarris, A. Detection of exposed and subsurface archaeological remains using multi-sensor remote sensing. *J. Archaeol. Sci.* **2007**, *34*, 795–803. [CrossRef]
23. Lasaponara, R.; Masini, N. QuickBird-based analysis for the spatial characterization of archaeological sites: Case study of the Monte Serico Medioeval village. *Geophys. Lett.* **2005**, *32*, L12313. [CrossRef]
24. Lasaponara, R.; Masini, N. Identification of archaeological buried remains basedon Normalized Difference Vegetation Index (NDVI) from Quickbird satellite data. *IEEE Geosci. Remote Sens. Lett.* **2006**, *3*, 325–328. [CrossRef]
25. Altaweel, M. The use of ASTER satellite imagery in archaeological contexts. *Archaeol. Prospect.* **2005**, *12*, 151–166. [CrossRef]
26. Beck, A. Archaeological site detection: The importance of contrast. In Proceedings of the Annual Conference of the Remote Sensing and Photogrammetry Society, Newcastle Upon Tyne, UK, 11–14 September 2007.
27. Laet, V.; Paulissen, E.; Waelkens, M. Methods for the extraction of archaeologicalfeatures from very high-resolution Ikonos-2 remote sensing imagery, Hisar (southwestTurkey). *J. Archaeol. Sci.* **2007**, *34*, 830–841. [CrossRef]
28. Noviello, M.; Ciminale, M.; Pasquale, V. Combined application of pansharpening and enhancement methods to improve archaeological cropmark visibility and identification in QuickBird imagery: Two case studies from Apulia, Southern Italy. *J. Archaeol. Sci.* **2013**, *40*, 3604–3613. [CrossRef]

29. Cowley, D.C.; Stichelbaut, B.B. Historic aerial photographic archives for European archaeology. *Eur. J. Archaeol.* **2012**, *15*, 217–236. [CrossRef]
30. Forte, M.; Campana, S.; Liuzza, C. *Space, Time, Place*; Archaeopress: Oxford, UK, 2010.
31. Orlando, P.; Villa, B.D. Remote sensing applications in archaeology. *Archeol. E Calc.* **2011**, *22*, 147–168.
32. Aminzadeh, B.; Samani, F. Identifying the boundaries of the historical site of Persepolis using remote sensing. *Remote Sens. Environ.* **2006**, *102*, 52–62. [CrossRef]
33. Oltean, I.A.; Abell, L.L. High-Resolution Satellite Imagery and the Detection of Buried Archaeological Features in Ploughed Landscapes. In *Satellite Remote Sensing: A New Tool for Archaeology*; Springer: Berlin/Heidelberg, Germany, 2011; pp. 291–305.
34. Bini, M.; Isola, I.; Zanchetta, G.; Ribolini, A.; Ciampalini, A.; Baneschi, I.; Mele, D.; D'Agata, A.L. Identification of Leveled Archeological Mounds (Höyük) in the Alluvial Plain of the Ceyhan River (Southern Turkey) by Satellite Remote-Sensing Analyses. *Remote Sens.* **2018**, *10*, 241. [CrossRef]
35. Rayne, L.; Donoghue, D. A Remote Sensing Approach for Mapping the Development of Ancient Water Management in the Near East. *Remote Sens.* **2018**, *10*, 2042. [CrossRef]
36. Ur, J. CORONA Satellite Imagery and Ancient Near Eastern Landscapes. In *Mapping Archaeological Landscapes from Space: In Observance of the 40th Anniversary of the World Heritage Convention*; Comer, C.D., Harrower, J.M., Eds.; Springer: New York, NY, USA, 2013; pp. 19–29.
37. Megarry, W.; Davis, S. Beyond the Bend: Remotely Sensed Data and Archaeological Site Prospection in the Boyne Valley, Ireland. In *Mapping Archaeological Landscapes from Space: In Observance of the 40th Anniversary of the World Heritage Convention*; Comer, C.D., Harrower, J.M., Eds.; Springer: New York, NY, USA, 2013; pp. 85–95.
38. Verhoeven, G.; Vermeulen, F. Engaging with the Canopy—Multi-Dimensional Vegetation Mark Visualisation Using Archived Aerial Images. *Remote Sens.* **2016**, *8*, 752. [CrossRef]
39. Traviglia, A.; Torsello, A. Landscape Pattern Detection in Archaeological Remote Sensing. *Geosciences* **2017**, *7*, 128. [CrossRef]
40. Verhoeven, J.G. Near-Infrared Aerial Crop Mark Archaeology: From its Historical Use to Current Digital Implementations. *J. Archaeol. Method Theory* **2012**, *19*, 132–160. [CrossRef]
41. Hejcman, M.; Smrž, Z. Cropmarks in stands of cereals, legumes and winter rape indicate sub-soil archaeological features in the agricultural landscape of Central Europe. *Agric. Ecosyst. Environ.* **2010**, *138*, 348–354. [CrossRef]
42. Czajlik, Z.; Árvai, M.; Mészáros, J.; Nagy, B.; Rupnik, L.; Pásztor, L. Cropmarks in Aerial Archaeology: New Lessons from an Old Story. *Remote Sens.* **2021**, *13*, 1126. [CrossRef]
43. Šošić Klindžić, R.; Šiljeg, B.; Kalafatić, H. Multiscale and Multitemporal Remote Sensing for Neolithic Settlement Detection and Protection—The Case of Gorjani, Croatia. *Remote Sens.* **2024**, *16*, 736. [CrossRef]
44. Winterbottom, S.J.; Dawson, T. Airborne multi-spectral prospection for buried archaeology in mobile sand dominated systems. *Archaeol. Prospect.* **2005**, *12*, 205–219. [CrossRef]
45. Aqdus, S.A.; Hanson, W.S.; Drummond, J. The potential of hyperspectral and multi-spectral imagery to enhance archaeological cropmark detection: A comparative study. *J. Archaeol. Sci.* **2012**, *39*, 1915–1924. [CrossRef]
46. Tabbagh, A. Prospections thermiques et électriques. *Gall. Préhistoire* **1977**, *20*, 271–280. [CrossRef]
47. Fourteau, A.M.; Tabbagh, A. Parcellaire fossile et prospection thermique, resultants des recherches à Lion en Beauce (Loiret). *ArchéoSciences Rev. D'archéométrie* **1979**, *3*, 115–123. [CrossRef]
48. Lunden, B. Aerial thermography: A remote sensing technique applied to detection of buried archaeological Remains at a site in Dalecarlia, Sweden. *Geogr. Anna. Phys. Geogr.* **1985**, *67*, 161–166.
49. Bellerby, M.N.; Branigan, K. A thermal method for archaeological prospection: Preliminary investigations. *Archaeometry* **1990**, *32*, 191–203. [CrossRef]
50. Colomina, I.; Molina, P. Unmanned aerial systems for photogrammetry and remote sensing: A review. *ISPRS J. Photogramm. Remote Sens.* **2014**, *92*, 79–97. [CrossRef]
51. Stott, D.; Boyd, D.S.; Beck, A.; Cohn, A.G. Airborne LiDAR for the detection of archaeological vegetation marks using biomass as a proxy. *Remote Sens.* **2015**, *7*, 1594–1618. [CrossRef]
52. James, K.; Nichol, C.J.; Wade, T.; Cowley, D.; Gibson Poole, S.; Gray, A.; Gillespie, J. Thermal and Multispectral Remote Sensing for the Detection and Analysis of Archaeologically Induced Crop Stress at a UK Site. *Drones* **2020**, *4*, 61. [CrossRef]
53. Challis, K.; Forlin, P.; Kincey, M. A generic toolkit for the visualization of archaeological features on airborne LiDAR elevation data. *Archaeol. Prospect.* **2011**, *18*, 279–289. [CrossRef]
54. Crow, P.; Benham, S.; Devereux, B.J.; Amable, G.S. Woodland vegetation and its implications for archaeological survey using LiDAR. *For. Int. J. For. Res.* **2007**, *80*, 241–252. [CrossRef]
55. Doneus, M.; Briese, C.; Fera, M.; Janner, M. Archaeological prospection of forested areas using full-waveform airborne laser scanning. *J. Archaeol. Sci.* **2008**, *35*, 882–893. [CrossRef]
56. Johnson, K.M.; Ouimet, W.B. Rediscovering the lost archaeological landscape of southern New England using airborne light detection and ranging (LiDAR). *J. Archaeol. Sci.* **2014**, *43*, 9–20. [CrossRef]
57. Risbøl, O.; Langhammer, D.; Schlosser Mauritsen, E.; Seitsonen, O. Employment, Utilization, and Development of Airborne Laser Scanning in Fenno-Scandinavian Archaeology—A Review. *Remote Sens.* **2020**, *12*, 1411. [CrossRef]
58. Chen, F.; You, J.; Tang, P.; Zhou, W.; Masini, N.; Lasaponara, R. Unique performance of spaceborne SAR remote sensing in cultural heritage applications: Overviews and perspectives. *Archaeol. Prospect.* **2018**, *25*, 71–79. [CrossRef]

59. Balz, T.; Caspari, G.; Fu, B.; Liao, M. Discernibility of Burial Mounds in High Resolution X-Band SAR Images for Archaeological Prospections in the Altai Mountains. *Remote Sens.* **2016**, *8*, 817. [CrossRef]
60. Moore, E.; Freeman, A.; Hensley, S. Spaceborne and airborne radar at Angkor: Introducing new technology to the ancient site. In *Remote Sensing in Archaeology*; Wiseman, J., El-Baz, F., Eds.; Springer: New York, NY, USA, 2007; pp. 185–216.
61. Stewart, C.; Oren, E.D.; Cohen-Sasson, E. Satellite Remote Sensing Analysis of the Qasrawet Archaeological Site in North Sinai. *Remote Sens.* **2018**, *10*, 1090. [CrossRef]
62. Jiang, A.; Chen, F.; Masini, N.; Capozzoli, L.; Romano, G.; Sileo, M.; Yang, R.; Tang, P.; Chen, P.; Lasaponara, R.; et al. Archeological crop marks identified from Cosmo-SkyMed time series: The case of Han-Wei capital city, Luoyang, China. *Int. J. Digit. Earth* **2017**, *10*, 846–860. [CrossRef]
63. Zaina, F.; Tapete, D. Satellite-Based Methodology for Purposes of Rescue Archaeology of Cultural Heritage Threatened by Dam Construction. *Remote Sens.* **2022**, *14*, 1009. [CrossRef]
64. Negula, D.; Moise, C.; Lazăr, M.A.; Rișcuța, C.N.; Cristescu, C.; Dedulescu, L.A.; Mihalache, E.C.; Badea, A. Satellite Remote Sensing for the Analysis of the Micia and Germisara Archaeological Sites. *Remote Sens.* **2020**, *12*, 2003. [CrossRef]
65. Kaimaris, D.; Georgiadis, C.; Patias, P.; Tsioukas, V. Aerial and Remote Sensing Archaeology. *Int. J. Comput. Methods Herit. Sci.* **2017**, *1*, 58–76. [CrossRef]
66. Vilbig, M.; Sagan, V.; Bodine, C. Archaeological surveying with airborne LiDAR and UAV photogrammetry: A comparative analysis at Cahokia Mounds. *J. Archaeol. Sci. Rep.* **2020**, *33*, 102509. [CrossRef]
67. Nikolakopoulos, G.K.; Soura, K.; Koukouvelas, K.I.; Argyropoulos, G.N. UAV vs classical aerial photogrammetry for archaeological studies. *J. Archaeol. Sci. Rep.* **2017**, *14*, 758–773. [CrossRef]
68. Field, S.; Waite, M.; Wandsnider, L. The utility of UAVs for archaeological surface survey: A comparative study. *J. Archaeol. Sci. Rep.* **2017**, *13*, 577–582. [CrossRef]
69. Robert, M.; Evelyn, B. Low altitude unmanned aerial photography to assist in rock art studies. *SAA Archaeol. Rec. Spec. Issue Drones Archaeol.* **2016**, *16*, 14–16.
70. Meyer, E.D.; Lo, E.; Afshari, S.; Vaughan, A.; Rissolo, D.; Kuester, F. Utility of low-cost drones to generate 3D models of archaeological sites from multisensor data. *SAA Archaeol. Rec. Spec. Issue Drones Archaeol.* **2016**, *16*, 22–24.
71. Harrison-Buck, E.; Willis, M.; Walker, C. Using drones in a threatened archaeological landscape rapid survey, salvage, and mapping of the Maya site of Saturday Creek, Belize. *SAA Archaeol. Rec. Spec. Issue Drones Archaeol.* **2016**, *16*, 30–35.
72. Verhoeven, G.J.J. Providing an archaeological bird's-eye view—An overall picture of ground-based means to execute low-altitude aerial photography (LAAP) in archaeology. *Archaeol. Prospect.* **2009**, *16*, 233–249. [CrossRef]
73. Masinia, N.; Marzo, C.; Manzari, P.; Belmonte, A.; Sabia, C.; Lasaponara, R. On the characterization of temporal and spatial patterns of archaeological crop-marks. *J. Cult. Herit.* **2018**, *32*, 124–132. [CrossRef]
74. Stek, D.T. Drones over Mediterranean landscapes. The potential of small UAV's (drones) for site detection and heritage management in archaeological survey projects: A case study from Le Pianelle in the Tappino Valley, Molise (Italy). *J. Cult. Herit.* **2016**, *22*, 1066–1071. [CrossRef]
75. Adamopoulos, E.; Rinaudo, F. UAS-Based Archaeological Remote Sensing: Review, Meta-Analysis and State-of-the-Art. *Drones* **2020**, *4*, 46. [CrossRef]
76. Materazzi, F.; Pacifici, M. Archaeological crop marks detection through drone multispectral remote sensing and vegetation indices: A new approach tested on the Italian pre-Roman city of Veii. *J. Archaeol. Sci. Rep.* **2022**, *41*, 103235. [CrossRef]
77. Hill, C.A.; Laugier, J.E.; Casana, J. Archaeological Remote Sensing Using Multi-Temporal, Drone-Acquired Thermal and Near Infrared (NIR) Imagery: A Case Study at the Enfield Shaker Village, New Hampshire. *Remote Sens.* **2020**, *12*, 690. [CrossRef]
78. Carmona, S.A.J.; Quirós, E.; Mayoral, V.; Charro, C. Assessing the potential of multispectral and thermal UAV imagery from archaeological sites. A case study from the Iron Age hillfort of Villasviejas del Tamuja (Cáceres, Spain). *J. Archaeol. Sci. Rep.* **2020**, *31*, 102312.
79. Uribe, P.; Angás, J.; Pérez-Cabello, F. Aerial mapping and multi-sensor approaches from remote sensing applied to the Roman Archaeological Heritage. *Int. Arch. Photogramm. Remote Sens. Spat. Inf. Sci.* **2015**, *XL-5/W4*, 461–467.
80. Brooke, C.; Clutterbuck, B. Mapping Heterogeneous Buried Archaeological Features Using Multisensor Data from Unmanned Aerial Vehicles. *Remote Sens.* **2019**, *12*, 41. [CrossRef]
81. Koucká, L.; Kopăcková, V.; Fárová, K.; Gojda, M. UAV Mapping of an Archaeological Site Using RGB and NIR High-Resolution Data. In Proceedings of the 2nd International Electronic Conference on Remote Sensing 2018, Thessaloniki, Greece, 22 March–5 April 2018.
82. Lehmann, J.R.K.; Smithson, K.Z.; Prinz, T. Making the invisible visible: Using UAS-based high-resolution color-infrared imagery to identify buried medieval monastery walls. *J. Unmanned Veh. Syst.* **2015**, *3*, 58–67. [CrossRef]
83. VanValkenburgh, P.; Cushman, K.C.; Butters, L.J.C.; Vega, C.R.; Roberts, C.B.; Kepler, C.; Kellner, J. Lasers, Without Lost Cities: Using Drone Lidar to Capture Architectural Complexity at Kuelap, Amazonas, Peru. *J. Field Archaeol.* **2020**, *45*, S75–S88. [CrossRef]
84. Ronchi, D.; Limongiello, M.; Barba, S. Correlation among Earthwork and Cropmark Anomalies within Archaeological Landscape Investigation by Using LiDAR and Multispectral Technologies from UAV. *Drones* **2020**, *4*, 72. [CrossRef]
85. Moriarty, C.; Cowley, D.C.; Wade, T.; Nichol, C.J. Deploying multispectral remote sensing for multi-temporal analysis of archaeological crop stress at Ravenshall, Fife, Scotland. *Archaeol. Prospect.* **2019**, *26*, 33–46. [CrossRef]

86. Fiz, J.I.; Martín, P.M.; Cuesta, R.; Subías, E.; Codina, D.; Cartes, A. Examples and Results of Aerial Photogrammetry in Archeology with UAV: Geometric Documentation, High Resolution Multispectral Analysis, Models and 3D Printing. *Drones* **2022**, *6*, 59. [CrossRef]
87. Poirier, N.; Hautefeuille, F.; Calastrenc, C. Low altitude thermal survey by means of an automated unmanned aerial vehicle for the detection of archaeological buried structures. *Archaeol. Prospect.* **2013**, *20*, 303–307. [CrossRef]
88. Thomas, H. Some like it hot: The impact of next generation FLIR Systems thermal cameras on archaeological thermography. *Archaeol. Prospect.* **2018**, *25*, 81–87. [CrossRef]
89. Waagen, J.; Sánchez, J.G.; van der Heiden, M.; Kuiters, A.; Lulof, P. In the Heat of the Night: Comparative Assessment of Drone Thermography at the Archaeological Sites of Acquarossa, Italy, and Siegerswoude, The Netherlands. *Drones* **2022**, *6*, 165. [CrossRef]
90. Cao, W.; Qing, H.; Xu, X.; Liu, C.; Chen, S.; Zhong, Y.; Liu, J.; Li, Y.; Jiang, X.; Gao, D.; et al. Pre-Archaeological Investigation by Integrating Unmanned Aerial Vehicle Aeromagnetic Surveys and Soil Analyses. *Drones* **2022**, *6*, 243. [CrossRef]
91. Fuldain González, J.J.; Varón Hernández, F.R. NDVI Identification and Survey of a Roman Road in the Northern Spanish Province of Álava. *Remote Sens.* **2019**, *11*, 725. [CrossRef]
92. Šedina, J.; Housarová, E.; Raeva, P. Using RPAS for the detection of archaeological objects using multispectral and thermal imaging. *Eur. J. Remote Sens.* **2018**, *52*, 182–191. [CrossRef]
93. Shin, Y.H.; Shin, S.Y.; Rastiveis, H.; Cheng, Y.-T.; Zhou, T.; Liu, J.; Zhao, C.; Varinlioğlu, G.; Rauh, N.K.; Matei, S.A.; et al. UAV-Based Remote Sensing for Detection and Visualization of Partially-Exposed Underground Structures in Complex Archaeological Sites. *Remote Sens.* **2023**, *15*, 1876. [CrossRef]
94. Cowley, D.C.; Moriarty, C.; Geddes, G.; Brown, G.L.; Wade, T.; Nichol, C.J. UAVs in Context: Archaeological Airborne Recording in a National Body of Survey and Record. *Drones* **2018**, *2*, 2. [CrossRef]
95. Casana, J.; Laugier, J.E.; Hill, C.A.; Reese, M.K.; Ferwerda, C.; McCoy, D.M.; Ladefoged, T. Exploring archaeological landscapes using drone-acquired lidar: Case studies from Hawai'i, Colorado, and New Hampshire, USA. *J. Archaeol. Sci. Rep.* **2021**, *39*, 103133. [CrossRef]
96. Li, Z. New opportunities for archaeological research in the Greater Ghingan Range, China: Application of UAV LiDAR in the archaeological survey of the Shenshan Mountain. *J. Archaeol. Sci. Rep.* **2023**, *51*, 104182. [CrossRef]
97. Index DataBase. A Database for Remote Sensing Indices. Available online: https://www.indexdatabase.de/db/a-single.php?id=1 (accessed on 6 June 2024).
98. Spectral Indices by Type. Available online: https://awesome-ee-spectral-indices.readthedocs.io/en/latest/list.html#soil (accessed on 6 June 2024).
99. Broadband Greenness. Available online: https://www.nv5geospatialsoftware.com/docs/BroadbandGreenness.html#Green3 (accessed on 6 June 2024).
100. Franzini, M.; Ronchetti, G.; Sona, G.; Casella, V. Geometric and Radiometric Consistency of Parrot Sequoia Multispectral Imagery for Precision Agriculture Applications. *Appl. Sci.* **2019**, *9*, 5314. [CrossRef]
101. Rouse, J.W.; Haas, R.H.; Schell, J.A.; Deering, D.W. Monitoring Vegetation System in the Great Plains with ERTS. In Proceedings of the Third Earth Resources Technology Satellite-1 Symposium, Greenbelt, MD, USA, 10–14 December 1974.
102. Gitelson, A.A.; Kaufman, Y.J.; Merzlyak, M.N. Use of a green channel in remote sensing of global vegetation from EOS-MODIS. *Remote Sens. Environ.* **1996**, *58*, 289–298. [CrossRef]
103. Barnes, E.M.; Clarke, T.R.; Richards, S.E.; Colaizzi, P.D.; Haberland, J.; Kostrzewski, M.; Waller, P.; Choi, C.; Riley, E.; Thompson, T.; et al. Coincident detection of crop water stress, nitrogen status and canopy density using ground based multispectral data. In Proceedings of the Fifth International Conference on Precision Agriculture, Bloomington, MN, USA, 16–19 July 2000; pp. 1–15.
104. Gitelson, A.A.; Merzlyak, M.N. Spectral reflectance changes associated with autumn senescence of *Aesculus hippocastanum* L. and *Acer platanoides* L. leaves. Spectral features and relation to chlorophyll estimation. *J. Plant Physiol.* **1994**, *143*, 286–292. [CrossRef]
105. Huete, A.R. A soil-adjusted vegetation index (SAVI). *Remote Sens. Environ.* **1988**, *25*, 295–309. [CrossRef]
106. Peter, G.B.; Messina, P.J.; Carroll, W.J.; Zhi, J.; Chimonyo, V.; Lin, S.; Snapp, S. Multi-Spatial Resolution Satellite and sUAS Imagery for Precision Agriculture on Smallholder Farms in Malawi. *Photogramm. Eng. Remote Sens.* **2020**, *86*, 107–119. [CrossRef]
107. Gitelson, A.A.; Kaufman, Y.J.; Stark, R.; Rundquist, D. Novel algorithms for remote estimation of vegetation fraction. *Remote Sens. Environ.* **2002**, *80*, 76–87. [CrossRef]
108. Hunt, E.R.; Cavigelli, M.; Daughtry, C.S.T.; McMurtrey, J.E.; Walthall, C.L. Evaluation of digital photography from model aircraft for remote sensing of crop biomass and nitrogen status. *Precis. Agric.* **2005**, *6*, 359–378. [CrossRef]
109. Savopoulou, T.; Giannakis, I. Highlighting the archaeological site of Europe: From planning to implementation. In Proceedings of the 26th Archaeological Work in Macedonia and Thrace, Thessaloniki, Greece, 10–12 May 2011.
110. Tsokas, G.N.; Giannopoulos, A.; Tsourlos, P.; Vargemezis, G.; Tealby, J.M.; Sarris, A.; Papazachos, C.B.; Savopoulou, T. A large scale geophysical survey in the archaeological site of Europos (northern Greece). *J. Appl. Geophys.* **1994**, *32*, 85–98. [CrossRef]
111. Oikonomidis, D.; Karamitrou, A.; Tsokas, G.N.; Astaras, T. Combined use of Satellite Remote Sensing, GIS, and Geophysical Data to Archaeological Research in Europos Area, Macedonia (Northern Greece). In Proceedings of the 31st EARSeL Symposium Remote Sensing and Geoinformation Not Only for Scientific Cooperation, Prague, Czech Republic, 30 May–2 June 2011.

112. WingtraOne GEN II Drone, Technical Specifications. Available online: https://wingtra.com/wp-content/uploads/Wingtra-Technical-Specifications.pdf (accessed on 6 June 2024).
113. Miyoshi, G.T.; Imai, N.N.; Tommaselli, A.M.G.; Honkavaara, E.; Näsi, R.; Moriya, E.A.S. Radio-metric block adjustment of hyperspectral image blocks in the Brazilian environment. *Int. J. Remote Sens.* **2018**, *39*, 4910–4930. [CrossRef]
114. Guo, Y.; Senthilnath, J.; Wu, W.; Zhang, X.; Zeng, Z.; Huang, H. Radiometric calibration for multispectral camera of different imaging conditions mounted on a UAS platform. *Sustainability* **2019**, *11*, 978. [CrossRef]
115. Mafanya, M.; Tsele, P.; Botai, J.O.; Manyama, P.; Chirima, G.J.; Monate, T. Radiometric calibration framework for ultra-high-resolution UAS-derived orthomosaics for large-scale mapping of invasive alien plants in semi-arid woodlands: *Harrisia pomanensis* as a case study. *Remote Sens.* **2018**, *39*, 5119–5140. [CrossRef]
116. Honkavaara, E.; Khoramshahi, E. Radiometric correction of close-range spectral image blocks captured using an unmanned aerial vehicle with a radiometric block adjustment. *Remote Sens.* **2018**, *10*, 256. [CrossRef]
117. Kaimaris, D. Utilization of different sensors in UAV for the detection and optimal visual observation of the marks over buried ancient remains. *Sci. Cult.* **2021**, *8*, 129–145.

land

Article

Is Historical Cartography a Useful Tool for Landscape Analysis? A Perspective from Inland Spain (Zorita de los Canes, Guadalajara) from the Middle Ages to the Present

Pilar Diarte-Blasco [1,*] and Manuel Castro-Priego [2]

[1] Department of Archaeology and Social Processes, Institute of History, Spanish National Research Council, C/Albasanz, 26, 28037 Madrid, Spain
[2] Department of History I and Philosophy, Alcalá University, C/Trinidad, 1, 28801 Alcala de Henares, Spain; manuel.castro@uah.es
* Correspondence: pilar.diarte@cchs.csic.es

Abstract: Historical cartography continues to be an essential resource in developing strategies for the analysis of anthropised landscapes. In recent years, the digitisation and conversion of a large number of pre-20th-century maps to modern geographic coordinate systems and data hierarchisation in GIS-integrated databases have opened up huge possibilities. In this paper, we highlight some of the advantages and issues that we observed in using historical cartography in the Iberian Peninsula heartlands by comparing archaeological data, textual sources and maps and various levels of information obtained for our area of study: the southeast of the present-day province of Guadalajara, Spain. Using the longue durée approach conceptualised by Fernand Braudel (1976), historical cartography enabled us to delve deeper into two essential aspects, land use and the road network, while providing elements of diachrony that suggest changes in the landscape in specific and disruptive periods.

Keywords: map regression analysis; medieval landscape; historical cartography; Reccopolis; longue dureé analysis

1. Introduction

Over the past decade, landscape archaeology has made considerable efforts to take the paradigm to the next level. As a result, there can now be no doubt that landscape archaeology is much more than archaeology, including, for example, geography, environmental studies, anthropology, history and cultural heritage management. However, it also combines methodologies and techniques such as archaeobiology, archaeometry, spatial analysis and isotope studies. Moreover, cutting-edge technologies, such as deep learning, are contributing to a new means of approaching research, although more 'traditional' tools, such as archaeological prospection, soil sampling and archival data (including maps and census data), are still crucial elements of analysis.

This paper provides an overview of the most recent results from spatial analysis based on one of landscape archaeology's traditional tools, namely historical cartography [1–3]. The aim is to assess its limits and potential as a viable tool for the 21st century. In the case of Spain, the data available for consultation and analysis have increased immeasurably over the last two decades [4–6], which increases the possibilities of designing more complex sampling strategies in relation to the area. From this perspective, we would like to reflect on the limits of such vast amounts of data by reconstructing a specific historical landscape from the Middle Ages to the present: the southeast of the present-day province of Guadalajara. Taking into account our proposal, our approach does not involve a closed model of analysis but the possibility of incorporating modifications, given that there is an emphasis on some aspects that may be key to the development of the methodology. The analysis also has a strong microspatial component. The study framework comprises an approximation to an area of ca. 10 km^2, which is a representative area of a historical space marked by important historical transformations.

Our territory of analysis is one of the most significant ones of the Early Medieval Ages of Ancient Hispania. It focuses on an area controlled by the city of Reccopolis (Figure 1), the only archaeologically attested city founded by the Visigoths in the Iberian Peninsula. Consequently, it is one of the most important locations for the study of the Visigothic and Early Islamic periods. Named by King Liuvigild in AD 578, the city played an important role in the landscape and remained the central hub of the area throughout the 8th and 9th centuries AD. Although the cartographical data for the period are limited and sometimes even absent (if we compare them with subsequent periods), our archaeological research over the past few years has enabled us to gain a clear picture of the area from its very beginnings. Hence, we have been able to trace the area's landscape evolution from then until now.

Figure 1. An image of Reccopolis and its surrounding area and a map of its location.

This study was made possible by gathering and processing important historical cartography data from the 17th century using geographical apps. Moreover, if we add the written sources that we obtained for the medieval period to the data, the dialogue between them is key to analysing road networks, settlement patterns and vegetation, including agriculture and water resources, as well as change processes versus a long duration in the formation dynamics of historical landscapes. The aim of this study, then, was to assess which natural and/or anthropic elements have been maintained in this landscape, and since when, and which features were distorted and/or transformed, if any.

2. Theoretical and Analytical Framework for an Early Middle Ages Landscape

With the fall of the Roman Empire, most of the Iberian Peninsula's rural area was once again restructured between the 6th and 7th centuries through various levels of settlement: villages and farms, intermediate centres of power such as the so-called hilltop sites, together with churches, monasteries and a limited number of aristocratic residences [7]. However, our case study is different given that it focuses on the foundation of a new settlement in an almost empty landscape or, at best, with no significant Roman habitats. There are three processes that clearly indicate the new organisation of the area: the transformation of the road network, the emergence of new settlements and, lastly, the modification of the agri-system [8,9].

The new road network is intrinsically linked to the other two elements, namely settlements and agriculture, given that it is a consequence of the foundation of new settlements dedicated to the exploitation and new organisation of the productive space for agriculture. However, the process involved two organisational levels. The first was the location of Reccopolis as a central hub of communication between the centre of the Iberian Peninsula and Levante, connecting the capital, Toledo, with the coast by following the left bank of the River Tagus. Second, in parallel to this reorganisation, the immediate area around Reccopolis was restructured through a series of radial pattern secondary roads that connected the main urban centre with secondary agglomerations and their associated land parcels, similar to other areas in Europe in the Early Middle Ages.

The organisation of the land was perhaps not only affected by the new road and settlement network but also by the need for a strategy to mitigate environmental difficulties, especially the aridity associated with the Dark Ages Cold Period [10]. The vegetation in the dry Mediterranean-style landscape was adapted to the steppe-like open spaces, similar to that described in nearby areas such as the Tablas de Daimiel (Ciudad Real) [11], the Castro de Peña Moñuz (Olmeda de Cobeta, Guadalajara) [12] and the Taravilla lagoon (Taravilla, Guadalajara) [13,14]. A shortage of water would have been a problem and must necessarily have affected agricultural production, which is attested to in the region by the transformation of Lake Somolinos (Guadalajara) into a wetland sometime between the 6th and 8th centuries [15] (p. 49).

The data obtained from the sampling performed in the area indicate that the Reccopolis landscape was characterised by being eminently open, given that the arboreal pollen did not exceed 40%. In addition to pines and conifers, which were the most represented taxa in the arboreal group, there was also the important presence of olive trees and trees belonging to the Quercus family (oaks, holm oaks, cork oaks), predominantly the evergreen varieties. A small number of hazelnut and walnut taxa were also documented, not exceeding 3% of the vegetation, which indicates the existence of species that had important productive and economic interest. There was also a riparian forest (ash and elm trees), representing 7% of the vegetation [9].

In relation to the steppe-like open landscape, of particular importance is the existence of the herbaceous group of the Aster family and grasses, especially Artemisia and Ephedra, which highlights the presence of pastures, as well as ruderal plants, which are also associated with cultivated fields. However, there is currently an evident absence of taxa associated with cereal crops, which is unsurprising if we consider that crops are self-pollinating species and that, in all likelihood, the fields were far from where the samples were taken.

Moreover, in connection with the foundation of the new settlement, it is highly likely that an organised process of deforestation was undertaken—possibly in large part due to the action of fires—precisely to favour the opening of new fields for agricultural and pastoral activity involving domestic livestock, with a clear predominance of ovicaprines (domestic sheep and goats). This would explain the development stage reached by the nitrophilous taxa (Plantago, Rumex, Urtica, Chenopodiaceae, Geraniaceae, Rubiaceae and Polygonaceae).

In addition to the area's agrarian characterisation, the use of remote sensing technologies (LiDAR, thermal photography), together with infrared aerial photography, has also been key in the study of the agri-systems around Reccopolis. LiDAR technology enabled us to identify various levels of data and design a specific digital surface model (DSM). This has been especially useful in areas where the microspatial parcel division is linked to the topographical organisation by means of terraces (near the settlement of Arroyo Badujo). Traditional aerial photography is not capable of providing precise visualisation of the structure of this type of space. However, microspatial research on the Arroyo Badujo valley, combining intensive prospecting with a specific DTM-LiDAR survey of the area, has identified the outline of an aqueduct and traces of three terraced curvilinear morphological units located between Reccopolis and La Paeriza (Figure 2), in an environment where the rectangular organisation of the landscape dominates. This curvilinear morphology was already evident in the Iron Age in areas of Central and Southern France [16] (p. 22), [17] (pp. 109, 384, 523). However, in the area between Reccopolis and La Paeriza, there is no evidence of habitation prior to the 6th century, which is when the new landscape structure appeared. As a result, the relationship between the curvilinear morphological units and the new road network created when the city was founded, with the characteristic crossroads as also seen in La Paeriza, and the new settlement structure constitute the elements that support the discussion of this hypothesis.

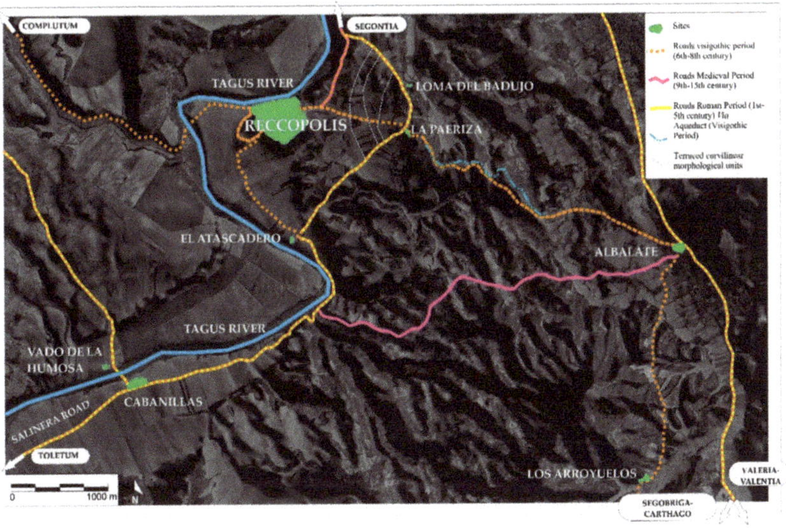

Figure 2. Reccopolis road reorganisation around the site, 6th-century settlement pattern.

3. Methodology

The longue durée analysis [18,19] was paramount to our approach, given that our aim was to examine transformations in a specific landscape in an extended period of time—from the Early Middle Ages to the present—and to draw conclusions in two ways. The first was in reference to historical trends and patterns, which usually occur slowly and are only perceived after a few generations but define the present-day landscape of the area. The second was whether it was possible to relate the information obtained from maps to our current knowledge of the landscape during the Visigothic period using landscape archaeology.

As noted in the Introduction, historical maps were the main source of interpretation used in this study. This enabled us to analyse historical processes from multiple perspectives [20]. We considered the area's landscape in the centre of the Iberian Peninsula with two clear parameters in mind: the present-day landscape and that created during the Visigothic period. The latter has already been reconstructed through archaeological prospection and excavation, remote sensing, archaeobiology, palynology and environmental analysis.

3.1. Study Area

Our basic unit of analysis is the so-called "Alfoz de Zorita" (Figure 3), an administrative district created in the 10th century. Although its etymology is clearly Arabic in origin, it remained in use throughout the Middle Ages and Modern Era. An "alfoz" is the rural area that surrounds a city or town and can comprise various population units, one of which acts as the capital, in this case Zorita, the nucleus that replaced Reccopolis from the 9th century onwards. The new al-Andalus settlement is in fact 800 m east of the Visigothic city. It should be noted that while we are familiar with the "alfoz" of the High Middle Ages (13th–14th century), we have to assume that it would be similar to that founded between the 9th and 10th centuries, given that Zorita maintained its undisputed status as the region's capital from its foundation until the 15th century.

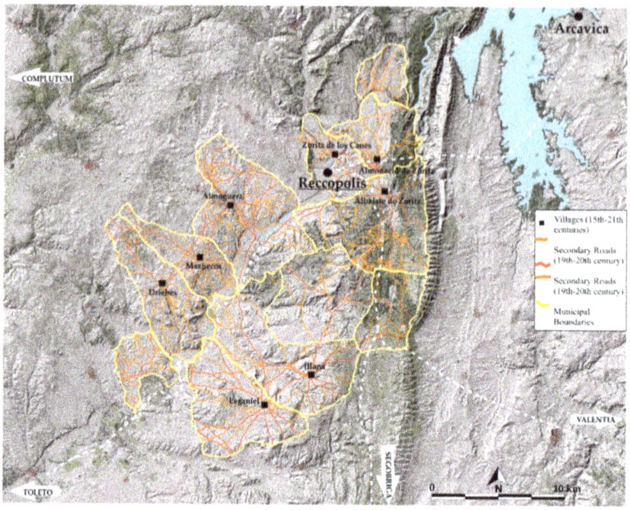

Figure 3. Alfoz de Zorita central area: territory and historical road network.

The first time that the Alfoz de Zorita demarcation was documented in written sources was in 1124 [21], which included the villages and settlements of the present-day towns of Zorita, Illana, Aldovera, La Bujeda, Albalate, Almonacid, Yebra, Sayatón, Pastrana, Escariche, Escopete, Loranca, Hontova, Hueva, Moratilla, Valdeconcha, Fuentelaencina, Alhóndiga, Auñón and Berninches.

3.2. *Sources of Information and Regression Map Analysis*

To approach a study of these characteristics, it is important to highlight that historical cartography offers very diverse information that is not only found on the map itself. On numerous occasions, as will be seen, the written information attached to maps is also very valuable for the interpretation and assessment of the map's extracted data. For this study, we performed a map regression analysis ranging from the 21st century to the medieval period. As already mentioned, although we did not use contemporary cartography and/or topographical data, we were familiar with the landscape due to previous studies. We worked backwards from more modern maps to earlier maps of the area to determine change and/or continuity and to locate former features. The entire data and indicator set was uploaded to an open-source GIS program (QGIS).

The cartographic data were processed in various ways. The data from 19th and 20th century maps were transformed from rasters into vectors. Priority was given to the construction of historical parcels, starting with the introduction of roads, including the footpath network. The 19th-century agricultural parcels were transformed into CAD format. The information provided by modern and medieval land registers was added to the cartographic base and supported with archaeological studies on specific areas, including abundant toponymic and parcel information. In some cases, however, it was not possible to perform a complete topographical reconstruction to reduce the data to the geographical locations of specific productive elements. The geographic dataset was linked to a specific MySql database. In fact, various data analysis models were designed in an ad hoc MySql structure. One of the most complex elements was managing non-quantitative data. In this regard, a first approximation was to construct a regression map model in which the data were contrasted with extensive archaeoenvironmental analyses. This enabled us to resolve questions relating to settlement patterns and the influence of the natural environment in a much broader sense.

We divided the information into six main data groups (Table 1): (1) 21st–20th century; (2) 19th century—mainly based on the "Diccionario geográfico-estadístico-histórico de España y sus posesiones de Ultramar"/Geographical-Statistical-Historical Dictionary of Spain and its overseas possessions [22]; (3) 18th century, based on the Catastro of Ensenada[1] [23], which is the first microspatial representation of the geography of Spain; (4) 16–17th century, using the Relaciones topográficas de los pueblos de España, hechas de orden de Felipe II/Topographic Relations of the Towns of Spain, commissioned by Philip II [24]; (5) the medieval period (11th–13th century), based on the Diplomática Arriacense [25]; and (6) Visigothic and Islamic written sources, which provide various levels of information.

Table 1. Documents and data used for the archaeogeographical analysis.

Document Source	Date	Name	Scale of Analysis	Data Spatiality (GIS)	Type of Information
1	1910–1970	Mapa Topográfico Nacional (Instituto Geográfico Nacional)	1:25,000/1:50,000	Yes	Topography, toponymy, roads, land use, water resources
1	1956–1957	Army Map Service (USA)	1:32,000	Yes	Aerial imagery
2	1890–1953	Planimetrías/Altimetrías (Instituto Geográfico Nacional)	1:25,000/1:50,000	Yes	Topography, toponymy, roads, water resources
2	1893–1953	Planos de Población (Instituto Geográfico Nacional)	1:2000/1:5000	Yes	Toponymy, roads, urban
2	1890–1953	Cuadernos de Campo (Instituto Geográfico Nacional)	1:2000/1:5000	Yes	Topography, toponymy, roads, water resources
2	1887	Mapa Militar (Instituto Geográfico Nacional)	1:200,000	Yes	Topography, toponymy, roads
2	1845–1850	Diccionario geográfico-estadístico-histórico de España y sus posesiones de Ultramar (Pascual Madoz)	-	No	Toponymy, roads, land use, water resources
3	1781	Castilla La Nueva. Cuenca y Guadalajara. Mapas generales. (Tomás López)	1:500,000	Yes	Toponymy, roads
3	1749–1759	Catastro de la Ensenada	-	No	Toponymy, roads, land use, agrarian information
4	1691	Averiguaciones de las antigüedades de Cantabria, enderezadas principalmente a descubrir las de Guipuzcoa, Vizcaya y Alava (Gabriel Henao)		No	Toponymy
4	1577	Crónica General de España (Ambrosio de Morales)		No	Toponymy
4	1575–1578	Relaciones topográficas de los pueblos de España, hechas de orden de Felipe II	-	No	Topography, toponymy, land use, agrarian information, production spaces
5	1156	Privilegio de Alfonso VII concedido a los mozárabes que fueron a poblar a Zorita (1156) Información toponímica, viaria, de paisaje y de usos productivos		No	Toponymy, land use, production spaces, roads
5	1132	Donación de Vállaga y Hueva, hecha por Don Alfonso el Emperador, a favor de Don Galindo, maestre de Santiago (1132). Información toponímica, agraria y de los usos productivos	-	No	Toponymy, land use, agrarian information
6	c. 1000	Ajbār mulūk Al-Andalus (Aḥmad al-Rāzi)	-	No	Toponymy
6	c. 590	Chronicon (Juan de Biclaro)		No	Toponymy

As explained, we decided to use a map regression analysis based on present-day landscape features to see whether it was possible to recreate the landscape of the Early Middle Ages and whether this type of analysis continues to be highly efficient in the 21st century or, in contrast, whether it has been completely superseded by more modern techniques and approaches.

The contemporary data were extracted from data available at the Spanish Geographic Institute, which reports to the Spanish Ministry of Public Works and Transport (www.cnig.es) (accessed on 13 June 2023). The advantages of using these data are numerous, including the availability of data collections and the fact that they are georeferenced in ETRS89 [26]. Freely accessible data have been available on a specific platform since 2010, which adheres to the EU INSPIRE directive, and can be combined to produce interoperable geographic information systems (GIS) and spatial data infrastructures (SDI) to provide a GIS node and geoportal and directly generate historical data with links to specific cartography from the 15th century.

The Diccionario geográfico-estadístico-histórico de España y sus posesiones de Ultramar is a geographical handbook of Spain, frequently referred to as "el Madoz", because it was edited by Pascual Madoz [22], a Spanish politician and statistician who held various senior political positions, such as Minister of Finance, during the "Bienio Progresista" (Progressive Biennium) during the reign of Queen Isabel II of Spain. Originally published in 16 volumes between 1845 and 1850, this reference work is outstanding for its sheer amount of data, its systematic organisation and for covering the entire area of Spain. In this study, we used Volume 7, which focusses on Faba in the province of Guadalajara. The handbook proves to be an interesting basis for the identification of ancient population centres, which have subsequently been identified as archaeological sites. It also highlights abandoned buildings and spaces for social use, which provides important information for the identification of cultural heritage.

However, the use of cartography to obtain cadastral information did not emerge until the late 19th century. In addition to "el Madoz", also particularly useful from the perspective of landscape reconstruction are the so-called "Minutas", which include a varied set of topologies: kilometre sheets, urban plots, altimetries and boundary drawings. Together with an initial survey of population centres, most of the "Minutas" for the study area were drawn up between 1870 and 1890 prior to industrialisation. The specific treatment of this data set, in which a spatial rather than a quantitative value predominates, as in the case of the so-called 'Bourbon cadastres', made it necessary to design a second database to include data on the administrative geographical boundaries, water resources and constructions relating to water and land use, and other structures, some linked to religious or even heritage buildings. In addition, one of the more unusual aspects is the inclusion of so-called production buildings (bee farms, mills, warehouses, etc.) in the database, as well as the articulation of the road network based on a distinction between primary and secondary roads.

In reference to the road network and based on the "Minutas" drafted in the late 19th century, we were able to analyse the area's road network. The development of roads enabled us to establish temporal superposition criteria closely linked to specific archaeological sites. From the analysis of the relationship between roads and sites, we observed a close relationship between medieval occupation and the organisation of the area, especially in the locations of settlements, which is particularly useful in the design of field sampling strategies. From the analysis of the relationship between roads and sites, based on the radial and linear configuration that was one of the central elements in the definition of Archaeogeography defended by the French School in the 1980s and the important work of various research groups in Spain, under the definition of Archaeomorphology [27], we were able to determine the structure of the area. In this regard, roads become a diachronic reality that can be defined as '[. . .] today's road systems are composed of an accumulation of strata that contain indications of their history' [28] (p. 69) within a model that operates 'at various spatial and temporal scales' [29,30] that have gradually been integrated into a reflection on the evolution of parcels of land as a reflection of archaeological materiality.

The analysis of the road network was neither simple nor homogeneous. A large part of the toponymic or cadastral references during the early Modern and Middle Ages lack specific spatiality. Moreover, we have data that are even more ephemeral for earlier periods.

For the 18th century, we used a large-scale census and statistical investigation from 1749 until 1756, commissioned by King Ferdinand VI of Spain and his minister the Marquis of Ensenada—known as the "Catastro de Ensenada" (Cadastre of Ensenada). It was conducted in the Crown of Castile, excluding the Basque Country, Navarre and the Crown of Aragon (fortunately, our study falls under the Crown of Castile). It includes not only geographical data, but also information about the population, territorial properties, buildings, cattle and all manner of revenue and trades. As a result, it is considered the best statistical register from the pre-statistical age of the Old Regime in Europe. This set of initiatives was maintained by successive Bourbon regimes throughout the 18th century, to which others were added, such as the "Censo del Conde de Aranda" (1768–1769) and the later "Censo de Floridablanca" (1787) [31]. Both primarily focus on demographic questions and provide fewer fiscal and cartographic data than the "Catastro de Ensenada". One of the key aspects of the latter was the design of a specific cartography, which included "agrimensores y geómetras" (surveyors and geometers) to draft individualised maps of each of the municipalities where the data were collected. However, the results are clearly heterogeneous. The lack of an extensive corps of geographers in the mid-18th century was a major obstacle to the development of the proposed objectives. In many cases, local surveyors had to be used to gather schematic data on municipal boundaries. Fortunately, the data are detailed enough to identify their representations as roads, toponyms and urban spaces.

It should be noted that this level of information was extremely important for our study given that we organised the data on the study area in a database based on the structure of 18th-century land registers. They primarily provide economic, land use and productivity data. Toponymic data are limited and primarily refer to geographical elements; therefore, they have limited capacity to provide information on human settlements. However, the wealth of data on agricultural use, geographical characteristics and landscape development provide an evolutionary nuance that is impossible to observe in other cartographies.

For example, we were unable to obtain this type of information for the 17th century and did not find a quality data record until the 16th century, such as the "Relaciones topográficas de los pueblos de España, hechas de orden de Felipe II" [24], which provides a detailed description of the settlements in Spain during the reign of Philip II, as well as the main characteristics of the ecosystem of the Crown's territories in the Iberian Peninsula in the 1570s. The "Relaciones topográficas" is perhaps one of the richest sources available for toponymic reconstruction, especially the "Aumentos" (a review and expansion of 16th-century data), which also undertakes minimal historical reconstruction, proposes hypotheses and mentions depopulated areas and even social and religious practices around certain spaces. It also indirectly mentions epigraphic elements, ancient roads and bridges. However, this work cannot be considered true cartography, given that it does not provide a representation of the geo/spatial reality.

However, in addition to the "Relaciones topográficas", one of the most interesting aspects of the medieval period is how various geographers, commissioned under the Habsburg dynasty, began to represent the Iberian Peninsula. In the first representations of Castile from 1606 to 1632[2] [32], Alfoz de Zorita is conspicuous by its absence, or has a very limited presence. Its absence is quite significant as it underlines a fundamental issue in the area: the loss of the region's importance during the transition to the Modern State as a result of the diminishing political and economic weight of the Order of Calatrava[3] [33,34], which, until then, had been the principal political body. As a result, the area was relegated to a mere transit zone between two large areas of the southern Iberian sub-plateau, crossing the Sierra de Altomira.

The "Diplomática Arriacense" [25] is a medieval document that provides important information on the toponymy and boundaries of the Alfoz de Zorita in the late 12th century. In certain cases, some of the population centres mentioned are considered uninhabited in "el Madoz" (19th century), which leads us to defend a model of occupation that was consolidated in the High Middle Ages and maintained until the Modern Era. These elements have been cross-referenced with the results of archaeological surveys and documentation on the Alfoz de Zorita as part of the Order of Calatrava in the second half of the 16th century over an area of 800 km^2. Although it can be a useful tool, direct toponymic identification entails its risks. Examples are the potential duplication of names throughout the same area and spatial ambiguity, reflected in the displacement of 'sites'.

4. Results and Discussion

The aim of this study was not to perform a detailed analysis of the evolution of a landscape from the Middle Ages to the present day, but rather to take a methodological approach to highlight some of the key elements in its constitution. As mentioned above, our study area is located in the southeast of the present-day province of Guadalajara and is characterised by its tabular relief, which is traversed by the River Tagus and some of its tributaries. The moorland is interspersed with watercourses, opened precisely by the action of water, whose substratum was formed fundamentally by a prolonged process of endorheic sedimentation of materials eroded from the Central System Mountain Range in the northwest and the Iberian System Mountain Range in the southeast [35]. This section of the "Meseta" or plateau is relatively high, ranging from 600 MASL in the valley bottoms to 1000 MASL in the highest foothills, such as the Sierra de Altomira.

Our study area is largely defined by the presence of the middle course of the River Tagus (Figure 4), which was dammed in the early 20th century. This has affected the sedimentation process and changes in the riverbed, as well as navigability. The analysis of the 19th-century documentation gives us a glimpse of small natural modifications in the course and structure of the meanders. The level of detail in the cartography prior to the Middle Ages does not allow us to accurately measure the oscillations. However, we know that this process must have been very intense, given that the geomorphological studies that we performed in the river's palaeo-course, in the west of the area immediately surrounding the Reccopolis site, showed that the course of the river has shifted by around 500 m (Figure 4) from the Early Middle Ages to the present day.

According to the Geological Map of Spain (1:50,000, IGME), it is important to note that the area is characterised by the presence of fluvial sediments composed of gravel, sand and silt. This fact, plus the absence of evident linearity in its modification, suggests that it is not an anthropic diversion or channelling (Figure 5). There is no defined pattern, so it seems evident that it would be a natural modification of the palaeochannel, which we were able to verify by comparing LiDAR data with earlier cartography.

Several data sources were combined for the study. On the one hand, a geomagnetic survey confirmed the fluvial origin of the palaeochannels, which was extended using geological sampling. The data were processed in QGIS, combined with a LiDAR flight specific to the area of study. Historical maps from the 19th century were also used to analyse the changes in the course of the River Tagus, which confirmed the frequent alteration of the channel as a result of floods.

One of the key elements in the study was the reconstruction of parcels and land use. In this regard, 19th-century maps made it possible to identify agricultural uses such as the traditional Mediterranean style of cultivation. In parallel, they also made it possible to determine exploitation strategies by identifying areas of wasteland and fallow land, allowing us to determine the rotation of crops. Although the medieval documentation is very generic, it did allow us to reconstruct a landscape composed of vineyards, olive trees and cereals for flour production, based on the mills on the River Tagus.

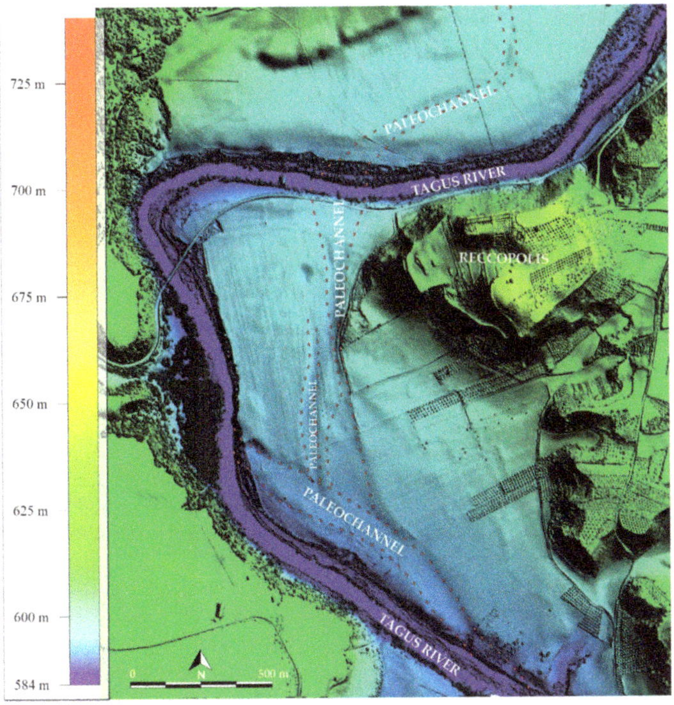

Figure 4. Modification of the course of the River Tagus as it passes through Reccopolis.

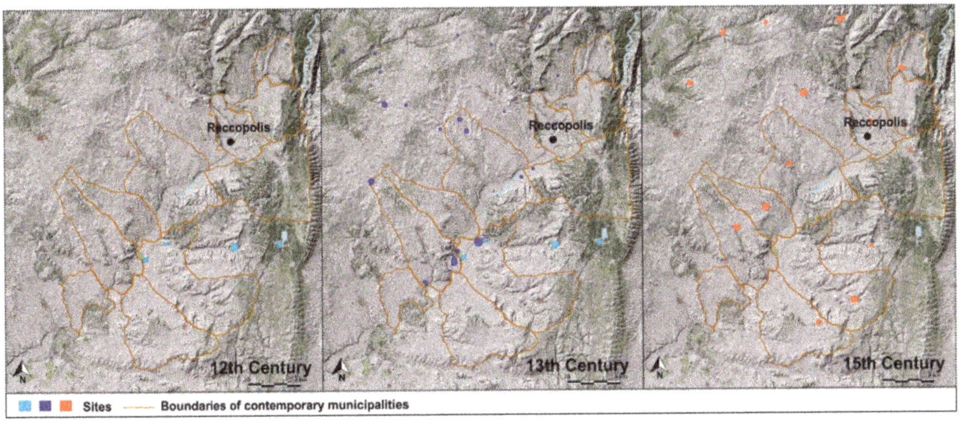

Figure 5. Evolution of settlement patterns (12th–15th century).

The reshaping of the landscape around the year 1880 was subsequently studied in a comparative analysis of cultivation and parcel dimensions using a panorama of land use provided by the first extensive aerial photography flights in the region, which started in 1956. In this brief window of time, less than 100 years, it is significant that a very intense process of deforestation can be observed, which suggests that the anthropogenic transformation of the ecosystem may have occurred relatively quickly and must have been very disruptive. Earlier, between 1790 and 1880, intense changes in the landscape were also evidenced through the decrease in forest cover, which was probably linked to the disentailment process. This was the auctioning of the lands belonging to the Order of

Calatrava to other landowners. However, with the exception of the logical transformation of land parcels and their ownership, even today, the predominant style of agriculture is still traditional Mediterranean cultivation.

The continuity of land use, excluding forest cover, was not extended to those settlements that underwent various stages of restructuring from the Visigothic period until the 15th century. However, from the 15th century onwards, as can be seen from historical maps and contemporary sources, most of the settlements were already established and have continued to the present day.

As already highlighted, the foundation of Reccopolis had an important impact both on the creation of new nuclei and on the new road network. The rural areas linked to the city were maintained until around the 8th–10th century, when they were abandoned, and the new "medina" of Zorita was founded. It is likely that this new city led to the creation of new rural settlements in the area, which are, however, currently undetectable in archaeological records.

From the 12th century onwards, a new social landscape was constructed as a consequence of the Castilian occupation of al-Andalus. New settlements arose that played a decisive role in the control of the area and its spatial hierarchy in opposition to pressure from the Islamic state, which turned the study area into a frontier zone until the 13th century. From then on, the area was predominantly used for agriculture, which lasted until the late 14th and early 15th centuries, when a significant decline in the number of settlements saw the disappearance of around 20–30% (Figure 5).

The evolution of settlement patterns analysis was the result of developing an extensive archaeological survey programme, reviewing textual sources and processing historical maps. The latter gave rise to high-resolution model designs up to the mid-19th century. From then on, cartographic construction, in which toponymy also played a prominent role, had to be supplemented with archaeological work in the field. This consisted of archaeological surveys and, occasionally, magnetometry research. As already mentioned, all the data were processed in QGIS, which included various levels of information—for example, land use.

One of the key elements in the organisation of settlement patterns is road networks. The reconstruction of the road network, at both a micro- and macrospatial level, formed part of the reorganisation of the area during the Visigothic period. However, the process had two organisational levels. The first was the location of Reccopolis as a central hub of communication between the centre of the Iberian Peninsula and Levante, connecting the capital, Toledo, with the coast by following the left bank of the River Tagus (Figure 6). Prospecting work in the area around Reccopolis has led to the discovery of a connection between Reccopolis-Complutum (an important nucleus in Roman times) and Toledo via a well-known Iron Age II road [36], which is still visible in 19th-century cartography. At that time, it was known as the Vía Salinera, an important communication intersection between the provinces of Cuenca, through the municipality of Illana (Guadalajara), to the present-day region of Madrid [36–38]. Second, in parallel to the general reorganisation process, the immediate area around Reccopolis was restructured (Figures 4 and 6) through a series of radial pattern secondary roads that connected the main urban centre with secondary agglomerations and their associated land parcels, similar to other areas in Europe in the High Middle Ages.

The road network endured over a long period of time, from the 13th century onwards. However, although the main roads were maintained, the network also became more complex as a result of the increase in the number of settlements. The structure was modified from the 16th century onwards, as a consequence of Madrid becoming the capital city and the fact that the area was becoming more marginalised, which can be seen in the cartography of the 19th and 20th centuries.

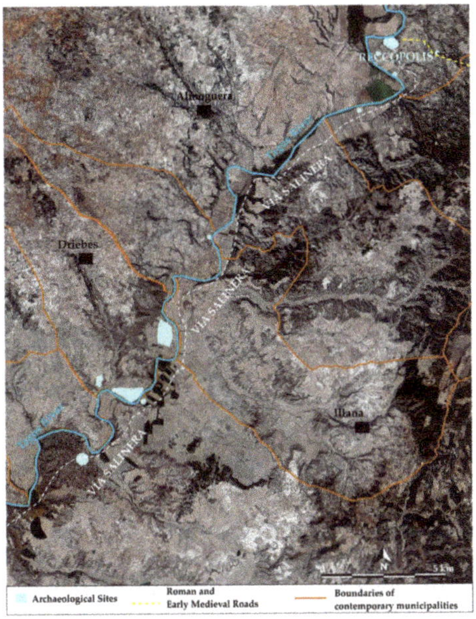

Figure 6. The Vía Salinera, the backbone of the medieval and Visigothic territory.

These are some of the main results of our investigation. As we have seen throughout the study, historical cartography has proven to be an extremely useful tool for the analysis of landscape archaeology. Indeed, it provides an invaluable source of geospatial information on what the landscape looked like in the past and how human groups were distributed, allowing us to better understand the distribution of settlements, infrastructure, communication routes and other relevant cultural and natural features. In fact, several elements have been identified that provide a significant improvement when carrying out an analysis within the discipline of landscape archaeology.

1. Temporal context: Historical maps show how the region was perceived in earlier times. By comparing modern maps or current satellite images with ancient ones, we have detected changes in the landscape, suggesting the presence of archaeological sites and socio-cultural changes.
2. Detection of missing features: Some structures and settlements may have disappeared over time due to erosion, abandonment or new urbanisation processes. Historical maps can reveal the locations of features that are no longer visible in today's landscape, which helps to direct archaeological investigations.
3. Identification of patterns and trends: By examining historical maps from different periods, we have been able to identify settlement patterns, areas of cultivation and other relevant data. These patterns have provided valuable information on how societies interacted with the landscape and with each other.
4. Planning archaeological research: Historical mapping can help archaeologists to identify areas potentially rich in archaeological remains, allowing them to plan excavations and field surveys more efficiently.
5. Cultural contextualisation: Historical maps can also provide information on toponymy, historical boundaries and other cultural aspects that may be essential in understanding the historical evolution of a territory.

All these elements show that historical cartography serves as a valuable form of documentation. By preserving the maps properly and combining them with remote sensing and georeferencing technologies, accurate records of the past landscape and its relationship with archaeological sites can be analysed and preserved.

5. Conclusions

This study emphasises the usefulness, even today, of historical cartography for landscape analysis, which, in recent years, has gradually been superseded by new remote sensing technologies and other cutting-edge predictive approaches. As we have seen, historical cartography is essential in performing a longue durée analysis, which enables us to have a global perspective to detect transformations and continuities in a specific historical space. However, it is difficult to obtain a precise temporal picture given that, in general, we move in broad chronological arcs. Notwithstanding, this type of analysis, in combination with other methods, is an essential basic step in performing a landscape archaeology study. In fact, the results obtained raise new questions about the importance of the earlier ecological substratum and historical heritage as indissoluble elements in the construction of a landscape.

In our case study, we were able to verify that specific socio-political actions, such as the foundation of new settlements, led to the new organisation of the productive space and the road network that connected them. The latter, in fact, has remained practically unchanged from the Middle Ages to the present day. We also observed that, due to rotation, the types of crops grown remained practically the same. The greatest fluctuations were found in relation to population centres before the 15th century, which varied in number and location as a result of social conflicts, epidemics and moments of growth and economic stability.

In summary, historical cartography, and especially the exponential growth of digitisation, is a significant starting point in identifying and analysing anthropised landscapes and, more specifically, the perception or continuity of long-term processes in the construction of historical landscapes.

Author Contributions: Conceptualization, P.D.-B. and M.C.-P.; methodology, P.D.-B. and M.C.-P.; software, M.C.-P. and P.D.-B.; validation, P.D.-B. and M.C.-P.; formal analysis, P.D.-B. and M.C.-P.; investigation, P.D.-B. and M.C.-P.; resources, M.C.-P. and P.D.-B.; data curation, M.C.-P.; writing—original draft preparation, P.D.-B. and M.C.-P.; writing—review and editing, P.D.-B. and M.C.-P.; visualization, M.C.-P. All authors have read and agreed to the published version of the manuscript.

Funding: This research was supported by the SUBRECC project (PID2022-136708NA-I00) and the Research Group Archaeology: Landscapes, Colonialism and Cultural Heritage, at the Universidad de Alcalá (Madrid, Spain).

Institutional Review Board Statement: Not applicable.

Informed Consent Statement: Not applicable.

Data Availability Statement: Not applicable.

Acknowledgments: Pilar Diarte-Blasco is currently a Ramón y Cajal Research Fellow (RYC2018-025646-I) at Spanish National Research Council in the Department of Archaeology and Social Processes (Instituto of History, Madrid), funded by the Ministry of Science and Innovation of Spain (MICINN). This research was supported by the SUBRECC project (PID2022-136708NA-I00) Manuel Castro Priego belongs to the Research Group Archaeology: Landscapes, Colonialism and Cultural Heritage, at the Universidad de Alcalá (Madrid, Spain), which has also contributed to this research. We are grateful for the collaboration of Professor Lauro Olmo-Enciso who initiated a first approach to the study area within the Spanish Ministry of Science and Innovation project HAR2009-11627 Dynamics and Construction of the Medieval Landscape.

Conflicts of Interest: The authors declare no conflict of interest.

Notes

1. https://pares.mcu.es/Catastro/.
2. https://centrodedescargas.cnig.es/CentroDescargas/index.jsp.
3. The Order of Calatrava was a military and religious order founded in the 12th century in the Kingdom of Castile, with the initial objective of protecting the town of Calatrava, Ciudad Real. In a relatively short time, its immense human and economic resources gave the Order sweeping political and military power, which lasted until the end of the Reconquista. The order owned lands and castles all along the border between Castile and Aragon and exercised feudal lordship over thousands of vassals [33,34].

References

1. Gillespie, S.D. Archaeological Drawing as Re-Presentations: The Maps of Complex A, La Venta, Mexico. *Lat. Am. Antiq.* **2011**, *22*, 3–36. [CrossRef]
2. Sobala, M. Do historical maps show the maximal anthropopressure in the Carpathians? *J. Mt. Sci.* **2021**, *18*, 2184–2200. [CrossRef]
3. Sánchez-Pinto, I. Historical Mapping vs. Archaeology: Rethinking Fort Sancti Spiritus (1527–1529). *Hist. Archaeol.* **2023**, *57*, 220–251. [CrossRef]
4. Quirós Castillo, J.A. Early medieval landscapes in north-wet Spain: Local power and communities, fifth-centuries. *Early Mediev. Eur.* **2011**, *19*, 285–311. [CrossRef]
5. Sánchez Pardo, J.C. Power and rural landscapes in early medieval Galicia (400–900 AD): Towards a re-incorporation of the archaeology into the historical narrative. *Early Mediev. Eur.* **2013**, *21*, 140–168. [CrossRef]
6. Fernández Mier, M. Peasant Communities and Landscape Change in North-West Iberia. In *Polity and Neighbourhood in Early Medieval Europe*; Escalona, J., Vésteinsson, O., Stuart Brookes, S., Eds.; Brepols: Turnhout, Belgium, 2019; pp. 57–82.
7. Diarte-Blasco, P. *Late Antique and Early Medieval Hispania: Landscapes without Strategy?* Oxbow Books: Oxford, UK, 2018.
8. Olmo-Enciso, L.; Castro Priego, M.; Diarte Blasco, P. Transformación social y agrosistema en el interior peninsular durante la Alta Edad Media (s. VI-VIII d. C.): Nuevas evidencias desde Recópolis (Zorita de los Canes, Guadalajara). *Lucentum* **2019**, *38*, 355–377. [CrossRef]
9. Olmo-Enciso, L.; Castro-Priego, M.; Ruiz-Zapata, B.; Gil-García, M.J.; Galindo-Pellicena, M.; Checa-Herráiz, J.; Gómez de la Torre-Verdejo, A. The Construction and Dynamics of Early Medieval Landscapes in Central Iberia. In *Mediterranean Landscapes in Post Antiquity: New Frontiers and New Perspectives*; Gelichi, S., Olmo-Enciso, L., Eds.; Archeopress: Oxford, UK, 2019; pp. 104–128. ISBN 978-1789691900.
10. Helama, S.; Jones, P.D.; Briffa, K.R. Dark Ages Cold Period: A literature review and directions for future research. *Holocene* **2017**, *27*, 1600–1606. [CrossRef]
11. Gil García, M.J.; Ruiz Zapata, M.B.; Santisteban, J.I.; Mediavilla, R.; Lopez-Parno, E.; Dabrio, C.J. Late holocene environments in Las Tablas de Daimiel (south central Iberian peninsula, Spain). *Veg. Hist. Archaeobot.* **2007**, *16*, 241–250. [CrossRef]
12. Ruiz Zapata, M.B.; Gil García, M.J.; Arenas Esteban, J.; Martín Arroyo, T. El impacto del asentamiento celtibérico de Peña Moñuz (cuenca superior del Tajo) sobre el paisaje vegetal: Análisis polínico. *Geogaceta* **2014**, *55*, 83–86.
13. Moreno, A.; Valero-Garcés, B.; González-Sampériz, P.; Rico, M. Flood response to rainfall variability during the last 2000 years inferred from the Taravilla Lake record (Central Iberian Range, Spain). *J. Paleolimnol.* **2008**, *40*, 943–961. [CrossRef]
14. Moreno, A.; Pérez, A.; Frigola, J.; Nieto-Moreno, V.; Rodrigo-Gámiz, M.; Martrat, B.; González-Sampériz, P.; Morellón, M.; Martín-Puertas, C.; Corella, J.P.; et al. Medieval Climate Anomaly in the Iberian Peninsula reconstructed from marine and lake records. *Quat. Sci. Rev.* **2012**, *43*, 16–32. [CrossRef]
15. Currás, A.; Zamora, L.; Reed, J.M.; García-Soto, E.; Ferrero, S.; Armengol, X.; Mezquita-Joanes, F.; Marqués, M.A.; Riera, S.; Julià, R. Climate change and human impact in central Spain during Roman times: High-Resolution multi-proxy analysis of a tufa lake record (Somolinos, 1280 m asl). *Catena* **2012**, *89*, 31–53. [CrossRef]
16. Chouquer, G. Transmissions et transformations dans les formes parcellaires en France. Esquisse d'un schéma général d'interprétation. In *Actes du Colloque "Forêt, Archéologie et Environnement" 14–16 Décembre 2004*; Dupouey, J.L., Dambrine, E., Dardignac, C., Georges-Leroy, M., Eds.; l'Office National des Forêts, l'Institut National de la Recherche Agronomique et la Direction Regionale des Affaires Culturelles de Lorraine: Nancy, France, 2007; pp. 21–33.
17. Watteaux, M. *La Dynamique de la Planimétrie Parcellaire et des Réseaux Routiers en Vendée Méridionale. Études Historiographiques et Recherches Archéogéographiques*; Université Panthéon-Sorbonne: Paris, France, 2009; p. 571.
18. Braudel, F. Histoire et sciences sociales: La longue durée. *Ann. Hist. Sci. Soc.* **1958**, *13*, 725–753. [CrossRef]
19. Braudel, F. *El Mediterráneo y el Mundo Mediterráneo en la Época de Felipe II*, 2nd ed.; Monteforte Toledo, M.; Roces, W.; Simón, V., Translators; Fondo de Cultura Económica: Madrid, Spain, 1976.
20. Porro, J.M. La defensa y consolidación de las fronteras en el Septentrión novohispano: Geografía y desarrollos cartográficos (1759–1788). *Anu. Estud. Am.* **2011**, *68*, 19–50. [CrossRef]
21. Ballesteros-San José, P. Hermandades en Zorita, Almoguera y sus tierras. *España Mediev.* **1985**, *7*, 973–990.
22. Madoz, P.; Sánchez Zurro, D.; Sánchez Sánchez, I. *Diccionario Geográfico-Estadístico-Histórico de España y sus Posesiones de Ultramar*; Ámbito: Valladolid, Spain, 1985; ISBN 84-8604-790-0.
23. Camarero Bullón, C. *Madrid y su Provincia en el Catastro de Ensenada. Los Pueblos de Madrid 1750–1759*; Ediciones del Umbral: Madrid, Spain, 2001; ISBN 978-8495-457-56-1.
24. Catalina-García, J. *Relaciones Topográficas de los Pueblos de España, Hechas de Orden de Felipe II, Guadalajara*; Aache Ediciones: Guadalajara, Spain, 2002; ISBN 8495179792.
25. Pareja Serrada, A. *Diplomática Arriacense. Colección de Algunos Documentos, Publicados unos, Inéditos otros, que Pueden Servir para Planear o Ilustrar una "Historia de Guadalajara y su Provincia"*; Taller Tipográfico de la Casa de Expósitos: Guadalajara, Spain, 1921.
26. Potti Manjavacas, H.; Sevilla Sánchez, C.; Villalón Esquinas, M.; Sánchez Fanjul, J. SignA, el geoportal del IGN de España: Explotando lo mejor de las IDE y los SIG. *Rev. Cart.* **2017**, *95*, 111–128. Available online: https://www.revistasipgh.org/index.php/rcar/article/view/278/275 (accessed on 13 June 2023). [CrossRef]

27. Palet Martínez, J.M.; Arnau, G.M.; Hèctor, A.O.R.; Alamino, T.P. Els espais altimontans pirenaics orientals a l'Antiguitat: 10 anys d'estudis en arqueologia del paisatge del GIAP-ICAC/The high mountain areas in the Eastern Pyrenees in Antiquity: 10 years of studies in landscape archaeology from GIAP-ICAC. *Treb. D'arqueologia* **2017**, *21*, 77–97. [CrossRef]
28. Vion, É. L'analyse archéologique des réseaux routiers: Une rupture méthodologique, des réponses nouvelles. *Paysages Découverts* **1989**, *1*, 67–99. Available online: https://www.formesdufoncier.org/pdfs/vion.pdf (accessed on 26 April 2016).
29. Watteaux, M. The Road Network in the Longue Durée: A Reading Key of the History of Territories. *Open Archaeol.* **2017**, *3*, 149–174. [CrossRef]
30. Watteaux, M. Medieval Agrarian Landscapes seen by French School of Archaeogeography. In *People and Agrarian Landscapes. Approaches and Methods for an Archaeology of Peasantry in the Western Mediterranean*; Quirós Castillo, J.A., Narbarte Hernández, J., Eds.; Archeopress: Oxford, UK, 2022; pp. 15–51. ISBN 978-1803-274-386.
31. Instituto Nacional de Estadística (INE). *Censo de 1787. "Floridablanca". Tomo II Comunidades Autónomas de la Submeseta sur*; I.N.E. Artes Gráficas: Madrid, Spain, 1987; NIPO WEB: 729-16-002-2.
32. Hernando, A. *El mapa en España. Siglos XV–XVIII*; Centro Nacional de Información Geográfica: Madrid, Spain, 1995; ISBN 84-7819-067-8.
33. Ayala Martínez, C. La Orden de Calatrava, problemas de organización territorial y militar (Siglos XII-XIII). *Arqueol. Territ. Mediev.* **2003**, *10*, 157–179. [CrossRef]
34. Madrid Medina, A.; Villegas Díaz, L.R. *El Nacimiento de la Orden de Calatrava. Primeros Tiempos de Expansión (Siglos XII y XIII)*; Instituto de Estudios Manchegos: Granada, Spain, 2009; ISBN 978-84-9915-121-2.
35. González-Martín, J.A.; Vázquez González, A. *Guía de los Espacios Naturales de Castilla-La Mancha*; Junta de Comunidades de Castilla-La Mancha: Toledo, Spain, 1991; ISBN 84-7788-181-2.
36. Abascal-Palazón, J.M. *Vías de Comunicación Romanas de la Provincia de Guadalajara*; Diputación de Guadalajara: Guadalajara, Spain, 1982; ISBN 843-006-3781.
37. Palomero Plaza, S. *Las Vías Romanas En La Provincia De Cuenca*; Diputación Provincial de Cuenca: Cuenca, Spain, 1987; ISBN 84-505-6258-9.
38. Álvarez González, Y.; Palomero Plaza, S. Las vías de comunicación en Madrid desde época romana hasta la caída del reino de Toledo. In *Madrid del siglo IX al XI*; Real Academia de Bellas Artes de San Fernando: Madrid, Spain, 1990; pp. 41–64; ISBN 84-451-0243-5.

land

MDPI

Article

Landscape Exploitation and Biotic Resource Management at the Tossal de la Vila Hillfort through the Long Durée

Marta Pérez-Polo [1,2,*], Joan Negre [3], Ferran Falomir [4], Guillem Pérez-Jordà [5], Sonia de Haro [6] and Gustau Aguilella [4]

1 Departamento de Prehistoria, Historia Antigua y Arqueología, Facultad de Geografía e Historia, Universidad de Salamanca, Calle Cervantes, 2, 37008 Salamanca, Spain
2 Dipartamento di Storia Culture Civiltà, Università di Bologna, Piazza S. Giovanni in Monte, 2, 40124 Bologna, Italy
3 Institut Català d'Arqueologia Clàssica, Plaça d'en Rovellat, s/n, 43003 Tarragona, Spain
4 Servei d'Investigacions Arqueològiques i Prehistòriques, Diputación de Castellón, Av. dels Germans Bou, 28 (Museu), 12003 Castelló de la Plana, Spain
5 Departamento de Prehistoria, Arqueología e Historia Antigua, Universidad de Valencia, Avda. Blasco Ibáñez, 28, 46010 Valencia, Spain
6 Independent Researcher, Cl. Enseñanza, 11, 12001 Castellón, Spain
* Correspondence: mppolo@usal.es

Abstract: This paper focuses on the anthropic dynamics of environmental transformation and natural resource management in the specific case of the archaeological site of Tossal de la Vila in Castelló, Spain, a hillfort located at the end of the Eastern Iberian Cordillera. It presents two phases of occupation determined by multiple radiocarbon dating analyses: the first phase during the late Bronze Age (8th–7th centuries BCE) and the second at the beginning of the al-Andalus period (8th–10th centuries CE). The results of the comparison of the subsistence strategies and the biotic configuration of the natural environment in the same place at two different times have shown us that in the protohistoric period, the diversity of economic activities represented in the archaeobiological record is certainly heterogeneous, while in the al-Andalus period the complexity of these options is reduced to much less significant limits, possibly due to the different functionality of this space in both periods.

Keywords: Tossal de la Vila; hillfort; archaeobiological; economic strategies; late Bronze Age; early Middle Ages

1. Introduction

In this study, we present the most important archaeobiological results from recent research performed at the Tossal de la Vila archaeological site, with the aim of highlighting the economic strategies developed at the settlement during its two phases and the occupants' relationship with the surrounding biotic resources. We used a comparative approach between its only two moments of occupation: first, as a hamlet from the late Bronze Age to the early Iron Age (8th–7th centuries cal BCE), and second, as a hillfort from the early Andalusi period (8th–10th centuries CE) (Tables 1 and 2) [1–5].

The Tossal de la Vila archaeological site is located on a wide rocky outcrop, on one of the highest peaks (954 m above sea level) in the south of the Serra d'En Galceran, a mountain range that delimits two of the main communication routes in the central area of the province of Castelló, Spain (Figures 1 and 2). This gives rise to two very important natural corridors to the east and west of the mountain range, used by the many human communities that have inhabited the area since prehistoric times. To the west runs the Carbonera wadi, a tributary of the river Millars, which connect the route between the Mediterranean coastal area and the inland regions of El Maestrat. To the east is the corridor between Cabanes and Sant Mateu, a natural and traditional passage between the sub-region of La Plana de l'Arc, to the south, and the coastal plains formed by the rivers Cervol, Sénia and Ebre, to the north.

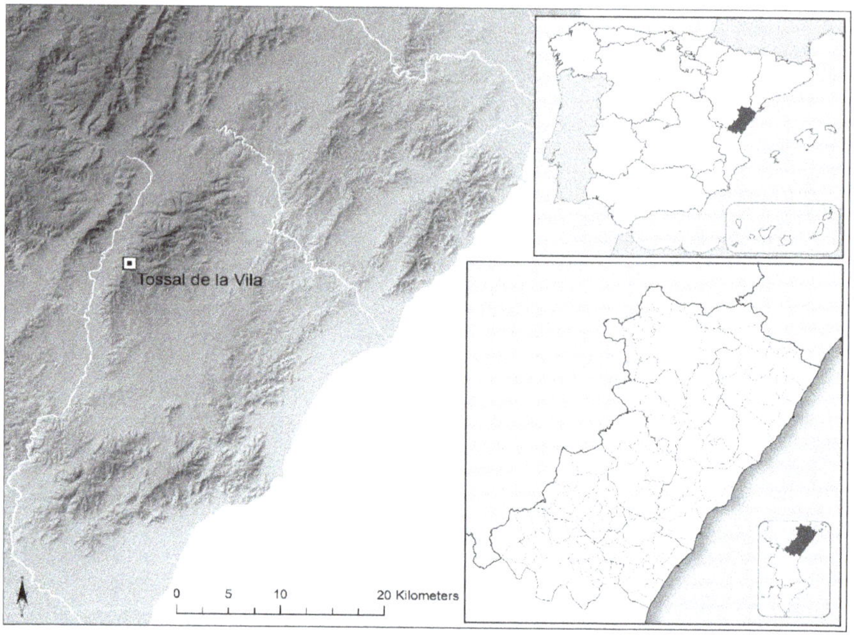

Figure 1. Location of the Tossal de la Vila (Serra d'en Galceran, Castellón).

Figure 2. Aerial photograph of Tossal de la Vila.

Table 1. Periods of occupation at Tossal de la Vila.

Period	Phase	Chronology	Dating Criteria
Late Bronze Age	Previous occupations	8th century BCE or earlier	Ceramics and stratigraphic
Late Bronze Age	Construction and use	8th–7th centuries BCE	TV-1; TV-2
Late Bronze Age early Iron Age	Destruction	First half of the 7th century BCE	TV-3
Emirate Andalusi period	Construction	8th century CE	Ceramics and stratigraphic
Emirate Andalusi period	Use	8th–10th centuries CE	TV-4; TV-5; TV-6
Emirate–Caliphate Andalusi period	Abandonment	10th century CE	TV-7; TV-8

Table 2. Absolute dating of Tossal de la Vila.

ID	Laboratory Code	Sample Type	Conventional Radiocarbon Age (BP)	Calibrated Age (1σ 68.2%)	Calibrated Age (2σ 95.4%)	Calibration
TV-1	Beta-336276	Charred material	2520 ± 30	800–770 cal BCE	800–760 cal BCE 680–670 cal BCE	IntCal09
TV-2	Beta-480635	Bone collagen	2460 ± 30	751–683 cal BCE 591–509 cal BCE 668–637 cal BCE 623–616 cal BCE 497–495 cal BCE	672–429 cal BCE 758–678 cal BCE	IntCal13
TV-3	Beta-393531	Charred material	2360 ± 30	540–410 cal BCE	745–685 cal BCE 665–645 cal BCE 550–400 cal BCE	IntCal13
TV-4	Beta-588420	Charred material	1220 ± 30	784–835 cal CE 844–877 cal CE 774–778 cal CE	770–888 cal CE 686–742 cal CE	IntCal20
TV-5	Beta-503127	Charred material	1200 ± 30	788–872 cal CE 774–779 cal CE	765–895 cal CE 714–744 cal CE 928–940 cal CE	IntCal13
TV-6	Beta-639687	Bone collagen	1160 ± 30	920–955 cal CE 870–898 cal CE 828–861 cal CE 776–787 cal CE	820–978 cal CE 772–790 cal CE 804–810 cal CE	IntCal20
TV-7	Beta-588421	Bone collagen	1130 ± 30	915–976 cal CE 889–902 cal CE	875–994 cal CE 830–851 cal CE 776–786 cal CE	IntCal20
TV-8	Beta-419185	Charred material	1050 ± 30	900–925 cal CE 945–990 cal CE	890–1015 cal CE	IntCal13

1.1. Theoretical Framework

The late Bronze Age and Medieval archaeology performed throughout the territory of Castelló still very much concentrate their attention on the traditional archaeological record, i.e., ceramics and constructive remains, while many other materials, such as archaeobiological remnants, are identified only occasionally. Nevertheless, the use of a multi-proxy perspective that includes the study of faunal, carpological and anthracological leavings, among other information sources, is increasingly considered as a more appropriate way to approach the historical analysis of economic strategies and palaeoenvironmental conditions in past societies [6–9]. In this sense, drawing on multiple lines of evidence allows us to better evaluate the conditions under which different communities would have developed broad subsistence and exploitation plans in the same place at different periods in time. Moreover, this viewpoint shows how helpful multi-proxy proposals can be for a refined understanding of past human–land interactions and the way in which political configuration could have affected that association and its correlation.

This assumption arises from the idea that different social actions of production, consumption and distribution cause variations in the archaeobiological record recovered during excavation and that these differences can be explained in terms of group survival decisions. In other words, consumption remains could be considered the material signature of different activities destined to fulfil the specific needs of a community, which in turn would indicate a specific pattern of behaviour towards the exploitation of different resources. Therefore, this paper presents a first approximation of Tossal de la Vila anthropic activity markers, the models connecting those archaeological proxies, such as archaebotanical or zooarchaeological remains, with the specific activities or strategies inferred from different information sources and interpretive paradigms coming from archaeological discussion [10–13].

Lastly, this text bases its discussion on the potential of archaeological comparative analysis, highlighting similarities and discrepancies in the management of the same agroecosystem in two different historical moments. In that regard, it is necessary to identify and reproduce the palaeoecological and historical setting of each one of them in order to evaluate the influence of environmental, social, economic and political causes on the inferred activities developed within the site. One of the most important issues confronting archaeological studies is the extent to which human behaviour is shaped by factors that operate cross-culturally throughout time, as opposed to those that are unique to particular societies [14] (p. 3). Lamentably, most common approaches to comparative analysis in archaeology focus on synchronic comparisons, due to a complex set of reasons, such as the interest in the study of historical questions regarding a particular society or time and the very nature of time and social change [15–17].

Our work operates in this scenario only, studying the material record of each phase to explore variation over time that could explain the causes that led these societies to adapt to a particular area. These kinds of comparisons, typically focused on major questions in cultural transformation such as the origins of agriculture and the emergence of states, could also benefit new robust comparative claims applied to case studies such as the one that is presented here [18–20]. As of today, unfortunately, and despite some remarkable early works on this topic from other domains [21], work in this area is still poorly developed and archaeology has yet to progress very far, with rigorous and relevant comparisons where the spatial settings remain stable, while varying the chronological focus.

1.2. Objectives and Hypothesis

In this paper, we provide an exhaustive analysis of archaeobiological records, which are currently unique in their respective fields of study, with the objective of determining two forms of distinct occupation. Based on the carpological, anthracological and zoological remains recovered during the archaeological excavations of the last ten years, the aim is to shed light on the economic strategies that both enclosures developed in relation to the biotic resources available in the environment. The detailed study of the patterns of production and consumption of plant species, in terms of food and the supply of raw materials for construction and fuel, in both periods of occupation, contributes to a substantial increase in knowledge of the relationship between the inhabitants and their sources of supply, both produced and exploited. The analysis of livestock management methods is also important for understanding the economic choices made, both in relation to the survival of the group and patterns more typical of a hierarchical society with an exploited workforce.

The archaeobiological materials recovered over the years have allowed us to sketch in this article a first approach to the subsistence strategies followed during the two periods of occupation. The main objective therefore is to confront these strategies against the drawn hypothesis about the general setting and functions of the two occupation phases. We look forward to refining and consolidating the image presented here, with future studies of new samples recovered from the site. At the moment, anthracological and archaeozoological information have been recovered from the two periods, while the carpological record is restricted to the first phase of occupation due to taphonomical and preservation problems. An initial carpological outline has already been presented [1], although in the following pages, we will detail and expand on some aspects (see Table A1). A draft of the initial archaeofaunal information from the Andalusi phase was also presented in the II Encuentro de Zooarqueología Ibérica (2021), of which the publication is under review [22]. Notwithstanding, this paper regroups this information, as well as summarising the zooarchaeological record of the first occupation and the entire anthracological record, both of which are unpublished.

Regarding our initial hypothesis or interpretation of the archaeological contexts, in the first period of occupation, in the 8th and early 7th centuries BCE, Tossal de la Vila consisted of a small settlement. Based on archaeological excavations in the region, subsistence economy strategies seem to have been the norm for the community. There is evidence of livestock farming, agricultural activity, gathering and the use of forestry resources. Hunting would also supplement the economic spectrum.

In the case study of the early Andalusi period of occupation, judging by the records recovered, the structured, hierarchical, militarised and ritualised nature of the site, the population has been identified as a garrison of early Islamic troops. Their main functions, therefore, would include road and coastal surveillance, as well as territorial control of the limits between different domains. The strong focus of this particular function would naturally lead to the detriment of other basic functions such as those linked to the maintenance and reproduction of social life. This context led to the exploitation of the closest forestry resources, with a preference for species that would ignite rapidly for fuel and have a short useful burn time. Exploitation of the wealth and heterogeneity of ligneous species would be limited by the restricted diversity of the activities performed. Regarding the specific register linked to food in this particular context, both plant and animal species conformed to the social norms marked by certain religious precepts—such as the prohibition of the consumption of suidae—and the patterns typical of these types of militarised groups, primarily a very limited diversity of species and the exploitation of their secondary products.

2. Materials and Methods

2.1. The Site within the Context

During the late Bronze Age and the early Iron Age, Tossal de la Vila was enclosed by a perimeter wall located on the north, west and south sides of the settlement. Various rectangular-shaped spaces were built against the wall that shared party walls and were accessed from the central interior space (Figure 3). At present, excavations have identified 17 of the spaces that are located along the entire length of the walled perimeter. The eastern side of the settlement was naturally defended by escarpments. This distribution delimited a large, central plateau-like space, largely free of structures and with limited sedimentation, from where herds could be controlled, or other activities performed.

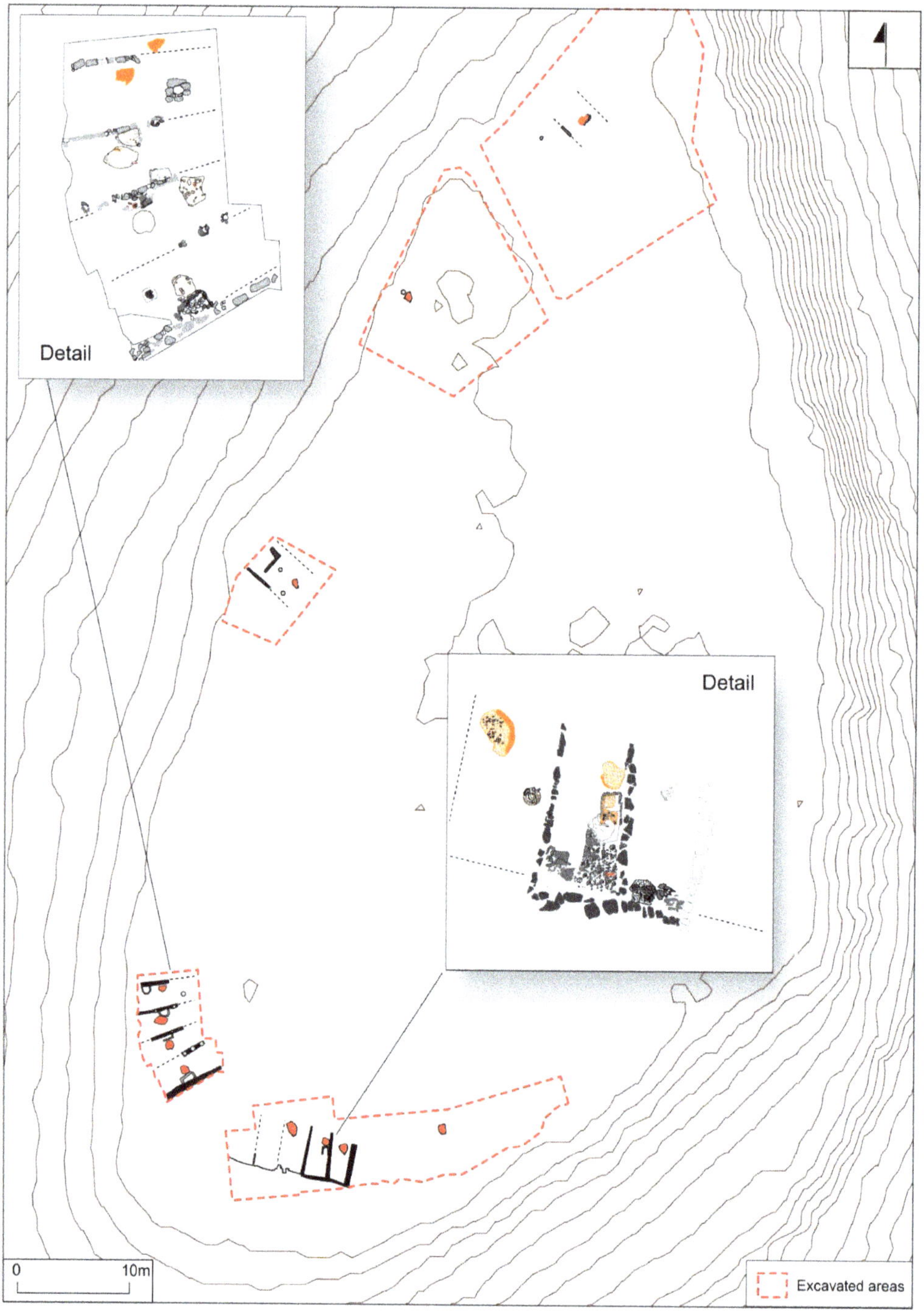

Figure 3. Occupation plan of the protohistoric phase. Although the remains of this phase have been greatly altered by the early medieval occupation, rectangular domestic spaces and various associated structures are documented, as shown in the plan.

Inside the different spaces, the excavations identified hearth and storage structures, as well as the remains of domestic activity, such as the processing and storage of foodstuffs (Figure 4). The archaeological record of the interior suggests certain functional differences; however, at present, features associated with social hierarchies have not been observed. A theoretical calculation based on replicating the module of the spaces on the perimeter wall gave a maximum of 56 possible areas. We do not believe that each one was associated with a family unit, but for estimation purposes, if we consider each pair of areas for a family of 3–4 individuals, it gives a demographic range between 84 and 112 individuals.

Figure 4. Details of domestic and combustion structures (**a**) and a storage space (**b**).

Among the construction techniques that characterise this period of occupation, there is a clear distinction between the large perimeter wall, which was built using double masonry held together with earth, and the internal partitions or walls separating dwellings, which were wooden frames covered in earth and vegetation raised on stone foundations. This same technique was also used for roofs, while floors took advantage of the geological level, which, if required, would be levelled with gravel and earth [1,23].

Regarding the assemblage of artifacts, handmade pottery is of particular interest as it represents a fairly broad repertoire of forms that can be dated to the 8th century BCE. However, some elements can be dated to the late 8th and, in particular, 7th century BCE. This is the case with certain forms and decorations on ceramic vessels, and also with various amphora fragments from Phoenician trade. Other elements that highlight trade contacts are metals, such as various iron objects, including two knives, and other copper-based objects.

Due to its general state of conservation and the lack of other comparable sites, Tossal de la Vila could be considered a model of a late Bronze Age III settlement in the northern region of Valencia. It features proto-urban characteristics, of which the origins can be traced to other sites in southern Catalonia with a slightly older chronology.

In the case study of the emirate Andalusi period of occupation, similar to the protohistoric period, most of the structures built in the interior are completely connected to the wall itself and are arranged longitudinally. Based on the floor area accounted for each residential unit identified in the site, we suggest that the settlement was inhabited by a small group of no more than fifty individuals (Figure 5).

Figure 5. Occupation plan of the Andalusi phase showing the fortification system and the different types of buildings described in the text: (a) simple modulations (a), multi-cellular building (b), privileged residential complex (c) and a mosque (d).

The easily accessible western front is defended with square or circular towers and bastions, together with a formidable wall with an average width of 3 m. The construction technique used consists of two solid dry-stone masonry walls, with the space between them filled with blocks and gravel (Figure 2). The layout was also conditioned by the intermittent nature of its function. Successive functions repeatedly repaired and modified the defensive structures, of which the intensity and functionality are still being analysed [3].

The entire central space of the settlement is free of buildings, with the exception of two free-standing structures, one described as a 'privileged' residence and the other as a mosque for communal use [5]. With regard to the different areas excavated protected by the wall, up to three different forms of modulation were identified. From the northern sector of the site to approximately half of the total length of the wall, a series of simple modules were identified, the basic residential units documented, each of which contained at least one hearth. The only multi-cellular building system discovered also shows a clear evolution from a simple module. Its development can be seen within different additions to the complex, which are not always orthogonal to the initial space and eventually form a clearly hierarchical space. We also identified a communal activity area attached to the complex with an independent entrance. The area had a raised oven built with slabs surrounded by various hearth structures, which we have identified as a small kitchen. Lastly, in the southern sector of the site, we identified a new simple module of completely different proportions and morphology to those of the residential modules. In this case, this was associated with various enclosures possibly made of perishable materials, which we believe to be a space for stables and storage (Figure 6).

Figure 6. Details of the different sectors from the Andalusi period, including an area for stabling and storing (**a**); a small communal kitchen (**b**); a mosque; (**c**) and a privileged residential complex (**d**).

The construction techniques associated with the buildings are homogeneous. However, two development phases of the settlement's urban fabric can be differentiated via the quality of the workmanship. The orthogonal walls of the residential modules are arranged against the main wall on stone foundations, built using double masonry filled with gravel and earth, possibly using a frame. Earth might also have been used as mortar and even cladding. The masonry is much more elaborate in the two free-standing buildings. The layout of the walls and partitions was built directly on natural rock during the hillfort's foundation phase, and on a relatively uniform layer of gravel and earth during the later renovation period. The roofs were probably built using plant-based materials in the form of flat slabs. Thoroughfares would have been built either on the calcareous geological base itself or on the protohistoric building materials which, if taken in order of collapse, would constitute levels of use in the al-Andalus period.

The archaeological records have made it possible to establish a solid scenario in terms of the periodisation of the site's architectural development, the spatial organisation of activities and the identification of the occupants and the functions they would have performed [3]. Ceramic materials mostly appear in a secondary position and are highly fragmented, due to the planned and ordered abandonment of the settlement. However, the analysis of the finds led to the initial morphotypological identification, which primarily comprises local production together with other elements from different parts of the Iberian Peninsula. This indicates the presence of occupants from different parts of the al-Andalus domains. In turn, the metal record gave rise to the recovery of a small set of military panoply, as well as other elements linked to domestic use. Other unique objects were also found, such as pieces carved in bone and game boards carved in stone.

Eleven excavations have been performed at the site since 2012, funded by the Serra d'en Galceran Town Council and the Castellón Regional Council. These campaigns have gone through different phases in terms of execution and objectives, and research is currently focused on obtaining a complete understanding of the phases of occupations from the perspective of chronological precision, functional distribution and construction of the space. It also aims to open the focus towards a better contextualisation of the territorial scope within the chronological context of each of the two periods.

2.2. Characterisation and Methodology of the Carpological Material Analysis

All the samples considered correspond specifically to the protohistoric levels [1]. The methodology for taking and obtaining samples was also replicated in the stratigraphy of the al-Andalus chronology. However, at present, no positive results have been obtained.

An intensive sampling of the sediment was performed, which included a 1 m^2 grid collection, with the aim of determining the possible spatial distribution of the remains. In some cases, seeds and fruit were collected manually during the excavation process.

The sediment was treated in a flotation machine with a 1 mm interior mesh and a 0.25 mm mesh, where the floating material was collected. Once the residue was dry, it was analysed in the Prehistory and Archaeology Department laboratory at the University of Valencia. With the aid of a stereoscopic microscope, the remains were selected and identified based on a comparison with a collection of present-day remains.

2.3. Characterisation and Methodology of the Anthracological Material Analysis

For the anthracological analysis, 46 samples were examined. Exhaustive sampling was performed to recover charcoal, and the sediment was processed through dry sieving and flotation. The latter facilitates the recovery of practically all the plant macro remains, thus providing comprehensive information on the content of each sample [24]. Other larger charcoal fragments were collected individually by hand during excavations, in order to prevent fragmentation and avoid the over-representation of taxa.

For the identification of flora, a Zuzi 10× metallographic microscope with 4×, 10×, 40× and 100× objectives was used. This facilitated the detailed observation of different types of wood cells that, when burnt, retain their anatomical structure, by which plants can be identified. In order to guarantee identification, the samples were checked against recent carbonised wood and various specialised plant anatomy atlases. The taxa were identified with varying degrees of success, ranging from precise to less precise identification at species level; genera that have been catalogued mention 'sp'.

Where the charcoal could not be identified between two genera of the same family, they were separated with a hyphen (-). Lastly, the *Quercineae* were grouped into two taxa that were easily distinguishable using a microscope: deciduous *Quercus*, including species such as oaks, and evergreen *Quercus*, including species such as holm oak/kermes oak/cork oak. Indeterminable taxa have also been included where it was impossible to observe the anatomical properties of fragments to determine their taxonomy, due to their small size or deformation.

At an anthracological level, the data obtained at Tossal de la Vila can be interpreted from either an ecological or an ethnological perspective, depending on the provenance and spatial distribution of the charcoal [25,26]. As a result, we correlated the charcoal samples to their spatial location and provenance in the settlement's two phases of occupation in order to be able to interpret them one way or another.

2.4. Characterisation and Methodology of the Archaeozoological Material Analysis

In both periods, the faunal elements correspond to food remains. This interpretation is based on the fact that the animal remains appear together with other materials, such as pottery, charcoal and seeds, and there are anthropic alterations such as butchery marks and there are no articulated portions or complete skeletons. The selection of stratigraphic units and sectors was motivated by the quality of the data: units derived from domestic residue, undisturbed units, sectors with significant assemblages, etc.

The faunal assemblage under study was collected manually during the excavation process. Although teeth, smaller animals such as leporids and a large number of bone fragments are recorded, we are aware that there may be a bias against smaller fragments.

The analysis of the sample was performed using comparative osteology. Given that, at present, no significant differences were found by breaking down the data by differentiated sectors, the total quantification for both periods are provided. In this article we work with the number of identified specimens (NISP, NISP%), and in the tables, we provide the minimum number of elements (MNE, MNE%) as a completment [27,28].

Sheep and goat bones and teeth were differentiated according to the criteria of Boessneck et al. [29], Payne [30], Halstead et al. [31], Zeder and Lapham [32] and Zeder and Pilaar [33]. However, the significant bulk of remains were included in the mixed category of caprinae (OC). When it was not possible to identify the remains at a species level due to the state of conservation, we quantified them osteologically by size, i.e., macro (cattle, horse) and meso-mammal (in which we grouped medium-sized animals, such as caprinae and suidae) categories. Small splinters with no osteological or taxonomic provenance were counted as unidentified remains. Unidentified avian remains were included in the general category of birds.

Regarding the frequency of the anatomical parts of the skeleton of each species, this was of each species, this has been evaluated taking into account the MNE, from which we have obtained the minimum number of animal units (MAU) [34]. MNE and MAU results have been grouped from larger units: head, axial skeleton, forelimb, hind limb and appendicular limbs.

The presence of equidae is low in both phases, only two remains could be attributed to horse (early medieval phase) and the rest were included in the *Equus* sp. category, because morphological and biometric criteria did not help us to make a specific identification. The size of the cattle remains leads us to rule out the presence of *Bos primigenius* for the time being. The differentiation between the domestic pig and its wild type has been established by metrical criteria. This differentiation was not always possible due to the presence of small fragments, and it is possible that some wild boar remains have been included in the category of domestic swine. In Appendix A (Table A9), we include some of the measurements (we have followed the method of Von den Driesch [35]) that have helped us in these differentiations. For the moment, there are only a few measurable data, and we hope to increase in the next years the corpus of biometric data important to know the skeletal variations of different species, among other questions.

Age was determined by dental eruption and wear [36–38], and the level of detail of the appendicular skeleton [39,40].

In general, the data extracted from the protohistoric period are more limited than that of the early Middle Ages, and therefore only enables us to sketch a vague outline of the management and exploitation of the enclosure's faunal resources. The low percentage of fragments identified at the species level and their state of conservation limits the informative value of some of the classic analyses performed in archaeozoology. In this article, we will not dwell on the taphonomic results and other alterations, such as butchery marks, which may have affected the final preservation of the complex (for more information on the Islamic occupation, see Pérez-Polo et al. [22]). However, we would like to point out that the early medieval occupation razed a large part of the preceding structures, which, together with other post-depositional alterations, has conditioned the quality and quantity of the faunal remains from the protohistoric phase.

3. Results

3.1. The Protohistoric Settlement

3.1.1. The Protohistoric Carpological Record

The carpological study led to the discovery of a remarkable set of seeds and fruit, the results of which were included in a previous publication [1]. All the remains are charred, which must be related to the fire that seems to have led to the abandonment of the protohistoric settlement.

At present, the analysis has not provided any relevant conclusions about the spatial distribution of the carpological samples. The flora remains recovered from the levels of use show that the agricultural activity of the settlement seems to be mainly characterised by the cultivation of two taxa: durum wheat (*Triticum aestivum-durum*) and domestic barley (*Hordeum vulgare* subsp. *vulgare*). The presence of domestic wheat is marginal and there is no evidence to suggest that they were domestic crops. There was also evidence of weeds (such as *Lolium temulentum*, *Galium* sp.) in the same fields as the two cereals.

There is no evidence that other cereals, such as millet, were cultivated, although there is evidence in the chronologies at nearby sites, such as Los Morrones [41] and at several other sites in Catalonia [42]. In turn, it is not currently possible to confirm the cultivation of legumes.

Accordingly (Figure 7; Table A1), the agricultural model can be defined as exclusively based on annual crops (wheat and barley). Added to this is the systematic use of wild plant resources. Acorns (*Quercus* sp.) were particularly important, either for animal or human consumption, and, to a lesser extent, blackthorns (*Prunus spinosa*).

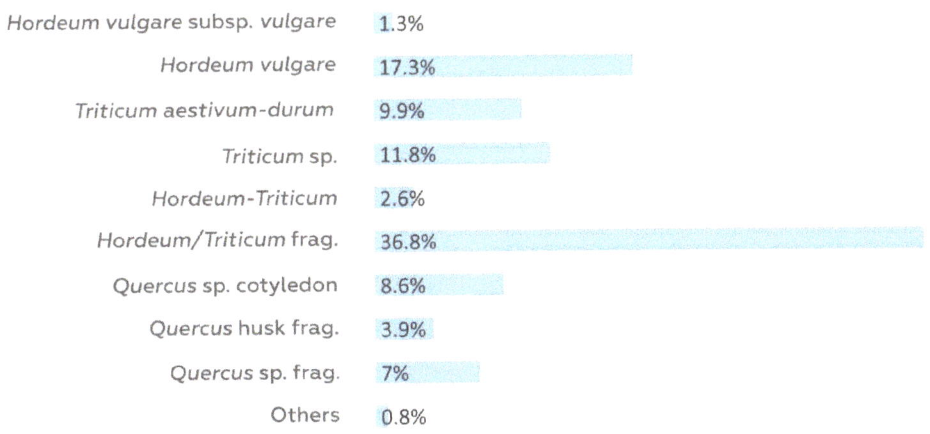

Figure 7. Protohistoric carpological record: relative frequency (%) of the carpological remains recovered. Data derived from Table A1 (Appendix A).

The collection and use of *Quercus* fruit is very common in the prehistory of the Iberian Peninsula, especially in the Bronze and the Early Iron Age [43]. This is documented in Tossal de la Vila by the constant and abundant presence of carbonised cotyledons in all the protohistoric areas excavated, including those in which the corresponding carpological study is not yet complete.

3.1.2. The Protohistoric Anthracological Record

A total of 2754 charcoal fragments were analysed in the anthracological study of the protohistoric occupation, corresponding to the construction and use (Phase I; 1362 fragments), and destruction and abandonment (Phase II; 1392 fragments) of the settlement.

In Phase I (Figure 8; Table A2), the charcoal samples appear either dispersed in the sediment or in carbonaceous concentrations not directly associated with any structure. In general, the samples have been interpreted as residue from cleaning out hearth structures that were in use over time and, therefore, are good ecological indicators. From this perspective, the samples analysed from the dispersed charcoal reveal the presence of evergreen *Quercus* in all but one of them, although with very variable percentages. There are also other taxa, most in negligible percentages, such as 'strawberry tree' (*Arbutus unedo*), legumes, gorse-type legume, ivy, rosacea, rosemary and heather. Five taxa were identified in the sample with the greatest variety. The most abundant was deciduous *Quercus*, followed by *Arbutus unedo*, evergreen *Quercus*, rosemary and heather.

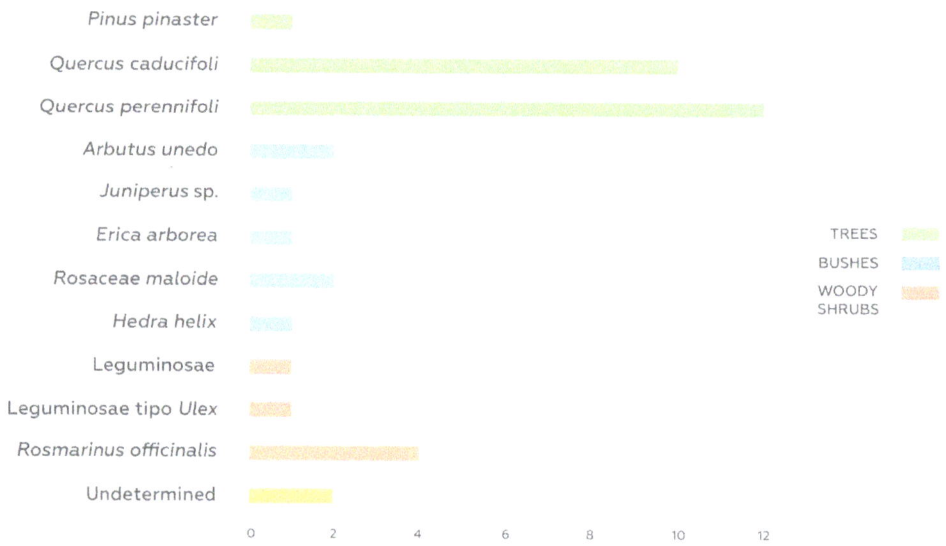

Figure 8. Protohistoric anthracological record: number of stratigraphic units in which different anthracological species have been recorded (Phase I). Data derived from Table A2 (Appendix A).

The charcoal found in carbonaceous concentrations was sometimes associated with building material, and in others, it was found next to or inside ceramic remains. These samples did not show significant variability, and most were monospecific such as evergreen and deciduous *Quercus*, except in one of the samples, where *Pinus pinaster* (maritime pine) and rosemary were also identified.

In the levels of abandonment or destruction (Phase II) (Figure 9; Table A3), most of the charcoal analysed was found at the level where the building structures had collapsed as a result of the fire. In several post holes, charcoal fragments of varying sizes were identified, the main taxa being oak and kermes oak. Oak seems to have been used more abundantly in construction, although holm oak was also used.

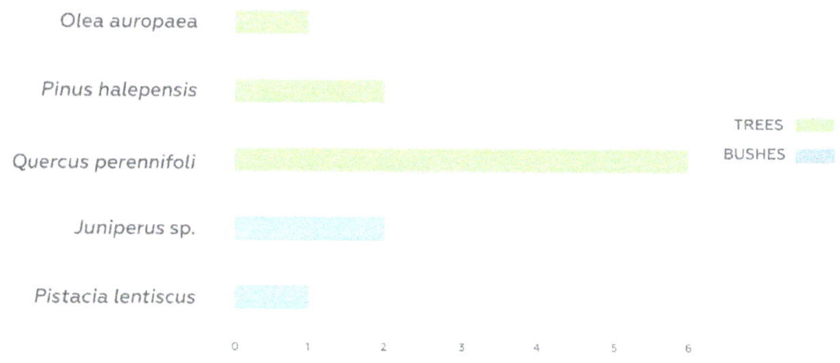

Figure 9. Protohistoric anthracological record: number of stratigraphic units in which different anthracological species have been recorded (Phase II). Data derived from Table A3 (Appendix A).

Charcoal was also found dispersed over these levels in the sediment, and the most abundant taxa found were also deciduous and evergreen *Quercus*. This suggests that they are the remains of the wooden structures that were destroyed, although they might be mixed with remains from different origins. As a result, there seems to be no difference between their use as fuel in hearths and as structural fixings (posts/roofs).

The results from the protohistoric period indicate that the plant cover was dominated by a well-preserved mixed forest, with dominant tree layers of oak and kermes oak together with juniper or scale-leaf juniper, heather and 'strawberry trees', an approximation of what would have been the area's potential flora. The inhabitants of the settlement would have used these resources to obtain firewood for their homes, mainly holm oak, and used oak to build structures, which, due to its characteristics, would have been used for beams and other architectural features. The differentiation between charcoal intended for constuction material or fuel has been established by the species identified and by the location of each example within the settlement (within a structure, post hole, etc.). Other nearby protohistoric sites where anthracological studies have been performed, such as Els Estrets in Vilafamés [44] and El Cormulló dels Moros in Albocàsser [45], also document the use of holm oak and oak as construction material and fuel.

3.1.3. The Protohistoric Archaeozoological Record

The faunal assemblage analysed from the late Bronze to the late Iron Age comprises 1476 remains (Figure 10; Table A5), of which only 512 were identified at the species level, representing 34.6% of the total. This percentage highlights the assemblage's poor state of conservation and the over-representation of unidentified remains including small splinters without osteological or taxonomic provenance. However, it is indicative of an exhaustive manual collection.

The majority of the remains identified, 89.1%, were from domestic animals compared to 10.9% from hunted animals. Caprinae dominate (caprinae + sheep + goats; 64.9%), while domestic suidae (11.8%) and cattle (10.8%), the second most represented species, had similarly low values. Lastly, equidae remains (1.6%) were negligible (Figure 8; Table A5). In turn, malacofauna remains were token and reduced to five fragments corresponding to marine bivalves.

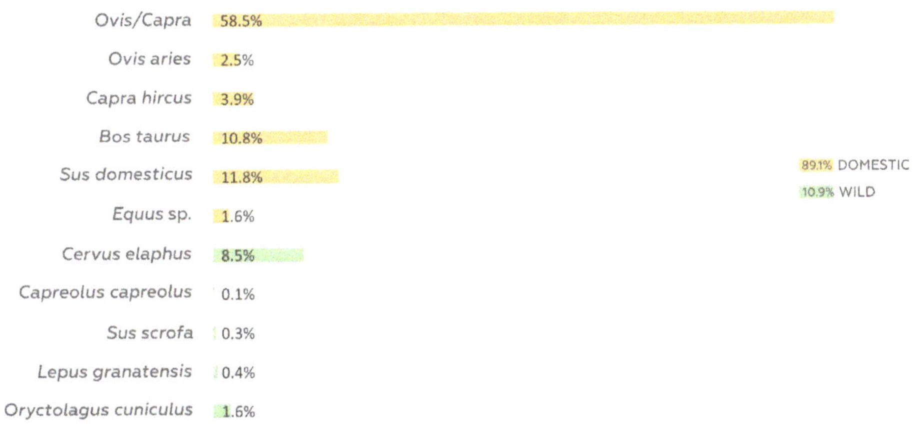

Figure 10. Protohistoric archaeozoological record: relative frequency (%) of faunal remains identified to species level recovered. Data derived from Table A5 (Appendix A).

The most recorded hunted species is red deer (8.5%). Other species documented in this group are less common and include: roe deer (0.1%), wild boar (0.3%), hare (0.3%) and rabbit (1.6%).

The differentiation between sheep and goats was not easy due to the state of conservation and the usual problems of identification between the two species. The bulk of the remains were included in the mixed category caprinae (OC; 58.5%). The limited number of sheep (2.5%) and goat (3.9%) remains do not support any particular hypothesis.

However, for the moment, the data indicate a similar weight of both species, with a slight predominance of goats.

The limited set of remains conditions representativeness and makes it impossible to establish the age at slaughter of caprinae, cattle and suidae, those with the highest rates recorded. Of the limited number of caprinae remains from which data on the state of epiphyseal fusion could be extracted, 30% derive from juvenile individuals, with preferential age of slaughter being between 1.5 and 2.5, and 2.5 and 3.5 years old. Examining the eruption and wear of the mandibular teeth to estimate the age of slaughter, only three mandibles belonged to sub-adult individuals (slaughtered between the second and third year of life, phases D and E), while the third belonged to an adult (>3 years old). In general, examining each tooth individually, the results coincide with the analysis of dental wear stages, given that most of the pieces (N: 23), except for one that shows the first stages of dental wear, show little wear (<9). In the case of cattle and suidae, the data are even less representative. In the former, there was one juvenile and one adult individual, while in the latter there were two sub-adult individuals.

Regarding anatomical distribution, the best represented parts of the caprinae according to the MAU analysis (Table A7) are the elements of the head, the rest being shown in a balanced way. As for the rest of the domestic species, the analysis yields insufficient data, so we point out, without further precision, the apparent presence of elements derived from all anatomical regions. Although the data are also scarce for the deer, the best represented wild species, we interpret the presence of all the anatomical parts as a consequence of a contribution of the whole animal to the settlement, and its hunting in the vicinity of the settlement. On the other hand, the axial skeleton is under-represented in all species, since due to the level of fragmentation, most of the remains were included in the meso-macromammal categories.

3.2. *The Andalusi Emirate Hillfort*

3.2.1. The Andalusi Carpological Record

A carpological analysis could not be performed, as no samples were found, despite several campaigns of exhaustive sample collection in multiple strata to determine hearth structures, ashes and levels of use. The question arises as to whether the uneven conservation between the protohistoric and early Medieval carpological contexts is due to a sampling problem or, in contrast, to taphonomy and the settlement's abandonment conditions at the end of their respective periods of occupation. In this regard, the fire that marked the settlement's abandonment in the early Iron Age gave rise to the exceptional conservation of seed and fruit remains from the period. In turn, archaeological records of the abandonment of the settlement during the first half of the 10th century CE highlight how the occupants of the hillfort removed all their possessions, including foodstuffs, in an orderly manner. Anything that was left behind disappeared due to decomposition, which only leaves the possibility of recovering contexts from hearths or middens, which are being thoroughly processed despite negative results.

3.2.2. The Andalusi Anthracological Record

A total of 687 charcoal fragments were analysed in the anthracological study of the Andalusi occupation (Figure 11; Table A4), corresponding to the use (Phase III; 610 fragments) and abandonment (Phase IV; 77 fragments) of the settlement.

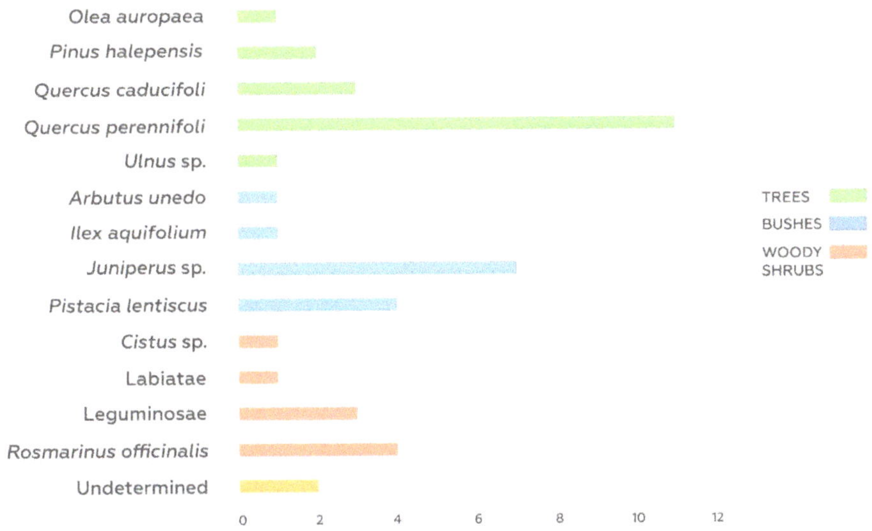

Figure 11. Andalusi anthracological record: Number of stratigraphic units in which different anthracological species have been recorded. Data derived from Table A4 (Appendix A).

The charcoal concentrated in hearth structures reflects their last level of use. The best represented the taxa of evergreen *Quercus*, together with other taxa such as juniper, mastic, *Cistus*, legumes and labiatae such as rosemary. In the dispersed charcoal, the most abundant taxon was still evergreen *Quercus*, together with *Juniperus* and *Rosmarinus*.

At the abandonment level, we worked with a sample of 77 fragments of concentrated charcoal, all identified as deciduous *Quercus*.

Similar to the protohistoric period, the charcoal recovered from the early Middle Ages is indicative of the forest that provided wood, especially holm oak that was used as the primary fuel, together with a high variability of species of shrubs and bushes that grew in low areas and on the sunny slopes of the sierra.

Although the identification of oak is practically non-existent in this period, it does not imply that it disappeared from the natural environment. Most of the charcoal recovered is the result of the last fire tended in the structure, which would have been the last firewood collected, and would not be a valid interpretation from an ecological perspective. However, the study of these levels has provided a diverse anthracological record that complements the data obtained from the dispersed charcoal. By determining which species would best suit specific needs, we were able to determine that they generally formed part of the flora ecosystem close to the site, which therefore demonstrates the use of plant resources from all the biotopes susceptible to exploitation.

3.2.3. The Andalusi Archaeozoological Record

The emirate faunal assemblage analysed to date comprises 3046 remains (Figure 12; Table A6), of which 1281 were identified at the species level, representing 42% of the total [22].

Figure 12. Andalusi archaeozoological record: relative frequency (%) of faunal remains identified at the species level recovered. Data derived from Table A6 (Appendix A).

The study revealed that the majority of the remains, 94.9%, belonged to domestic animals, compared to 5.1% from hunted species, which represent a marginal resource. Caprinae (caprinae + sheep + goats; 54.3%) and chickens (36%) dominate compared to the token presence of cattle (2.9%), pigeons (1%) and suidae (0.5%), with even lower values for horses (0.1%) and canidae (0.1%). The minor presence of hunted animals is represented by the remains of red deer (0.5%), foxes (0.1%), lagomorpha—both hares (0.6%) and rabbits (3.6%)—and partridges (0.3%) (Figure 10; Table A6).

The higher incidence of caprinae and chickens enabled us to extract a greater volume of data for both species. In the former, the two main species appear in balanced proportion, with 105 sheep compared to 88 goat remains. Examination of the state of eruption and wear of caprinae mandibular teeth revealed the dominance of sub-adult individuals slaughtered between the second and third year of life. Additionally, the presence of adult individuals over three years old is noteworthy, as well as the absence of immature young animals (phases A–C). In general terms, the results coincided with the analysis of dental wear stages, examining each tooth individually, as most of the teeth recorded were in the advanced wear stage (>9). Lastly, the state of fusion of the long bones confirmed the same trend: only 34.4% of the finds correspond to juvenile individuals, which indicates a preferential slaughter age of between two and a half and three and a half years.

Regarding chickens, the overwhelming proportion of remains with formed epiphyses indicates the presence of adult specimens, while only 5.5% of the sample belongs to juvenile individuals.

In terms of anatomical distribution, caprinae and chickens are the species that provide us with the most information (Table A8). According to the MAU analysis, the best represented anatomical region within caprinae is the forelimb, although it is fairly balanced with respect to the hindlimb and appendicular limbs, while the anatomical elements of the head reflect a lower index. As for the gallinaceous species, all anatomical regions are represented except the cranial region, possibly due to the higher rate of fragmentation in this area. There is an imbalance between the leg and the wing, with the latter being less represented. For the rest of the species, we can only point out that there is representation of the different anatomical regions. As in the protohistoric phase, the axial skeleton presents a marginal incidence, since most of these remains, due to their level of fragmentation, have been included in the meso-macromammal group.

4. Discussion

4.1. Implications of the Results for the Protohistoric Historiographical Debate

Tossal de la Vila's first period of occupation took place in the general archaeological context of the Iberian Peninsula's Mediterranean seaboard during the late Bronze Age (ca. 1300/1100–750/700 cal BCE) and the early Iron Age (750/700–550 cal BCE). The first phase is characterised by what is known as the north–eastern Urn Fields [46–50], which had an uneven impact in the regions of Aragon, Navarra and Valencia [51–63], while the second, that of the destruction and abandonment, coincided with the Phoenician colonial trade period [64–75].

For a clearer image of the occupation, the recovery of a good sample of archaeobiological remains in the excavated areas provided a fairly specific picture of the economic subsistence and exploitation activities performed by the inhabitants of Tossal de la Vila. Thus, through seed studies, we were able to determine an agricultural model based on the cultivation of wheat and barley with an annual cycle and a systematic exploitation of wild plant resources, especially acorns. The result is in line with other known chronologically related contexts and is characteristic of a phase that precedes the appearance of fruit trees in the early Iron Age [23,76].

In relation to the use of trees, woody shrubs and bushes, the anthracological study reveals that oak/holm oak was primarily used for construction and, by extension, as fuel for households, with a certain homogeneity between the different sectors. The use of woody plants, such as rosemary, seems to be very limited. Comparison with other settlements of a similar chronology suggests that the use of plant species as construction elements and fuel may have been selective, subject to availability in the surrounding environment [77].

The location of Tossal de la Vila deserves a separate mention given that the environment seems somewhat unsuitable for cultivating crops. This suggests that the selection of the location was not based on an agricultural economy as a priority, and other reasons must be considered, such as a preference for the exploitation of woodlands, animal grazing and perhaps the advantages derived from its inaccessibility.

The faunal study shows a clear preference for domestic species over wild species. In the former, caprinae dominate the sample and a mixed herd of goats and sheep were identified from the analysis. However, in general terms, there was a diversified faunal exploitation where suidae, cattle and game, mainly red deer, constitute the bulk of the total. Other species, such as equidae, have a low incidence, with values similar to those found at other sites in the region of Valencia [78,79]. Some researchers [78] suggest that some of the horses exploited for meat were wild.

The reduced incidence of hunting is common to various chronologically comparable contexts in the region of Valencia (Racó dels Cantos I [80]; Torrelló del Boverot, Los Villares and La Frontera [78]; Tossal del Mortòrum [79] and Los Morrones [81]). The values represented in Tossal de la Vila are along the same lines, with hunted animals only representing 10.9% of the total.

At present, it is difficult to determine the exploitation pattern of the most representative herds in Tossal de la Vila (caprinae, cattle and suidae) from the protohistoric period. Only the herd of caprinae enabled us to draw a tentative outline of their age at slaughter and, therefore, of livestock management strategies. However, this must be understood with due caution owing to the low representativeness of the samples. At the moment, the analyses suggest a mixed and relatively balanced exploitation of animals in an optimal state of maturity for meat production and secondary products throughout their lives.

In general terms, the livestock of the late Bronze to late Iron Age settlements in the region of Valencia was based on caprinae, suidae, cattle and equidae, the latter to a lesser extent. The diet was also supplemented with wild animals. The production strategy would be mixed, structured around the production of meat and secondary products, as well as using livestock as transport, pack and draught animals [82]. Despite these commonalities, there is variability in the percentages of the species represented in the different settlements [78,79,81,82], which cannot always be explained or directly attributed to environmental factors [82]. Other conditioning factors, such as the strategies and needs of each individual group, as well as their corresponding relationships with surrounding groups, must be taken into account when explaining the differences documented between different settlements.

4.2. Implications of the Results for the Andalusi Historiographical Debate

Tossal de la Vila's Andalusi period of occupation falls within the general historiographical question that studies the occupation and fortification of certain high-ground enclaves from late Antiquity to the early Middle Ages, a phenomenon that has clear parallels in the western Mediterranean [83–88]. The site is an example of complex historical dynamics that correspond to diverse social and political realities that are rooted in a myriad of causes [84,89–92]. We owe the definition of this specific case in the extreme north of Šarq al-Andalus to André Bazzana [91,93–95], even though it has not been developed further since the last years of the 20th century.

Notwithstanding, recent and systematic analysis of Arabic written sources and wide advances in archaeological excavations in several of the sites within this territory have made it possible to establish a hierarchy in those settlements not always coincident with Bazzana's first hypotheses. In this sense, sites such as Tossal de la Vila have been identified as fortifications with signs of proto-urbanism, although still without clearly defined functions and economic strategies. In turn, other examples could be linked to enclosures, called *maràbit* in the Arabic sources, large spaces surrounded by dry masonry walls, identical to those of fortresses, in which no permanent settlement areas can be observed. Consequently, we suggest that they were used to install mobile garrisons to control the coast and roads [3,5,96,97] (pp. 223–227; 404–405).

Regarding our site and taking into consideration the anthracological contexts recovered, of note is the strong homogeneous use of forestry resources by the occupants of the emirate hillfort. In this regard, the charcoal recovered from the different hearth structures identified as domestic hearths shows a balanced composition in the elements burnt. Thus, woody bushes and plants such as juniper, scale-leaf juniper, mastic, 'strawberry tree' and rosemary would be used as ignition fuels, while holm oak and cork trunks would provide sufficient raw material to generate long-lasting embers. The conservation of charcoal and ashes is exceptional, as they were recovered in their primary position owing to the orderly abandonment of the settlement, and, in one case, they were still covered by a stone slab.

In contrast, the deterioration levels of the houses, of which the layers of collapse include roof and building materials, show a higher proportion of deciduous *Quercus*, which provides denser and more resistant timber than holm oak or cork oak. There is a notable absence of pine in the samples, both in combustion and construction. This seems to indicate a type of flora that is in strong contrast to the area's current configuration, where pine has acquired a major presence, to the detriment of holm oak and oak groves. The latter is due to the fact that, in general, there has been a change in the landscape in the historical period due to deforestation, fires, etc. and, above all, more recently, to the intense repopulation of pine trees in the area. It is worth highlighting one last important detail: the recovery of charred fragments of olive trees not linked to houses or other living spaces. Although this does not provide a great deal of precision for the analysis of the use of wood resources, it helps to confirm the importance of olive crops throughout the al-Andalus period.

Lastly, the analysis of the faunal contexts from the emirate occupation period reveals consumption patterns based mainly on domestic species, especially through rearing small flocks of sheep, goats and chickens. Regarding caprinae, which is the taxon that provided the greatest volume of data, the analyses show a mixed and balanced flock of sheep and goats, in which the use of secondary products, such as wool and milk, was prioritised over the production of meat, as can be seen from the pattern of slaughter. This same pattern was observed in chicken remains, which revealed an overwhelming majority of adult individuals. Consequently, meat consumption was secondary to egg production. In general terms, it shows a subsistence-oriented production system, similar to those documented in other contemporary al-Andalus rural contexts [98,99], although with specific characteristics depending on the functional significance of the site.

Within the group of domestic animals, the minuscule proportion of suidae remains in the archaeofaunal assemblage is particularly noteworthy, accounting for around 0.5% of the sample. This situation can easily be explained by the presence of a human community that had adopted dietary guidelines incorporating certain Islamic religious precepts, such as the prohibition of eating pork and wild boar. From the early development of Islam, this was one of the most basic symbols of belonging to the new community of believers. Islam spread progressively in al-Andalus during a long transformation process that developed through the interaction between the different groups settled in the Iberian Peninsula [99] (pp. 663–664). This transformation, which is known generically as the Islamisation of al-Andalus led, in terms of food, to the adoption of a certain orthopraxis specific to medieval Islam. Tossal de la Vila is one of the earliest examples of this transformation, given its chronology and peripheral location.

The archaeozoological studies performed in al-Andalus contexts also highlight the secondary or insignificant role of other domestic species such as cattle [98,99], a situation that is repeated in the case of Tossal de la Vila. However, the question arises whether the very low values reflected in the analysis of this species at the site, even lower than those found at other contemporary sites, could be linked to the military function of the settlement.

In this regard, the evidence that indicates the presence of an armed group at the site, that would have performed tasks such as surveillance and territorial control, could be the reason for the lack of diversity in the herd maintained, as it would have been easier to reduce husbandry to a minimum.

The presence of wild fauna is practically marginal, and only rabbits are significantly represented in consumption remains. Although there is evidence of rabbits breeding in al-Andalus from early times [99], there are reasons that lead us to include them in the hunting category. One reason is their scarce representation in the food remains, and another, the isolation and difficulty to access the nature of the settlement.

The faunal data published on the territorial scope of Šarq al-Andalus in the early Islamic period are limited to those presented here [22] and to those published for Molón de Camporrobles [98]. Therefore, only new, accurately processed data from other rural enclaves in the region can shed light on these and other questions. At present, only more questions are emerging from these initial studies.

In summary, the archaeobiological evidence as a whole reinforces the initial hypothesis that the settlement was a hillfort occupied by a small human community that would have performed surveillance and territorial control functions together with subsistence activities. In this regard, the anthracological record shows homogeneous patterns of exploitation of forestry resources. The diversity of species used is typical of a close supply. The majority of the surrounding arboreal flora, comprising various species of evergreen *Quercus*, was effectively combined with various woody shrubs and bushes to feed the hearths of the various living areas of the hillfort, while the more resistant deciduous varieties, such as oak, were used for roofs and other construction material. As previously highlighted, the sphere of action for supply is relatively close to the settlement, which also seems to indicate that the inhabitants made careful choices aimed at avoiding using supplies from more distant resources.

The archaeozoological evidence also seems to highlight the same careful choices through the intensive exploitation of animals and their by-products, such as milk and eggs as a source of animal protein. Bearing in mind the highly developed level of Islamisation of this population in terms of dietary orthopraxis, and taking into account the geographical and chronological context of the site, the absence of suidae in the diet did not seem to lead to an increase in hunting activities. This seems to indicate that such activities in the settlement were reduced to a minimum. The evidence seems to suggest that economic strategies favoured the simplification of the work processes linked to subsistence. This would allow for a greater effort to be devoted to the military functions that were expected of a fortification of this type.

5. Conclusions

Owing to our extensive research, we are now in a position to offer concrete answers to the question of why two different human communities, separated by almost two millennia, chose to occupy the same site. In both cases, the underlying cause for choosing the site is linked to subsistence reasons, based on economic strategies that prioritised interest in nearby resources, especially timber. Another common aspect will be the keeping of mixed flocks of sheep and goats, although the landscape of the immediate surroundings seems particularly suitable for the latter. The analysis of the anatomical distribution has allowed us to explore the caprinae and chickens (the latter within the Andalusi phase) with a greater precision, but at a general level we can highlight that the presence of the remains in all the anatomical regions of the different species is indicative of the complete processing of a large number of animals within the settlement within both phases of occupation. However, this is where the coincidences seem to end. In the protohistoric period, the archaeobiological record reveals heterogeneous, diverse economic activities, while evidence from the al-Andalus period seems to suggest that complexity was reduced to much less significant limits.

Thus, the carpological analysis of the plant resources used for food in the protohistoric period highlights a cereal diet based on annual cycle crops, such as durum wheat and domestic barley, although heavily supplemented with the collection of acorns, whether for animal or human use. The considerable distance between the settlement and the probable cereal fields points either to the existence of an exchange relationship between enclaves or the existence of complementary settlements in the valley floor, of which the function would be to cultivate and store crops. Unfortunately, we are not yet in a position to clarify this aspect, although future survey work might shed light on key questions for understanding the survival of the settlement during the late Bronze Age. Similarly, the al-Andalus scenario remains a complete mystery, as no carpological remains have been recovered to date, probably due to the poor conservation conditions of differential taphonomy in the formation of its levels of abandonment and deterioration.

In turn, the livestock strategies between the two periods show some significant differences, which leads us to suggest two clearly divergent options. Hunting did not seem to account for an excessive volume of animal protein intake of either group, although hunting activity was clearly more relevant in the protohistoric period, in which red deer seems to play an important role in the food supply. It seems, therefore, that a conscious and programmed dedication to hunting can be identified, which would supplement a predominant domestic diversified herd, caprinae, the majority, and well-represented suidae and cattle. The age of slaughter, which identified a large number of juveniles, suggests a well-structured livestock scenario, in which breeding and consumption coexisted with other secondary productions.

The al-Andalus scenario is very different in terms of livestock strategies, with an absolute dominance of caprinae and chickens, taxa that were slaughtered at an advanced age. This highlights an economic activity that was totally secondary to other functions and ensured a constant supply of animal protein in the form of milk and eggs, as well as the sporadic consumption of meat from animals that were no longer productive. These characteristics, together with the token hunting of various species and the anecdotal presence of other animals in the domestic herd, lead us to consider that the subsistence practices of the fortified enclave were never considered priority economic activities.

Therefore, when it comes to the reasons that led these two human communities to settle on top of a rocky outcrop at an altitude of almost a thousand metres, a distinction can be made between economic and strategic reasons. Thus, the protohistoric settlement seems to have chosen the enclave primarily for its accessibility to optimal forest resources, the possibility of keeping a small herd of livestock and the development of other supplementary subsistence activities, such as hunting and gathering. In contrast, the installation and fortification of the enclave in the Andalusi period was due to well-founded military reasons, as it was located at the crossroads between the great emirate cities of Valencia, Tortosa and Zaragoza. The economic activities documented in its interior are the minimum needed for the subsistence of the group, linked practically and exclusively to the supply of firewood for the living spaces and the maintenance of a small group of sheep, goats and chickens to provide permanent food for the garrison. This subsistence strategy was not based on economic design. In fact, it most probably had to be supplemented by contributions of other food sources from nearby peasant communities, through mechanisms that we are not yet able to define precisely.

Therefore, and by way of conclusion, it is worth highlighting the multiple causal structures that arise when faced with the decision of occupying a high-altitude enclave. On the one hand, the example of a protohistoric settlement built in a location that provided certain characteristics for the development of the economic activities contemplated when the settlement was founded. In this case, the biotic resources around the site played a key role in the choice of location. The settlement's survival was based on the occupants' correct evaluation of the sources of supply and their capacity to exploit them. The set of tasks performed, including livestock husbandry, gathering, agriculture and domestic activities, seem to have revolved around varied and well-defined economic strategies, typical of settlements that characterise the late Bronze Age communities. However, it was an economic model that was not perpetuated in the new phase that marked the end of the settlement, the early Iron Age.

In contrast, the al-Andalus hillfort did not seem to take these conditions into account in the choice of the site, which responds to military needs and is an optimal location from a strategic perspective. The needs of its occupants seem to have played a secondary role during the construction of the fortification, with no nearby food sources and not even cisterns for storing water, basic elements for the permanent establishment of a human community. The strategies used to solve these shortcomings must have been different to those employed during the protohistory period, possibly linked to the ability to extract resources from the surrounding territory by mechanisms that are still under study. All the indicators highlight activities oriented towards more than meagre subsistence requirements, with no apparent economic vocation.

Author Contributions: All authors contributed equally; funding acquisition, M.P.-P., J.N. and G.P.-J. All authors have read and agreed to the published version of the manuscript.

Funding: This research received funding from the following projectsand and contracts: M.P.-P has carried out this work within the contract as a postdoctoral researcher at the University of Salamanca. This contract is part of the Margarita Salas grant, financed by the Ministry of Universities (Spain) with Next Generation funds from the European Union. J.N. has worked within the framework of ENALFE: Entre al-Ándalus y la feudalidad. Poderes territoriales y desarrollo de sistemas defensivos altomedievales en el nordeste peninsular (PID2020-114484GB-I00), funded by the Ministry of Science and Innovation; G.P.-J. has worked within the framework of FRUITCOM: Un nuevo modelo agrícola: frutales, hierro y comercio durante el I milenio A.E. (CIDEGENT/2019/003), funded by the Generalitat Valenciana (Regional Government of Valencia).

Data Availability Statement: The results of this research will be compiled from new and published sources. All data are available in the research cited and in this manuscript.

Acknowledgments: The authors would like to thank the Castellón Provincial Council and the Serra d'en Galceran Town Council for their constant support (financial and logistical) both in the development of the archaeological excavation and in the research plan. To M. Moreno-García, head of the Laboratorio de Arqueozoología del Instituto de Historia (IH/CCHS-CSIC), and his team, E. López-Romero and E. Checa, we are very grateful for allowing us to consult the reference collection of this centre. Finally, we would like to thank the reviewers who contributed constructively with critical comments on the final version of the article.

Conflicts of Interest: The authors declare no conflict of interest.

Appendix A

Table A1. Protohistoric carpological record: number of remains (NR.) and relative frequency (%) of recovered species.

	Total NR.	%
Hordeum vulgare subsp. *vulgare*	126	1.3
Hordeum vulgare	1723	17.3
Triticum aestivum-durum	987	9.9
Triticum sp.	1177	11.8
Triticum cf. *diccoccum*	5	0.05
Triticum cf. *Monococcum*	1	0.01
Triticum dicoccum spikelet	1	0.01
Hordeum-Triticum	265	2.6
Hordeum/Triticum frag.	3673	36.8
Fallopia convulvulus	7	0.07
Galium sp.	1	0.01
Leguminosae	1	0.01
Lolium temulentum	15	0.1
Poacea	2	0.02
Prunus spinosa	3	0.03
Quercus sp. cotyledon	865	8.6
Quercus sp. cupula base	10	0.1
Quercus husk frag.	386	3.9
Quercus sp. frag.	701	7
Undetermined fruit	2	0.02
Undetermined mass	29	0.3
Total	9980	100

Table A2. Protohistoric anthracological record: percentages (%) of occupation levels (phase I). Concentrated and dispersed charcoal.

Sector	UE	Charcoal Distribution	*Pinus pinaster*	*Quercus caducifoli*	*Quercus perennifoli*	*Arbutus unedo*	*Juniperus* sp.	*Erica arborea*	*Rosaceae maloide*	*Hedera helix*	Leguminosae	Leguminosae tipo*Ulex*	*Rosmarinus officinalis*	Undetermined
3	3018	Dispersed		100%										
6	6018			50%	16.7%	20.4%		5.6%					5.6%	1.9%
1	10013				37.2%	2.3%			2.3%	4.7%	9.3%	9.3%	34.9%	
6	6014				45.8%		54.1%							
6	6043				100%									
3	3029-E4-PI02				100%									

Table A2. *Cont.*

Sector	UE	Charcoal Distribution	*Pinus pinaster*	*Quercus caducifoli*	*Quercus perennifoli*	*Arbutus unedo*	*Juniperus* sp.	*Erica arborea*	*Rosaceae maloide*	*Hedera helix*	Leguminosae	Leguminosae tipo*Ulex*	*Rosmarinus officinalis*	Undetermined
3	3009-23			100%										
3	3010-12			100%										
3	3010-28			100%										
3	3010-27				100%									
3	3010-35			100%										
3	3008-39			100%										
3	3008-20	Concentrate	15.8%	47.4%	10.5%								26.3%	
3	3010-01				50%								50%	
3	3009-43				100%									
3	3009-06													100%
3	3046-FP1				100%									
3	3010-09				100%									
7	7006-83P			100%										
3	3030P			97.2%	0.9%				1.9%					
				Trees				Bushes				Woody Shrubs		

Table A3. Protohistoric anthracological record: percentages (%) of occupation levels (phase II). Concentrated and dispersed charcoal.

Sector	UE	Charcoal Distribution	*Olea europaea*	*Pinus halepensis*	*Quercus perennifoli*	*Juniperus* sp.	*Pistacia lentiscus*
4	4012		13.1%		9.8%	11.5%	1.6%
3	3008-A2						
3	3008-A1				28.6%		
3	3023	Dispersed			58.3%		
47	47103-95						
47	47113-FP21						
47	47103-96			99.6%			
47	47103				100%		
47	47121 FP13				100%		
6	6002-50						
6	6002-49	Concentrate					
6	6034 FP5				100%		
5	5115-91					100%	
1	10017V			100%			
				Trees		Bushes	

Table A4. Andalusi anthracological record: percentages (%) of occupation and abandonment levels (phases III-IV). Concentrated and dispersed charcoal.

Level	Sector	UE	Charcoal Distribution	*Olea europaea*	*Pinus halepensis*	*Quercus caducifoli*	*Quercus perennifoli*	*Ulmus* sp.	*Arbutus unedo*	*Ilex aquifolium*	*Juniperus* sp.	*Pistacia lentiscus*	*Cistus* sp.	Labiatae	Leguminosae	*Rosmarinus officinalis*	Undetermined
Occupation	11	11002	Dispersed			51%	47.1%										0.019%
	14	14126			2.3%		59.1%				38.6%						
	1	10014 (PI01)					88.2%				5.9%					5.9%	
	5	5117					100%										
	5	5017 LL11	Concentrate			3.8%	4.1%		7.6%	1.9%	6.7%	5.7%			6.7%	16.2%	4.8%
	5	5101 LL17					12.5%				25%	37.5%			12.5%	12.5%	
	5	5009 LL11					80%				20%		20%	1%			
	5	5079 LL14					21.4%	7.1%			14.3%				7.1%	50%	
	2	2010 LL04					33.3%				11.1%	55.6%					
	5	5091 LL16					80%					20%					
	4	4003-45		74.1%	24.1%		1.7%										
Abandonment	47	47108-94	Concentrate			100%											
						Trees				Bushes				Woody shrubs			

Table A5. Protohistoric archaeozoological record: number of identified specimens (NISP, NISP%) and minimum number of elements (MNE, MNE%). The category "caprinae + sheep + goats" includes the remains identified at a specific level and at the *Caprinae* subfamily level (these values are not included in the total quantifications).

	NISP	%	MNE	%
Caprines (*Ovis/Capra*)	299	58.5%	208	52.6%
Sheep (*Ovis aries*)	13	2.5%	12	3%
Goat (*Capra hircus*)	20	3.9%	17	4.3%
Caprinae + sheep + goat	332	64.9%	237	60%
Cattle (*Bos taurus*)	55	10.8%	49	12.4%
Domestic suidae (*Sus domesticus*)	61	11.8%	56	14.1%
Equidae (*Equus* sp.)	7	1.6%	5	1.3%
Total domestic	455	89.1%	347	87.9%
Red deer (*Cervus elaphus*)	44	8.5%	35	8.9%
Roe deer (*Capreolus capreolus*)	1	0.1%	1	0.2%
Wild boar (*Sus scrofa*)	2	0.3%	2	0.5%
Hare (*Lepus granatensis*)	2	0.4%	2	0.5%
Rabbit (*Oryctolagus cuniculus*)	8	1.6%	8	2%
Total wild	57	10.9%	48	12.1%
Total Identified	512	100%	395	100

Table A6. Andalusi archaeozoological record: number of identified specimens (NISP, NISP%) and minimum number of elements (MNE, MNE%). The category "caprinae + sheep + goats" includes the remains identified at a specific level and at the Caprinae subfamily level (these values are not included in the total quantifications).

	NISP	%	MNE	%
Caprines (*Ovis/Capra*)	501	39.2%	345	37%
Sheep (*Ovis aries*)	105	8.2%	95	10.2%
Goat (*Capra hircus*)	88	6.9%	79	8.4%
Caprinae + sheep + goat	694	54.3%	519	55.5%
Cattle (*Bos taurus*)	36	2.9%	30	3.2%
Domestic suidae (*Sus domesticus*)	7	0.5%	6	0.6%
Horse (*Equus caballus*)	2	0.1%	2	0.2%
Dog (*Canis familiaris*)	1	0.1%	1	0.1%
Chicken (*Gallus* dom.)	461	36%	322	34.4%
Pigeon (*Columba palumbus*)	14	1%	3	0.3%
Total domestic	1215	94.9%	883	94.5%
Red deer (*Cervus elaphus*)	7	0.5%	6	0.6%
Fox (*Vulpes vulpes*)	1	0.1%	1	0.1%
Hare (*Lepus granatensis*)	8	0.6%	5	0.5%
Rabbit (*Oryctolagus cuniculus*)	45	3.6%	38	4.1%
Partridge (*Alectoris rufa*)	5	0.3%	1	0.1%
Total wild	66	5.1%	51	5.5%
Total Identified	1281	100%	934	100%

Table A7. Protohistoric archaeozoological record: anatomical distribution (MNE and MAU; species with extremely low values in the sample are not represented).

Caprinae + Sheep + Goat + Cattle + Domestic Suidae+Red Deer										
Caprines	**MNE**	**MAU**	**Sheep**	**MNE**	**MAU**	**Goat**	**MNE**	**MAU**	**Total MNE**	**Total MAU**
Head	103	16.5	Head	2	0.6	Head	10	2.9	115	20
Axial skeleton	21	1.7	Axial skeleton			Axial skeleton	1	1	22	2.7
Forelimb	18	9	Forelimb	3	1.5	Forelimb	4	2	25	12.5
Hind limb	22	11	Hind limb	3	1.5	Hind limb	1	0.5	26	13
Appendicular limb	44	11	Appendicular limb	4	2	Appendicular limb	1	0.5	49	13.5
Total	**208**		**Total**	**12**		**Total**	**17**		**237**	
Cattle	**MNE**	**MAU**	**Domestic Suidae**	**MNE**	**MAU**	**Red Deer**	**MNE**	**MAU**		
Head	14	2.4	Head	27	5.1	Head	6	3		
Axial skeleton			Axial skeleton	1	1	Axial skeleton	1	1		
Forelimb	8	4	Forelimb	9	4.5	Forelimb	7	3.5		
Hind limb	5	2.5	Hind limb	5	2.5	Hind limb	7	3.5		
Appendicular limb	22	6.5	Appendicular limb	14	3.1	Appendicular limb	14	4.4		
Total	**49**		**Total**	**56**		**Total**	**35**			

Table A8. Andalusi archaeozoological record: anatomical distribution (MNE and MAU; species with extremely low values in the sample are not represented).

Caprinae + Sheep + Goat + Cattle + Rabbit + Chicken										
Caprines	**MNE**	**MAU**	**Sheep**	**MNE**	**MAU**	**Goat**	**MNE**	**MAU**	**Total MNE**	**Total MAU**
Head	124	24.7	Head	8	3.1	Head	8	2.3	140	30.1
Axial skeleton	36	7.6	Axial skeleton			Axial skeleton			36	7.6
Forelimb	51	25.5	Forelimb	33	16.5	Forelimb	27	13.5	111	55.5
Hind limb	59	29.2	Hind limb	13	6.5	Hind limb	10	5	82	40.7
Appendicular limbs	75	18.7	Appendicular limbs	41	14.1	Appendicular limbs	34	12.1	150	44.9
Total	**345**		**Total**	**95**		**Total**	**79**		**519**	
Cattle	**MNE**	**MAU**	**Rabbit**	**MNE**	**MAU**	**Chicken**	**MNE**	**MAU**		
Head	23	3.1	Head	5	2	Head				
Axial skeleton	1	1	Axial skeleton	2	0.07	Axial skeleton	29	14.1		
Forelimb	2	1	Forelimb	6	3	Wing	96	48		
Hind limb	1	0.5	Hind limb	18	9	Leg	197	89.5		
Appendicular limb	3	0.75	Appendicular limb	6	0.6	**Total**	**322**			
Total	**30**		**Total**	**37**						

Table A9. Protohistoric and Andalusi archaeozoological record: metric data in mm [35].

			Bos taurus				
Late Bronze Age—Early Iron Age	P4 lower	L		A			
	UE: 10013-S.1		19.1		13		
	M1 lower	L		A			
	UE: 3036-S.3		19.52		13.5		
	M3 upper	L		A			
	UE: 4004-S.4		27.12		21.57		
	M3 upper	L		A			
	UE: 6043-S.6		26.84		17.82		
	M3 lower	L		A			
	U3: 3010-S.3		33.1		14.5		
	Metatarsus	Bp					
	UE: 6014-S.6		44.2				
	Astragalus	GLI					
	UE: 3010-S-6		63.51				
	PH1	Bp		Bd		GL	
	UE: 10013-S.1		25.32		22.9		47.94
	UE: 6035-S.6		31.4		28.21		55.6
	PH2	Bp		Bd		GL	
	UE: 6018-S.6		24.6		19.83		35.73
	PH3	Bp					
	UE: 3009-S.3		23.12				

Table A9. *Cont.*

	Sus domesticus				
Late Bronze Age—Early Iron Age	M2 lower	L	A		
	UE: 3009-S.3		19.4	11.9	
	M3 lower	L	A		
	UE: 10013-S.1		28.2	15.5	
	M3 lower	L	A		
	UE: 10024-S.1		29.4	14.9	
	Scapula	SLC			
	UE: 6021-S.3		22.3		
	Radio	Bp			
	UE: 6018.S.6		26.1		
	Radio	Bp			
	UE: 3009-S.3		26.81		
	Ulna	BCP			
	UE: 4004-S.4		19.6		
	PH1	Bp	Bd	GL	
	UE: 3009-S.3		16.2	15.4	32.9
	Sus scrofa				
Late Bronze Age—Early Iron Age	M3 lower	L			
	UE: 3010-S.3		41.1		
	Astragalus	GIL	GLm	Bd	
	UE: 3009-S.3		50.2	46.9	31
	Sus domesticus				
Early Middle Ages	M3 lower	L	A		
	UE: 5078-S.5		31.5	15.6	

References

1. Aguilella, G.; Falomir, F.; Pérez, G.; Laguna, M.C.; García, D.A.; Arquer, N. Tossal de la Vila (la Serra d'en Galceran, Castellón). Un asentamiento en la transición del Bronce Final al Hierro Antiguo. Primeros resultados de las campañas 2012–2014. *Quad. Prehist. Arqueol. Castelló* **2016**, *34*, 5–41.
2. Falomir, F.; Negre, J.; Aguilella, G.; Arquer, N. Noves perspectives en l'estudi del període andalusí als districtes meridionals de Turtusa. Resultats preliminars als jaciments castellonencs del Tossal de la Vila (serra d'en Galceran), Xivert i Santa Llúcia (Alcalà de Xivert). In *Arqueologia Medieval: Fortaleses a la Vall de l'Ebre (Segles VII–XI)*; Brufal, J., Negre, J., Sabaté, F., Eds.; Agira: Lleida, Spain, 2020; pp. 331–354.
3. Negre, J.; Pérez-Polo, M.; Falomir, F.; Aguilella, G.; Medina, P.; Blasco, M. Una lectura contextual del recinto emiral del Tossal de la Vila (Castelló): Algunas reflexiones sobre el origen, morfología y funciones de los asentamientos en altura en el extremo septentrional del Šarq al-Andalus". In *El Sitio de las Cosas: La Alta Edad Media en Contexto*; Publicacions de la Universitat d'Alacant: Alacant, Spain, 2020; pp. 195–217.
4. Negre, J.; Falomir, F.; Pérez-Polo, M.; Aguilella, G. Poliorcética, morfología edilicia y técnicas constructivas en el Tossal de la Vila, un recinto fortificado de época emiral en el extremo septentrional del Šarq al-Andalus. In *Defensive Architecture of the Mediterranean X*; Navarro, J., García-Pulido, L.J., Eds.; Universidad de Granada: Granada, Spain; Universitat Politècnica de València: Valencia, Spain, 2020; pp. 155–162. [CrossRef]
5. Negre, J.; Falomir, F.; Pérez-Polo, M.; Aguilella, G. La mezquita del fortín emiral del Tossal de la Vila (Castellón). Secuencia estratigráfica y diseño arquitectónico de un edificio religioso rural en los albores del islam andalusí. *Lucentum* **2022**, *XLI*, 301–319. [CrossRef]
6. Connolly, R.; Jambrina-Enríquez, M.; Herrera-Herrera, A.V.; Vidal-Matutano, P.; Fagoaga, A.; Marquina-Blasco, R.; Marin-Monfort, M.D.; Ruiz-Sánchez, F.J.; Laplana, C.; Bailon, S.; et al. A multiproxy record of palaeoenvironmental conditions at the Middle Palaeolithic site of Abric del Pastor (Eastern Iberia). *Quat. Sci. Rev.* **2019**, *225*, 106023. [CrossRef]

7. Fuks, D.; Dunseth, Z.C. Dung in the dumps: What we can learn from multi-proxy studies of archaeological dung pellets. *Veg. Hist. Archaeobot.* **2021**, *30*, 137–153. [CrossRef]
8. Scherer, S.; Höpfer, B.; Deckers, K.; Fischer, E.; Fuchs, M.; Kandeler, E.; Lechterbeck, J.; Lehndorff, E.; Lomax, J.; Marhan, S.; et al. Data from: Middle Bronze Age land use practices in the north-western Alpine foreland—A multi-proxy study of colluvial deposits, archaeological features and peat bogs. *Soil* **2021**, *7*, 269–304. [CrossRef]
9. Eixea, A.; Bel, M.Á.; Carrión, Y.; Ferrer-García, C.; Guillem, P.M.; Martínez-Alfaro, Á.; Martínez-Varea, C.M.; Moya, R.; Rodrigues, A.L.; Días, M.I.; et al. A multi-proxy study from new excavations in the Middle Palaeolithic site of Cova del Puntal del Gat (Benirredrà, València, Spain). *Comptes Rendus Palevol* **2023**, *10*, 159–200. [CrossRef]
10. Rondelli, B.; Lancelotti, C.; Madella, M.; Pecci, A.; Balbo, A.; Ruiz Pérez, J.; Inserra, F.; Gadekar, C.; Cau Ontiveros, M.A.; Ajithprasad, P. Anthropic activity markers and spatial variability: An ethnoarchaeological experiment in a domestic unit of Northern Gujarat (India). *J. Archaeol. Sci.* **2014**, *41*, 482–492. [CrossRef]
11. Barrachina, A. L'edat del bronze. Cent anys d'evolució dels estudis castellonencs. In *En els Problemes Arqueològics de la Província de Castelló: 100 Anys Després de Pere Bosch Gimpera*; Oliver, A., Ed.; Diputació de Castelló: Castelló, Spain, 2021; pp. 47–57.
12. Falomir, F. L'edat del ferro als territoris de Castelló. In *En els Problemes Arqueològics de la Província de Castelló: 100 Anys Després de Pere Bosch Gimpera*; Oliver, A., Ed.; Diputació de Castelló: Castelló, Spain, 2021; pp. 59–69.
13. Negre, J. La formació d'al-Andalus des de l'arqueologia: Un debat historiogràfic amb arrels castellonenques. In *En els Problemes Arqueològics de la Província de Castelló: 100 Anys Després de Pere Bosch Gimpera*; Oliver, A., Ed.; Diputació de Castelló: Castelló, Spain, 2021; pp. 91–109.
14. Trigger, B.G. *Understanding Early Civilizations: A Comparative Study*; Cambridge University Press: New York, NY, USA, 2003.
15. Smith, M.E. Braudel's Temporal Rhythms and Chronology Theory in Archaeology. In *Annales, Archaeology, and Ethnohistory*; Knapp, A.B., Ed.; Cambridge University Press: New York, NY, USA, 1992; pp. 23–34.
16. Bailey, G.N. Time Perspectives, Plimpsests, and the Archaeology of Time. *J. Anthropol. Archaeol.* **2007**, *26*, 198–223. [CrossRef]
17. Holdaway, S.; LuAnn, W. *Time in Archaeology: Time Perspectivism Revisited*; University of Utah Press: Salt Lake City, UT, USA, 2008.
18. Smith, M.E. Sprawl, Squatters, and Sustainable Cities: Can Archaeological Data Shed Light on Modern Urban Issues? *Camb. Archaeol. J.* **2010**, *20*, 229–253. [CrossRef]
19. Feinman, G. Comparative Frames for the Diachronic Analysis of Complex Societies: Next Steps. In *The Comparative Archaeology of Complex Societies*; Smith, M., Ed.; Cambridge University Press: Cambridge, UK, 2011; pp. 21–43. [CrossRef]
20. Drennan, R.; Peterson, C. Challenges for Comparative Study of Early Complex Societies. In *The Comparative Archaeology of Complex Societies*; Smith, M., Ed.; Cambridge University Press: Cambridge, UK, 2011; pp. 183–191. [CrossRef]
21. Bartolini, S. On Time and Comparative Research. *J. Theor. Politics* **1993**, *5*, 131–167. [CrossRef]
22. Pérez-Polo, M.; Falomir, F.; Negre, J.; Aguilella, G. Lectura arqueozoológica del fortín emiral del Tossal de la Vila (Castellón): Primeros datos. *Archaeofauna* **2023**. *submitted*.
23. García, D.; Laguna, M.C.; Aguilella, G.; Falomir, F. Arquitectura de terra i pedra en la transició del Bronze final al Ferro antic (segles VIII-VII a.C.). Un cas d'estudi en l'assentament del Tossal de la Vila (La Serra d'en Galceran). In *Actes de les XXI Jornades Culturals de la Plana de l'Arc: (21, 22 i 23 d'octubre de 2016)*; Associació Cultural la Balaguera: La Pobla de Tornesa, Spain, 2018; pp. 7–16.
24. Buxó, R.; Piqué, R. *Objetivos y Propuestas Metodológicas: La Gestión de los Recursos Vegetales y la Transformación del Paleopaisaje en el Mediterráneo Occidental*; Museu d'Arqueologia de Catalunya: Barcelona, Spain, 2003.
25. Chabal, L. Pourquoi et comment prélever les charbons de bois pour la période antique, les méthodes utilisé es sur le site de Lattes (Hérault). *Lattara* **1988**, *1*, 187–222.
26. Grau, E. El uso de la Madera en Yacimientos Valencianos de la Edad del Bronce a Época Visigoda. Datos Etnobotánicos y Reconstrucción Ecológica Según la Antracología. Ph.D. Thesis, Universitat de València, València, Spain, 1991.
27. Lyman, R.L. *Vertebrate Taphonomy*; Cambridge University Press: Cambridge, UK, 1994.
28. Lyman, R.L. *Quantitative Paleozoology*; Cambridge University Press: Cambridge, UK, 2008.
29. Boessneck, J.; Müller, H.-H.; Teichert, M. Osteologische Unterscheidungsmerkmale zwischen Schaf (*Ovis aries* Linne) und Zeige (*Capra hircus* Linne). *Kühn Archiv.* **1964**, *78*, 1–129.
30. Payne, S. Morphological distinctions between the mandibular teeth of young sheep, *Ovis*, and goats, *Capra*. *J. Archaeol. Sci.* **1985**, *12*, 139–147. [CrossRef]
31. Halstead, P.; Collins, P.; Isaakidou, V. Sorting the sheep from the goats: Morphological distinctions between the mandibles and mandibular teeth of adult *Ovis* and *Capra*. *J. Archaeol. Sci.* **2002**, *29*, 545–553. [CrossRef]
32. Zeder, M.A.; Lapham, H.A. Assessing the reliability of criteria used to identify postcranial bones in sheep, *Ovis*, and goats, *Capra*. *J. Archaeol. Sci.* **2010**, *37*, 2887–2905. [CrossRef]
33. Zeder, M.A.; Pilaar, S.E. Assessing the reliability of criteria used to identify mandibles and mandibular teeth in sheep, *Ovis*, and goats, *Capra*. *J. Archaeol. Sci.* **2010**, *37*, 225–242. [CrossRef]
34. Binford, L.R. *Nunamiut Ethnoarchaeology*; Academic Press: New York, NY, USA, 1978.
35. Driesch, A.V.D. *A Guide to the Measurement of Animal Bones from Arcaheological Sites*; Peabody Museum, Harvard University: Cambridge, UK, 1976.
36. Payne, S. Kill-off patterns in sheep and goats: The mandibles from Asvan Kale. *Anatol. Stud.* **1973**, *23*, 281–304. [CrossRef]

37. Grant, A. The use of tooth wear as a guide to the age of domestic ungulates. In *Ageing and Sexing Animal Bones from Archaeological Sites*; Wilson, B., Grigson, C., Payne, S., Eds.; BAR British Series: Oxford, UK, 1982; pp. 91–108.
38. O'connor, T. Process and terminology in mammal carcass reduction. *Int. J. Osteoarchaeol.* **1993**, *3*, 63–67. [CrossRef]
39. Silver, I.A. The ageing of domestic animals. In *Science in Archaeology*; Brothwell, D., Higgs, E.S., Eds.; Thames and Hudson: London, UK, 1969; pp. 283–302.
40. Zeder, M.A. Reconciling rates of long bone fusion and tooth eruption and wear in sheep (*Ovis*) and goat (*Capra*). In *Recent Advances in Ageing and Sexing Animal Bones*; Ruscillo, D., Ed.; Oxbow Books: Oxford, UK, 2006; pp. 87–118.
41. Pérez-Jordà, G. Estudio Arqueobotánico de Los Morrones. In *Los Morrones de Cortes de Arenoso. Un Poblado de la Edad del Hierro en el Alto Mijares. Castellón*; Barrachina, A., Ed.; SIAP. Diputació, Monografies de Prehistòria i Arqueologia Castellonenques: Castelló, Spain, 2020; Volume 15, pp. 183–191.
42. Albizuri, S.; Alonso, N.; López Cachero, F.J. Economia i Canvi Social a Catalunya Durant l'edat del Bronze i la Primera Edat del Ferro. *Arqueomediterrània* **2011**, *12*, 11–36.
43. García Gómez, E.; Pereira Sieso, J. El uso de la bellota en la Prehistoria en la Península Ibérica. In *Cambio Climático y Cultural en la Península Ibérica: Una Perspectiva Geohistórica y Paleoambiental*; Pérez-Díaz, S., Ruiz-Fernández, J., López-Sáez, J.A., García-Hernández, C., Eds.; Servicio de Publicaciones de la Universidad de Oviedo: Oviedo, Spain, 2017; pp. 169–183.
44. Barrachina, C.; De Haro, S. Contribució a l'estudi del medi ambient a partir de les restes de fusta recuperades al jaciment dels Estrets de Vilafamés. In *XII Jornades Culturals de la Plana de L'Arc*; Ajuntament de VilafamésPublisher: Vilafamés, Spain, 2007.
45. Espí, I.; Iborra, M.; De Haro, S. El área de almacenaje del poblado Ibero-Romano del Cormulló dels Moros (Albocàsser, Castelló). *Saguntum* **2000**, *3*, 147–152.
46. Ruiz Zapatero, G. Los Campos de Urnas del NE de la Península Ibérica. Ph.D. Thesis, Universidad Complutense de Madrid, Madrid, Spain, 1983.
47. Ruiz Zapatero, G. Los Campos de Urnas. In *Protohistoria de la Península Ibérica: Del Neolítico a la Romanización*; Almagro, M., Ed.; Fundación Atapuerca: Burgos, Spain, 2014; pp. 195–215.
48. Ruiz Zapatero, G. Bronce Final-Hierro: La naturaleza de los Campos de Urnas. In *XV Coloqui Internacional d' Arqueología de Puigcerdà. La Transició Bronze Final—1a Edat del Ferro en els Pirineus i Territoris Veïns*; Institut d' Estudis Ceretans: Puigcerdà, Spain, 2014; pp. 635–658.
49. López Cachero, J. Sociedad y economía durante el Bronce final y la Primera edad del Hierro en el noreste peninsular: Una aproximación a partir de las evidencias arqueológicas. *Trab. Prehist.* **2007**, *64*, 99–120.
50. Rafel, N. El Bronce Final y la primera Edad del Hierro en la fachada oriental peninsular y las Baleares. In *La Protohistoria en la Península Ibérica*; Ediciones Akal: Madrid, Spain, 2017; pp. 341–440.
51. Ruiz Zapatero, G. Las penetraciones de campos de urnas en el País Valenciano. *Cuad. Prehist. Arqueol. Castellon.* **1978**, *4*, 243–256.
52. González Prats, A. El proceso de formación de los pueblos ibéricos en el Sureste y Levante de la Península Ibérica. *Complutum* **1992**, *2*, 137–150.
53. Mata, C.; Martí, M.A.; Iborra, M.P. El País Valencià del Bronze recent a l'Ibèric antic: El procés de formació de la societat urbana ibérica. In *Models D'ocupació, Transformació i Explotació del Territori Entre el 1600 i el 500 A.N.E. a la Catalunya Meridional i Zones Limítrofes de la Depressió de l'Ebre*; Rovira, J., Ed.; Museu Nacional d´ Art de Catalunya: Barcelona, Spain, 1996; pp. 183–217.
54. Picazo, J. El poblamiento en el Valle Medio del Ebro durante la Prehistoria reciente: Zonas y procesos. *Rev. d ´ Arqueol. Ponent* **2005**, *15*, 97–118.
55. Picazo, J.; Rodanés, J.M. *Los Poblados del Bronce Final y Primera Edad del Hierro: Cabezo de la Cruz, La Muela, Zaragoza*; Gobierno de Aragón: Zaragoza, Spain, 2009.
56. Fatás, L. La Edad de Hierro en el valle del Matarraña (Teruel). Las investigaciones del Institut d'Estudis Catalans en el Bajo Aragón. *Caesaraugusta* **2016**, *85*, 2016.
57. Rodanés, J.M.; Picazo, J. Interaction and Interchange. The Genesis of the Late Bronze and Early Iron Age in the Middle Ebro Valley. In *Interchange in Pre and Protohistory. Case Studies in Iberia, Romania, Turkey an Israel*; Cruz, A., Gibaja, J.F., Eds.; BAR Publishing: Oxford, UK, 2018; pp. 161–175.
58. Jover, F.J.; Lorrio, A.J.; Díaz, M.Á. El Bronce Final en el levante de la península Ibérica: Bases arqueológicas y periodización. *Complutum* **2016**, *27*, 81–108. [CrossRef]
59. Oliver, A. Aproximación al poblamiento del Hierro antiguo en Castellón. *Trab. Var. SIP* **1992**, *89*, 29–38.
60. Oliver, A. *El Poblado Ibérico del Puig de la Misericordia de Vinaròs*; Associació Cultural Amics de Vinaròs: Vinaròs, Spain, 1994.
61. Gusi, F.; Barrachina, E. L'evolució dels grups culturals del bronze final i del ferro al País Valencià. Estat de la qüestió i problemàtica. In *Actes del XIII Congrés Internacional de Puigcerda Món Ibèric als Països Catalans. Homenatge a Josep Barberà i Farràs (Puigcerdà, 2003)*; Institut d´Estudis Ceretans: Puigcerdà, Spain, 2006; pp. 95–116.
62. Vives-Ferrándiz, J. Negociando encuentros. Situaciones coloniales e intercambios en la costa oriental de la Península Ibérica (s. VIII-VI a.C.). *Cuad. Arqueol. Mediterr.* **2005**, *12*, 15–269.
63. Barrachina, A.M. *Indesinenter: Permanencia y Cambio. El Pic dels Corbs como Modelo de Interpretación de la Edad del Bronce en el Norte del País Valenciano*; Servei d'Investigacions Arqueològiques i Prehistòriques: Castelló, Spain, 2012.
64. Oliver, A.; Gusi, F. *El Puig de la Nau. Un Hábitat Fortificado Ibérico en el Ámbito Mediterráneo Peninsular. Monografies de Prehistòria i Arqueologia Castellonenques 4*; Diputació de Castelló: Castelló, Spain, 1995.

65. Oliver, A.; García Fuertes, J.M.; Moraño Poblador, I. *El Castellet, Castelló de la Plana. Yacimiento Emblemático en la Historiografía de la Edad del Bronce Peninsular*; Fundación Dávalos-Fletcher: Castellón, Spain, 2005.
66. Oliver, A.; Falomir, F.; Aguilella, G. El Puig de la Misericordia, Vinaròs. In *Jornades d'Arqueologia de la Comunitat Valenciana 2016-2017-2018*; Ponce, M., Tendero, F., Alamar, Y., Alapont, L., Eds.; Generalidad Valenciana, Direcció General de Cultura i Patrimoni: València, Spain, 2020; pp. 61–70.
67. Oliver, A. *El Puig de la Nau, Benicarló*; Museu de Bellas Artes de Castelló: Castelló, Spain, 2006.
68. Hernández, F.J.; Cabanes, S. *En Balaguer I (Portell de Morella, Castellón). La Evolución de una Comunidad Rural Desde la Edad del Hierro Hasta la Romanización*; Generalitat Valenciana, Renomar, EIN Mediterráneo: València, Spain, 2010.
69. Aguilella, G. Primeros datos sobre el asentamiento del Hierro Antiguo de Santa Llúcia (Alacalà de Xivert, Castellón). Sector 1, fase 2. *Quad. Prehist. Arqueol. Castelló* **2016**, *34*, 105–118.
70. Aguilella, G. (Ed.) *Tossal del Mortorum. Un Assentament de l'Edat del Bronze i del Ferro Antic a la Ribera de Cabanes (Castelló). Monografies de Prehistòria i Arqueologia Castellonenques 12*; Diputació de Castelló: Castelló, Spain, 2017.
71. Vicente, M.; Pérez Milián, R.; Arquer, N.; Aguilella, G.; Mas, P.; Pérez Jordà, G. El Mas de Fabra (Benicarló, Castellón). Un asentamiento del hierro antiguo en la llanura litoral del Baix Maestrat. *Quad. Prehist. Arqueol. Castelló* **2016**, *34*, 79–104.
72. Aguilella, G.; García-Vuelta, O.; Montero-Ruiz, I.; Vila López, J.M. Oro y bronce en el yacimiento de Santa Llúcia (Alcalà de Xivert-Alcossebre, Castellón): Reflexiones sobre la actividad metalúrgica en los asentamientos del Hierro Antiguo del Levante peninsular. *Quad. Prehist. Arqueol. Castelló* **2020**, *38*, 51–70.
73. Oliver, A. (Ed.) *El Puig de la Misericordia y los Inicios de la Arquitectura de Prestigio en el Llano Litoral de Vinaròs. Un Ámbito Especializado en el Edificio Fortificado del Hierro Antiguo*; Diputació de Castelló: Castelló, Spain, 2021.
74. Barrachina, A. (Ed.) *Los Morrones de Cortes de Arenoso. Un Poblado de la Edad del Hierro en el Alto Mijares*; Monografies de Prehistòria i Arqueologia Castellonenques 15; Diputació de Castelló: Castelló, Spain, 2020.
75. Mateu, R.; Aguilella, G. El Tossal de Subarra (Benlloc, Castelló). L'ocupació de l'assentament entre el Bronze final i el Ferro antic. *Quad. Prehist. Arqueol. Castelló* **2021**, *39*, 13–23.
76. Pérez Jordà, G. La Agricultura en el País Valenciano Entre el VI y el I Milenio a.C. Ph.D. Thesis, Universidad de Valencia, Valencia, Spain, 2013.
77. De Haro, S. Estudio de los carbones vegetales. In *Los Morrones de Cortes de Arenoso. Un Poblado de la Edad del Hierro en el Alto Mijares. Castellón*; Barrachina, A., Ed.; SIAP. Diputació, Monografies de Prehistòria i Arqueologia Castellonenques: Castelló, Spain, 2020; Volume 15, pp. 176–181.
78. Iborra, P. *La Ganadería y la Caza Desde el Bronce Final Hasta el Ibérico Final en el Territorio Valenciano. Serie de Trabajos Varios 103*; SIP: Valencia, Spain, 2004.
79. Tormo, C. La fauna del Tossal del Mortórum. Campañas de excavación 2005–2010. In *Tossal del Mortòrum un Assentament de l ´Edad del Bronze i del Ferro Antic a la Ribera de Cabanes (Castelló)*; Aguilella, G., Ed.; SIAP. Diputació, Monografies de Prehistòria i Arqueologia Castellonenques: Castelló, Spain, 2017; Volume 12, pp. 113–128.
80. Sanchis, A.; Tormo, C. Estudio de la fauna, El Racó dels Cantos I. Un asentamiento de la Edad del Hierro. In *Paisaje y Arqueología en la Sierra de la Menarella. Estudios Previos del Plan Eólico Valenciano. Zona II: Refoies y Todolella*; Vizcaíno, D., Ed.; Renomar: Valencia, Spain, 2007; pp. 118–124.
81. Navarrete, V. Los restos de fauna del yacimiento de Los Morrones. In *Los Morrones de Cortes de Arenoso. Un Poblado de la Edad del Hierro en el Alto Mijares. Castellón*; Barrachina, A., Ed.; SIAP. Diputació, Monografies de Prehistòria i Arqueologia Castellonenques: Castelló, Spain, 2020; Volume 15, pp. 192–199.
82. Barrachina, A.; Sanchis, A. Valoración diacrónica de un modelo económico de la Edad del Bronce: La fauna del poblado del Pic dels Corbs, Sagunt (València). *Quad. Prehist. Arqueol. Castelló* **2008**, *26*, 43–94.
83. Brogiolo, G.P.; Gelichi, S. *Nuove Ricerche sui Castelli Altomedievali in Italia Settentrionale*; All'Insegna del Giglio: Florència, Italy, 1996.
84. Schneider, L. Oppida et castra tardo-antiques. À propos des établissements de hauteur de la Gaule Méditerranéenne. In *Les Campagnes de la Gaule à la Fin de l'Antiquité*; Ouzoulias, P., Pellecuer, C., Rayunaud, C., van Ossel, P., Garmy, P., Eds.; APDCA: Antibes, France, 2001; pp. 433–448.
85. Gutiérrez González, J.A. La fortificación pre-feudal en el norte peninsular: Castros y recintos campesinos en la Alta Edad Media. In *Mil anos de Fortificaçoes na Península Ibérica e no Magreb (500–1500): Actas do Simposio Internacional sobre Castelos*; Ferreira, I.C., Ed.; Edições Colibri, Camara Municipal de Palmela: Liboa, Portugal, 2002; pp. 19–28.
86. Catalán, R.; Fuentes, P.; Sastre, J.C. *Fortificaciones en la Tardoantigüedad: Élites y Articulación del Territorio (Siglos V-VIII d.C.)*; La Ergástula: Madrid, Spain, 2014.
87. Quirós, J.A.; Tejado, J.M. (Eds.) *Los Castillos Altomedievales en el Noroeste de la Península Ibérica*; Euskal Herriko Unibertsitatea: Bilbao, Spain, 2012.
88. Gibert, J. *La fi del Món Antic i els Inicis de l'Edat Mitjana a la Catalunya Central. Economia, Societat i Territori entre els Segles V i VIII*; ICAC: Tarragona, Spain, 2018.
89. Castellanos, S.; Martín-Viso, I. The local articulation of central power in the north of the Iberian Peninsula, 500–1000. *Early Mediev. Eur.* **2005**, *13*, 1–42. [CrossRef]
90. Martín, I. Espacios sin estado: Los territorios occidentales entre el Duero y el Sistema Central (siglos VIII-IX). In *¿Tiempos Oscuros? Territorios y Sociedad en el Centro de la Península Ibérica (Siglos VII–X)*; Martín, I., Ed.; Sílex: Madrid, Spain, 2010; pp. 107–135.

91. Bazzana, A. *Maisons d'Al-Andalus. Habitat Médiéval et Structures du Peuplement dans l'Espagne Orientale*; Casa de Velázquez: Madrid, Spain, 1992.
92. Acién, M. De nuevo sobre la fortificación del emirato. In *Mil Anos de Fortificaçoes na Península Ibérica e no Magreb (500–1500): Actas do Simposio Internacional sobre Castelos*; Ferreira, I.C., Ed.; Edições Colibri, Camara Municipal de Palmela: Liboa, Portugal, 2002; pp. 59–75.
93. Bazzana, A.; Guichard, P. Investigaciones arqueológicas en hábitats medievales de la provincia de Castellón de la Plana. *Cuad. Prehist. Arqueol. Castellon.* **1976**, *3*, 297–300.
94. Bazzana, A.; Guichard, P. Campaña 1977 de investigación arqueológica en yacimientos medievales de la provincia de Castellón (La Magdalena; Monte Mollet; Zufera). *Cuad. Prehist. Arqueol. Castellon.* **1977**, *4*, 333–350.
95. Bazzana, A. Investigaciones de arqueología medieval en la provincia de Castellón. Campañas de 1978 y 1979. *Cuad. Prehist. Arqueol. Castellon.* **1979**, *6*, 307–312.
96. Negre, J.; Suñé, J. Territorio, fiscalidad y actividad militar en la formación de un espacio fronterizo. La consolidación de Tortosa como límite extremo del al-Andalus omeya. *Anu. Estud. Mediev.* **2019**, *49*, 705–740. [CrossRef]
97. Negre, J. *En els Confins d'al-Andalus. Territori i Poblament Durant la Formació d'una Societat Islàmica a les Terres de l'Ebre i el Maestrat*; Onada Edicions: Benicarló, Spain, 2020.
98. Lorrio, A.; Almagro-Gorbea, M.; Sánchez de Prado, M.D. *El Molón (Camporrobles, Valencia). Oppidum Prerromano y Hîsn Islámico*; Ayuntamiento de Camporrobles: Valencia, Spain, 2009.
99. García-García, M. Explotación y Consumo de los Animales en el Sudeste de la Península IBÉRICA durante la Alta Edad Media (Siglos VII–XII): Perspectivas Históricas y Arqueozoológicas. Ph.D. Thesis, Universidad de Granada, Granada, Spain, 2019.

land

MDPI

Article

Continuity, Resilience, and Change in Rural Settlement Patterns from the Roman to Islamic Period in the Sicani Mountains (Central-Western Sicily)

Angelo Castrorao Barba [1,*], Carla Aleo Nero [2], Giuseppina Battaglia [2], Luca Zambito [3], Ludovica Virga [3], Alessandra Messina [3], Marco Cangemi [4] and Giuseppe Bazan [5]

[1] Institute Archaeology and Ethnology, Polish Academy of Sciences, 00-140 Warsaw, Poland
[2] Soprintendenza BB.CC.AA. di Palermo, 90133 Palermo, Italy; carla.aleonero@regione.sicilia.it (C.A.N.); giuseppina.battaglia@regione.sicilia.it (G.B.)
[3] Independent Researcher, 90100 Palermo, Italy; luca.zambit@gmail.com (L.Z.); ludovica.virga@live.it (L.V.); ale.mess96@gmail.com (A.M.)
[4] Department of Cultures and Societies, University of Palermo, 90128 Palermo, Italy; marco.cangemi05@you.unipa.it
[5] Department of Biological, Chemical and Pharmaceutical Sciences and Technologies (STEBICEF), University of Palermo, 90123 Palermo, Italy; giuseppe.bazan@unipa.it
* Correspondence: castroraobarba@gmail.com or a.barba@iaepan.edu.pl

Abstract: This study aims to analyze the dynamics of change in settlement models from the Roman, late antique, and Byzantine periods, focusing on how these transformations influenced the formation of Islamic societies in the rural landscapes of western Sicily. The study is centered around the territory of Corleone in the Sicani Mountains (central-western Sicily). This region, strategically located between the significant cities of Palermo on the Tyrrhenian Sea and Agrigento on the Strait of Sicily, has been pivotal in the communication network spanning from the Roman era to the Middle Ages and beyond. The area has been subject to extensive surveys and excavations, revealing diverse dynamics of continuity, resilience, and innovation in settlement patterns from the Roman to the Islamic periods. Beyond presenting the results of archaeological fieldwork, this study employs GIS-based spatial and statistical analyses and utilizes a range of topographic (elevation, slope, aspect, topographic position index (TPI), and distance to water sources) and ecological factors (vegetation series). These analyses aim to assess the evolving relationships and site positioning within the territory over time. Combining archaeological data with topographic and ecological landscape analysis, this integrated approach elucidates the complex transition dynamics from the Roman settlement system to the Islamic age's landscape formation in western Sicily's rural areas. The study thereby contributes to a deeper understanding of the intricate interplay between historical developments and environmental factors in shaping rural settlement patterns.

Keywords: landscape archaeology; landscape ecology; archaeological survey; settlement patterns; Roman and Islamic pottery; GIS; spatial analysis; vegetation series; late antiquity and early middle ages; Islamic Sicily; Corleone

1. Introduction

The complexity of the archaeological debate on the development of Islamic rural settlement patterns in Sicily following the 827 CE conquest [1,2] is unparalleled when compared to regions like al-Andalus in the Islamic West [3]. Initial archaeological investigations of medieval rural Sicilian settlements by the G.R.A.M. ("Gruppo Ricerche di Archeologia Medievale") [4,5] and French scholars [6,7] did not ignite substantial debate or lead to the establishment of a specialized field of medieval archaeology in Sicily [8,9]. Comprehensive scholarly works on late antique Sicilian settlements are notably absent, as highlighted by

Wilson's 1990 volume on Roman Sicily [10], which only marginally addresses the 5th–6th centuries CE. Maurici's 1992 book [11] remains the primary synthesis of Byzantine, Islamic, and Norman settlement patterns in Sicily, yet it relies heavily on written sources and offers limited material evidence, particularly in the realm of stratigraphical excavations. However, a study of Norman records and current place names reveals the presence of *rahl/qarya/manzil* as open settlements with distinct Islamic roots [1]. Archaeological breakthroughs, such as in the Islamic village that evolved over the Late Roman villa of Casale in Piazza Armerina (Enna) [12,13], demonstrate the feasibility of identifying such settlements archaeologically [14]. Excavations at various sites, like the large Roman-origin villages linked to the road system network, including, for example, Sofiana/Philosophiana (Caltanissetta), Casale San Pietro (Palermo), and Carini/*Hykkara* (Palermo), have uncovered material evidence that demonstrates an enduring occupation stretching into the Early Middle Ages [15,16].

The evolution of Byzantine forts (*castra*) after the Islamic conquest is an area where archaeological insights are scant [17]. A historical source by al-Nuwairī states that in 967 CE, Fatimid caliph al-Mu'izz instructed Sicilian Emir 'Aḥmad to construct fortified urban centers with *ǧami* mosques in each district (*iqlīm*), centralizing populations and discouraging rural sprawl [1]. This directive suggests that the *ḥuṣun* constructions were possibly part of a broader caliphal strategy [18] or a state endeavor to oversee and modify existing centers, incorporating essential features like fortifications and *ǧami* mosques [19]. The archaeological evidence for Islamic hilltop sites or fortresses is still largely uncharted [19,20]. The discovery of material culture artifacts, especially pottery from Palermo, is shedding light on the initial stages of Islamic presence, particularly during the late 9th and early 10th centuries CE [21,22].

In recent years, interest in early medieval Sicily has inspired two European projects on the Mediterranean Mountainous Landscape (MEMOLA, https://cordis.europa.eu/project/id/613265, accessed on 10 February 2024), with a case study in Sicily in the Trapani Mountains and on the archeology of changing regimes from the Byzantine to Norman eras (SICTRANSIT, https://cordis.europa.eu/project/id/693600/it/, accessed on 10 February 2024), mainly focused on the territory of Castronovo di Sicilia in the province of Palermo. Furthermore, specific projects on suburban and rural contexts [23] in the province of Palermo, such as Bagni di Cefalà Diana [24], San Giovanni dei Lebbrosi [25], or Santa Maria di Campogrosso [26], have provided new archaeological data on the transition between the Islamic and Norman periods and on that between the Byzantine and Islamic periods, as in the case of Contrada Castro [27–30]. The IS_LANDAS project "Islamic landscapes in southern Andalusia and western Sicily: patterns of change in settlements and rural communities between Late Antiquity and the Islamic age" (MSCA COFUND PASIFIC No. 847639) was conceived to understand, also from a comparative perspective and with an interdisciplinary approach, the dynamics of change of the Roman–late antique–Byzantine landscapes with the formation of Islamicate societies. As part of this project, the territory of Corleone was chosen to address the theme of the development of new settlement models in the Islamic era, as well as the pre-existing population and the characteristics of the landscape.

In the framework of recent archaeological research on the Roman and medieval [23,31–35] periods in western Sicily, the Corleone area has lacked systematic research efforts. Sporadic reconnaissance efforts revealing material evidence have identified several sites mainly related to the prehistoric and pre-Roman periods [36,37] and less so to the Roman and Middle Ages [38,39] (Table S1). More comprehensive surveys have been confined to neighboring areas such as Contessa Entellina in the southwest [40] and certain regions in the southern part of the municipality of Monreale [41]. Brief excavation campaigns were occasionally conducted, notably near Cozzo Spoletino [42] and Montagna Vecchia [43,44].

In 2015, a comprehensive archaeological and ecological study project commenced. It focused on a section of the southern part of the municipality of Corleone, specifically in the Castro and Giardinello districts [29,45]. This project encompassed the archaeological excavation (2017–2021) of the rural early medieval site of Contrada Castro [27,28,30].

The study at the Contrada Castro site expanded beyond a site-centric view, exploring the dynamics of human–environment interaction across a broader landscape scale [46].

This study introduces new data, highlighting continuity, resilience, and transitional settlement patterns from Roman to Islamic periods in an area of the hinterland between Palermo and Agrigento. This area serves as a paradigm for understanding the formation of Islamic settlements in Sicilian landscapes and its relationship with environmental factors.

2. Study Area

The study area is located in central-western Sicily in the province of Palermo. Some of the prominent elevations include Rocca Busambra, at 1613 m a.s.l.; Mt. Barraù, reaching 1420 m a.s.l.; and Mt. Cardellia, at 1266 m a.s.l. (Figure 1).

The lithological formations, regardless of their stratigraphic order and based on all the characteristics (lithology, petrography, sedimentology, structure, texture, erodibility, etc.) that may have shaped the geomorphological configuration of the landscape can be grouped into the following seven lithological complexes: clastic continental deposition complex; Plio–Pleistocene sandy–calcarenite complex; clayey–marly complex; evaporitic complex; conglomeratic–sandstone complex; sandy–clayey–limestone complex; and carbonate complex [47]. The lithostratigraphic attributes of the geologic formations of this region delineate a terrain marked by two distinctly different relief patterns. Intermittently, carbonate rock ridges rise amidst a terrain dominated by gentle clayey hills. The carbonate elevations vary in magnitude compared to the uniform clay hills. These elevations typically display uneven profiles, featuring moderately sloped, solid, and rocky sides, and are often fringed with layers of debris. In certain instances, these elevations possess plateaued summits encircled by cliffs, leading to their local designation as "Rocche", an Italian term for a fortress. This name underscores their visual similarity to majestic and dilapidated bastions, as described by Sestini [48].

The climatic data recorded in the thermo-pluviometric station of Corleone (594 m a.s.l.) show average temperatures (in °C) of 15.8 and annual precipitation (in mm) of 747.2 [49], with a dry summer period typical of the Mediterranean climate. According to the bioclimatic classification proposed by Rivas-Martínez [50], the region is characterized by a Mediterranean pluviseasonal oceanic bioclimate [51]. This bioclimatic framework can be segmented into several distinct belts. The thermo-Mediterranean belt fluctuates from an upper dry ombrotype, predominantly found on the drier southern and western inclines, to a lower subhumid bioclimate, typical in valley basins up to altitudes of approximately 450–550 m a.s.l. The meso-Mediterranean belt shifts from a lower subhumid ombrotype, generally seen at elevations spanning 550 m to 900 m a.s.l., to an upper subhumid ombrotype evident on the tallest mountain peaks exceeding 900 m in elevation. The supra-Mediterranean belt predominantly exhibits an upper subhumid ombrotype observed in the summit areas of mountains such as Rocca Busambra [52].

The landscape predominantly features arable land, vast crops, uncultivated areas, and pastures. The agricultural landscape of this area is typical of the Sicilian hinterland, characterized by cultivated fields, mainly of wheat in rotation with legumes, vineyards, and olive groves, sometimes characterized by old olive trees [53]. Grazing by sheep, goats, and cattle has led to the development of seminatural environments in the hilly and sub-mountainous areas characterized by prairies with *Ampelodesmos mauritanicus* and, in marginal areas, by shrublands dominated by spiny Rosaceae and, in some cases, brooms [54]. Natural woods, frequently surrounded by man-made forest plantations, are typically restricted to fringe areas or higher altitudes mainly characterized by evergreen and deciduous oaks dominated by *Quercus ilex* L., *Quercus suber* L., and *Quercus pubescens* Willd. subsp. *pubescens*, mainly associated with *Acer campestre* L., *Fraxinus ornus* L., *Fraxinus angustifolia* Vahl, *Ulmus minor* Mill. subsp. *minor*, *Sorbus torminalis* L. (Crantz), and, sporadically, *Ostrya carpinifolia* Scop. [55,56].

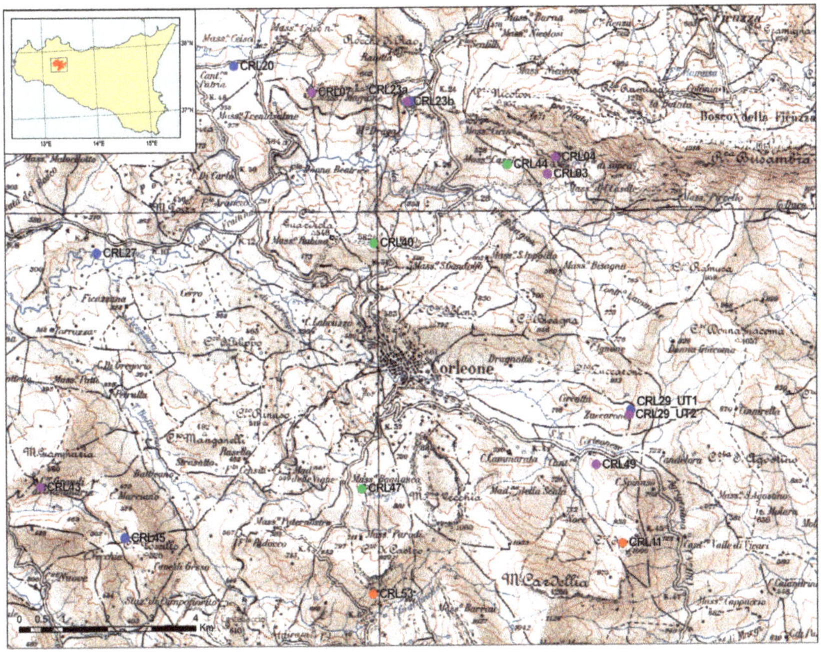

Figure 1. Location of the study area and topographic map from the Italian Geographic Military Institute [57] and sites relevant for settlement dynamics across the ages: Roman (blue); Byzantine (green); Byzantine–Islamic (red); Islamic (purple).

3. Historical Context

The archaeological research on this territory was undertaken due to the significant historical dimension that Corleone [58,59] had in the history of the population of the Sicilian hinterland between Antiquity and the Middle Ages. The location near Corleone of the ancient settlement of *Schera*/Σχήρα [60], mentioned by ancient sources such as Ptolemy [Ptol., 3, 4, 14], Cicero [Cic., 2, *Verr.*, 3, 103, 43], and Pliny [Plin., *n.h.*, 3, 91], was already advanced by Clüver in the 17th century [61], while a more precise identification of the presence of an ancient city on the plateau of "La Vecchia" or "Old Mountain", south-east of current Corleone, was already hypothesized by the discovery of coins and ceramics by Jean Houel in the 18th century [62]. Although the identification of the city of *Schera* is still debated [63], archaeological research has nevertheless confirmed the existence of a large center from the Sican period, perhaps occupied sporadically until the 3rd century BCE and, only later, partially reoccupied in the 13th century CE [43,44]. However, researchers have not excluded the possibility of prior human habitation in the Byzantine and Islamic eras near the calcarenitic plateau of Montagna Vecchia (maximum elevation of 1081 m a.s.l.), which stands isolated in the valley of the Corleone River (Figure 2). In Roman times, it was only an unverified hypothesis that the site had already moved to the site of the current town of Corleone [64]. The exceptional document of the Roman milestone of the consul C. Aurelius Cotta comes from the territory (Contrada Zuccarone) (252–248 BCE) [65,66]. It is the only Roman milestone found in Sicily. We have no mention of Corleone in the Byzantine era, while the Arab chroniclers of the conquest of Sicily (Ibn al-Athīr [67] and al-Nuwairī [68]) attest that in 839–840 CE, Corleone (*Q.r.lūn*) was one of the various fortresses (called *ḥuṣūn* in Arabic sources) of the Island that made a surrender agreement with the Muslims. In the 10th century, *Qurliyūn*/Corleone was cited as a chief town (*mudun*) by al-Muqaddasi [68,69], while in 1154, Idrîsi describes it as a fortified small town ("une bourgade fortifièe" (*ḥiṣn*) [70]), well defended and inaccessible, with solid and high fortifications, contiguous agricultural crops, and a nearby river of the same name [67]. In documents from the Norman era (late 12th century), the *iqlīm* of *κουρουλλιούνη*/*qurliyūn* or an administrative district center is attested [71,72].

Figure 2. The right peak of the southwest side of Montagna Vecchia, where the medieval settlement was located.

4. Materials and Methods

4.1. Archaeological Survey

Two archaeological survey campaigns were carried out in the summer of 2022 and 2023. In this phase of the research, the investigation strategy focused on verifying the many sites with possible phases from the Roman, late ancient, Byzantine, Islamic, and Norman periods (3rd century BCE–12th century CE), which had been sporadically reported over the years. In this phase of the research, therefore, an extensive investigation of portions of the territory was not carried out field by field, but we opted for an intra-site survey to understand the extension of the sites and, above all, to verify the existence of ceramic materials on the surface. The sites were mapped with mobile GPS, and the georeferenced data (points and tracks) were inserted into a GIS platform integrated with information sheets on the individual sites and the material culture connected to them. According to the directions of the Sicilian Region Archaeological Service (Superintendence BB.CC.AA. of Palermo) responsible for authorizing research, it was decided not to carry out a total collection of the material on the surface. This choice was also dictated by the limited possibilities of the conservation of the materials and to avoid completely erasing the traces on the ground of the presence of sites, thereby protecting the areas for future investigations. Therefore, we carried out an autopsy evaluation of the areal dispersion and density of the archaeological material present on the surface, and only some diagnostic materials were collected, intended to act as fragments for the reconstruction of a shape (e.g., edges, bottoms, and handles), fragments with particular surface treatments (e.g., black gloss, ARS, and painted pottery), or significant fabrics (e.g., calcitic ware). In this way, we tried to obtain a significant sample of materials useful for the chronological definition of the occupation of the various sites. In some cases, different areas of concentration or "topographic units" (UT) were identified within a site, while, in other cases where the density of the material was particularly high, it was decided to divide the areas into squares within which the material was quantified in situ before proceeding with the selective collection of diagnostic materials square by square. Finally, in some cases which will not be covered in this study, only structural evidence relating to medieval and post-medieval watermills was mapped. To address the theme of the dynamics of continuity, resilience, and change between the settlement dynamics of the Roman and late antique periods and those of the Byzantine and Islamic periods, a subset of sites was created that presented phases of occupation falling within these chronological areas.

4.2. GIS-Based Spatial Analysis

A GIS-based analysis was conducted to elucidate the relationship between archaeological sites and landscape characteristics during the Roman and Islamic periods. The study encompassed 17 sites categorized into three macro-historical periods: (1) Roman, (2) Byzantine, and (3) Islamic.

The area of concentration of ceramic materials for each site, detected in the field with GPS, was mapped in GIS as polygons. Each polygonal feature was then used to perform zonal statistics on the raster layers, which returned the statistical values of five landscape parameters: elevation (meters above sea level), slope (degrees), aspect (degrees to the north), topographic position index (TPI), and water distance (meters).

The processing was carried out using both Esri ArcGIS 10.2 and QGIS 3.28 for all terrain covariates. The raster layers were processed starting from the LIDAR-DTM of the Sicilian region with a cell size of 2 m [73].

TPI values, as formulated by Weiss [74], facilitated the classification of the landscape into six slope position categories: (1) valley, (2) lower slope, (3) flat slope, (4) middle slope, (5) upper slope, and (6) ridge.

For the distance from water sources, the nearest distance was calculated from the hydrographical network obtained from the regional technical map of Sicily at a scale of 1:10,000 [73]. Given that it is a hilly and mountainous area, the morphological dynamics of the watersheds, as well as the other topographical factors considered, are not affected by significant natural changes in spatial terms over the time scale of the last two thousand years, such that they would influence the results of the analyses considered.

The extracted values were then recorded in a table, which was used for statistical analyses in R [75]. Given the limited size of the dataset, an exploratory data analysis was performed to understand the value distribution better. A boxplot was created to visually represent the data distribution across different periods to determine the differences between groups.

To understand and interpret the relationships between the topographical variables considered (elevation, slope, aspect, TPI, and water distance) and visualize how the sites group together, a principal component analysis (PCA) was carried out, with the results displayed as a biplot. The PCA is a technique widely used in exploratory data analysis, specifically with multivariate data [76]. The biplot visually represents the relationships between the variables and their impact on different archaeological sites and periods.

To interpret the distribution of the sites in relation to the ecological characteristics of the territory, a vegetation series map was created according to the methodology proposed by Bazan et al. [46]. The vegetation series (or sigmetum) depicts a set of plant communities within homogeneous environmental areas (tessella or ecotope), which are stages of the same ecological succession process, and the series serves as synthetic indicators of all environmental characteristics (climate, lithology, and landforms). As ecological indicators, vegetation series are also indicators of potential land use suitability. The fact that it is based on factors whose change generally takes tens of thousands of years [77] allows the use of vegetation series as indicators of potential land use, even in retrospective terms. The frequency of different vegetation series was analyzed within a 250 m buffer around each site using R.

5. Results

In total, 53 sites from different periods were identified through bibliographic research and field surveys and mapped in a GIS environment (Table S2, Figure S1). From the general dataset of the sites, a group of sites relevant from the point of view of the settlement dynamics of the Roman and Islamic eras was identified. In particular, these are sites falling into four thematic categories: (1) topographical continuity of Roman sites during the Islamic period; (2) new settlement patterns during the Byzantine period; (3) early medieval reoccupation of pre-Roman sites; and (4) new Islamic settlements.

5.1. Topographical Continuity of Roman Sites during the Islamic Period

5.1.1. Contrada Zuccarone (CRL29)

The Contrada Zuccarone is located about 5 km southeast of Corleone and about 1 km north of the "Strada Statale 118 Corleonese-Agrigentina", between the local road Agnone-Zuccarone (to the north) and the "Regia Trazzera N. 81, Bivio Amendola—Portella dei Giudei—Bivona (east-west)" and the "Regia Trazzera N. 481, Bivio Cicio—Bivio Cappuccio (north-south)". The only Roman milestone found in Sicily was discovered in Contrada Zuccarone in 1954. The milestone bore the inscription [*C. A]VRELIVS/[L.F.]COTTAS/[C]ONSOL/↓VII*, in reference to the consul *C. Aurelius Cotta* (252, 248 BCE) [65,66]. It is still debated whether the milestone was connected with the construction of an official road network in the 3rd century BCE or whether this milestone was a trophy or a terminal stone [65] connected to the phases of the conquest of the territory by the Romans during the First Punic War (264–241 BCE). Despite this, the discovery is certainly connected to the proximity of the Contrada Zuccarone to a road route that was connected with a stretch of the Palermo–Agrigento road mentioned in the *Itinerarium Antonini* [78]. During our archaeological survey, we had the opportunity to systematically document, for the first time, an archaeological site located in the area of the discovery of the milestone and, in particular, on a small hill of approximately 800 m above sea level. This site was approximately 80 m higher than the access road below, which was near a watering trough. In a vineyard (CRL29_UT1) with an eastward slope, we identified a dense concentration area (about 4100 m^2) of ceramic materials (Figure 3). Given the considerable concentration of materials, we divided the area into 35 squares, according to the rows of vineyards, and quantified the material in situ with a selective collection of diagnostic material for each square (Table S3).

According to diagnostic materials (Figure 4), the site seems to have been frequented, starting from the middle Hellenistic age, for about one or two centuries, after which there seems to have been a hiatus until the Early Imperial age when the frequentation resumed. Continuous ensuing habitation from that time was demonstrated until Late Antiquity.

No valuable elements have been found on the site, and it is not easy to propose a univocal interpretation of this site; perhaps it was a villa, which, in any case, would seem to be well connected to regional and extra-regional commercial circuits until the end of Late Antiquity.

Despite this, the occupation of this area continued beyond Late Antiquity. In fact, in the field adjacent to the southern side of the vineyard (currently uncultivated but with soil devoid of surface vegetation), another area of concentration of materials was identified (CRL29_UT2), covering approximately 4900 m^2, mainly from the Islamic era. In fact, among the diagnostic materials found, we noted the characteristic tiles with vegetal inclusions, some of which also had engraved streaks, which, in association with painted and glazed ceramics, allow us to hypothesize human presence in the Islamic era (Figure 5).

Figure 3. Map of the density of ceramic fragments per m^2 in the two topographic units (UT1 and UT2) identified in the Contrada Zuccarone site (CRL29).

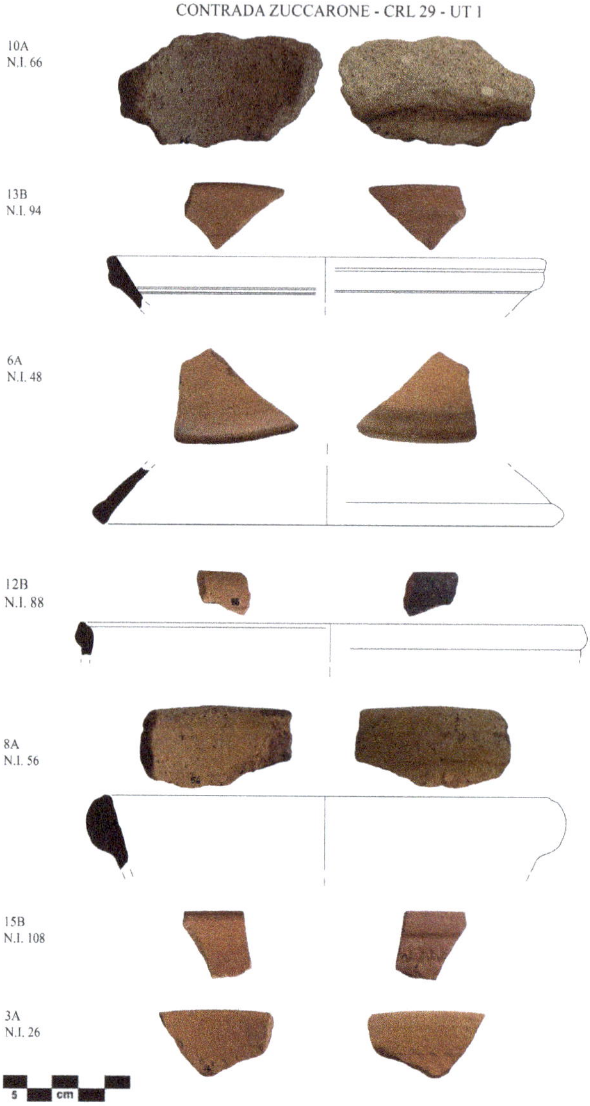

Figure 4. Hellenistic and Roman pottery from the Contrada Zuccarone site (CRL29_UT1). 10A, N.I. 66: worn rims referable due to their texture (numerous volcanic inclusions of Tyrrhenian origin) and typology to MGS amphorae and probably dating back to between the 4th and 3rd centuries BCE [79]. 13B, N.I. 94: fragment of an African Red Slip A cup, Hayes 8B form, with an undecorated strip (2nd century CE) [80]. 6A, N.I. 48: a fragment of rim of plate lid with cinerish patina of Hayes 196 form/Bonifay types 11.5 (2nd–5th century) [80,81]. 12B, N.I. 88: almond-shaped rim and vertical body of a North African cooking pot (casserole) related to the Hayes 197 form (4th–5th century) [80]. 8A, N.I. 56: a fragment of the rim of a cooking pan of Pantellerian ware (4th–5th century) [82,83]. 15B, N.I. 108: a fragment of a rim of an African Red Slip D cup with "nailed" decoration on the external body of a Hayes 81 A/Bonifay type 43–44 form (half/second half of 5th century) [80,81]. 3A, N.I. 26: a fragment of a rim of Hayes 81A/Bonifay type 43–44 form (half/second half of 5th century) [80,81].

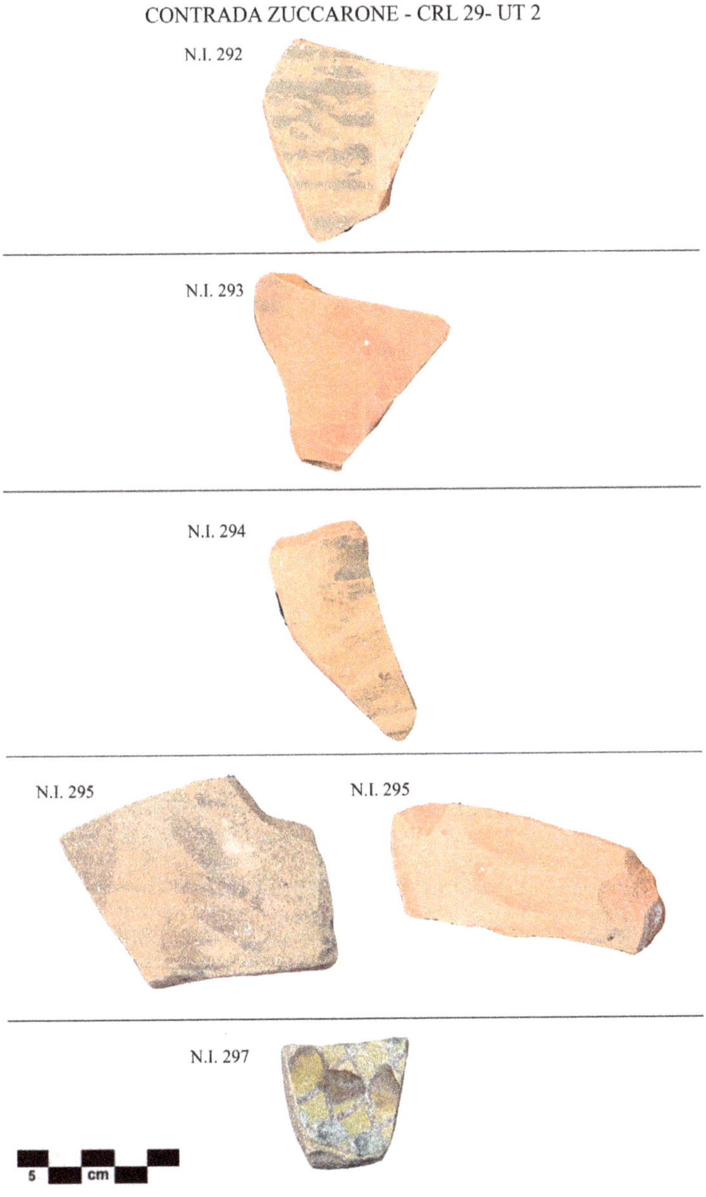

Figure 5. Islamic pottery from the Contrada Zuccarone site (CRL29_UT2). N.I. 292–295: painted amphorae/jugs. N.I. 297: Islamic glazed pottery.

In particular, numerous fragments of the walls of amphorae/jugs were found alongside a modest quantity of handles. Many of the walls are painted in both red and brown: among the most common decorative motifs, we found the broad band motif (Figure 5, N.I. 292), a sinusoidal motif with vertical lines alternating with sinusoidal ones rendered with brushstrokes of the same thickness, which appears in the contexts of the end of the 9th century. During the first half of the following century, the sinusoidal lines were gradually modified to cursive lines until the motif transformed into the variant with parallel bands and detached sections, also, in this case, possessing lines of the same thickness (mid-10th–mid-11th century). In the Zuccarone district site, in particular, the most widespread decorative motifs are the sinusoidal motif (Figure 5, N.I. 293) and the motif with parallel detached lines (Figure 5, N.I. 294–295); the cursive sinusoidal motif is less attested (Figure 5, N.I. 296) [84]. In the same UT, fragments of glazed ceramics were also found: for example, a fragment pertaining to an open shape with green and brown decoration (Figure 5, N.I. 297) was found, comparable to a piece found in urban and suburban excavations in Palermo in contexts dated to the second half of 10th/mid-11th century such as in the Gancia [85,86] and in the cesspits/silos of Via Imera [87].

5.1.2. Case Scalilli–Gole del Drago (CRL23)

In the northern part of the municipality of Corleone, there is a promontory of calcarenite rock (max. elevation 498 m a.s.l.), stretched in a NW–SE direction, bordered on the west by the Frattina torrent, from which the Belice Sinistro River originates and which, at this point, forms the natural waterfalls named Gole del Drago. This promontory is located 350 m east from State Road 118 "Corleonese-Agrigentina". In this area, 18 different topographical units (UTs) of concentration of archaeological materials were identified (Table S4) UT14, UT15). These materials were found around the Case Scalilli–Gole del Drago, located east of the promontory (UT1, UT2, UT7, UT16, and UT17), and in the olive grove north of the promontory, beyond the access road to the site (UT18) and about 90 m southeast of the Case Scalilli–Gole del Drago (UT8) (Figure 6).

Figure 6. (**a**) Topographic units identified in the Case Scalilli–Gole del Drago site (CRL23); (**b**) the rocky plateau at the top of the isolated relief at the Case Scalilli–Gole del Drago site.

In the lower eastern part of the limestone promontory and near the Case Scalilli–Gole del Drago, areas were identified (UT1, UT2, UT7, and UT16) in which it was possible to recover material, mainly from the Roman period (Figure 7).

The site presents an important phase dating back to the Early Imperial age, which was followed by survival (possibly with contraction) until the early Byzantine age. It would seem to be one of the sites where activity began at the end of the 1st century BCE, perhaps in relation to the colonial deduction of which we have traces in literary sources.

In the Islamic period, the occupation of the site moved to the upper part of the limestone promontory on the Dragon Gorges. The topographic units refer to this occupation (UT3, UT4, UT5, UT6, UT9, UT10, UT11, UT12, UT13, UT14, and UT15), while the area downstream of the northern promontory (UT18) appears to be connected to more sporadic visits and the south (UT8).

This medieval occupation is evidenced by the presence of fragments of polychrome glazed ceramic, such as the fragment of a ring-bottom found with a phytomorphic/vegetal motif under a transparent glaze (Figure 8, UT 14, N.I. 187) dating back to the first half of the 11th century or the mid-11th century [88]. Other fragments found in the various UTs belong to the same class, but the decoration is not always present or identifiable: for example, a fragment of a wall belonging to an open form with a transparent window and a nonreadable decoration in green and brown are cited (Figure 8, UT 15, N.I. 288), as is a ring foot fragment with green glaze (Figure 8, UT 15, N.I. 188). The discovery of several fragments of amphorae/jugs was also reported. Many of these are painted both red and brown (Figure 8, UT 15, N.I. 291). The most frequently identified decorative motif with parallel detached lines (Figure 8, UT 14, N.I. 286–287) appeared around the mid-10th century and remained unchanged until the middle of the following century. During the 12th century, the oblique strokes tended to become thinner and more spaced out (Figure 8, UT 15, N.I. 290) [84]. There is no shortage of evidence of other types of materials, as evidenced by the fragment of the edge of a circular plate lamp with a small dome reservoir (Figure 8, UT 15, N.I. 289), a type that appears in contexts from the second half of the 9th century but which persisted at least until the second half of the following century [89]. The fragments of green monochrome glazed ceramic (Figure 8, UT 14, N.I. 185–186), internally bearing a furrowed continuous ribbon decoration with parallel transversal sections, are, however, to be referred to a later occupation of the site (12th century) [90,91].

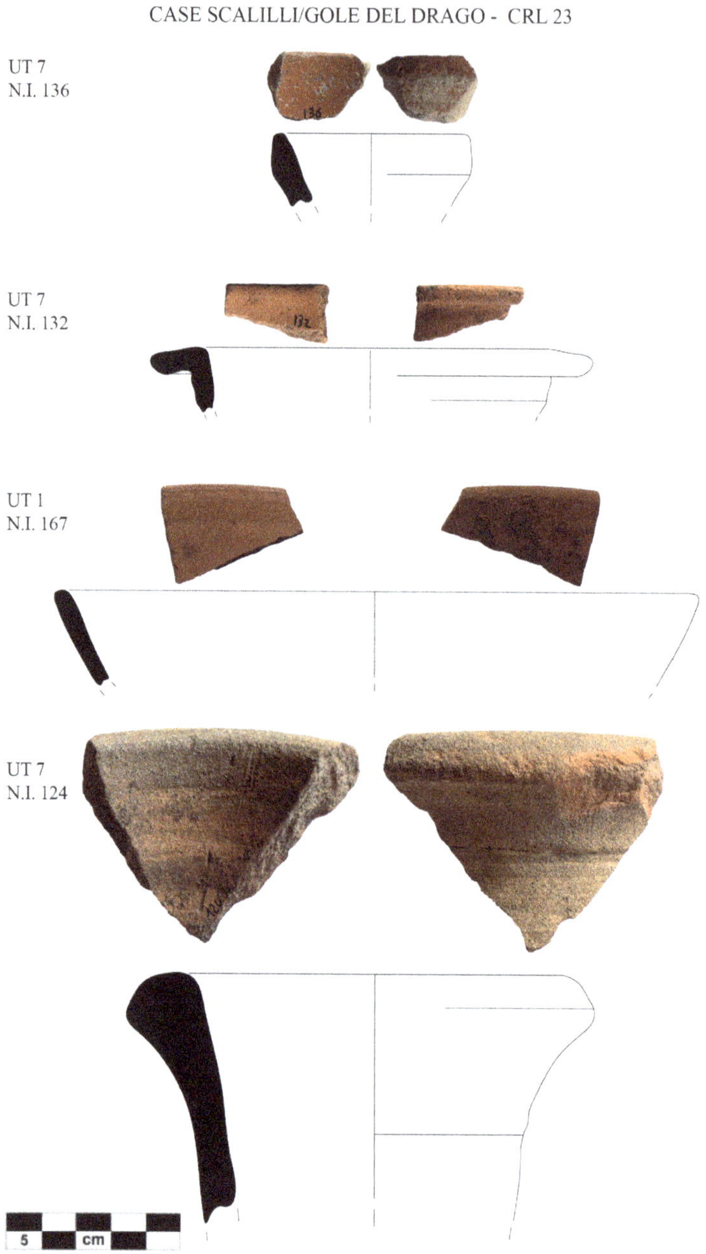

Figure 7. Roman pottery from the Case Scalilli–Gole del Drago site (CRL23). UT7, N.I. 136: a small cup in *terra sigillata*, approximately 15 BCE—50 CE [92]. UT7, N.I. 132: a cup in African red slip A with everted rim and subhorizontal brim, Hayes 32 form [80] (1st–2nd century CE). UT7, N.I. 167: a bowl in African Red Slip F [93] with an indistinct rim and vertical body may be similar to Hayes 50 (mid-5th century CE) [80]. UT7, N.I. 124: an amphora Africana IIIA (4th century) [56].

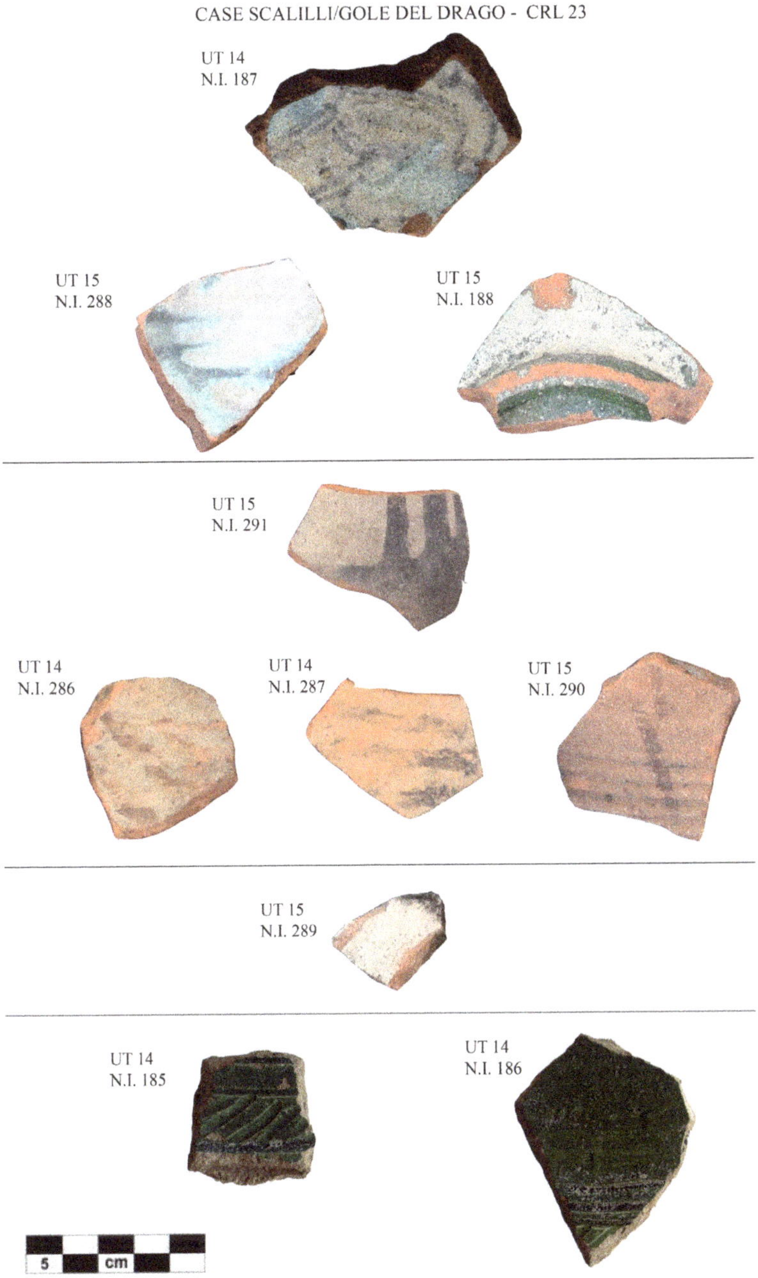

Figure 8. Shards of Medieval pottery from the Case Scalilli–Gole del Drago site (CRL23). UT 14, N.I. 187; UT 15, N.I. 188; UT 15, N.I. 288: glazed pottery. UT 15, N.I. 291; UT 14, N.I. 286; UT 14, N.I. 287; UT 15, N.I. 290: painted amphorae/jugs. UT 15, N.I., 289: lamp. UT 14, N.I. 185; and UT 14, N.I. 186: glazed pottery with furrowed decoration.

5.2. New Settlement Patterns during the Byzantine Period

Small Byzantine Sites: Costa Rubina (CRL40), Contrada Casale (CRL44), and Case Bingo (CRL47)

Regarding the period of occupation of this territory during the late antique and Byzantine periods, we do not have much definite information on the settlement typologies if we exclude sites with long continuity until the 5th–6th century, such as the aforementioned cases of Contrada Zuccarone and Case Scalilli–Gole del Drago. From irregular excavations between the 1950s and 1970s, some ceramic and glass materials from 5th–6th-century grave goods come from the Bisagna and Carrubba districts [39,94], while a group of 6th–7th-century grave goods appear to come from the Palastanga locality (not identifiable with certainty). Golden basket earrings and ceramics characterize these and include a jug decorated on the body with two overlapping irregular painted bands of vertical stripes in red, a lanceolate shape in the upper part, a comma shape in the lower part, and with a thin horizontal band on the edge [95] (Figure 9).

From our archaeological survey, we identified sites of limited extension at Costa Rubina (CRL40, area size of 6245 m^2), Contrada Casale (CRL44, area size of 1744 m^2), and Case Bingo (CRL47, area size of 2931 m^2). These sites are united by the low density of fragments of ceramic pottery and characterized by the almost exclusive presence of fragments of striped/combed tiles (Figure 10), a typical indicator of the Sicilian contexts during the Byzantine era between the late 6th and 7th centuries [96].

Figure 9. Earrings (**a**,**b**) and painted ceramic jugs (**c**,**d**) from grave goods of the Byzantine period (6th–7th century) from the Palastanga area in the Corleone area (reworked from Dannheimer [97] to Metaxas [98]).

Figure 10. Byzantine striped/combed tiles from the Costa Rubina (CRL40), Contrada Casale (CRL44), and Case Bingo (CRL47) sites.

5.3. Back to the Hilltop: Early Medieval Reoccupation of Pre-Roman Site

Contrada Castro (CRL53) and Cozzo Spolentino (CRL11)

The site of Contrada Castro (CRL53) was identified following systematic reconnaissance during 2015–2016 [99]. It is located on a high plateau south of the rocky slope of the highest relief of Pizzo Castro/Rocche di Mezzogiorno in the southern part of the municipal territory of Corleone, almost on the border with the municipality of Campofiorito. From 2017 to 2021, several excavation campaigns were carried out [27,28,30], which led to the identification of an interesting sequence of occupation divided into four main periods of the pre-Roman period and Byzantine–Islamic transition age: (1) late 6th–4th century BCE; (2) late 7th–early 8th century CE; (3) late 8th–9th century CE; and (4) 10th–11th century CE. The first phase of occupation of the site dates back to the late archaic/classical period (late 6th–4th century BCE), to which several walls probably connected to residential units. After a long hiatus, the site was reoccupied in the early Middle Ages. The first evidence of this re-frequenting is children's tombs dated by radiocarbon analyses between the end of the 7th century and the beginning of the 8th century CE. In the late 8th–9th century, the site of Contrada Castro saw a building intensification documented by the construction, in the two excavation areas investigated, of a warehouse/deposit with a deep silo inside. At the end of the 9th century, these buildings were obliterated, and from the beginning of the 10th century CE, other buildings (houses) were built with a completely divergent orientation compared to that of the previous phase. This last phase, heavily eroded over the centuries by subsequent plundering and agricultural activities, can be seen within the full Islamic era (10th–11th century), after which the site was only sporadically frequented between the end of the 11th and the beginning of the 12th century. This site has been the subject of multidisciplinary analyses focused on the study of human–environment interactions, which have highlighted—through zooarchaeological, archaeobotanical, and phytosociological analyses [100]—a long-lasting continuity in the exploitation of environmental and animal resources [101]. The ceramic record demonstrates a good level of economic complexity in this inland area, in which the presence of amphorae produced in Palermo testifies to an exchange connection between the city and the countryside [102].

A similar dynamic of reoccupation of a high-altitude site (about 950 m above sea level) from the pre-Roman era during the early Middle Ages could be that of Cozzo Spolentino (CRL11). At this site, emergency excavations in the 1990s brought to light layers of use with ceramic materials, still unpublished, which could refer to the early Middle Ages [42].

5.4. New Islamic Sites

5.4.1. Casale di Sotto (CRL3), Casale di Sopra (CRL4)

Starting from the full Islamic age (at least during the 10th century), new sites emerged in lowland areas connected to roads and on high ground. The most interesting case is represented by the evidence in Contrada Casale (Municipality of Monreale), located on the southern slope of the Rocca Busambra mountain, on the border between the southern part of the Municipality of Corleone and the Municipality of Monreale, and delimited to the south by the Bicchinello stream, a tributary of the Frattina stream. In the 1980s, the first identification of two sites was carried out [103]: one in the Casale di Sotto area (CRL3), on the plateau (558 m above sea level, below the rocky part of this side of the Rocca Busambra), and another in the Casale di Sopra area (CRL4), on the top of a high plateau (826 m above sea level) located on the offshoot of the limestone massif of Rocca Busambra and accessible from the south through the Piano Pilato plateau (794 m above sea level) (Figure 11a).

The Casale di Sopra site is defined by a fortification wall, which is preserved for a length of 243 m. In some places, it is often almost 2 m high and constructed of freshly rough-hewn stones laid dry without the use of mortar binder (Figure 11b,c). This line of wall encloses the plateau on three sides: north, east, and west (where a small door opens). To the south, the site is naturally defended by the wall of the promontory, which has a practically sheer morphology. Overall, the surface area enclosed by the fortification is equal to approximately 30,903 m^2.

Despite the strong erosion of the site, which in some places has outcropping rock, five topographical units have been identified (Figure 11d; Table S5); these have yielded a large number of ceramic finds.

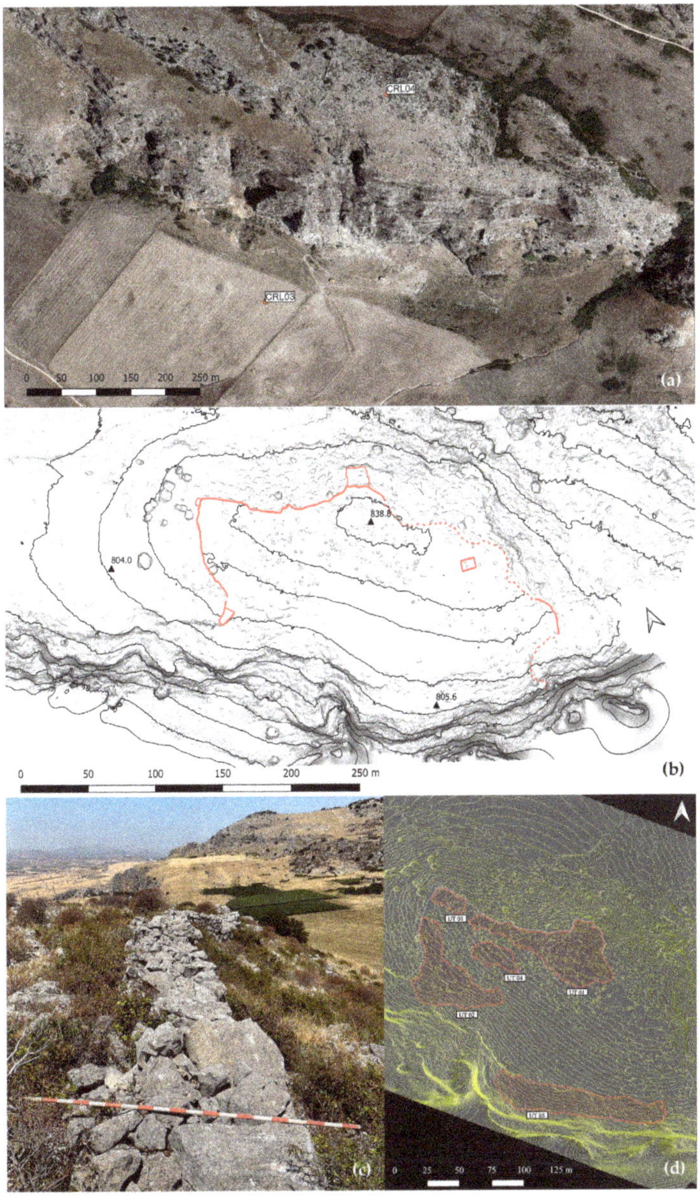

Figure 11. (**a**) Topographic location of the sites of Casale di Sotto (CRL3) and Casale di Sopra (CRL4); (**b**) map of the fortification line of Casale di Sopra; (**c**) the fortification wall of Casale di Sopra; (**d**) topographic units of concentration of pottery inside the wall of Casale di Sopra.

The most frequently observed class was that of vacuolated tiles with vegetable inclusions. Some fragments relating to amphorae (Figure 12) are equally interesting, and they often have a decorative motif of wide brown bands on the external surface (Figure 12, UT 2,

N.I. 281). Among the rims within the same class, examples were identified consisting of a rim with a triangular section (Figure 12, UT 1, N.I. 2) or a pendulous triangular section (Figure 12, UT 2, N.I. 239–240; UT4, N.I. 242), likely related to Palermo production, which dates back chronologically to between the second half of the 10th and the beginning of the 11th century [104]. The other fragment of an almond-shaped rim (Figure 12, UT 5, N.I. 244) also refers to Palermo production and is enlarged with two evident concentric grooves on the outside. This type of amphora, dated to the 11th century, belongs to containers with a wide mouth [84,105].

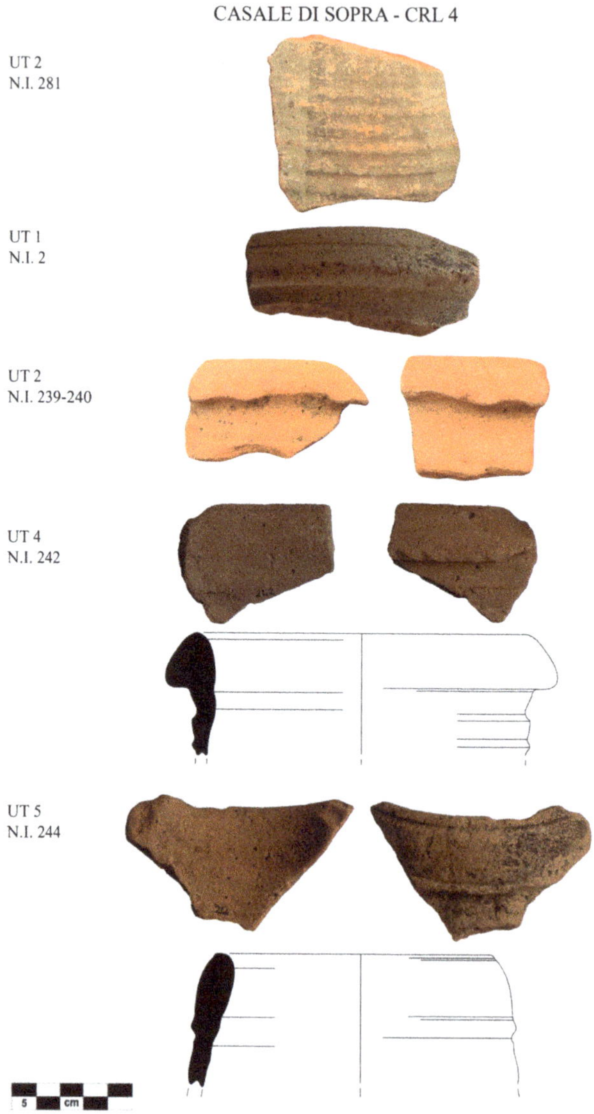

Figure 12. Islamic amphoras from Casale di Sopra (CRL4). UT 2, N.I. 281: painted amphora. UT 1, N.I. 2; UT 2, N.I. 239–240; UT 4 N.I. 242; and UT 5, N.I. 244: amphoras.

Regarding glazed pottery (Figure 13), among the different types identified, a fragment of a ring foot was found. This was of an open form, with a transparent glaze and balanced leaf decoration in green and brown (Figure 13, UT4, N.I. 243) [88,106]. The source of its production is currently uncertain, and it can be dated to between the second half of the 10th and the first half of the 11th century. In addition, a fragment of a thickened rim with traces of green glaze was identified (Figure 13, UT3, N.I. 241). This, most likely, was produced later (12th century CE). In fact, the shape seems to recall the so-called keeled basins with a lowered cap, such as those found in the exhaust of the kiln of Palazzo Lungarini [107] and in the workshop at the Santa Cecilia theater, both in Palermo [108]. Based on the study of the finds, it is clear that the site of Casale di Sopra seems to have had evident use during the Islamic period (second half of the 10th–11th century), with some sporadic traces of use in the Norman age.

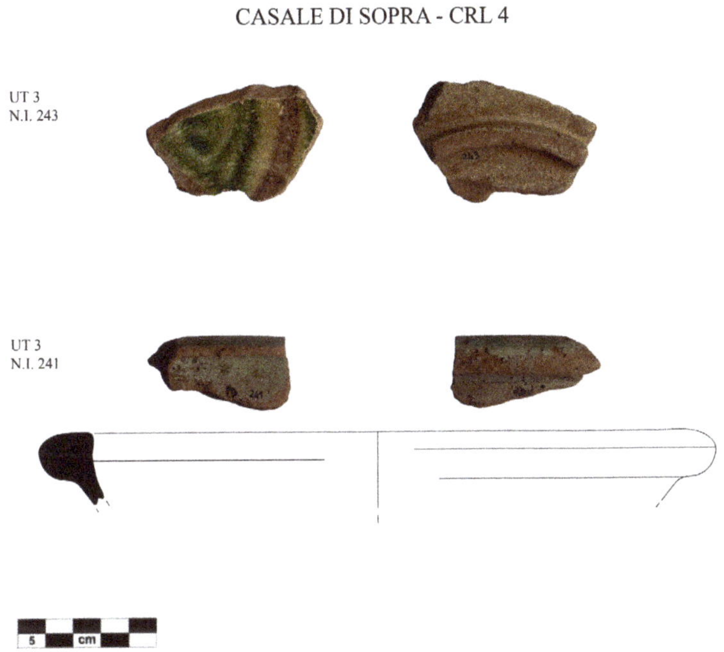

Figure 13. Medieval glazed pottery from Casale di Sopra (CRL4): UT 3, N.I. 243; UT 3, and N.I. 241.

At the foot of the steep wall, towards the south, medieval materials were collected in the Casale di Sotto site area. This was previously reported in the surveys of the 1980s [103].

5.4.2. Masseria Magione (CRL7), Case Grandi di Giammaria (CRL43), and Contrada Carrubba (CRL49)

Furthermore, materials from the end of the 9th to the 11th century were found in other open sites on the valley floor, such as those at Masseria Magione (CRL7), Case Grandi di Giammaria (CRL43), and Contrada Carrubba (CRL49). Of the three sites, those that have returned the most diagnostic material are Masseria Maggione and Contrada Carrubba (Figure 14). The most significant and numerically largest finds belong to the class of amphorae, likely also, in this case, as in the Casale di Sopra site of Palermo production. Among the most recurring painted decorative motifs, we found the sinusoidal motif (Figure 14, N.I. 292, late 9th–early 10th century) and the sinusoidal cursive motif (Figure 14, N.I. 283, first half of the 10th century) with wide bands of both brown and red (Figure 14, N.I. 284) at the site of Masseria Maggione. At Contrada Carruba, we also identified the motif of parallel detached lines in brown (NI 261–262, mid-10th–11th century) [84]. The only fragment of the rim of an amphora from Masseria Maggione (Figure 14, N.I. 194) refers to that of the same type found on the site of Casale di Sopra and dated to the 11th century (see above

Figure 12, UT5, N.I. 244). Finally, a fragment of an olla (cooking pot) rim with an everted and rounded almond-shaped brim (Figure 14, N.I. 192), identified as fire pottery, was found. The rim, probably of Palermo production, dates back to between the end of the 9th and the first half of the 10th century [109].

Figure 14. Shards of Islamic painted amphoras and cooking pots from the Masseria Magione (CRL7) and Contrada Carrubba (CRL49) sites. CRL 7, N.I. 282; CRL 7, N.I. 283; CRL 7, N.I. 284; CRL 49, N.I. 261; and CRL 49, N.I. 262: painted amphorae/jugs. CRL 7, N.I. 194: amphora. CRL 7, N.I. 192: cooking pot.

5.5. Landscape Position Analysis

GIS-based analyses focusing on settlement dynamics between the Roman and Islamic periods were conducted on a collection of 17 sites. This included those previously described with Islamic era phases and sites exclusively occupied during the Roman period.

The GIS analysis of the sites' positions in relation to the topographic characteristics of the landscape first resulted in the creation of GIS maps depicting the elevation, slope, aspect, TPI, and water distance (Figures 15 and S2–S6). The mean value of each parameter for each site, calculated using zonal statistics, is shown in Table 1.

Table 1. Topographical parameter of analyzed sites.

ID	Site Names	Period *	Elevation [m]	Slope (°)	Aspect (°)	TPI	Water Dist. [m]
CRL03	Casale di Sotto	3	558	7.4	214.4	4	244
CRL04	Casale di Sopra	3	824	16.3	207.4	6	472
CRL07	Masseria Magione	3	476	10.6	185.7	3	396
CRL11	Cozzo Spolentino	2, 3	952	18.1	210.7	6	346
CRL20	Masseria Patrìa	1	367	8.2	252.5	3	300
CRL23a	Case Scalilli/Gole del Drago	1	483	5.8	90.2	4	115
CRL23b	Case Scalilli/Gole del Drago	3	489	10.8	200.8	5	111
CRL27	Strasatto/Case Scuderi	1	254	5.2	173.1	3	54
CRL29_UT1	Zuccarone	1	798	9.7	82.1	4	444
CRL29_UT2	Zuccarone	3	804	10.8	101.2	3	500
CRL40	Costa Rubina	2	565	9.3	210.3	4	503
CRL43	Case Grandi di Giammaria	3	455	12.8	177.5	5	60
CRL44	Contrada Casale	2	531	9,9	266.2	3	19
CRL45	Casa Paternostro	1	465	7.9	99.7	4	109.
CRL47	Case Bingo	2	709	6.6	250.8	4	234
CRL49	Contrada Carrubba	3	733	1.7	69.4	3	231
CRL53	Contrada Castro	2, 3	712	4.5	209.2	6	262

* (1) Roman; (2) Byzantine; and (3) Islamic.

The analysis of the distribution of variable elevation across different periods (Figure 16a) shows that the Roman sites have a median altitude value of 465 m above sea level, which is lower than those for the Byzantine period (709 m above sea level) and the Islamic period (711 m above sea level). However, within the group of Roman sites, there is a site (CRL29_UT1) with an "outlier" elevation value of 798.15 m above sea level. Similarly, in the group of the Byzantine period, there is a site (CRL11) with an elevation value of 952.08 m above sea level, which significantly differs from the average. An interesting observation is that, in general, the altitude of the Roman sites tends not to overlap with the higher altitudes of the sites from the Byzantine and Islamic periods. From a morphological point of view, the medians of the slopes (Figure 16b) increase from the Roman period (7.9°) to the subsequent Byzantine (9.3°) and Islamic periods (10.9°). In general, these median values correspond to two slope gradient classes: "moderately sloping" (4.7°–9°) and "strongly sloping" (9.1°–16.7°), according to the USDA [110].

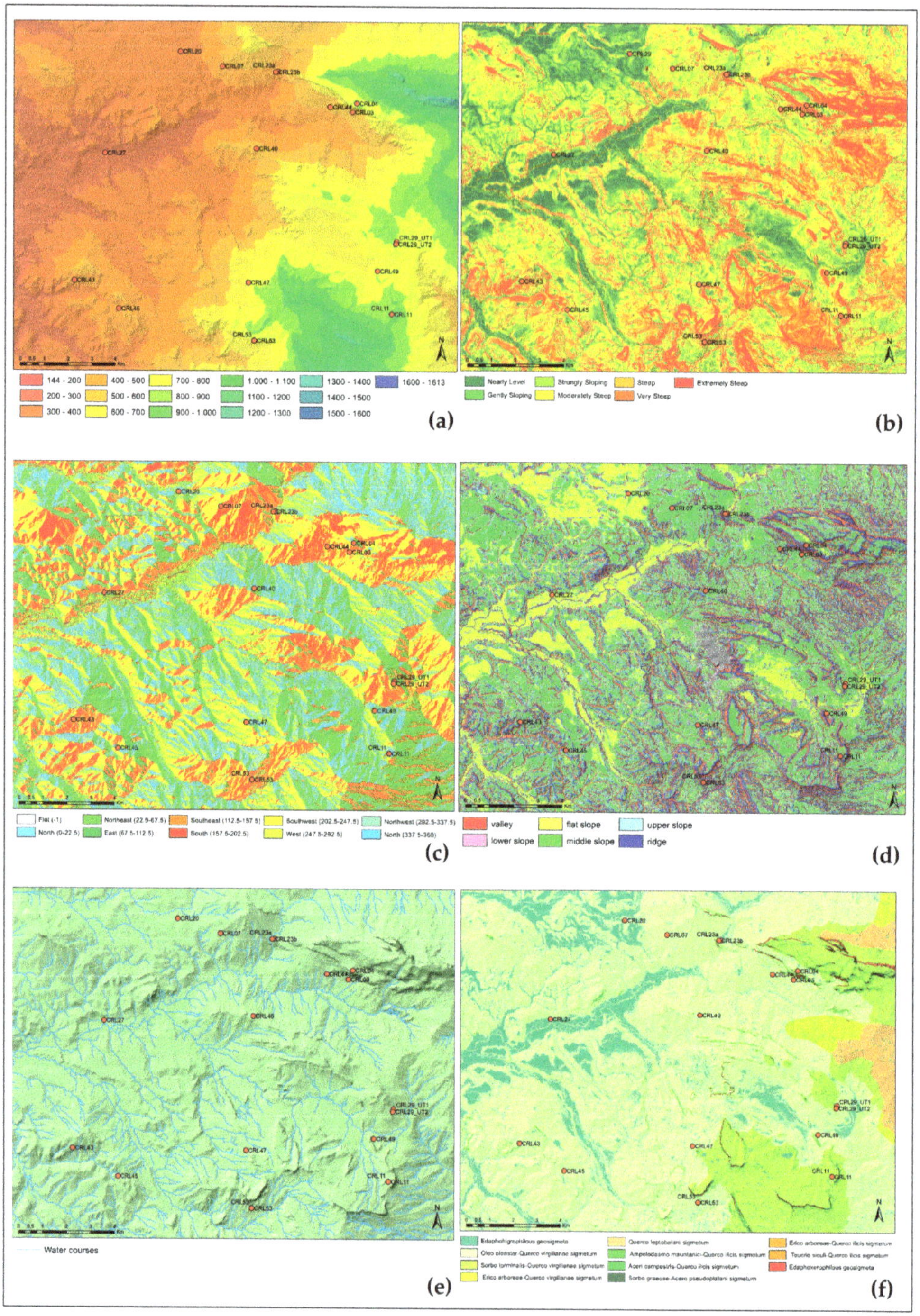

Figure 15. Maps of GIS-based spatial analyses of key environmental variables: (**a**) elevation [m], (**b**) slope (°), (**c**) aspect (°), (**d**) TPI, (**e**) water distance [m], and (**f**) vegetation series. Red circles are archaeological sites.

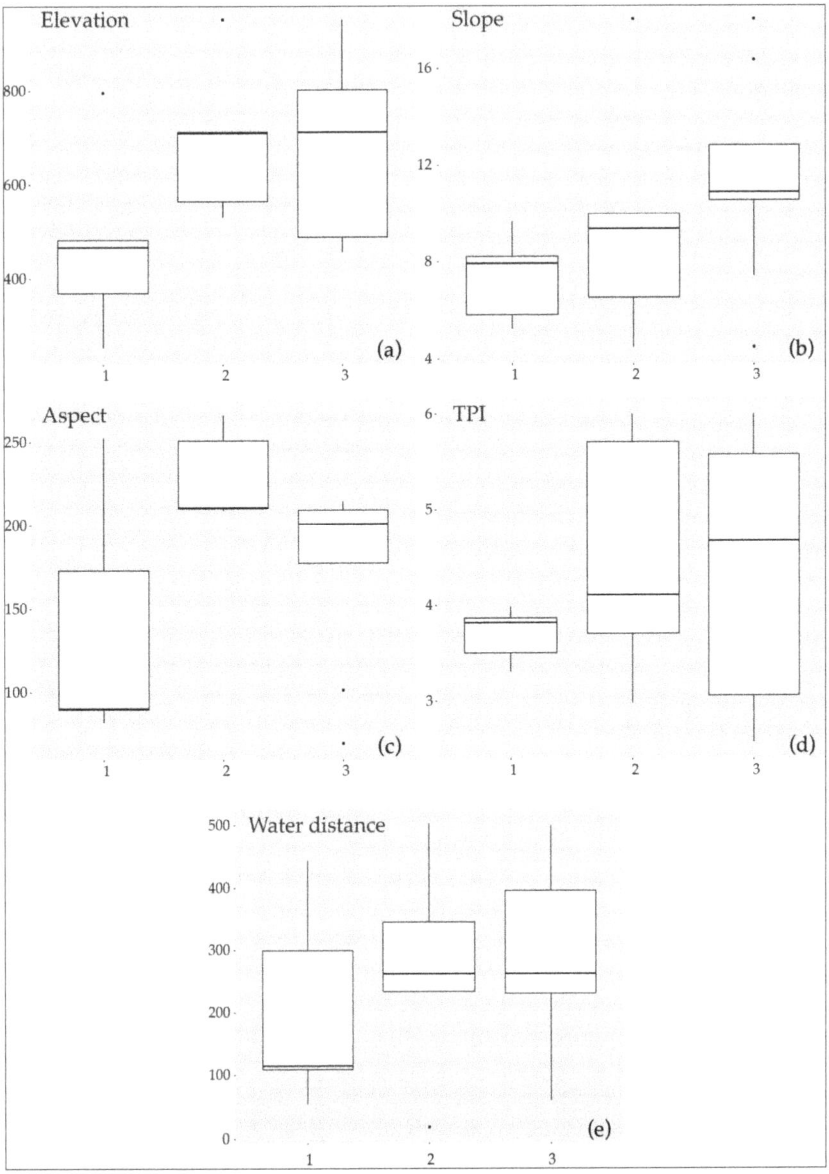

Figure 16. Boxplots representing the distribution of key environmental variables across different periods at archaeological sites. Each plot corresponds to a unique variable: (**a**) elevation [m], (**b**) slope (°), (**c**) aspect (°), (**d**) TPI, and (**e**) water distance [m]. The boxplots are categorized by the Roman (1), Byzantine (2), and Islamic (3) periods and illustrate the range, interquartile range, median, and potential outliers within each period.

However, there are some outliers among the sites from the Byzantine and Islamic periods, which are located in "gently sloping" (CRL53) and "steep" (CRL11) positions. Regarding the aspect variable (Figure 16c), the Roman sites have median values that indicate an eastward aspect, while the Byzantine and Islamic sites tend to face southwest. The most frequent TPI class (Figure 16d), based on the medians for all three periods, is the middle slope (class 4). The values of the median distances from watercourses per period are as follows: 115 m for the Roman period and 262 m for the Byzantine and Islamic periods (Figure 16e).

A PCA biplot reveals a data distribution pattern along a gradient of topographic factors consistent with the data described in the boxplot (Figure 17). The PCA identifies the existence of at least three clusters of sites exhibiting similar topographic characteristics. The first cluster comprises six sites, including four from the Roman period (CRL20, CRL23a, CRL27, and CRL45) and two from the Islamic period (CRL03 and CRL07). These sites have an average elevation of 434 m above sea level, an average slope of 7.5°, an average aspect of 167° (S), a TPI of 3–4 (flat slope/middle slope), and a water distance of 203 m. The second cluster consists of three sites occupied during the Islamic era (CRL23b, CRL43, and CRL53), featuring average elevation values of 552 m above sea level, an average slope of 9.4°, an average aspect of 196° (S), a TPI of 5 (upper slope), and a water distance of 144 m. The final cluster includes two sites also occupied in the Islamic era [CRL11, CRL04], with average elevation values of 888 m above sea level, an average slope of 17.1°, an average aspect of 209° (SW), a TPI of 5–6 (upper slope/ridge), and a water distance of 409 m.

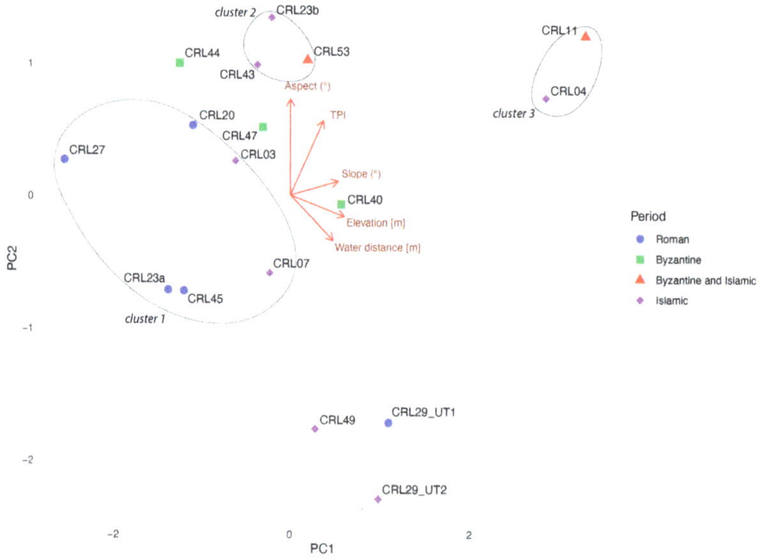

Figure 17. PCA biplot depicting the distribution of archaeological sites categorized by period. Each point represents a site positioned according to the first two principal components (PC1 and PC2) derived from environmental variables. The red arrows indicate the direction and strength of the variables' association with the principal components. Sites are visually distinguished by period: Roman (1), Byzantine (2), and Islamic (3).

The ecological analysis of the landscape conducted in terms of vegetation series led to the creation of a map in which 11 types have been identified (Figure S7).

The zonal statistic of the area surrounding the sites (250 m buffer) indicates that the landscape is predominantly characterized by two calcicolous series of downy oak (Figure 18). For CRL_11, due to its elevated location (>900 m asl) within the bioclimatic zone of the meso-supramediterranean subhumid ombrotype, the vegetation series is almost entirely (91.7%) represented by the *Sorbus-Quercus virgilianae* sigmetum. The other sites, which fall within the thermo-meso-Mediterranean subhumid bioclimatic zone, are primarily characterized (over 80% in 11 sites) by the *Olea-Quercus virgilianae* sigmetum series. In both instances, the remaining forest vegetation is comprised of woodlands dominated by *Quercus pubescens*, located in mountainous and hilly areas with soils of varying depths. This pedological feature provides this ecological context with good productive potential, especially for cereal cultivation, which has been the predominant type of land use in the hilly areas for centuries. In three Roman period sites (CRL20, CRL23a, and CRL27), the vegetation context is partially influenced by a sub-flat terrain with alluvial soils where the potential vegetation is represented by edaphohygrophilous sigmeta (27%, 19%, and 40%, respectively, for the three sites). From an agricultural standpoint, these areas with a good productive capacity are used for the cultivation of cereals and produce from vineyards.

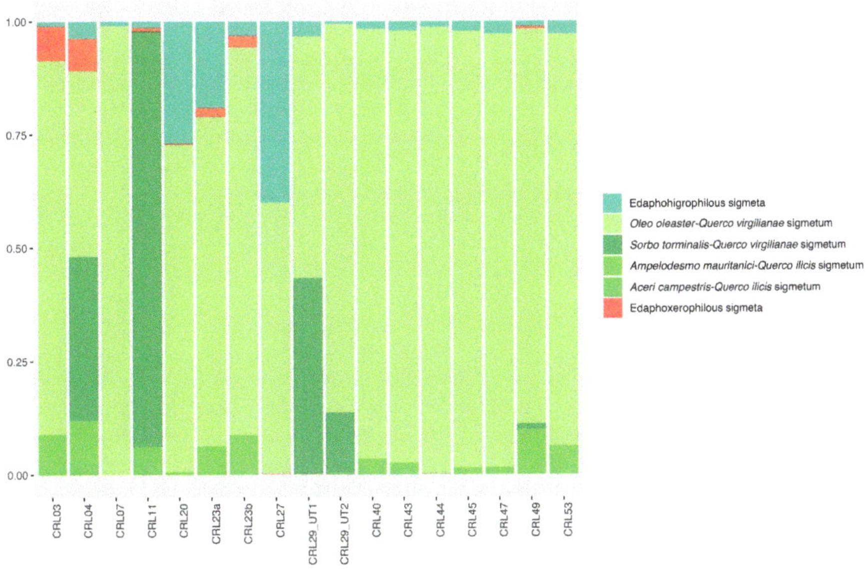

Figure 18. Frequency of surface assigned to a different vegetation series in the surrounding area of each archaeological site.

6. Discussion and Conclusions

The territory of Corleone reveals itself as a crucial territory for the analysis of the settlement dynamics in the transition from the Roman to the Byzantine and Islamic periods. In fact, being in a strategic position for the connection axis between two important cities such as *Panormus/Balarm* (Palermo) and *Agrigentum/Kirkant-Jirjant* (Agrigento), it lends itself to this type of diachronic analysis of the change and resilience of settlement patterns. We have seen how, in Roman times, the sites mainly occupied locations at the bottom of valleys with a close connection to the road axes and water resources. The connection with the road network of the Roman era is demonstrated by the notable quantity and variety of products that were imported, as evidenced by amphorae, ARS, and fire ceramics, which, until Late Antiquity, reached the territory of Corleone from North Africa and Pantelleria. The significant presence of Pantellerian ware from the 4th to 5th century CE provides evidence of the latter. These ceramic imports reveal a significant connection with extra-regional trade flows, a characteristic that defines the entire internal region of the Sicani Mountains. In the adjacent territory of Contessa Entellina, bordering the Municipality of Corleone to the west, decades of meticulous research by the survey have uncovered similar patterns [40]. This is particularly noted in the discovery of materials in Contessa Entellina that closely resemble those found in the Corleone area, indicating a shared belonging to similar long-distance exchange networks.

Even in the Islamic era, the territory seems to continue to be connected to an active road network; although, during this era, it is within a changed economic condition that no longer sees massive extra-regional imports but a dynamic of city–countryside exchange attested to by the presence of glazed ceramics, which seem to link this agricultural territory with the urban markets and port of the capital, Palermo. Evidence from the Byzantine era, to this day, still appears incomplete due to the scarcity of documentation, and it is still to be assessed whether, after a reorganization of the properties after the end of the villa system [14], there was a rearrangement of the population according to new subdivisions of the property and new management of the earth through the system of *massae fundorum* [111, 112]. Uncovering small Byzantine sites in the Corleone area, marked solely by roof tiles decorated with combed impressions, might suggest a hierarchical structure of 6th-7th century settlements featuring sites of diverse sizes and purposes. A similar pattern emerged in the adjacent area of Contessa Entellina [40], where larger, material-rich sites containing imported goods and fine tableware ceramics coexisted alongside smaller sites, identifiable only by comber/striped tiles akin to those in the Corleone territory. This phenomenon extends to other Sicilian regions, like the Alesa area in the northern coast's hinterland, where sites bearing these characteristics have been identified and interpreted as temporary structures linked to combined agricultural and pastoral uses [113].

Another unresolved problem is that of the changes that occurred during the constitution of the theme (Byzantine military province [114]) of Sicily—an important administrative reform of the Byzantine Empire—between the end of the 7th and 8th centuries and, above all, during the process of "militarization" [16,17,20,115] that characterized the island in the previous decades and included the landing of the Aghlabid army in Mazara in 827 CE [2]. It was perhaps precisely during this period of the late 8th and early 9th centuries that the pre-Roman highland sites that had been abandoned during the previous centuries began to be reoccupied, as demonstrated by the example of Contrada Castro. The hills became attractive for settlement in the Islamic era. This occurred in parallel with a new population settling in the valley bottom, both in places not previously settled and in proximity to settlements from the Roman era, demonstrating a new productive boost in the agricultural economy of the Sicilian hinterland. The case of the Islamic era occupation of Casale di Sopra, in particular, opens up interesting perspectives on the types of occupation of the hills due to the presence of fortified enclosures connected either for defensive purposes/control of the territory or as occupation nuclei also connected to the shelter of the livestock. All of these hypotheses could be verified and analyzed with archaeological excavations. The fact of being able to trace a settlement equipped with a fortification enclosure to the Islamic era remains important, given that there are quite a few examples in Sicily of fortified sites dated, with certainty, to the Islamic period, such as the fortified granary of Pizzo Monaco [116], or to the phases of the 10th–11th century of the castles of Calathamet [117] or Calatubo [118] in the province of Trapani.

The GIS approach to the relationship between site positioning and environmental factors has not been widespread in the study of settlement dynamics in post-classical rural landscapes in Sicily. Some applications have been created to understand the relationships between water sources and Islamic sites in the province of Trapani [119] or the distribution of medieval fortified sites in Sicily based on their viewsheds [120]. Our study proposed an original GIS-based approach to the transition between Roman and Islamic settlement patterns.

The landscape position analysis contributed to our understanding of the settlement dynamics of this territory and the relationships of topographic and ecological factors with political, social, and economic changes over time. The statistical analysis (PCA) of the topographic factors (elevation, slope, aspect, TPI, and water distance) made it possible to highlight clusters that show coherent topographic characteristics in the positioning of the settlements. The integration of spatial information with historical information highlighted common characteristics that would otherwise not have emerged from a simple statistical analysis or descriptive historical observation.

The first cluster is made up of four sites from the Roman era (CRL20, CRL23a, CRL27, and CRL45) and two from the Islamic era (CRL03 and CRL07); these are located at the bottoms of valleys, at low altitudes, and in slightly steep positions consistent with the agricultural exploitation of the territory and likely connected with the road network. Notably, this cluster includes the majority of Roman sites and two sites from the Islamic period, which present particular characteristics. The Casale di Sotto site is located in an area close to the passage of a road axis, evidenced by the nearby presence of an old bridge named Ponte del Casale, now out of use, which permits the Fosso del Bicchinello stream to be crossed.

This site has been interpreted as the *Casal Buchinene/Bū Kināna* [33,103] mentioned in written sources from the Norman era of the late 12th century [121,122]. The connection with an ancient road axis could be related to the hypothesis, cited in the *Itinerarium Antonini* [78], that the Roman *statio Pirama* was located in this area 4 km north of the *miliarium* of Zuccarone (i.e., at mile LVII) of the *Panormus-Agrigentum* road (24 miles north of Casale San Pietro a Castronovo di Sicilia (*statio Petrina*) and 24 miles south of Palermo), although hypotheses are still unresolved regarding the location of this Roman *statio*. It may have been located in the area between Pizzo Nicolosi [123] or in the aforementioned site of Casale di Sotto or further north, therefore, in contradiction to the distance of the *miliarium* of Zuccarone near the Nor-man hospital of *Sancte Agnes* [124], located at the site of Contrada Sant'Agata in Piana degli Albanesi, where a large necropolis of Late Antiquity was identified [125]. The continuity of use, in medieval times, of a road axis connecting Palermo and Corleone was witnessed in the sources of the Norman era by a road (*via*) *quae ducit in Corilione ad Panormum* [124], near which the aforementioned Norman *hospitalis* of Sant'Agnese stood and the *publica* via *que ducit a panormo ad briacam* that corresponds to a stretch of road located further south of Contrada Casale and which connected Corleone to Prizzi [124].

The site of Masseria Magione, where 10th–11th-century materials have been identified, currently appears as a fortified grange in which a probable residence of the commander of the *Haiarzeneto* fiefdom, which came into possession of the Order of the Teutonic Knights, was located as of the 1240s [126,127]. Therefore, this site was strategically positioned close to a road axis, functional for the management of agricultural resources and rural infrastructures such as water mills (Primary source "Archivio di Stato di Palermo, Tabulario della Magione": March 1243, ASP TM, 43; 10 May 1260, ASP TM, 87; and 23 March 1293, ASP TM, 258).

These two sites reveal how, even in the Islamic period and, subsequently, between the 12th and 13th centuries, topographical proximity to the road axes and centrality with respect to the agricultural spaces of the valley floor continued to represent decisive factors in the occupation of the territory.

The second cluster is made up of three sites from the Islamic era (CRL23b, CRL43, and CRL53), which are all located in flat terraces in small reliefs towering over the surrounding territory and which, therefore, allow both good defensibility and favorable visibility to control the territory. In this regard, it is interesting to underline the strategic position controlling the Palermo–Corleone road system and the Frattina/Belice Sinistro River of the Islamic site near Case Scalilli/Gole del Drago. This is probably the place, which falls within the *Haiarzeneti/ḥajar al-Zanāta* estate (the stone of the *Zanāta*), to which the *Raalginet/raḥl al-Zanāta* (the village of the *Zanāta*) referred, and which was an object of donation by the Norman kings to the hospital of San Giovanni dei Lebbrosi, is located in the southern suburbs of the city of Palermo [25].

The particular conformation of this limestone promontory standing out in the middle of the valley defines this place as a real landmark, perhaps identifiable as *ḥajar*/stone, while the presence of materials dating back to the Islamic period makes it plausible to identify a connection between the site and a settlement connected to the Berber tribe named *Zanāta* [128,129].

The last cluster also refers to the Islamic period and is made up of two sites [CRL11 and CRL04], which are located in decidedly elevated and isolated positions compared to the valley bottoms and which, therefore, are characterized by the manner in which they perch upon the hills. The characteristics of a fortified site on high ground and the location on the southern slopes of Mount Rocca Busambra suggest that the site of Casale di Sopra can be identified more so with the *Qal^c at Bū Samra/Kala Busamara* [122,124] mentioned in the Norman boundary register (*jarīda*) of the late 12th century [121] than with the category of the few fortified sites that include Arab personal names [130], i.e., *Abū Samra,* which likely reference commanders of the Aghlabid *jund* (Islamic army) who were assigned to defensible places and related surrounding areas [128]. This hypothesis is compelling, although we cannot dismiss an alternative interpretation of the toponym *Qal^c at Bū Samra*. It might not refer to the castle of *Abū Samra* but to a "very dark" caste [131], considering that *sam(a)rah* could be a regional variant of samrā', signifying "brown", or it could mean *sumrah,* which translates to "brown color" [132].

However, the possibility that this site could also refer to the *Qal^c at 'at Tarîq* (fortress/castle/fortress of the road/way) mentioned by al-Idrisi in the mid-12th century [1] cannot be excluded. Idrisi had described the site as distant from Corleone, towards the north, nine Arabian miles away, which would correspond to three Frankish miles [70]. This would correspond to 17.748 km (if 1 Arabic mile = 1.972 m, [133]) or 17.694 km (if 1 Arabic mile = 1.966 m [134]). These attributions with places reported in 12th-century sources for a site already in use in the Islamic era remain quite problematic and not without risks. Significantly, these sources documented the presence of materials from the Islamic period on the site. The hypothesis that this site served as an enclosure-refuge for people and animals [11] is confirmed and supported by a more precise chronology. This kind of site may be similar to *ḥisn-albacar* [135]; this last term is derived from the Arab word *baqqár,* literally meaning "cattle guar" [136]. These settlements were found in al-Andalus from the 9th century onwards [137]. Investigations of excavations could better clarify the sequence of use of this site and provide further information on its function. It may have been a fortified settlement or an enclosure for less stable occupation.

In the Corleone territory, the Islamic-era settlement dynamics, succinctly put, unveil a compelling juxtaposition of hilltop settlements (exemplified by Casale di Sopra, Contrada Castro, and arguably Case Scalilli/Gole del Drago) alongside valley/flatland settlements established close to Roman/late ancient sites such as Contrada Zuccarone, or in areas previously uninhabited, like Masseria Magione, Case Grandi di Giammaria, and Contrada Carrubba. This diversified landscape shows a complexity where the hilltop vs. valley–flatland dichotomy should not be overly emphasized. More intriguing would be exploring the functions and interconnections among sites within the same region, as well as examining the presence or absence of internal social hierarchies in these settlements, something seemingly nonexistent based on recent archaeological findings [138], including those from the Islamic phase of Contrada Castro, with the exception of Calathamet. This case aligns with Islamic and Norman written sources, illustrating a collective political dynamism in rural as well as urban settlements [138]. The pottery data from sites of the Islamic era unveil another interesting trend. As already substantiated by petrographic/archaeometric analyses of pottery from Contrada Castro [102], certain types of painted amphoras indicate a significant shift in trade patterns. In rural areas, unlike the Roman/late antique period's extra-regional scale, trade operated at a mid-range scale in Islamic period, especially along a city (Palermo)/countryside axis.

An ecological analysis of the territory, through the vegetation series, provided the basic knowledge to evaluate the relationship between the sites' catchment area and the territory's agricultural potential. All the sites are located in a territorial context where the oak series (*Oleo oleaster-Querco virgilianae* sigmetum and *Sorbo torminalis-Querco virgilianae* sigmetum) represent the dominant typologies. The plant communities that make up these two series are indicators of environmental contexts with more or less deep and evolved soils. In particular, in the Corleone area, vertexes and vertisols are distributed on brown soils, which have good agronomic potential, especially with regard to the cultivation of cereal [139] but also for the produce of vineyards, olive groves, and orchards. Over the centuries, the high agricultural suitability of the ecotopes of these series has led to the disappearance of the related forest formations (of which only the potential remains), which cultivated fields have replaced. All the sites analyzed, regardless of the topographic position of the hill or valley, are located within contexts with high potential for agricultural productivity. This agricultural potential has been supported over the centuries by the efficient management of agricultural space. An interesting case is represented by the documentary evidence from the Norman era of the *ḫandaq* (ditch)/*ḫandaq al-ġarīq* (literally a submerged ditch that may also mean an "underwater", sunken, or waterfilled ditch) [140] system, probably widespread in Sicily, along with other methods of water management, during the Islamic period [141]. These methods consisted of creating canalizations to regulate surface water, as reported at the *Haiarzeneti*/*ḥajar al-Zanāta* estate area a few kilometers north of Corleone. Additional material evidence of the historical agricultural productivity of the Corleone area is the presence of a high number of water mills along the Frattina/Belice Sinistro stream, some dating back to the 12th century, and the existence of areas of medieval grain pits [142].

The integration of the archaeological survey with spatial analyses of the characteristics of the landscape allows us to propose a more accurate vision of the positioning of the sites and the choice of settlement in the various periods. The enlargement of the sample and the application of the same modeling techniques to other contexts could be an important tool to provide a spatial, quantitative, and ecological dimension to the phenomena of change, resilience, and innovation in territorial occupation between the Roman and Islamic periods.

Supplementary Materials: The following supporting information can be downloaded at: https://www.mdpi.com/article/10.3390/land13030400/s1, Figure S1: GIS map of archaeological sites; Figure S2: Elevation map of the study area with the analyzed archaeological sites; Figure S3: Slope map of the study area with the analyzed archaeological sites; Figure S4: Aspect map of the study area with the analyzed archaeological sites; Figure S5: TPI map of the study area with the analyzed archaeological sites; Figure S6: Water distance map of the study area with the analyzed archaeological sites; Figure S7: Vegetation series map of the study area with the analyzed archaeological sites; Table S1: Sicilian timeline from Roman age to Norman age; Table S2: Table of sites mapped in the study area located in the municipality of Corleone and surrounding areas; Table S3: Table of the sampling grid squares of the concentration area of archaeological materials and its density at the Contrada Zuccarone site (CRL29_UT1); Table S4: Table of the sampling grid squares of the concentration area of archaeological materials and its density at the Case Scalilli–Gole del Drago site (CRL23a); Table S5: Table of the sampling grid squares of the concentration area of archaeological materials and its density at the Casale di Sopra site (CRL4).

Author Contributions: Conceptualization, A.C.B.; methodology, A.C.B. and G.B. (Giuseppe Bazan); archaeological investigation, A.C.B., M.C., A.M., L.V. and L.Z.; historical study, A.C.B.; vegetation study, G.B. (Giuseppe Bazan); data analysis, G.B. (Giuseppe Bazan); GIS mapping, G.B. (Giuseppe Bazan) and M.C.; pottery illustration, A.M., L.V. and L.Z.; writing—original draft preparation, A.C.B. and G.B. (Giuseppe Bazan); writing—review and editing, A.C.B. and G.B. (Giuseppe Bazan); archaeological supervision, C.A.N. and G.B. (Giuseppina Battaglia); project administration, A.C.B.; funding acquisition, A.C.B. All authors have read and agreed to the published version of the manuscript.

Funding: This paper is part of Angelo Castrorao Barba's MSCA COFUND project "IS_LANDAS Islamicate landscapes in Southern Andalusia and Western Sicily: patterns of change in settlements and rural communities between Late Antiquity and the Islamic age" within the PASIFIC program (Polish Academy of Sciences' Individual Fellowships: Innovation & Creativity). This program has received funding from the European Union's Horizon 2020 research and innovation program under the Marie Skłodowska-Curie grant agreement No. 847639 and from the Ministry of Education and Science.

Data Availability Statement: The original contributions presented in the study are included in the article/Supplementary Material, further inquiries can be directed to the corresponding author.

Acknowledgments: Since 2022, a new archaeological survey project has been underway in the territory of Corleone and its surrounding areas. This project is part of the MSCA COFUND PASIFIC project "IS_LANDAS Islamicate landscapes in Southern Andalusia and Western Sicily: patterns of change in settlements and rural communities between Late Antiquity and the Islamic age" (IP, Angelo Castrorao Barba, Institute of Archeology and Ethnology, Polish Academy of Sciences), in collaboration with the Superintendency BB.CC.AA. of Palermo. We thank the Superintendency BB.CC.AA. of

Palermo for having stipulated an agreement (prot. 32036/2022) for survey archaeologists as part of the "IS_LANDAS" project. We thank the municipality of Corleone for paying attention to this research and for good intentions in supporting future investigations. We also thank the agricultural company Bona Furtuna LLC and Pasquale Marino for supporting research on the Contrada Castro site over the past few years. We thank Filippo Pisciotta for suggestions regarding Late Roman and Islamic pottery.

Conflicts of Interest: The authors declare no conflicts of interest.

References

1. Amari, M. *Storia Dei Musulmani Di Sicilia*; Romeo Prampolini: Catania, Italy, 1933.
2. Granara, W. *Narrating Muslim Sicily: War and Peace in the Medieval Mediterranean World*; The Early and Medieval Islamic World; I.B. Tauris: London, UK, 2019; ISBN 978-1-78831-306-3.
3. Molinari, A. 'Islamization' and the Rural World: Sicily and al-Andalus. What Kind of Archaeology? In *New Directions in Early Medieval European Archaeology: Spain and Italy Compared*; Haut Moyen Âge; Gelichi, S., Hodges, R., Eds.; Brepols Publishers: Turnhout, Belgium, 2015; Volume 24, pp. 187–220. ISBN 978-2-503-56520-0.
4. Giunta, F. A Proposito Di Archeologia Medievale in Sicilia. *Sicil. Archeol.* **1983**, *51*, 11–12.
5. Bresc, H. Carmelo Trasselli Pioniere Dell'archeologia Medievale. *Mediterranea* **2022**, *54*, 196–206.
6. Pesez, J.-M. (Ed.) *Brucato. Histoire et Archéologie d'un Habitat Médiéval en Sicile*; École Française de Rome: Rome, Italy, 1984.
7. Aymard, M.; Bresc, H. Problemi di storia dell'insediamento nella sicilia medievale e moderna, 1100–1800. *Quad. Stor.* **1973**, *8*, 945–976.
8. Maurici, F. Archéologie Médiévale «en Sicile» et Archéologie Médiévale «sicilienne». Depuis Les Fouilles de Calathamet Jusqu'à Aujourd'hui: Quelques Considerations. In *Calathamet: Archéologie et Histoire d'un Château Normand en Sicile*; Lesnes, É., Poisson, J.-M., Eds.; École Française de Rome: Rome, Italy; Officina di Studi Medievali: Palermo, Italy, 2013; pp. 405–416.
9. Gelichi, S. I quarant'anni di Archeologia Medievale e l'archeologia in Italia negli ultimi quarant'anni. In *Archeologia Medievale: Cultura Materiale, Insediamenti, Territorio: 2014, Numero Speciale*; All'Insegna del Giglio: Sesto Fiorentino, Italy, 2014. [CrossRef]
10. Wilson, R.J.A. *Sicily under the Roman Empire: The Archaeology of a Roman Province, 36 BC–AD 535*; Archaeologists' Guides to the Roman Empire; Aris and Phillips: Warminster, UK, 1990; ISBN 978-0-85668-160-8.
11. Maurici, F. *Castelli Medievali in Sicilia. Dai Bizantini Ai Normanni*; Sellerio: Palermo, Italy, 1992.
12. Pensabene, P.; Barresi, P. *Piazza Armerina, Villa del Casale: Scavi e Studi nel Decennio 2004–2014*; Bibliotheca Archaeologica; "L'Erma" di Bretschneider: Roma, Italy, 2019; ISBN 978-88-913-1870-1.
13. Bonanno, C. (Ed.) *Piazza Armerina: L'area Nord Dell'insediamento Medievale Presso la Villa del Casale: Indagini Archeologiche 2013–2014*; Access Archaeology; Archaeopress Publishing Ltd.: Oxford, UK, 2020; ISBN 978-1-78969-461-1.
14. Castrorao Barba, A. Sicily before the Muslims. The Transformation of the Roman Villas between Late Antiquity and the Early Middle Ages, Fourth to Eighth Centuries CE. *J. Transcult. Mediev. Stud.* **2016**, *3*, 145–189. [CrossRef]
15. Molinari, A. Sicily between the 5th and the 10th Century: Villae, Villages, Towns and beyond. Stability, Expansion or Recession? In *The Insular System of the Early Byzantine Mediterranean*; BAR International Series; Michaelides, D., Pergola, P., Zanini, E., Eds.; BAR: Oxford, UK, 2013; pp. 97–114.
16. Cacciaguerra, G.; Castrorao Barba, A. The Sicilian Countryside during the Byzantine Period. Archaeological Perspectives on Settlement Patterns. In *Perspectives on Byzantine Archaeology*; Castrorao Barba, A., Castiglia, G., Eds.; Brepols Publishers: Turnhout, Belgium, 2022; pp. 141–163. ISBN 978-2-503-59684-6.
17. Arcifa, L. Byzantine Sicily. In *A Companion to Byzantine Italy*; Brill's Companions to the Byzantine World; Cosentino, S., Ed.; Brill: Leiden, The Netherlands, 2021; pp. 472–495. ISBN 978-90-04-30770-4.
18. Vanoli, A. *La Sicilia Musulmana*; Universale Paperbacks Il Mulino; Il Mulino: Bologna, Italy, 2012; ISBN 978-88-15-23779-8.
19. Molinari, A. *Fortified and Unfortified Settlements in Byzantine and Islamic Sicily: In Fortified Settlements in Early Medieval Europe*; Defended Communities of the 8th–10th Centuries; Christie, N., Herold, H., Eds.; Oxbow Books: Oxford, UK, 2016; pp. 320–332. ISBN 978-1-78570-235-8.
20. Cacciaguerra, G.; Castrorao Barba, A. Dopo i paesaggi delle ville: Nuove dinamiche insediative e siti d'altura nel mondo rurale della Sicilia tardoantica, bizantina ed islamica (V–X/XI Sec.). In *Perchement et Réalités Fortifiées en Méditerranée et en Europe, Vème-Xème Siècles*; Fortified Hilltop Settlements in the Mediterranean and in Europe (5th–10th Centuries); Pergola, P., Castiglia, G., Hanna, E.E.K., Martinetto, I., Segura, J.-A., Eds.; Archaeopress: Oxford, UK, 2023; pp. 310–323. ISBN 978-1-80327-241-2.
21. Ardizzone, F.; Pezzini, E.; Sacco, V. *Chapter 18 Aghlabid Palermo: Written Sources and Archaeological Evidence1*; Brill: Leiden, The Netherlands, 2018; pp. 362–381. ISBN 978-90-04-35604-7.
22. Sacco, V. The palermitan ceramic production from the ninth to the eleventh century. *Libyan Stud.* **2020**, *51*, 74–86. [CrossRef]
23. Castrorao Barba, A.; Mandalà, G. (Eds.) *Suburbia and Rural Landscapes in Medieval Sicily*; Limina/Limites: Archaeologies, Histories, Islands and Borders in the Mediterranean (365–1556), 14; Archaeopress Archaeology: Oxford, UK, 2023; ISBN 978-1-80327-545-1.
24. Bagnera, A.; Nef, A. (Eds.) *Les Bains de Cefalà (Xe–XIXe Siècle): Pratiques Thermales D'origine Islamique dans la Sicile Médiévale = I Bagni di Cefalà (Secoli X–XIX): Pratiche Termali D'origine Islamica nella Sicilia Medievale*; Collection de l'École Française de Rome, 538; École Française de Rome: Rome, Italy, 2018; ISBN 978-2-7283-1250-4.

25. Mandalà, G.; de los Utrero Agudo, M.Á. The King's Hospital in Norman Palermo: San Giovanni dei Lebbrosi in context. In *Suburbia and Rural Landscapes in Medieval Sicily*; Limina/Limites: Archaeologies, Histories, Islands and Borders in the Mediterranean (365–1556); Castrorao Barba, A., Mandalà, G., Eds.; Archaeopress Archaeology: Oxford, UK, 2023; pp. 12–29. ISBN 1803275464.
26. Moździoch, S.; Szubert, B.; Moździoch, E. The process of the creation and decline of the local religious and economic centres of Medieval Sicily: A case study of the Santa Maria di Campogrosso monastery. In *Suburbia and Rural Landscapes in Medieval Sicily*; Limina/Limites: Archaeologies, Histories, Islands and Borders in the Mediterranean (365–1556); Castrorao Barba, A., Mandalà, G., Eds.; Archaeopress Archaeology: Oxford, UK, 2023; pp. 30–43.
27. Castrorao Barba, A.; Miccichè, R.; Pisciotta, F.; Speciale, C.; Aleo Nero, C.; Marino, P.; Bazan, G. The settlement of Contrada Castro (Corleone, Palermo) between the Byzantine and Islamic periods (7th–11th c. ad). In *Suburbia and Rural Landscapes in Medieval Sicily*; Castrorao Barba, A., Mandalà, G., Eds.; Limina/Limites: Archaeologies, Histories, Islands and Borders in the Mediterranean (365–1556); Archaeopress Archaeology: Oxford, UK, 2023; pp. 77–93. ISBN 1803275464.
28. Castrorao Barba, A.; Miccichè, R.; Pisciotta, F.; Speciale, C.; Nero, C.; Marino, P.; Bazan, G. The rural settlement of Contrada Castro (Sicily) between Antiquity and the Early Middle Ages. In *Archaeology of the Mediterranean during Late Antiquity and the Middle Age*; Castrorao Barba, A., Tanasi, D., Miccichè, R., Eds.; University Press of Florida: Gainesville, FL, USA, 2023; pp. 240–265. ISBN 0-8130-7045-7.
29. Castrorao Barba, A.; Miccichè, R.; Pisciotta, F.; Marino, P.; Bazan, G. Archeologia nel Territorio dei Monti Sicani (Harvesting Memories Project). L'insediamento di Lunga Durata di Contrada Castro (Corleone, Palermo). Prima Campagna Di Scavo 2017. *FOLD&R* **2018**, *416*, 1–21.
30. Castrorao Barba, A.; Miccichè, R.; Pisciotta, F.; Speciale, C.; Aleo Nero, C.; Vassallo, S.; Marino, P.; Bazan, G. Nuovi Dati Sull'occupazione Altomedievale Dell'insediamento di Contrada Castro (Corleone, Palermo) Nell'area dei Monti Sicani. *FOLD&R* **2020**, *468*, 1–15.
31. Rizzo, M.S. *L' Insediamento Medievale nella Valle dei Platani*; Studi e Materiali/Dipartimento di Beni Culturali, Sezione Archeologica; Università di Palermo, L'Erma di Bretschneider: Roma, Italy, 2004; ISBN 978-88-8265-316-3.
32. Molinari, A. La Sicilia islamica: Riflessioni sul passato e sul futuro della ricerca in campo archeologico. In *MEFRM: Mélanges De L'école Française De Rome: Moyen-Âge*; L'école Française De Rome: Rome, Italy, 2004. [CrossRef]
33. Maurici, F. *L'insediamento Medievale nel Territorio della Provincia di Palermo: Inventario Preliminare degli Abitati Attestati dalle Fonti D'archivio (Secoli 11–16)*; Regione Siciliana: Palermo, Italy, 1998.
34. Nef, A.; Ardizzone, F. (Eds.) *Les Dynamiques de l'Islamisation en Méditerranée Centrale et en Sicile: Nouvelles Propositions et Découvertes Récentes = Le Dinamiche dell'Islamizzazione del Mediterraneo Centrale in Sizilia: Nuove Proposte e Scoperte Recenti*; Collection de l'École Française de Rome, Edipuglia: Bari, Italy, 2014; ISBN 978-88-7228-735-4.
35. Arcifa, L. *La Sicilia e il Mediterraneo dal Tardoantico al Medioevo: Prospettive di Ricerca tra Archeologia e Storia: Atti del Convegno Internazionale di Studi Dedicato a Fabiola Ardizzone lo Bue*; Palermo University Press: Palermo, Italy, 2022; ISBN 978-88-550-9403-0.
36. Spatafora, F. Ricerche e prospezioni nel territorio di Corleone: Insediamenti preistorici e centri indigeni. In *Atti delle Seconde Giornate di Studi sull'Area Elima (Gibellina 1994)*; Scuola Normale Superiore di Pisa, Comune di Gibellina: Pisa, Italy, 1997; pp. 1273–1286.
37. Scuderi, A.; Tusa, S.; Vintaloro, A. *La Preistoria e Protostoria del Territorio di Corleone nel Quadro della Sicilia Occidentale*; Città di Corleone/Archeoclub di Corleone: Corleone, Italy, 1996.
38. Vintaloro, A.; Orlando, M. *Corleone Archeologica III*; Città di Corleone/Archeoclub di Corleone: Corleone, Italy, 2019.
39. Vintaloro, A.; Scuderi, A. *Corleone Archeologica I: Studio Storico-Archeologico sul Territorio Corleonese*; Città di Corleone/Archeoclub di Corleone: Corleone, Italy, 1995.
40. Corretti, A.; Facella, A.; Gulletta, M.I.; Michelini, C.; Vaggioli, M.A. (Eds.) *Entella II: Carta Archeologica del Comune di Contessa Entellina dalla Preistoria al Medioevo*; Scuola Normale Superiore di Pisa: Pisa, Italy, 2021; ISBN 978-88-7642-700-8.
41. Johns, J. The Monreale Survey: Indigenes and invaders in Medieval West Sicily. In *Papers in Italian Archaeology IV. The Cambridge Conference, Part IV: Classical and Medieval Archaeology*; Malone, C., Stoddart, S., Eds.; BAR International Series; BAR: Oxford, UK, 1985; pp. 215–224.
42. Spatafora, F. Tracce di culto nell'entroterra sicano: Il santuario extraurbano di Cozzo Spolentino (Palermo). In *L'Occident Grec de Marseille à Mégara Hyblaea*; Bouffier, S., Hermary, A., Eds.; Publications du Centre Camille Jullian: Aix-en-Provence, France, 2013; pp. 277–294. ISBN 978-2-87772-553-8.
43. D'Angelo, F.; Spatafora, F. La Vecchia di Corleone. In *Federico e la Sicilia, dalla Terra alla Corona. Archeologia e Architettura*; Di Stefano, C.A., Cadei, A., Eds.; Ediprint: Siracusa, Italy, 1995; Volume 1, pp. 173–175.
44. Spatafora, F. La Vecchia di Corleone. *Archeol. Viva* **1996**, *59*, 34–40.
45. Castrorao Barba, A.; Speciale, C.; Miccichè, R.; Pisciotta, F.; Aleo Nero, C.; Marino, P.; Bazan, G. The Harvesting Memories Project: Historical Ecology and Landscape Changes of the Sicani Mountains in Sicily. *Ecocycles* **2022**, *8*, 51–60. [CrossRef]
46. Bazan, G.; Castrorao Barba, A.; Rotolo, A.; Marino, P. Vegetation series as a marker of interactions between rural settlements and landscape: New insights from the archaeological record in Western Sicily. *Landsc. Res.* **2020**, *45*, 484–502. [CrossRef]
47. Liguori, V. Geologia, Geomorfologia, Idrologia. In *Linee Guida del Piano Territoriale Paesistico Regionale*; Regione Siciliana: Palermo, Italy, 1996.
48. Sestini, A. *Il Paesaggio*; Conosci l'Italia; Touring Club Italiano: Milano, Italy, 1963; Volume 7.

49. Duro, A.; Piccione, V.; Scalia, C.; Zampino, S. Precipitazioni e temperature medie mensili in Sicilia relative al sessantennio 1926–1985. In *Proceedings of the Atti 5° Workshop Progetto Strategico Clima, Ambiente e Territorio nel Mezzogiorno CNR*; ISHS: Leuven, Belgium, 1996.
50. Rivas-Martínez, S.; Rivas-Saenz, S.; Penas, A. Worldwide Bioclimatic Classification System. *Glob. Geobot.* **2002**, *1*, 1–638. [CrossRef]
51. Gianguzzi, L.; Bazan, G. The Olea Europaea L. Var. Sylvestris (Mill.) Lehr. Forests in the Mediterranean Area. *Plant Sociol.* **2019**, *56*, 3–34. [CrossRef]
52. Bazan, G.; Marino, P.; Guarino, R.; Domina, G.; Schicchi, R. Bioclimatology and vegetation series in Sicily: A geostatistical approach. *Ann. Bot. Fenn.* **2015**, *52*, 1–18. [CrossRef]
53. Schicchi, R.; Speciale, C.; Amato, F.; Bazan, G.; Di Noto, G.; Marino, P.; Ricciardo, P.; Geraci, A. The monumental olive trees as biocultural heritage of Mediterranean landscapes: The case study of Sicily. *Sustainability* **2021**, *13*, 6767. [CrossRef]
54. Marino, P.; Guarino, R.; Bazan, G. The Sicilian taxa of genista sect. voglera and their phytosociological framework. *Flora Mediterr.* **2012**, *22*, 169–190. [CrossRef]
55. Marino, P.; Castiglia, G.; Bazan, G.; Domina, G.; Guarino, R. ertiary relict laurophyll vegetation in the madonie mountains (Sicily). *Acta Bot. Gall.* **2014**, *161*, 47–61. [CrossRef]
56. Bazan, G.; Castrorao Barba, A.; Rotolo, A.; Marino, P. Geobotanical approach to detect land-use change of a mediterranean landscape: A case study in Central-Western Sicily. *GeoJournal* **2019**, *84*, 795–811. [CrossRef]
57. Ministero dell'Ambiente Geoportale Nazionale. Distributed under the Terms and Conditions of the Creative Commons Attribution (CC BY-NC) License. Available online: https://gn.mase.gov.it/portale/servizio-di-consultazione-wms (accessed on 2 February 2022).
58. Bejor, G. Corleone. In *Bibliografia Topografica della Colonizzazione Greca in Italia e nelle Isole Tirreniche*; Scuola normale superiore: Pisa, Italy; École Française de Rome: Rome, Italy, 1987; Volume 5, pp. 418–420.
59. Colletto, G. *Storia della Citta di Corleone*; Helix Media: Palermo, Italy, 1992.
60. Gulletta, M.I. Schera. In *Bibliografia Topografica della Colonizzazione Greca in Italia e nelle Isole Tirreniche*; Scuola normale superiore: Pisa, Italy; École Française de Rome: Rome, Italy, 2010; Volume 18, pp. 448–457.
61. Clüver, P. *Philippi CluverI Sicilia Antiqua; Cum Minoribus Insulis, Ei Adjacentibus. Item, Sardinia et Corsica. Opus Post Omnium Curas Elaboratissimum; Tabulis Geographicis, Aere Expressis, Illustratum*; Ex Officina Elseviriana: Lugduni Batavorum, The Netherlands, 1619.
62. Houel, J. *Voyages Pittoresques des Îles de Sicile, de Lipari et de Malte Où l'on Traite des Antiquités Qui s'y Trouvent Encore; Des Principaux Phénomènes que la Nature y Offre; Du Costume des Habitants, & de Quelques Usages*; Imprimerie de Monsieur: Paris, France, 1782.
63. Trasselli, C. Schera-Corleone o Monte dei Cavalli. *Sicil. Archeol.* **1969**, *7*, 19–28.
64. Badami, A.; Carta, M. *Storia Urbanistica della Città di Corleone*; Dipartimento Città e Territorio: Palermo, Italy, 1993.
65. Prag, J. Il miliario di Aurelius Cotta (ILLRP n.1277): Una lapide in contesto. In *Guerra e Pace in Sicilia e nel Mediterraneo antico (VIII-III sec.a.C.). Arte, Prassi e Teoria della Pace e della Guerra (Atti V Giornate Internazionali (Erice 2003)*; Ampolo, C., Ed.; Edizioni Della Normale: Pisa, Italy, 2006; Volume 2, pp. 733–744.
66. Di Vita, A. Una recente nota e la datazione del miliario siciliano del console C. Aurelio Cotta. *Latomus* **1963**, *22*, 478–488.
67. Amari, M. *Biblioteca Arabo-Sicula*; E. Loescher: Torino, Italy, 1880; Volume I.
68. Amari, M. *Biblioteca Arabo-Sicula*; E. Loescher: Torino, Italy, 1881; Volume II.
69. Al-Muqaddasī, M.I.-A.; Collins, B.A.; al-Muqaddasī, M.I.-A. *The Best Divisions for Knowledge of the Regions: A Translation of Ahsan al-Taqasim Fi Maʿrifat al-Aqalim; Great Books of Islamic Civilization*, 1st ed.; Garnet: Reading, UK, 1994; ISBN 978-1-873938-14-0.
70. Bresc, H.; Nef, A. (Eds.) *Idrīsī. La Première Géographie de l'Occident*; GF Flammarion: Paris, France, 1999; ISBN 978-2-08-071069-7.
71. Johns, J. Sulla condizione dei musulmani di Corleone sotto il dominio normanno nel XII secolo. In *Bizantino-Sicula IV. Atti del I Congresso Internazionale di Archeologia della Sicilia Bizantina, Palermo, (Corleone, 28 Luglio–2 Agosto 1998)*; Bonacasa Carra, R.M., Ed.; Istituto Siciliano di Studi Bizantini e Neoellenici: Palermo, Italy, 2002; pp. 275–294.
72. Bover Fonts, I. L'iqlîm di Corleone: Studio del territorio e della sua popolazione durante l'epoca musulmana. *Ann. Dell'istituto Orient. Di Napoli* **1996**, *56*, 257–264.
73. Regione Siciliana S.I.T.R—Sistema Informativo Territoriale Regionale. Available online: https://www.sitr.regione.sicilia.it/ (accessed on 2 February 2024).
74. Weiss, A. Topographic Position and Landforms Analysis. In Proceedings of the Poster Presentation, ESRI User Conference, San Diego, CA, USA, 9–13 July 2001; Volume 200.
75. R Core Team. *R Foundation for Statistical Computing R: A Language and Environment for Statistical Computing*; R Core Team: Vienna, Austria, 2015.
76. Jolliffe, I.T. *Principal Component Analysis*; Springer Series in Statistics; Springer: New York, NY, USA, 2002; ISBN 978-0-387-95442-4.
77. Klijn, F.; De Haes, H.A.U. A hierarchical approach to ecosystems and its implications for ecological land classification. *Landsc. Ecol.* **1994**, *9*, 89–104. [CrossRef]
78. Uggeri, G. *La viabilità della Sicilia in età romana*; Journal of Ancient Topography—Rivista di Topografia Antica Supplemento II; Congedo Editore: Galatina, Italy, 2004; ISBN 88-8086-559-5.
79. Cibecchini, F.; Capelli, C. Nuovi dati archeologici e archeometrici sulle anfore greco italiche: I relitti di III secolo del Mediterraneo occidentale e la possibilità di una nuova classiicazione. In *Itinéraires des Vins Romains en Gaule IIIe–Ier Siècles Avant J.-C. Confrontation de Faciès, Actes du Collo-que Européen Organisé par l'UMR 5140 du CNRS Lattes, 30 Janvier–2 Février 2007*; Olmer, F., Ed.; Publication de l'UMR 5140 du CNRS: Lattes, Italy, 2013; pp. 423–451.

80. Hayes, J.W. *Late Roman Pottery*; British School at Rome: London, UK, 1972; ISBN 978-0-904152-00-5.
81. Bonifay, M. *Etudes sur la Céramique Romaine Tardive d'Afrique*; BAR International Series; Archaeopress: Oxford, UK, 2004; ISBN 978-1-84171-651-0.
82. Fiertler, G. La Pantellerian Ware dal quartiere ellenistico-romano di Agrigento: Aspetti della problematica e proposta per una tipologia. In *Archeologia del Mediterraneo: Studi in Onore di Ernesto de Miro. Bibliotheca Archaeologica 35*; "L'Erma" di Bretschneider: Roma, Italy, 2003; ISBN 978-88-913-0221-2.
83. Baldassari, R. La Produzione della Ceramica a Pantelleria e la sua Circolazione in età Tardo Antica. Ph.D. Thesis, Università degli Studi di Sassari, Sassari, Italy, 2014.
84. Sacco, V. Produzione e circolazione delle anfore palermitane tra la fine del IX e il XII secolo. *Archeol. Mediev.* **2018**, *45*, 175–192. [CrossRef]
85. Sacco, V. Le ceramiche invetriate di età islamica a Palermo: Nuovi dati dalle sequenze del quartiere della Kalsa. *Archeol. Mediev.* **2017**, *44*, 337–366. [CrossRef]
86. Ardizzone, F.; Pezzini, E.; Sacco, V. Lo scavo della chiesa di Santa Maria degli Angeli alla Gancia: Indicatori archeologici della prima età islamica a Palermo. In *Les Dynamiques de l'Islamisation en Méditerranée Centrale et en Sicile: Nouvelles Propositions et Découvertes Récentes = Le Dinamiche dell'Islamizzazione del Mediterraneo Centrale in Sizilia: Nuove Proposte e Scoperte Recenti*; Nef, A., Ardizzone, F., Eds.; École Française de Rome: Roma, Italy, 2014; pp. 197–223. ISBN 978-88-7228-735-4.
87. Ardizzone, F.; Agrò, F. Una discarica di età islamica fuori le mura della città di Palermo. *Archeol. Postmedievale Soc. Ambiente Produzione* **2012**, *2012*, 16. [CrossRef]
88. Molinari, A.; Meo, A. (Eds.) *Mazara/Māzar: Nel Ventre della Città Medievale (Secoli VII–XV): Edizione Critica degli Scavi (1997) in via Tenente Gaspare Romano*; Biblioteca di Archeologia Medievale 32; All'insegna del Giglio: Firenca, Italy, 2021; ISBN 978-88-928-5046-0.
89. Aleo Nero, C. Lucerne medievali a Palermo e nel territorio, nuove scoperte e vecchi dati a confronto. In Proceedings of the Atti del XLIX Convegno Internazionale della Ceramica (Savona 2016), Firenze, Italy, 27–28 May 2017; pp. 319–336.
90. Rizzo, M.S.; Fiorilla, S.; Guzzetta, G. Nuove osservazioni sulle produzioni delle fornaci di S. Lucia ad Agrigento. *Cronache Di Archeol.* **2021**, *40*, 435–460.
91. Agrò, F. A proposito dell'invetriata verde con decorazione solcata di produzione siciliana. Nuovi dati e vecchie linee di ricerca. In *Proceedings of the Amalfi e la Sicilia nel Medioevo: Uomini, Commerci, Culture: Atti del Convegno, Amalfi, Italy, 13–14 Dicembre 2019*; Centro di Cultura e Storia Amalfitana: Amalfi, Italy, 2022; pp. 303–320.
92. Olcese, G. Terra sigillata italica a roma e in area romana: Produzione, circolazione e analisi di laboratorio (Con Il Contributo Di M. Picon). *Rei Cretariae Rom. Fautorum Acta Roma* **2003**, *38*, 11–26.
93. Malfitana, D.; Bonifay, M. (Eds.) *La Ceramica Africana nella Sicilia Romana*; Monografie dell'Istituto per i beni Archeologici e Monumentali; IBAM/CCJ: Catania, Italy, 2016; ISBN 978-88-89375-13-6.
94. Di Stefano, C.A. Il territorio della provincia di Palermo tra la tarda età romana e l'età bizantina. problemi aperti e nuove acquisizioni. In *Proceedings of the Bizantino-Sicula IV. Atti del I Congresso Internazionale di Archeologia della Sicilia Bizantina, Palermo, Italy, 28 July–2 August 1998*; Bonacasa Carra, R.M., Ed.; Istituto Siciliano di Studi Bizantini e Neoellenici: Palermo, Italy, 2002; pp. 307–322.
95. Puglisi, M.; Sardella, A. Ceramica locale in Sicilia Tra VI e VII Secolo d.C. Situazione attuale e prospettive future della ricerca. In *Proceedgins of the Ceramica in Italia: VI–VII Secolo, Atti del Convegno in Onore di John, W. Hayes, Roma, Italy, 11–13 Maggio 1995*; All'Insegna del Giglio: Rome, Italy, 1998; pp. 777–785. ISBN 88-7814-128-3.
96. Arcifa, L. Indicatori archeologici per l'alto medioevo nella sicilia orientale. In *Piazza Armerina: Villa del Casale e la Sicilia tra Tardoantico e Medioevo*; Pensabene, P., Ed.; Studia Archaeologica, L'Erma" di Bretschneider: Roma, Italy, 2010; pp. 105–128.
97. Dannheimer, H. *Byzantinische Grabfunde aus Sizilien Christliches Brauchtum im Frühen Mittelalter*; Ausstellungskataloge der Archäologischen Staatssammlung, Prähistorische Staatssammlung: München, Gemany, 1989; ISBN 978-3-927806-03-0.
98. Metaxas, S. *Die Materielle Kultur des byzantinischen Sizilien (6–10. Jahrhundert)*; Universität Wien: Vienna, Austria, 2009.
99. Castrorao Barba, A.; Rotolo, A.; Bazan, G.; Marino, P.; Vassallo, S. Long-Term human occupation of a Rural Landscape in Central-Western Sicily (Castro/Giardinello Valley and Mt Barraù): Harvesting Memories Project case study. *ArkeoGazte* **2017**, *7*, 175–192.
100. Bazan, G.; Speciale, C.; Castrorao Barba, A.; Cambria, S.; Miccichè, R.; Marino, P. Historical Suitability and Sustainability of Sicani Mountains Landscape (Western Sicily): An Integrated Approach of Phytosociology and Archaeobotany. *Sustainability* **2020**, *12*, 3201. [CrossRef]
101. Castrorao Barba, A.; Speciale, C.; Miccichè, R.; Pisciotta, F.; Aleo Nero, C.; Marino, P.; Bazan, G. The Sicilian Countryside in the Early Middle Ages: Human–Environment Interactions at Contrada Castro. *Environ. Archaeol.* **2023**, *28*, 240–255. [CrossRef]
102. Montana, G.; Gasparo Morticelli, M.; Bazan, G.; Pisciotta, F.; Aleo Nero, C.; Marino, P.; Castrorao Barba, A. Sources of Geomaterials in the Sicani Mountains during the Early Middle Ages: A Case Study of Contrada Castro, Central Western Sicily. *Geoarchaeology* **2022**, *37*, 617–636. [CrossRef]
103. Maurici, F.; Vassallo, S. Due siti medievali nel territorio Santa Maria la Nuova di Monreale. *Sicil. Archeol.* **1987**, *20*, 13–28.
104. Sacco, V. L'islamizzazione a Palermo attraverso due contesti di palazzo Bonagia (scavi di Stefano). In *Les Dynamiques de l'Islamisation en Méditerranée Centrale et en Sicile: Nouvelles Propositions et Découvertes Récentes = Le Dinamiche Dell'islamizzazione del Mediterraneo Centrale in Sizilia: Nuove Proposte e Scoperte Recenti*; Nef, A., Ardizzone, F., Eds.; École Française de Rome: Rome, Italy, 2014; pp. 225–231. ISBN 978-88-7228-735-4.

105. Ardizzone Lo Bue, F. *Anfore in Sicilia VIII–XII Sec. dC.*; Torri del Vento: Palermo, Italy, 2012; ISBN 978-88-97373-29-2.
106. Ardizzone, F.; D'Angelo, F.; Pezzini, E.; Sacco, V. Ceramiche di età islamica provenienti da Castello della Pietra (Trapani). In *Atti del IX Congresso Internazionale sulla Ceramica Medievale nel Mediterraneo: Venezia, Scuola Grande dei Carmini, Auditorium Santa Margherita, 23–27 Novembre 2009*; Gelichi, S., Ed.; All'Insegna del Giglio: Firenze, Italy, 2012; pp. 167–172.
107. D'Angelo, F. Sicilia XII Secolo: Importazioni dal Mediterraneo orientale, importazioni dal Mediterraneo occidentale, produzioni locali. In *Proceedings of the Atti del IX Congresso Internazionale sulla Ceramica Medievale nel Mediterraneo, Venezia, Italy, 23–27 November 2009*; Gelichi, S., Ed.; All'Insegna del Giglio: Firenze, Italy, 2012; pp. 178–180. ISBN 978-88-7814-540-5.
108. Spatafora, F.; Canzonieri, E.; Di Leonardo, L. Ceramica da mensa nella Palermo di XI secolo: Dalla fornace al butto. *Archeol. Postmedievale* **2012**, *16*, 23–33.
109. Sacco, V.; Pezzini, E. le produzioni da fuoco a Palermo (IX–XI Secolo). In Proceedings of the XIth Congress AIECM3 on Medieval and Modern Period Mediterranean Ceramics Proceedings, Antalya, Turkey, 19–24 October 2015; Yenişehirlioğlu, F., Ed.; Koc Universitesi VEKAM: Ankara, Turkey, 2018; pp. 347–356.
110. Edmonds, W.J.; Thomas, P.J.; Simpson, T.W.; Baker, J.C. *Land Judging and Soil Evaluation*; Virginia Polytechnic Institute and State University Department of Crop and Soil Environmental Sciences: Blacksburg, USA, 1998.
111. Arcifa, L. Dinamiche insediative e grande proprietà nella Sicilia bizantina. In *L'héritage Byzantin en Italie (VIIIe–XIIe Siècle)*; Martin, J.-M., Peters-Custot, A., Prigent, V., Eds.; Publications de l'École Française de Rome: Rome, Italy, 2017; pp. 237–267. ISBN 978-2-7283-1224-5.
112. Nef, A.; Prigent, V. Contrôle et exploitation des campagnes en Sicile: Le rôle du grand domaine et son évolution du VIe Siècle au XIe Siècle. In *Authority and Control in the Countryside. From Antiquity to Islam in the Mediterranean and Near East (6th–10th Century)*; Leiden Studies in Islam and Society; Delattre, A., Legendre, M., Sijpesteijn, P., Eds.; Brill: Leiden, Belgium, 2018; pp. 313–366. ISBN 978-90-04-38654-9.
113. Burgio, A. *Il Paesaggio Agrario nella Sicilia Ellenistico-Romana: Alesa e il Suo Territorio*; Studi e Materiali; "L'Erma" di Bretschneider: Rome, Italy, 2008; ISBN 978-88-8265-483-2.
114. Konstantin; Haldon, J.F. *The De Thematibus ('on the Themes') of Constantine VII Porphyrogenitus: Translated with Introductory Chapters and Notes*; Translated Texts for Byzantinists; Liverpool University Press: Liverpool, UK, 2021; ISBN 978-1-80085-998-2.
115. Molinari, A. *Sicily from Late Antiquity to the Early Middle Ages: In Change and Resilience*; The Occupation of Mediterranean Islands in Late Antiquity; Ontiveros, M.Á.C., Florit, C.M., Eds.; Oxbow Books: Oxford, UK, 2019; Volume 9, pp. 87–110. ISBN 978-1-78925-180-7.
116. Civantos, J.M.M.; García, M.T.B.; Corselli, R.; Jiménez, A.P.R.; Maestro, E.C. Arqueología del paisaje en Trapani: Primera campaña de excavación en elgranero colectivo fortificado de Pizzo Monaco (Custonaci-Trapani, Italia). *Inf. Trab. Excavaciones Exter.* **2016**, *14*, 131–151.
117. Lesnes, É.; Poisson, J.M. *Calathamet: Archéologie et Histoire d'un Château Normand En Sicile*; Collection de l'École Française de Rome; École Française de Rome, Officina di Studi Medievali: Rome, Italy, 2013; ISBN 978-88-6485-068-9.
118. Di Liberto, R. L'apporto dell'architettura normanna alla conoscenza dell'ars fortificatoria islamica in Sicilia: Il contributo dell'analisi stratigrafica delle murature. *Mélanges L'école Française Rome Moyen Âge* **2004**, *116*, 319–350. [CrossRef]
119. Rotolo, A. The Trodden Path: GIS-Analyses of Settlement and Mobility Patterns in Western Sicily during the Islamic Period. *JIA* **2016**, *3*, 109–136. [CrossRef]
120. Kirk, S. Sicilian Castles and Coastal Towers. *Open Archaeol.* **2017**, *3*, 21. [CrossRef]
121. Metcalfe, A. *Muslims and Christians in Norman Sicily: Arabic-Speakers and the End of Islam*; Culture and civilization in the Middle East; Routledge Curzon: London, UK; New York, NY, USA, 2003; ISBN 978-0-7007-1685-2.
122. Nef, A. *Conquérir et Gouverner la Sicile Islamique aux XIe et XIIe Siècles*; Bibliothèque des Écoles Françaises d'Athènes et de Rome; Publications de l'École Française de Rome: Rome, Italy, 2011; ISBN 978-2-7283-0889-7.
123. Vassallo, S. Pizzo Nicolosi. *Bibliogr. Topogr. Della Colon. Greca Ital. E Nelle Isole Tirreniche* **1996**, *14*, 32–34.
124. Cusa, S. *I Diplomi Greci ed Arabi di Sicilia Pubblicati nel Testo Originale, Tradotti ed Illustrati da Salvatore Cusa: I, 1*; Stabilimento Lipograf. Lao.: Palermo, Italy, 1868.
125. Greco, C. Un sito tardoromano sulla via Agrigentum-Panormus: Scavi nella necropoli in Contrada S. Agata (Piana degli Albanesi). *Kokalos* **1993**, *39–40*, 1143–1163.
126. Toomaspoeg, K. *Les Teutoniques En Sicile (1197–1492)*; Collection de l'École Française de Rome; École Française de Rome: Rome, Italy, 2003; ISBN 2-7283-0655-9.
127. Toomaspoeg, K. Base Économique de l'expansion des bourgs siciliens. Exemple des possessions de l'Ordre Teutonique dans la zone Corleone-Vicari-Castronovo, 1220–1310. In Proceedings of the Actas del XVII Congreso de Historia de la Corona de Aragón,"El món urbà a la Corona d'Aragó del 1137 als decrets de Nova Planta", Barcelona, Spain, 7–12 September 2000; Claramunt Rodríguez, S., Ed.; Universidad de Barcelona: Barcelona, Spain, 2003; Volume 1, pp. 595–604.
128. Metcalfe, A. *The Muslims of Medieval Italy*; The New Edinburgh Islamic Surveys; Edinburgh University Press: Edinburgh, UK, 2009; ISBN 978-0-7486-2007-4.
129. Chiarelli, L.C. *A History of Muslim Sicily*, 2nd ed.; Central Mediterranean Studies Series; Midsea Books: Venera, Malta, 2018; ISBN 978-99932-7-645-6.
130. Bresc, H. Nommer La Ville: L'exemple de La Sicile (Xe-XVIIIe Siècle). In *Compositions Urbaines au Moyen Âge. Actes du 137e Congrès National des Sociétés Historiques et Scientifiques (Tours, 2012)*; Editions du CTHS: Paris, France, 2015; pp. 5–20.

131. Caracausi, G. *Dizionario Onomastico della Sicilia: Repertorio Storico-Etimologico di Nomi di Famiglia e di Luogo*; l'Epos: Palermo, Italy, 1993; ISBN 978-88-8302-053-7.
132. Wehr, H.; Cowan, J.M. *A Dictionary of Modern Written Arabic: Arabic–English*, 4th ed.; Considerably Enl. and Amended by the Author; Harrassowitz: Wiesbaden, Germany, 1979; ISBN 978-3-447-02002-2.
133. Mercier, R.P. Geodesy. In *The History of Cartography, Vol.II, Book I: Cartography in the Traditional Islamic and South Asian Societies*; Harley, J.B., Woodward, D., Eds.; University of Chicago Press: Chicago, IL, USA; pp. 175–188.
134. Santagati, L. *Viabilità e Topografia della Sicilia Antica. Bd. II: La Sicilia Altomedievale ed Arabo-Normanna*; Edizioni Lussografica: Caltanissetta, Italy, 2013.
135. Bazzana, A.; Cressier, P.; Guichard, P. *Les Châteaux Ruraux d'al-Andalus: Histoire et Archéologie des Ḥuṣūn du Sud-Est de l'Espagne*; Publications de la Casa de Velázquez. Série Archéologie; Casa de Velázquez: Madrid, Spain, 1988; ISBN 978-84-600-5814-4.
136. Epalza, M. De funciones ganaderas de los albacares en las fortalezas musulmanas. *Sharq Al-Andalus* **1984**, *1*, 47–54. [CrossRef]
137. Bazzana, A. Les Structures: Fortification et Habitat. In *Proceedings of the Castrum I. Habitats Fortifiés et Organisation de L'espace en Méditerranée Médiévale. Table Ronde Tenue à Lyon les 4 et 5 mai 1982*; Maison de l'Orient et de la Méditerranée Jean Pouilloux: Lyon, France, 1983.
138. Wickham, C. *The Donkey and the Boat: Reinterpreting the Mediterranean Economy, 950–1180*, 1st ed.; Oxford University Press: Oxford, UK, 2023; ISBN 978-0-19-885648-1.
139. Fierotti, G.; Dazzi, C.; Raimondi, S. *Commento alla Carta dei Suoli della Sicilia*; Istituto di Agronomia Generale: Palermo, Italy, 1988.
140. Metcalfe, A. Dynamic Landscapes and Dominant Kin Groups: Hydronymy and Water-Management in Arab-Norman Western Sicily. In *Urban Dynamics and Transcultural Communication in Medieval Sicily*; Jäckh, T., Kirsch, M., Eds.; Brill: Schöningh, Germany, 2017; pp. 97–139. ISBN 978-3-506-78629-6.
141. Todaro, P.; Barbera, G.; Castrorao Barba, A.; Bazan, G. Qanats and Historical Irrigated Landscapes in Palermo's Suburban Area (Sicily). *Eur. J. Post-Class. Archaeol.* **2020**, *10*, 335–370.
142. Arcifa, L. Facere fossa et victualia reponere: La conservazione del grano nella Sicilia medievale. *Mélanges L'école Française Rome Moyen Âge* **2008**, *120*, 39–54. [CrossRef]

land

MDPI

Article

Exploring the Dynamics of Occupation between Resilience and Abandonment in Two Post-Classic Rural Landscapes on the Iberian Peninsula

Julia Sarabia-Bautista

University Institute for Research in Archeology and Historical Heritage (INAPH), University of Alicante, 03690 Alicante, Spain; julia.sarabia@ua.es

Abstract: In this paper, we present a comparison of two rural landscapes in the southeast of the Iberian Peninsula, where the dynamics of occupation have differed since the end of the ancient world in terms of both the degree of resilience of settlements and the land use. Our purpose was to explore the social, political, economic, and environmental factors that could explain why there has been a long-term cross-cultural occupation of some resilient sites and landscapes for almost a millennium, while there have been only very specific temporary occupations in other areas. The first part of this paper describes the archaeological investigations carried out by means of intensive survey methods, geophysics, and some excavations in peripheral and peri-urban spaces. In the second part, we reflect on whether the use of the same methodology in all cases allows us to compare and understand what makes societies sustainable (or not) over time through their archaeological record.

Keywords: landscape archaeology; off-site surveys; historical land uses; agrarian practices; Late Antiquity; Middle Ages; Al-Andalus; "Campos de Hellín"; Balazote River Valley

1. Introduction

The research we have been carrying out over recent years has focused mainly on understanding and measuring interactions between societies and their environments across multiple timescales (especially from the Roman period to the Middle Ages) and spatial scales (from a regional framework, centred on the southeast of the Iberian Peninsula) [1,2]. The general objective is to reconstruct the historical socioecological systems in different landscapes using archaeological materiality, paying special attention to scenarios in which we have detected patterns of sustainability and historical resilience. In this article, we focus on presenting two case studies from the "Campos de Hellín" and the Balazote River Valley, which are located in the current province of Albacete (Spain). The enclaves of "El Tolmo de Minateda" (TM) and the "Camino Viejo de las Sepulturas" (CVS) stand out as the central places of these two regions, where the agrarian vocation remains the main economic base to this day (Figure 1). This has led to scarce urbanisation in their ecosystems. It is in these rural landscapes that we proposed to understand the causes of the permanence, transformation, or abandonment of the central places, rural settlements, and productive spaces in both areas from Late Antiquity to the Islamic period (5th to 10th centuries AD), when both territories were incorporated into the *Cora* or the province of *Tudmīr* from 713 [2].

In addition to their historical importance within the framework of our research, these landscapes were specifically selected due to the shape of the settlements and the morphology of their productive spaces, especially their agricultural systems, which are among the most homogeneous expressions of the interactions between humans and territory [3]. The application of the same methodology allowed us to contrast the data from both areas and identify any peculiarities and differences in terms of environmental patterns, social patterns, land-use management, etc., with the aim of revealing the causes of the greater or lesser resilience of the settlements and agricultural spaces analysed.

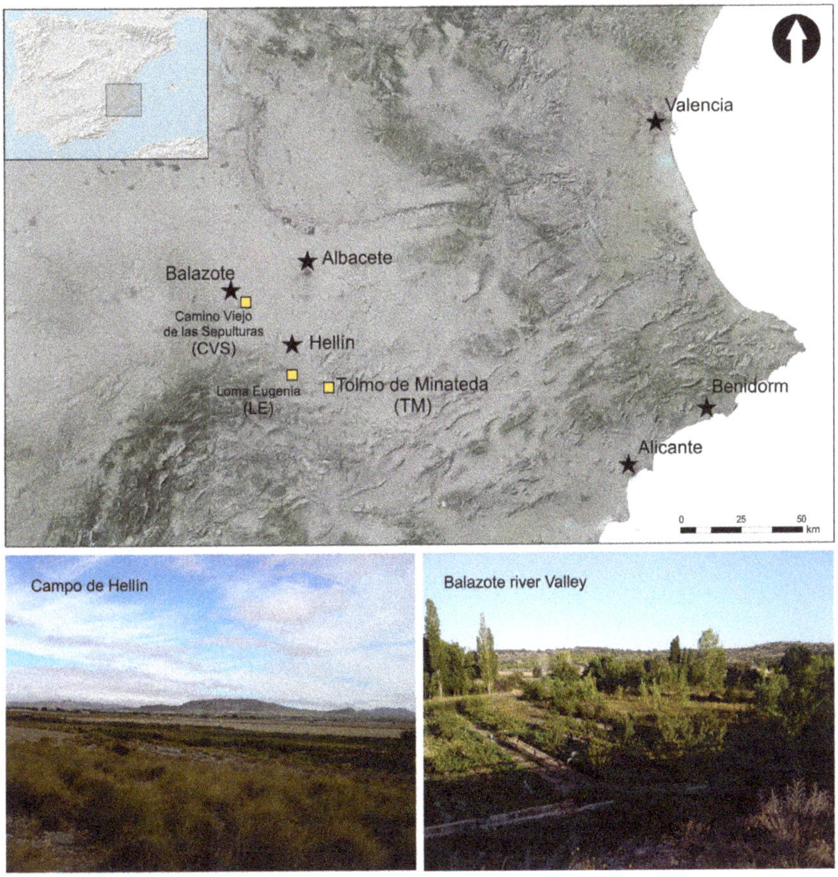

Figure 1. A location map of the study areas, showing the main settlements analysed and the views of the landscapes at the two study areas.

2. Methods and Context of Research

In both areas, we applied the same methodology of analysis after fusing all of the data, which allowed us to conduct a comparative study of the degree of continuity, transformation, or abandonment of rural structures since the end of the ancient world to the medieval Islamic period (5th to 10th centuries AD). The research was designed using a multiscale approach and incorporated multiple methodologies (Figure 2):

(1) Surface surveys with offsite strategy (Figure 2a),
(2) Ground-penetrating radar surveys (GPR) (Figure 2b),
(3) Analysis of surface material repertoires, filtering the surface diagnostic materials by periods and calculating their densities,
(4) Archaeological excavations to confirm chronological sequences in some key settlements (Figure 2c),
(5) Synthetic integration of large amounts of data into GIS platforms (Figure 2d).

During the first phases of the research, we carried out a general action to identify archaeological evidence in the different rural environments of TM and CVS, where we did not find many urban alterations. The selected transects were subjected to remote sensing and intensive full-coverage surface surveys, assisted by geospatial technologies. The data collected were integrated into a GIS environment with various cartographic sources and known archaeological sites from previous works [4–7].

Figure 2. Images of the different phases of research: (**a**) surface prospecting; (**b**) geophysical prospecting using GPR; (**c**) archaeological excavation; (**d**) data processing in GIS.

This first phase allowed us to obtain a detailed archaeological cartography with a general panorama at macro-spatial scale of the settlement patterns in Roman, Visigothic, and Islamic times, the preferred areas of human occupation, and the different traces left by those societies on the landscapes. The next objective was to transcend the point locations on the map and develop a detailed characterisation of the archaeological sites. This was a question of broadening our knowledge of the places that offered the most complex surface structures of remains and the settlements that showed possible evidence of a longer occupation or, conversely, a very short occupation. For this purpose, several geophysical GPR surveys and some excavations (still in progress) were carried out at the archaeological sites of Loma Eugenia (LE, Campos de Hellín) and CVS. The village in LE has never been excavated, whereas, in CVS, excavations were carried out in the 1970s but did not document the postclassical phases with the same intensity [8]. The aim of our research was to verify the possible existence of archaeological structures, gather direct information about buried deposits, and ultimately, characterise in detail the sites identified by the dispersion of materials on the surface.

Beyond the analysis of habitat zones, our interest also focused on the recognition of the complex human interactions with landscapes, especially land uses and agricultural structures, which are the fundamental basis of rural life. To this end, we carried out surface surveys of the main rural settlements in both areas using an off-site strategy [9–11]. Thanks to the registration of off-site materials, we were able to evaluate the intensity with which the plots were used in the different periods, demonstrating the clear differences in management, exploitation strategies, and land uses in both regions, which were certainly also reflections of the existence of different socioeconomic and probably political models of articulation.

3. Results: Settlements and Land Uses

In this section, we describe our results and the evidence collected from each of the investigated landscapes. We paid special attention to the concentrations of materials, which we interpreted as settlements of several different types. We also examined the fainter dispersions that were observed, which we interpreted as evidence related to the different uses of the land.

3.1. The Rural Landscapes around Tolmo de Minateda (TM) in the Late Antiquity and Early Middle Ages

The research carried out over the last 30 years in TM has shown the existence of a city from the Late Antiquity and Early Middle Ages, which was built on the remains of the ancient Roman municipality of *Ilunum* [4]. This city has also been identified as the Visigothic episcopal see of *Eio* or *Elo* and the later Islamic *madīna* of *Iyyuh*, both of which were of great importance within the geopolitical and historical context of Southeast Hispania between the 7th and 10th centuries [12,13]. Archaeological investigations have traditionally focused on this site because of the depopulated condition of the city and the fact that its archaeological records and transition contexts are so well preserved. For this reason, one of the main archaeological problems that has been faced in recent years is the recognition of the rural settlement patterns and landscape management strategies used in the different periods when TM headed this territorial system, as well as the degree of continuity of Roman structures in the Early Middle Ages [5,7,14,15].

The archaeological data that were available up to the beginning of our research were mainly related to material and structural remains from residences; therefore, they very focused on identifying the forms of settlements, without taking into account evidence from off-site records.

3.1.1. Characterisation of the Analysed Register: Concentrations of Materials without Dispersion Borders

Our work was concentrated in two areas of the TM suburban territory, within about 8–9 km to the north and west of the site. The selected study areas were large flat areas with some hills, which were surrounded by mountains and furrowed by small river branches that converged with the Mundo River. Within each study area, we chose certain spaces according to the topographic study, geomorphological criteria, and archaeological data from our previous work that could potentially specify the continuity of agricultural spaces and landscape management strategies from Roman times to Medieval times. We carried out a full-coverage survey with the aim of identifying and analysing the spatial structures of the surface records (Figure 3). The total area covered during the work was about 50 ha, with a geolocation of 1000 records (understood as every single piece of an archaeological object detected on the surface).

The strategy chosen to map the continuum of possible evidence was off-site registration; however, from the beginning, we could see that the surface records were concentrated in discrete and specific locations, with little dispersion around the sites and practically no records in the rest of the space. Therefore, we decided to reorientate our strategy and site concept, instead focusing our attention on places with potential house remains and verifying the existence of areas that were completely sterile.

The surface records presented a very clear pattern: dense concentrations of material and large intermediate voids. This picture clearly contrasted the records that we observed in the Balazote River Valley case study.

The spatial patterns were also recurrent: the fertile lands around rivers or small streams continued to be occupied, but the specific locations of rural sites did not since practically none of the Roman settlements survived beyond the 3rd and 4th century AD; furthermore, as we describe in the next section, new villages on small hills appeared at the end of the 6th century that replaced the previous settlements (Figure 4).

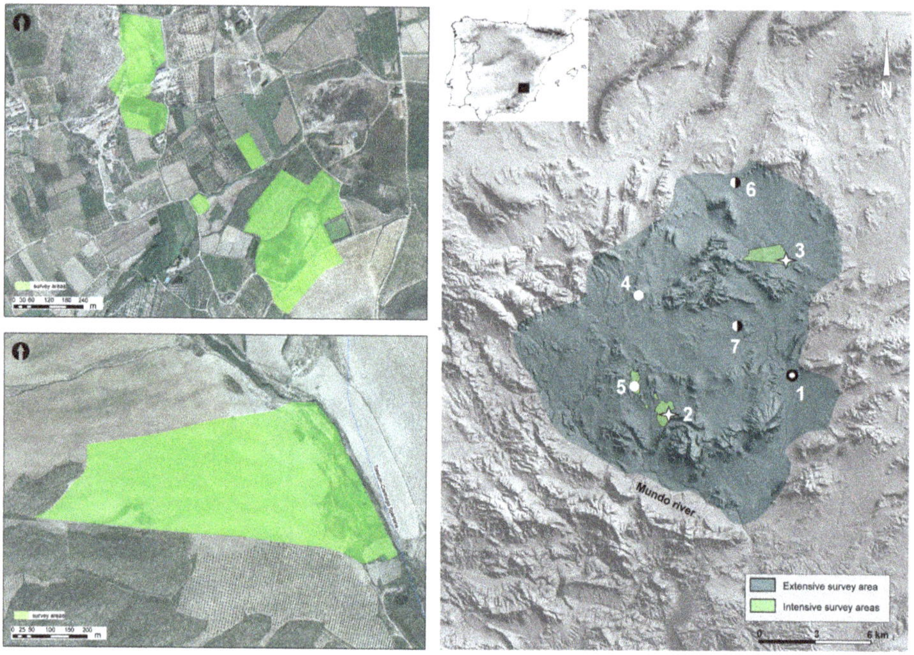

Figure 3. The surveyed areas in "Campos de Hellín" (**left**) and the main archaeological sites identified by the extensive surveys and some excavations (**right**): (1) *Eio/Iyyuh* city (TM); (2) Visigothic village of Loma Eugenia; (3) Visigothic village of Loma Lencina; (4) Roman *villa* of Hellín; (5) Roman *villa* of Agra; (6) Late Antiquity *castrum* of Alboraj; (7) Late Antiquity *castrum* of Torre Uchea.

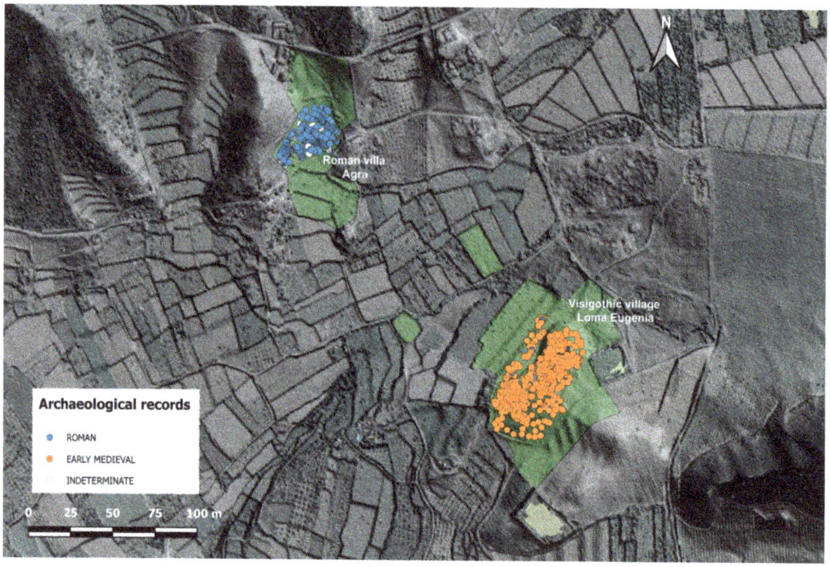

Figure 4. The LE study area, showing the individual localisations of records in the Roman villa of Agra and the Visigothic village of LE.

From the analysis of material concentrations, we observed how the dimensions of the new peasant villages in the Visigothic period (6th–8th centuries AD) were also similar in most cases. The enclaves located in the hills occupied an area between 0.5 and 1 ha. In all cases, the concentrations of ceramics coincided with the existence of structures, which could be recognised on the surface, in topographic anomalies, or from geophysical surveys. Therefore, the surface records seemed to indicate the zones of residence, transformation, storage, and necropolis, but not fields under exploitation.

3.1.2. The Materialisation of a Discontinuous Rural Population Model

Focusing on the settlement patterns identified from our archaeological data, we detected a Roman agrarian landscape in the *territorium* of *Ilunum* from around the 1st century AD, to which some farmhouses and numerous *villae* corresponded [16,17]. The establishments of the rural and suburban elites that articulated the space around the Roman city experienced different evolution patterns from the late ancient times. Some showed some continuity of space, but the most widespread pattern was that after the installation of the Visigothic episcopal see of *Eio–Elo* at the end of the 6th and the beginning of the 7th century, when a new landscape model was created. In it, in addition to other settlements, such as *castra* or *turris*, new forms of concentrated peasant settlements (villages) prevailed on small hills, which generally appeared in new habitat sites around the agricultural spaces that were exploited in Roman times. The occupation of these villages declined after the Islamic conquest, coinciding with the establishment of the Province of *Tudmīr* in 713 and the installation of the *madīna* of *Iyyuh* in TM (Figure 3) [5,7]. Therefore, we encountered some occupation dynamics that could be defined as discontinuous, with the clear abandonment of Roman *villae* although not of the agrarian spaces or land uses, as the new villages only appeared about 1–2 km away from the previous settlements and continued to exploit the same fields in the same way [7,16,17]. This is probably because they were still the most fertile lands, and the agro-systems were already articulated. The rupture of this pattern and the widespread abandonment of the agrarian landscapes that had historically been exploited occurred after the Islamic conquest of the Iberian Peninsula.

Some examples of this new pattern of occupation could be observed in the well-known villa of Hellín, which did not last beyond the 3rd century. From our surface surveys, geophysical surveys, and the excavation of some specific sites, we could more clearly characterise the creation of new rural settlements around the same time that the episcopal see was installed in TM. On the one hand, we found some examples of higher-ground settlements, such as the possible *castrum* of Alboraj, which presented some defensive sections, buildings of certain magnitudes, and surface materials with chronological forks between the 6th and the 10th centuries [7] (pp. 223–226).

On the other hand, some of the peasant settlements or villages emerged at the end of the 6th and the beginning of the 7th century, such as Loma Lencina (LL), which was established in a space that was not previously occupied and was probably built for the exploitation of lime. Another example is the case of Loma Eugenia (LE), which was located in an agrarian space that had already been exploited by the ancient Roman villa of Agra, although the village moved to a small hill on the other side of the river.

We especially focused the research on analysing the organisation of the peasant societies in this type of settlement and its relationship with the Visigothic city.

The case of LE was especially interesting because, after a one-off excavation of its necropolis in the 1990s [18] and the surveys carried out by us in recent years (both surface and GPR), a set of materials was recovered that established the provisional chronology from around the end of the 6th century to the mid-8th century [5] (pp. 184–187), [7] (pp. 221–223) and [19] (pp. 162–164).

According to the surface evidence and the results of our geophysical survey, the space occupied by the village did not seem to exceed 6000–9000 m^2; moreover, within the terrain, there were some masonry walls with vertical slabs that resembled the domestic Early Middle Age structures that have been documented in TM [5,7]. The residences seemed to be concentrated in the southern and central areas of the hill surface, while the necropolis area of the peasant community that lived in the village was located to the north. The cemetery included a series of burials with a west–east orientation. In the partial excavation in 1995, 33 graves were exhumed and were found to have been buried in earth, lined and covered with slabs and, in some cases, masonry to join the slabs and regularise the graves (Figure 5). We could not specify the number of individuals in each tomb, but they were mostly reused burials. Among the grave goods were two bronze rings, a possible leather tanner, and three belt buckles dated between the late 6th and 7th centuries [20] (p. 69). Absolute radiocarbon dating was also performed on one of the individuals found in one of these tombs (grave 6), which offered us a chronology of 676–779 cal AD (1274–1171 cal BP, Beta-572260). This necropolis was only partially investigated; however, according to the remains on the surface, it was a cemetery plot with a surface area of about 3000 m^2. Taking into account the excavated sector that had a surface area of about 1000 m^2, the necropolis could have held 100 tombs, which were probably reused on several occasions. Therefore, we identified a rural community of medium size, with a single necropolis detected so far, in which some items of personal clothing were found.

Figure 5. An orthophoto of the hill where the village of Loma Eugenia (LE) was installed. The right-hand side of the necropolis was partially excavated in 1995, which is where the belt buckles were found.

On the basis of all this evidence, we recently carried out excavations in two areas: one was located in the village and the other was located in the necropolis. The work carried out in the first campaign allowed us to detect a large room in the area of the village, where the sequence of occupation corresponded to the Visigothic period and did not extend beyond the middle of the 8th century. The room was very similar to those in other Visigothic villages that have been excavated in the centre of Madrid, such as the village of "Gózquez de Abajo", where domestic buildings have been found that included spaces for storage or production [21]. This could also be the case with the room documented in LE, which seems to include a fireplace or combustion structure (Figure 6).

As can be seen in Figure 7, the room also seemed to extend northwards; hence, this could corroborate the existence of a building with several rooms and walls of great technical quality and significant dimensions, including spaces for productive activities. In the necropolis, we also excavated one of the tombs that remained intact, which turned out to be a reused tomb containing three individuals. The last individual buried was a teenager with a radiocarbon dating of between 772–825 cal AD (1178–1025 cal BP, Beta-643123), under which was a pregnant woman with the remains of a foetus in her womb and another adult individual who had been placed in a secondary position when the other two individuals had been buried (Figure 7). Again, the configuration of the necropolis and the fact that the tombs were reused reinforced the idea that we had identified a peasant community who buried their dead in the necropolis for at least three generations.

Figure 6. Details of the evidence documented in the recent excavation of LE. The top panel shows a reused burial with three Visigothic individuals (the yellow circle marks the location of the grave that appears in the detail image); the bottom panel shows a room that was part of a larger building, which could have been a domestic space with several rooms (the yellow triangle marks the location of the structure that appears in the detail image).

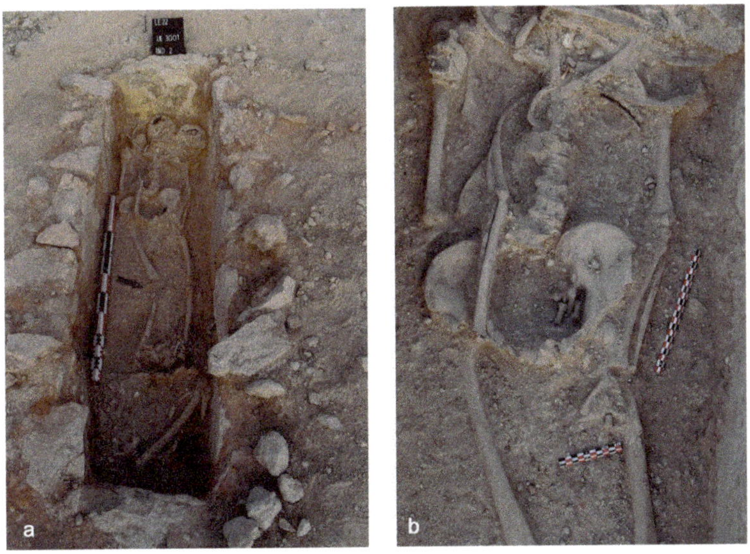

Figure 7. (**a**) The tomb excavated in 2022 in which three individuals were buried with a stone cist orientated from west to east. (**b**) Details of the pregnant woman and the remains of the foetus.

3.2. A Model of an Agrarian Settlement of Great Resilience: The Case of the Ancient Roman Villa of El Camino Viejo de las Sepulturas (CVS)

The Balazote River Valley is a wide endorheic area bounded to the east by the western edge of the "Llanos de Albacete" and to the west by the elevations of the "Campo de Montiel". The Balazote River, where small lagoons, wells, and water courses converge, has created a low-altitude (between 780 and 800 m) valley of alluvial soil that slopes slightly towards the river. These abundant water resources have made the land fertile and, together with the semi-arid continental Mediterranean climate, have helped to support intensive agriculture practices in the fertile plains, including extensive grain agriculture (currently mostly barley) and viticulture, along with an important livestock tradition on the grasslands of the nearby hills. Unlike the TM landscape, the surface records analysed after our survey in this valley showed extensive concentrations of materials from various periods with varying densities, which were concentrated in specific settlements or enclaves [2]. Beyond these sites and their agricultural spaces, there were areas that were empty of materials. For this reason, our research in this study area focused on thoroughly mapping and analysing the surface and subsurface evidence from one of these enclaves (CVS). In the 7.75 ha prospected using our intensive strategy (selected fields were used as sampling units to cover a large area), we geolocated more than 5000 items from different periods (Figure 8). More than 3000 items were diagnostic materials that showed a long sequence of uninterrupted occupation from the 1st century to the 11th century AD.

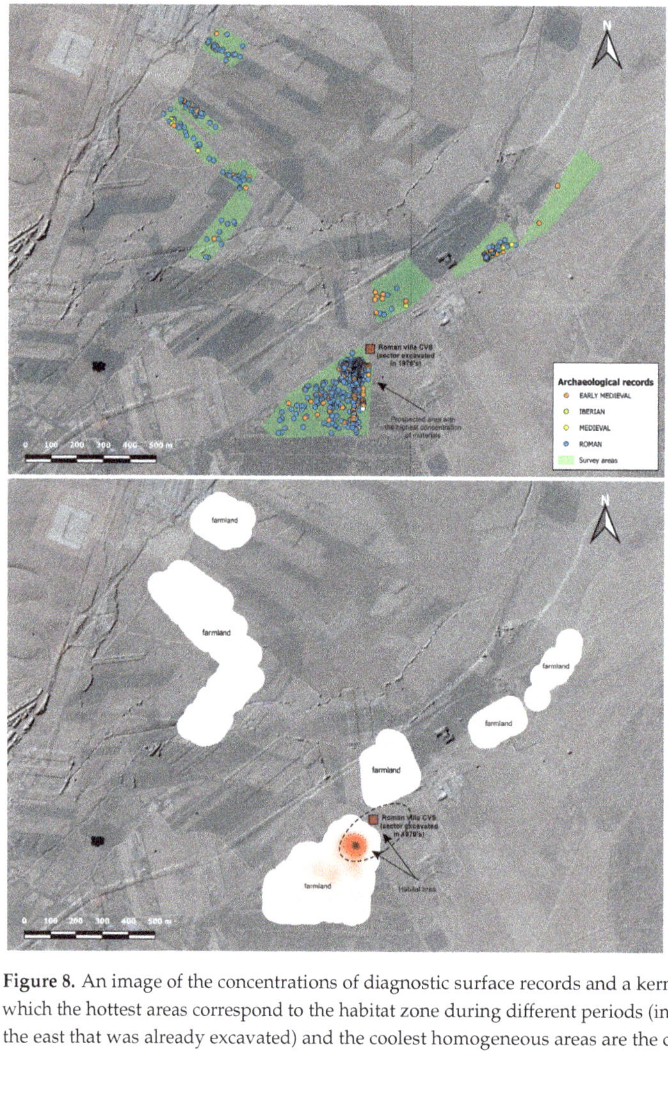

Figure 8. An image of the concentrations of diagnostic surface records and a kernel density map, in which the hottest areas correspond to the habitat zone during different periods (including the part to the east that was already excavated) and the coolest homogeneous areas are the cultivated fields.

3.2.1. Long-Term Occupation Dynamics at CVS

From the historical point of view, this region was linked to the *Colonia Libisosa Foroaugustana* in Roman times [22], which was a city that administered a large territory in which the establishment of *villae* and some *mansiones* was quite early [17]. We did not have many archaeological data on the evolution of these *villae* from the Late Antiquity and Early Middle Ages; however, after the abandonment of the city during the late Roman Empire, it seems that many of these settlements continued to be inhabited to a greater or lesser extent, albeit obviously losing their "aristocratic" character. At this time, the structures were transformed according to the new needs of the peasant communities that settled in the buildings after the widespread abandonment of the small and medium farmhouses located near of these *villae*. The goal was to seek security under the protection of these larger establishments.

This process was detected at the CVS site, where the excavation campaigns carried out in the 1970s exposed part of a large Roman villa, with several phases of occupation and transformation [23]. In addition to the structures discovered at that time, there were others found in specific interventions, which confirmed the extensive occupation of this area in the Roman, Late Antiquity, and the Early Middle Age periods [6,24,25]. In short, the origins of this settlement could be traced back to the 1st century AD, when colony status was granted to the city of *Libisosa* (located about 20 km away), to which administrative territory this rural complex belonged. This original phase was followed by a phase of development and monumentalisation in the Late Roman Empire, which provided a great number of remains during the excavations. After this period of prosperity, from the end of the 4th century and throughout the 5th century, some rooms began to be abandoned, and only those that could be used as domestic environments were occupied, in which cooking areas, storage/production areas, or even tombs were installed (Figure 9).

The phase of occupation in which the residential areas were reduced and transformed seemed to be linked to the installation of peasant populations in the surrounding areas of the site and when the old Late Roman elites abandoned these types of rural establishments as places of residence [26]. This new rural pattern gradually consolidated, as evidenced by the appearance of a community necropolis in the vicinity of the old villa, which implied the existence of a stable community that buried their dead for several generations. The radiocarbon dating of this necropolis ranged from the mid-7th century to at least the 11th century [27].

In order to sort all these partial data and begin to understand the dynamics of occupation of the ancient Roman villa of Balazote, we reconnoitred the evidence from the territory in which the settlement was located using surface surveys and geophysical surveys and proved the resilience of this fertile landscape, at least until the Islamic Caliphate period (10th–11th century) [2]. We began the research with a surface survey, from which the concentrations of materials allowed us to obtain a provisional chronology of post-classical contexts that oscillated between the middle of the 5th century and the 11th century. After this surface survey, we conducted geophysical prospecting using GPR in the areas with the highest concentrations of materials. The data extracted from this exploration showed structural testimonies in the subsoil, both to the west of the excavated spaces in the old roman villa and in slightly more distant areas, which supported our location hypothesis of a settlement in the Late Antiquity and Early Middle Ages. These structures were shallower and had a different orientation to those detected in the Roman villa (Figure 10). This settlement was related to the necropolis with Visigothic and Islamic tombs, which was located about 150 m to the northeast.

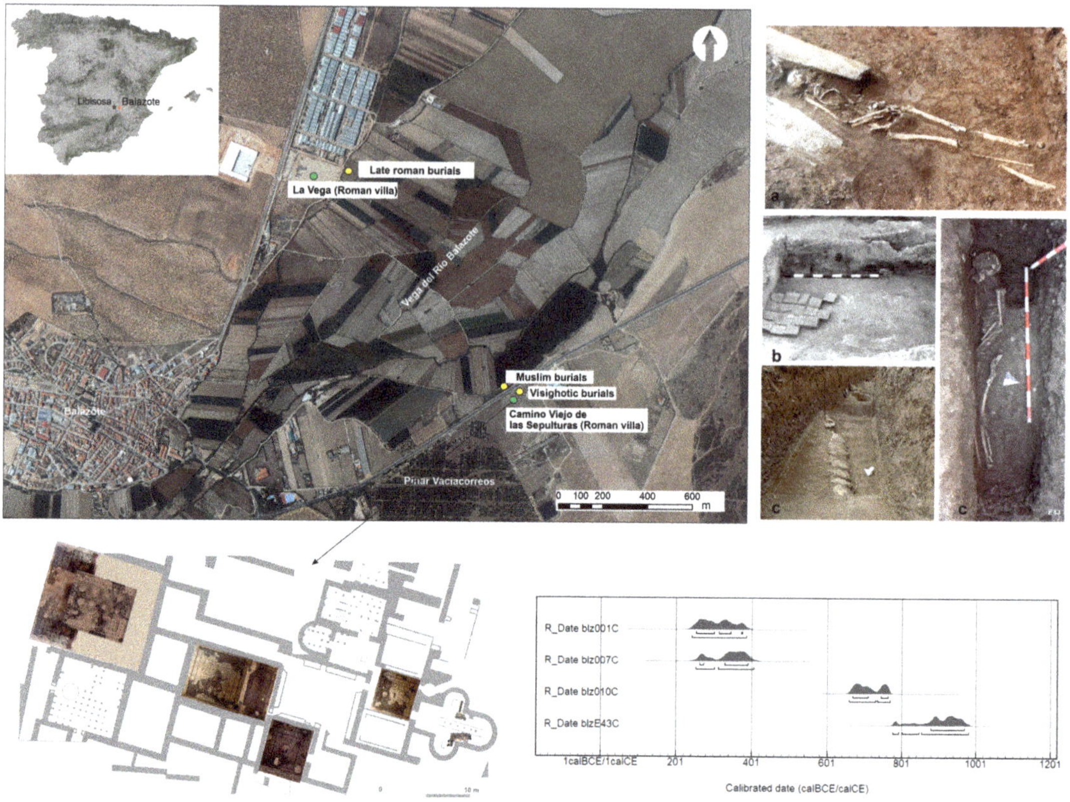

Figure 9. The locations of the different sectors of the CVS site. The right-hand panel shows details of the Late Antiquity (**a**) and Islamic (**c**) funerary uses and some combustion structures (**b**) that transformed the functionality of the original roman villa in the 5th century. The bottom panels show a plan of the *pars urbana* of the villa during its phase of splendour and the radiocarbon dating of several burial sites in the post-classical necropolis.

In order to verify and clarify the results obtained in the surveys, we began the excavation of a new sector to the west of the domestic spaces that had already been excavated in the 1970s. We documented new rooms related to Roman monumental housing, as well as new resurfacing and repairs from the late 4th and early 5th centuries, and evidence of use and construction from the occupation phases in the Late Antiquity and Medieval times. These later phases seemed to relate to the productive and domestic reoccupation of old rooms around the end of the 5th–6th centuries until the 10th century, with the construction of new walls that reduced and compartmentalised the rooms, silos for storage carved into the pavements of the *opus signinum* of the roman villa, and combustion structures, which demonstrated the installation of peasants in the ancient villa who buried their dead in the nearby necropolis that showed the same chronologies (Figure 11). This period of occupation related to a possible Visigothic village and a later Islamic settlement, which was followed by a phase of Medieval Islamic plundering before the definitive abandonment of the old village around the 11th century.

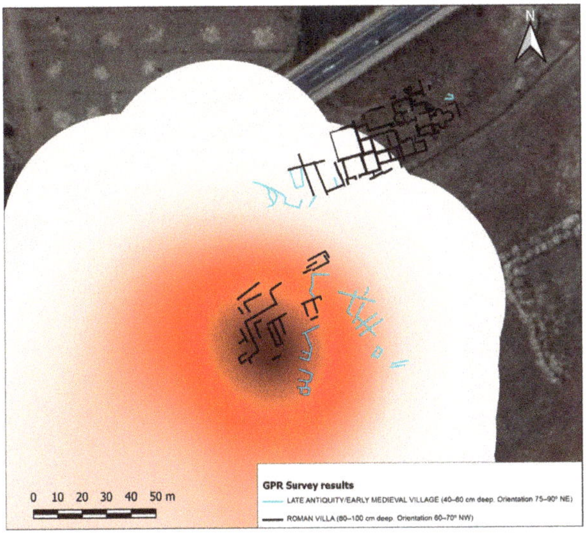

Figure 10. An image of the structural anomalies obtained by geophysical prospecting using GPR overlaid on the kernel density map interpolated from the concentrations of surface materials. On the basis of the depth and orientation of the anomalies, we differentiated remains from the Roman villa (black) and those from a possible Late Antiquity–Early Middle Ages village (blue).

Figure 11. An orthophoto of the excavation to the west of the area that was excavated in the 1970s, showing details of the Late Antiquity and Medieval reoccupation of an old building, where spaces had been compartmentalised, and where silos and other domestic structures had been built.

3.2.2. Land Use: Examples of Long-Term Sustainable Uses

Beyond these spaces that we interpreted as residential areas from different eras, we found fainter dispersions of materials in which ceramic fragments from ancient and Medieval times were intermingled with varying proportions, according to the counts of diagnostic pieces and the total counts from some plots. We interpreted these scattered remains as evidence of agricultural practices (the fields with the highest concentration of materials are interpreted as fertilisation remains with domestic waste), which allowed us to more precisely define the perimeters of cultivated fields during the two historical macro-periods (Roman and Medieval) that marked the transformation of the agrarian landscape of CVS.

According to our off-site registration, the agricultural areas located in the fertile river basin at the foot of the settlement had a perimeter between 200 and 230 ha in the ancient period. From these data and those from other nearby spaces, we proposed the existence of intensive agricultural practices according to the exploitation of plots or orchards that did not exceed 2–3 ha and were located near the settlement, which would work with other types of agricultural practices, such as the extensive cultivation of grains and pastures, more focused on the market needs of the people from the villa.

The off-site data corresponding to the Medieval period showed the continuity of the old intensive agricultural spaces, with a cultivated area of about 200 ha. However, at this time, it seems that the drylands from around the Early Middle Ages and Islamic settlement were reduced, which indicated a change in the model of the exploitation of agricultural landscapes that was probably connected to a change in the social organisation of the villa around the 5th century.

4. Discussion

4.1. Dynamics of Rural Occupation between Abandonment and Resilience: A Proposal of Determining Factors

As previously described, the surveys allowed us to identify and analyse the rural spaces of two different areas on the Iberian Peninsula, which had similar historical dynamics but markedly different socioecological systems. The application of identical methods allowed us to compare the different spatial structures of these two regions.

The results of this comparative exercise showed us clear differences in population patterns, as well as in agrarian practices and exploitation systems. Regarding settlement patterns, a similar typology of villages was found, although with significant variations in the duration of occupation. Compared to the model of the "Campos de Hellín", where rural villages rarely lasted more than one cultural phase, the occupation in the Balazote River Valley was much more resilient, occupying the same spaces during different historical phases (Figure 12).

The duration of occupation seemed to be affected by several factors that were opposed between the two rural landscapes.

4.1.1. Geopolitical and Historical Factors

In the case of "Campos de Hellín", the archaeological data showed territory occupation dynamics that were intricately linked to the city in TM. This city functioned as the epicentre from which a hierarchised rural landscape was developed in Roman times, which continued through Late Antiquity and the Early Middle Ages when it became a Visigothic episcopal see in the 7th century and an Islamic madīna in the 8th century [4]. From the episcopal see, the old and new rural settlements that were distributed within its diocese were managed and in Islamic times, it seems that the rural landscape was transformed definitively due to the abandonment of the old rural settlements from the Visigothic period [2,5,7,15]. The creation of new settlements was not documented until at least the 10th century, when the city was definitively abandoned.

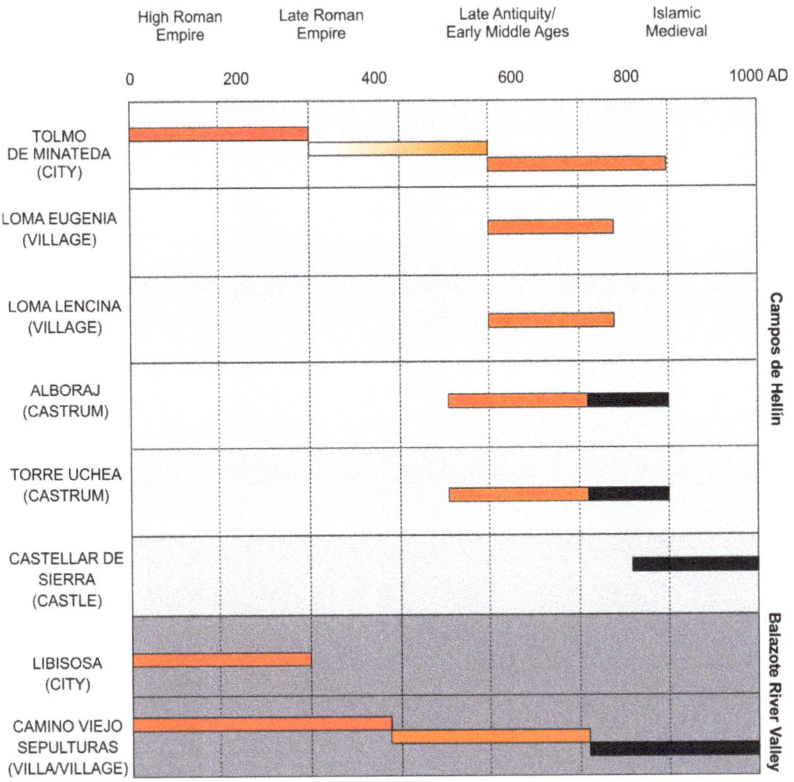

Figure 12. A diagram showing the chronology of the occupation of the settlements in both study areas, based on the diagnostic ceramic fragments that were recovered.

Therefore, from the geopolitical and historical points of view, the development and evolution of rural settlements in the countryside of TM were clearly linked to the different powers established in the city, from its foundation as a Roman municipality until its definitive abandonment at the beginning of the 10th century. A crucial fact was that the documented Late Antiquity villages (i.e., LL and LE, among others) were installed at the same time that the episcopal see was established in TM. This undoubtedly corresponded to the need for the bishopric of *Eio* to generate stable farming villages, in which productive agrarian activities could be concentrated and from which produce could be redistributed to rural markets and the city, probably via the payment of taxes. This hypothesis could explain the fact that, at the time of the conquest and after the configuration of the Islamic province of *Tudmīr* in 713, these villages began to be abandoned as the old fiscal role of the bishopric was lost, perhaps encouraging the population to settle in the new Islamic site of *Iyyuh*. The city at the end of the 8th and the beginning of the 9th century was fully Islamised and had a densified and rural urban fabric, with domestic spaces and other productive environments distributed around the surface of the hill, such as oil mills and corrals [13]. The abandonment of the agrarian settlement model around the mid-8th century was not detected in the examples of higher-ground settlements or *castra* that we found, which remained occupied until the beginning of the 10th century, as with the Islamic city itself. This reinforced the idea of the importance of the higher-ground settlements within the rural landscape of *Hispania* around the 5th and 6th centuries since various sources have cited that they, together with civitates, were key to the strategic and fiscal control of the territory. In fact, in 711, the *castra* constituted the main type of rural settlement for

Visigothic aristocracies, which had previously settled in episcopal sees and cities until the agreements of surrender with the Arabs had been established [28], undoubtedly leading to the durability of these settlements beyond the convulsive 8th century.

In contrast to this area, which was clearly linked to the historical development of the city in TM, in the case of the Balazote River Valley, we recognised the implementation of a Late Antiquity and Early Middle Ages rural settlement pattern, which had a high degree of continuity from the ancient Roman settlement [2]. Along with the other economic, environmental, and social factors that we describe below, we hypothesised that the continuity of these residential and agricultural spaces was significantly linked to the peripheral character of this region with respect to the nearest city because, after the abandonment of the Roman city of *Libisosa* around the 3rd century AD, the nearest centre was TM, which was located about 70 km away and, therefore, was outside regional control.

The archaeological data from CVS showed a long continuity in the sequence of the occupation of the ancient villa, albeit with different functionalities and physiognomies. The final moments of this settlement were when it was known as *Balāt al-Ṣūf* (the origin of the name Balazote), as cited by Arab sources [29] (p. 223), the first mention of which was during the Caliphate period (year 935) and which some authors have referred to as "Calzada de la Lana" (woollen driveway) [30] (p. 359). This could have been related to the important trade of this product in Islamic times, which, together with the intensive agricultural practices developed secularly in the fertile plains, resulted in the resilience and historical sustainability of this space. This settlement was abandoned during a time of territorial instability that was caused by the civil war or *Fitna* of Al-Andalus (1009–1031), which ended the Caliphate of Cordoba.

4.1.2. Economic Factors (Farm Models)

Beyond the archaeology of the settlements detected in both regions, we also examined the nature of land uses and agrarian practices using archaeological indicators that were related to the spatial structures of the surface records.

A whole series of geographical and anthropological studies on the investigation of Mediterranean landscapes have indicated the deployment of two economic strategies, according to the forms of land exploitation [31–33]. One strategy is defined by an extensive type of agriculture, which is based on the development of dryland and fallow farming, with some importance placed on grazing. The number of work hours required for this type of agriculture is not very high per unit of surface area, which allows it to be applied for large fields that practically do not need to be fertilised. In contrast, there is an intensive strategy, which is based on crop rotation and irrigation. This tillage requires constant work in the fields, especially for the maintenance of high levels of soil fertility through irrigation or the incorporation of organic fertilisers. Livestock farming is a fundamental complement to this type of agriculture and provides essential manure to maintain the fertility of the fields. In the case of intensive agriculture, proximity to the farmland facilitates this constant work, meaning that most of the peasant work is tied to agricultural practices and not displacement. Therefore, there is a clear correlation between management models and the locations of farmland [11].

The results of our fieldwork showed a clear contrast between the two case studies of Hellín and Balazote, which was undoubtedly linked to clearly different land uses and different forms of field management, as well as the presence or absence of off-site ceramic records. Due to the agricultural modalities offered by the absence of off-site registration around the villages of "Campos de Hellín", the economic orientation of peasant communities towards extensive grain agriculture and livestock seemed evident. Meanwhile, the archaeological indicators that showed the existence of an intensive agricultural model in Balazote were due to the detection of off-site ceramic materials, i.e., they were not directly associated with places of residence, according to the decomposition of the primary archaeological deposits.

We did not find any accurate archaeobotanical information in the landscape around TM, but the palynological analyses of nearby areas, as well as the indications of traditional uses prior to the mechanisation of farming in the 18th century, indicated the presence of crop fields intended for the cultivation of barley (*Hordeum vulgare*) and nude wheat (*Triticum aestivum/durum*), which are plants suitable for dry climates and saline soils [34,35]. These agricultural practices were combined with the cultivation of other complementary crops (such as cherries (*Prunus avium/cerasus*) or acorns (*Quercus*)), but no evidence was found of vegetable cultivation; however, evidence for livestock management was clear. Livestock farming was one of the main economic activities in this area as it was very suitable for this type of exploitation due to the suitability of the pastures, the abundant low vegetation, and the existence of numerous sources of fresh water, as well as the ease of access to salt in some salt marsh spaces (such as Alboraj-Cordovilla), which is an indispensable element in livestock diets [36].

For this region, the information provided by the Cadastre of Ensenada (18th century) was of great interest as it expressly states that, during its drafting, new plots were developed for cultivation: 14 plots were "taken from the meadow", 85 were in low mountain areas, and one was in a high–low mountain area. The surface area of cultivated land was only 17.83% of the Hellín area, according to the Cadastre [36] (p. 748). This suggested that, until the 18th century, land in the area was mostly low-lying vegetation, grasslands, a moderate degree of farmland, and vineyards or olive groves, along with trough spaces that were not occupied by stagnations [37]. The coexistence of the mountains (with conifers or scrub vegetation, depending on the area) and the small arable areas contributed to the main historical uses of these lands, which were mainly orientated towards livestock [38].

A few historical sources have Indicated that livestock has been considered one of the main resources of the area since at least the Middle Ages. The first known records were from 1576 (topographic listings for Philip II), when the authorities of the Hellín area made an agricultural catalogue at the request of the king, which included mulberry crops, raisin and wine vineyards, and some fruit trees [39] (pp. 40, 65). It has also been pointed out that the land was "mountainous and uncultivated due to its roughness, which is only good for pastures, and firewood". Referring to areas close to Hellín that were in the studied region, the catalogue sent to the king stated that the mountains "are very populated with firewood, pines, rosemary, sabina, junipers, and gorse ... There are also many pastures for cattle ... " [40] (p. 211). Due to the presence of grasslands, it was necessary to move livestock activity, which would target the barley crops, fodder, etc., and which, in the 18th and 19th century, included about 1800 heads of wool livestock [39] (pp. 65, 89).

In short, according to the agricultural modalities presented in all of these sources, as well as our archaeological data from "Campos de Hellín", it seemed that this region traditionally focused on extensive grain farming and, above all, the management of livestock herds, which is an area that is historically related to transhumance, where "Cañadas reales" or droves are even preserved to this day. These land uses leave little or no macroscopic archaeological footprints on the ground; hence, we did not find any off-site ceramic evidence on the surface, which was instead only found when the intensive land uses were fertilised with household waste. This type of activity allowed peasant communities to move to more favourable spaces or as indicated by local authorities because not much work was required for these extensive agricultural practices.

In contrast to the extensive model of the Hellín territory, which was based on dryland crops and livestock, in the case of the Balazote River Valley, we detected an intensive model of plots with contiguous tillage. Most likely, the high amounts of work required to build, condition, and maintain agricultural fields could explain the multigenerational links between certain groups and certain plots and, therefore, the permanence of some settlements, such as CVS, which experienced an uninterrupted occupation for a millennium (from the 1st to the 11th centuries AD), despite the change in the model of occupation around the 5th century (from a large Roman villa or hacienda to a Medieval peasant village).

This continuity of the CVS settlement indicated the clear fixation of the populations that occupied this area of the Balazote River Valley to the land and, therefore, the maintenance of the intensive agricultural management modality by these communities, which led them to systematically occupy the spaces near the river valley, where the potentially irrigated fields were located. Thus, there was clear evidence of the uses and management of the agricultural landscapes that have been maintained over the centuries, including the logical modifications of irrigation infrastructures and the morphologies of the fields.

During the Roman period, we hypothesised the existence of an intensive agriculture model that was based on the exploitation of small and medium plots and the extensive agriculture of grains, which probably occupied large areas, along with pastures. This agricultural model was most likely based on a mixture of grain and legume crops, as well as the cultivation of fruit trees, vines, and olives [35].

During the medieval period, the CVS off-site records showed the continuity of the ancient intensive farming spaces; however, at this time, it seemed that the fields destined for dryland farming or extensive cultivation were reduced. The amount of fertiliser available from the generation of household waste and animal manure was directly related to the size of the community and the available livestock herds. In the case of CVS, the more or less stable maintenance of intensive farming practices, both in Roman and in Medieval times, indicated the use of similar amounts of fertiliser and fairly large livestock herds, when taking into account the surface area (between 200–230 ha) of the intensive farming fields around the fertile plains near the site.

However, judging by the surface area of the residential zone and the dryland agricultural space of the settlement in Late Antiquity, we hypothesised that the post-classical village of CVS was formed by a medium-sized community that maintained small cereal plots for subsistence and produced surplus resources for local or regional exchange and for feeding livestock, which was demonstrated by the presence of silos inside domestic spaces during this late phase. The capacity of these silos ranged from 700 to 750 L.

The extraordinary attachment of the peasant communities that settled in this space to the land and intensive agricultural practices was undoubtable. These communities were first the dependents of Roman landowning families that managed large agricultural haciendas and large expanses of dryland and irrigated fields; then, from the beginning of the Middle Ages, rural communities with some autonomy settled in the medium-sized units of these settlements, such as the Visigothic village and the later Islamic village of *Balāt al-Ṣūf*, which had its own irrigated area of several hectares. This again demonstrated the existence of a special relationship between peasant settlements and the construction of work and residence spaces and the exploitation of the environment.

4.1.3. Social Factors

The different settlement patterns and agrarian models detected in both areas at different times were also related to other questions about the forms of social organisation and the legitimacy, ownership, and use of the land, which, although they have eluded archaeological reality, were related to other issues and could be identified using appropriate anthropological models. In this sense, the fact that certain human groups were linked to certain spaces for centuries could be due to the development of intensive farming practices, such as those detected in Balazote. This is because it is a strategy that involves high levels of dedication from farmers to their agricultural work in order to increase or maintain the production per unit of land [41]. This strategy also requires higher labour and/or capital costs [42]. The fertilisation of land and the construction of agro-engineering facilities, such as terraces or irrigation canals, are the most common forms of intensification [43]. The extension of the growing season, the introduction of new crops, and changes in the patterns of the organisation of community work are also considered to be forms of intensification [44]. Thus, agricultural intensification was defined in this study as any increase in the amount of labour per unit of land that increased production or minimised a decline in production [1]. Additionally, the investment of time and effort required by these practices meant that they developed over more or less prolonged periods of time, unlike other socioecological systems, such as those detected in "Campos de Hellín", where only extensive practices and livestock activities were identified.

The detailed analyses of the causes and different trajectories of intensification presented in other works allowed us to discuss its significance in relation to the forms of socioeconomic organisation among the communities installed secularly in CVS [45].

There are several theories to explain why communities begin to develop intensive farming practices and dedicate great amounts of effort to conditioning and maintaining fields. One of the main theories concerns population growth, which could lead to demographic pressure on resources and the need for agricultural intensification [43,46]. In this context, farmers reduce fallow land and maximise their land use or develop agro-engineering facilities. In our case study, it was unlikely that population growth was the cause of any intensification since the agrarian colonisation of the fertile plains around the Balazote River was only detected since the original installation of the Roman villa in the early imperial era; hence, demographic saturation could be ruled out. Other variables that play key roles in intensification have also been proposed, such as social demands and the payment of taxes related to political economies or trade. These variables could be related to the Roman phase of the settlement, when, in addition to intensive orchard farming with large concentrations of materials, other extensive land uses were detected that were undoubtedly related to the latifundia model, which is traditionally attributed to rustic Roman haciendas or *villae*.

However, this model of the socioeconomic organisation of agricultural work, in which the farmers who were dependent on the *possessores* of the villa carried out the intensive and extensive management of the fields, seemed to change following the dismemberment of the Roman State. From the 5th century onwards, we detected a clear reduction in dryland land uses and a concentration of agricultural work in intensive cultivation fields. From then on, we observed a continuity of the settlement, but with clear transformations in the type of settlement. There was a reduction in residential space and an increase in domestic and storage structures, which were linked to the installation of a peasant community that was probably generated after the integration of groups who had been installed in both the old Roman hacienda and the surrounding farmhouses that were abandoned [47]. This integration was a phenomenon that was widely observed in other regions of *Hispania* and could be explained by the instability and the lack of security caused by the crisis of the Roman Empire and the fall of classical territorial structures, among other things. These concentrated settlements or villages followed the social organisation model of communities based on nuclear or polygynic family groups, which are understood to be the social units that normally mobilise work activities, organise consumption, and exercise ownership over small plots of land that are intensively cultivated. In the case of CVS, the archaeological interventions were very recent and did not allow us to understand the extensions to the structure of the village; however, in other nearby cases, such as "Gózquez de Abajo" in Madrid, productive spaces, storage spaces, and plots for vegetable and cereal cultivation have been detected around each domestic unit [21]. These agricultural plots seem to have been distributed and assigned to each family unit within the community from the initial phase of the settlement until its abandonment. Trenches, roads, and aligned structures have allowed us to discern the shape and size of each plot. Each household occupied about 0.6 ha, comprising both land for the house and land for agricultural cultivation. Zooarchaeological and archaeobotanical analyses of samples from Gózquez have provided interesting data related to production and cultivation patterns, showing a balance between livestock and agricultural practices that, together with the systematic appearance of silos (each household had between four and six silos), certainly allowed for the production of a surplus that benefited both the family groups and the community. In this sense, Wickham [48] identified a peasant model of production among Early Middle Ages agrarian societies, which was based on the fact that the central element of the production process was the home, meaning that all members of the group worked but the accumulation of goods and the constitution of an estate were not essential concerns, nor was the response to the requests of a master, which differentiated this model from the slavery model of the Roman period and the feudal model of production. Additionally, he reported that there were no clear signs of social hierarchies or differences in wealth between different households. This social model best fitted our specific case study. Between the 5th and 8th centuries, the predominant role of the CVS village was due to the initiatives of small peasant groups that did not have strict dependences on external factors. Although these groups had connections to neighbouring regions through political structures and exchange networks, their economic activity was orientated towards satisfying their own needs and the perpetuation of their own communities through continuous investment.

This model of social organisation was maintained during the process of Islamisation that led to the Arab–Berber conquest of the Iberian Peninsula, throughout which the old Visigothic village continued to be in use and was probably occupied by the same family groups that adapted their socioecological systems to the new cultural and productive dynamics introduced by the Umayyads [2,19]. This could be assumed due to the archaeological records that were found in the necropolis located to the northeast of the settlement, where we detected the continued use of the same burial space from Late Antiquity to Medieval Islamic times. This continuity of the cemetery, together with the fact that stable isotope analyses carried out on individuals buried in the 7th and 9th centuries indicated similar consumption patterns (i.e., the preferential consumption of plants without the consumption of new crops introduced by Arab populations), demonstrated that different generations of the same family groups had become Muslim [27]. Agricultural work continued to be based on the intensive model, but the labour investment decreased with the installation of new hydraulic technologies that facilitated the fertilisation of the land [1].

5. Conclusions

As revealed in this paper, the investigations carried out so far on the selected areas have allowed us to identify and analyse the spatial structures of Late Antiquity and Medieval landscapes, including a city that functioned as the administrative head of a historical territory for several centuries and the peripheral area where urban dynamics did not directly influence the configuration and development of rural populations.

With regard to the settlement patterns that emerged from the 5th century, there were certain similarities in both cases since the type of rural settlement that was predominant from Late Antiquity was fundamentally the concentrated peasant settlement (village), which usually developed around ancient Roman *villae* or directly on them. In the case of the area near the city of TM, it seemed clear that these villages developed at the same time as the city became an episcopal see and, therefore, the epicentre of an administrative territory; furthermore, its occupation was dictated by the greater or lesser political importance of the city. This was why the villages around TM were abandoned at the same time as the city ceased to be a bishopric after the Islamic conquest and its incorporation as a territory of the *Tudmīr Cora*. Meanwhile, in the peripheral area of Balazote, the villages were established in ancient settlements and lasted for centuries, in more or less the same forms. The stability and fixation on the land shown by the peripheral peasant communities seemed to be explained by the models of agricultural exploitation that were deployed in those areas, in addition to the very condition of the urban peripheries of the territories they occupied.

Our analyses of the detailed records of the scattered remains in the fields near the CVS settlement in Balazote suggested the continued use of fertilisation practices related to intensive agriculture in those plots; hence, there were concentrations of materials from different periods that indicated the use of household waste to fertilise the fields. These peasant groups fundamentally based their productive structures on self-consumption and their operational structures on family labour. They deployed extraordinarily sustainable and autonomous land-use management strategies with respect to the control of regional or central powers, which remained stable from Roman times until at least the Middle Ages. Most likely, the high amounts of work required to build, condition, and maintain these intensive agricultural spaces could explain the multigenerational links between these groups and certain agrosystems.

In terms of this type of long-term settlement with extraordinary political and economic autonomy, we found evidence of the Late Antiquity and Medieval rural settlement model in TM, as the intensive surveys we conducted around the villages of LE and LL did not reveal any off-site records, meaning that the economies of these settlements were based mainly on the extensive cultivation of grain, which did not need to be fertilised, and livestock. There was also a clear fiscal and commercial relationship with the bishopric of *Eio*, to which these peasant communities supplied resources, at least until the middle of the 8th century, when they were abandoned. This also demonstrated a link to the religious power that administered these territories until the arrival of the Arab and Berber conquerors.

Author Contributions: J.S.-B. (University of Alicante) is a full lecturer of archaeology in the Department of Prehistory, Archaeology, Ancient History, and Greek and Latin Philology, as well as a researcher at the University Institute for Research in Archaeology and Historical Heritage (INAPH) at the University of Alicante. Her main research focuses on the organisation and transformation of cultural landscapes on the Iberian Peninsula and in northern Italy around Late Antiquity and the Early Middle Ages. She is currently conducting multidisciplinary research on several rural settlements in the southeast of Spain, where she is characterising new settlement patterns and production strategies that were developed after the end of the Roman villa model. The author read and agreed to the published version of the manuscript.

Funding: This research was funded by the "Ministerio de Ciencia e Innovación de España" (grant number: PID2019-108192GB-100/AEI/10.13039/501100011033). CONTEXT: El contexto como herramienta. Escalas de aplicación en los procesos de cambio en el Alta Edad Media.

Data Availability Statement: The data presented in this study are available upon request from the corresponding author. The data is not publicly available because it is currently being processed to include it in a public repository of the University of Alicante.

Acknowledgments: The research presented here was carried out between 2019 and 2022 by a team of researchers from INAPH, University of Alicante. The author would like to thank her colleagues who collaborated with her on the archaeological fieldwork. The author would also like to thank to the municipalities of Balazote and Hellín for the help they provided during this research.

Conflicts of Interest: The author declares no conflict of interest. The funders had no role in the design of the study, in the collection, analyses or interpretation of data, in the writing of the manuscript or in the decision to publish the results.

References

1. Grau Mira, I.; Sarabia-Bautista, J. Multiscaled Archaeological Survey in Eastern Iberia: Ancient Settlement Dynamics, Agrarian Practices, and Rural Landscapes. *J. Field Archaeol.* **2022**, *47*, 471–485. [CrossRef]
2. Sarabia-Bautista, J. El paisaje rural tardoantiguo y altomedieval en torno a la antigua villa romana de Balazote (Albacete, España): Detección y primeras reflexiones. In *Paisajes, Espacios y Materialidades: Arqueología Rural Altomedieval en la Península Ibérica*; Prata, S., Cuesta-Gómez, F., Tente, C., Eds.; Archaeopress Access Archaeology: Oxford, UK, 2022; pp. 152–165.
3. Fischer, J.; Gardner, T.A.; Bennett, E.M.; Balvanera, P.; Biggs, R.; Carpenter, S.; Daw, T.; Folke, C.; Hill, R.; Hughes, T.P.; et al. Advancing sustainability through mainstreaming a social–ecological systems perspective. *Curr. Opin. Environ. Sustain.* **2015**, *14*, 144–149. [CrossRef]
4. Abad Casal, L.; Gutiérrez Lloret, S.; Gamo Parras, B.; Cánovas Guillén, P. El Tolmo de Minateda (Hellín, Albacete, España): Un proyecto de investigación y puesta en valor del patrimonio. In *Documentos de Arqueología Medieval*; UPV: Bilbao, Spain, 2012; Volume 2, pp. 351–382.
5. Gutiérrez Lloret, S.; Grau Mira, I. El territorio tardoantiguo y altomedieval en el sureste de Hispania: Eio-Iyyuh como caso de estudio. In *Visigodos y Omeyas. El Territorio*; Caballero, L., Mateos, P., Cordero, T., Eds.; Anejos de AEspA-CSIC: Mérida, Spain, 2012; Volume 61, pp. 171–198.
6. Sarabia-Bautista, J. *La Villa de Balazote (Albacete): Un Ejemplo de la Vida en la Campiña Entre el Alto y el Bajo Imperio Romano*; Universidad de Alicante: Alicante, Spain, 2012.
7. Sarabia-Bautista, J. La transformación del paisaje rural tras la fundación del obispado de Eio-El Tolmo de Minateda (Hellín, Albacete, España): Siglos V al IX d.C. *HAM* **2014**, *20*, 216–231. [CrossRef]
8. de los Santos Gallego, S. Excavaciones en la villa romana de Balazote (Albacete). *NAH* **1977**, *5*, 249–254.

9. Bintliff, J.L. The concepts of 'site' and 'offsite' archaeology in surface artefact survey. In *Non-Destructive Techniques Applied to Landscape Archaeology*; Pasquinucci, M., Trement, F., Eds.; Oxbow Books: Oxford, UK, 2000; Volume 4, pp. 200–215.
10. Grau Mira, I.; Carreras Monfort, C.; Molina Vidal, J.; De Soto, P.; Segura Martí, J.M. Propuestas metodológicas para el estudio del paisaje rural antiguo en el área central de la Contestania. *Zephyrus* **2012**, *70*, 131–149.
11. Grau Mira, I.; Jiménez Vialás, H.; Sarabia-Bautista, J. Arqueología de los espacios y comunidades agrarias desde el registro superficial: Un análisis comparado de paisajes y prácticas rurales desde la Antigüedad al Medievo. In *Arqueología y Sociedad de los Espacios Agrarios: En Busca de la Gente Invisible a Través de la Materialidad del Paisaje*; Mayoral, V., Grau, I., Bellón, J.P., Eds.; Anejos de AEspA-CSIC: Mérida, Spain, 2021; Volume 91, pp. 27–46.
12. Abad Casal, L.; Cánovas Guillén, P.; Gamo Parras, B.; Gutiérrez Lloret, S. El Tolmo de Minateda: El camino desde el conocimiento hasta la divulgación. In *I Reunión Científica de Arqueología de Albacete*; Gamo Parras, B., Sanz Gamo, R., Eds.; I.E.A. Don Juan Manuel: Albacete, Spain, 2016; pp. 71–90.
13. Gutiérrez Lloret, S.; Abad Casal, L.; Gamo Parras, B. Eio, Iyyuh y el Tolmo de Minateda (Hellín, Albacete): De sede episcopal a Madîna islámica. In *Les Ciutats Tardoantigues d'Hispania: Cristianització i Topografia/VI Reunió d'Arqueologia Cristiana Hispànica*; Gurt, J.M., Ribera, A., Eds.; Institut d'Estudis Catalans: Barcelona, Spain, 2005; pp. 345–370.
14. Sarabia-Bautista, J. El paisaje rural y suburbano del Tolmo de Minateda (Hellín) durante la Antigüedad Tardía y la Alta Edad Media. In *I Reunión Científica de Arqueología de Albacete*; Gamo Parras, B., Sanz Gamo, R., Eds.; I.E.A. Don Juan Manuel: Albacete, Spain, 2016; pp. 723–744.
15. Sarabia-Bautista, J.; Gutiérrez Lloret, S.; Amorós Ruiz, V. The rural and suburban landscape of Eio-Iyyuh (Tolmo de Minateda, Hellín, Spain): New methodological approaches to detect and interpret its main generating elements. In *Mediterranean Landscapes in Post Antiquity. New Frontiers and New Perspectives*; Gelichi, S., Olmo-Enciso, L., Eds.; Archaeopress Access Archaeology: Oxford, UK, 2019; pp. 145–162.
16. Abad Casal, L.; Sanz Gamo, R. La comarca hellinera ante la romanización. *Ponen. Hist. Hellín* **1991**, *11*, 35–41.
17. Sanz Gamo, R. La distribución de las villas romanas en la provincia de Albacete. *AnMurcia* **2001**, *16–17*, 351–364.
18. Rico, M.T.; López Precioso, J.; Gamo, B. La Loma Eugenia. Noticia sobre un asentamiento rural visigodo en el campo de Hellín (Albacete). *Antigüedad Crist.* **1997**, *10*, 85–98.
19. Gamo Parras, B. *La Antigüedad Tardía en la Provincia de Albacete*; I.E.A. Don Juan Manuel: Albacete, Spain, 1998.
20. Ripoll López, G. *Toréutica de la Bética (Siglos VI y VII d.C.)*; Reial Acadèmia de Bones Lletres: Barcelona, Spain, 1998.
21. Vigil-Escalera, A. Granjas y aldeas altomedievales al norte de Toledo (450–800 d.C.). *AEspA* **2007**, *80*, 239–284.
22. Uroz Sáez, J. La colonia romana de Libisosa y sus precedentes. In *La Ciudad Romana de Castilla-La Mancha*; Carrasco, G., Ed.; Universidad de Castilla-La Mancha: Cuenca, Spain, 2012; pp. 87–130.
23. de los Santos Gallego, S. Excavaciones en la villa romana de Balazote (Albacete). In *Symposium de Arqueología Romana*; Instituto de Arqueología y Prehistoria, Universidad de Barcelona: Barcelona, Spain, 1977; pp. 367–370.
24. Sanz Gamo, R. El poblamiento rural del área de Balazote (Albacete) a la luz de las últimas investigaciones. In *Poblamiento Rural Romano en el SE de Hispania*; Noguera, J.M., Ed.; Universidad de Murcia: Murcia, Spain, 1995; pp. 339–356.
25. Sanz Gamo, R.; Gamo Parras, B. La Villa romana de Balazote (Albacete): Reflexiones para una revisión. In *Villas Tardoantiguas en el Mediterráneo Occidental*; Chavarría, A., Arce, J., Brogiolo, G.P., Eds.; Anejos de AEspA-CSIC: Mérida, Spain, 2006; Volume 39, pp. 153–171.
26. Chavarría Arnau, A. *El Final de las "Villae" en "Hispania (Siglos IV–VIII)*; Association Antiquité Tardive: Turnhout, France, 2007; Volume 7.
27. Sarabia-Bautista, J.; Bolívar, H.; Ureña, I. Who was buried there and what did they eat? Dietary study of the Balazote late Roman villa (Albacete, Spain). *J. Post-Class. Archaeol. (PCA)* **2022**, *12*, 135–162.
28. Díaz Martínez, P.C. El Parrochiale Suevum: Organización eclesiástica, poder político y poblamiento en la Gallaecia tardoangitua. In *Homenaje a José María Blázquez*; Ediciones Clásicas: Madrid, Spain, 1998; pp. 35–47.
29. Franco, F. *Vías y Defensas Andalusíes en la Mancha Oriental*; Instituto Alicantino de Cultura Juan Gil-Albert: Alicante, Spain, 1995.
30. Rubiera Mata, M.J. Los precedentes geopolíticos musulmanes del Señorío de Villena. In *Congreso de Historia del Señorío de Villena*; I.E.A. Don Juan Manuel: Albacete, Spain, 1987; pp. 357–360.
31. Halstead, P. Traditional and ancient rural economy in Mediterranean Europe: Plus ça change? *J. Hell. Stud.* **1987**, *107*, 77–87. [CrossRef]
32. Van Der Veen, M.; O'connor, T. The expansion of agricultural production in Late Iron Age and Roman Britain. In *Science in Archaeology. An Agenda for the Future*; Bayley, J., Ed.; English Heritage: London, UK, 1998; pp. 127–144.
33. Whitelaw, T.M. An ethnoarchaeology study of rural land use in North-West Keos. Insights and implications for the study of Past Aegean Landscapes. In *Structures Ruralesetsociétés Antiques: Actes du Colloque de Corfou (14–16 Mai 1992)*; Doukellis, P.N., Mendoni, L.G., Eds.; Collection de l'Institut des Sciences et Techniques de l'Antiquité: Paris, France, 1994; pp. 165–186.
34. Precioso Arevalo, M.L. El desarrollo de la agricultura en Lorca a través de los restos paleobotánicos. Estado de la cuestión. *AlbercA* **2004**, *2*, 27–38.
35. Peña-Chocarro, L.; Pérez-Jordà, G.; Alonso, N.; Antolín, F.; Teira-Brión, A.; Tereso, J.P.; Moya, E.M.M.; Reyes, D.L. Roman and medieval crops in the Iberian Peninsula: A first overview of seeds and fruits from archaeological sites. *Quat. Int.* **2019**, *499 Pt A*, 49–66. [CrossRef]

36. Abad Casal, L.; Sanz Gamo, R. En busca de los paisajes perdidos en época antigua. La cuenca baja del río Mundo. In *Libro Jubilar en Homenaje al Profesor Antonio Gil Olcina*; Universidad de Alicante: Alicante, Spain, 2016; pp. 741–764.
37. Lemeunier, G. Crecimiento agrícola y roturaciones en el antiguo marquesado de Villena (siglo XVIII). *Al-Basit* **1987**, *21*, 5–31.
38. Pretel Marín, A. *Hellín Medieval*; I.E.A. Don Juan Manuel: Albacete, Spain, 1988.
39. Rodríguez De La Torre, F.; Moreno García, A. *Hellín en Textos Geográficos Antiguos (Facsímiles y Transcripciones)*; I.E.A. Don Juan Manuel: Albacete, Spain, 1996.
40. Panadero Moya, M. Noticia geográfica de la villa de Tobarra en la segunda mitad del siglo XVIII. *An. Cent. Asoc. UNED Albacete* **1979**, *1*, 201–216.
41. Brookfield, H.C. Intensification and Disintensification in Pacific Agriculture. A Theoretical Approach. *Pac. Viewp.* **1972**, *13*, 30–48. [CrossRef]
42. Stone, G.D.; Netting, R.M.; Stone, M.P. Seasonality, Labor Scheduling, and Agricultural Intensification in the Nigerian Savanna. *Am. Anthropol.* **1990**, *92*, 7–23. [CrossRef]
43. Boserup, E. *The Conditions of Agricultural Growth. The Economics of Agrarian Change under Population Pressure*; University of Chicago: Chicago, IL, USA, 1965.
44. Netting, R.M. *Smallholders, Householders: Farm Families and the Ecology of Intensive, Sustainable Agriculture*; Stanford University Press: Stanford, CA, SUA, 1993.
45. Morrison, K.D. The Intensification of Production: Archaeological Approaches. *J. Archaeol. Method Theory* **1994**, *1*, 111–159. [CrossRef]
46. Boserup, E. *Population and Technological Change: A Study of Long-Term Trends*; University of Chicago: Chicago, IL, USA, 1981.
47. Bermejo Tirado, J.; Grau Mira, I. *The Archaeology of Peasantry in Roman Spain*; De Gruyter: Berlín, Germany, 2022.
48. Whickam, C. *Framing the Early Middle Ages. Europe and the Mediterranean, 400–800*; Oxford University Press: Oxford, UK, 2005.

land

MDPI

Article

Adaptive Dynamics of Settlement Models in the Urban Landscape of Termez (Uzbekistan) from c. 300 BCE to c. 1400 CE

Enrique Ariño [1,*], Paula Uribe [2], Jorge Angás [3], Raquel Piqué [4], Rodrigo Portero [1], Verónica Martínez-Ferreras [5] and Josep M. Gurt [5]

1 Department of Prehistory, Ancient History and Archaeology, University of Salamanca, C/Cervantes s/n, 37002 Salamanca, Spain; rodrigoportero@usal.es
2 Department of Antiquity Sciences, University of Zaragoza, c. Corona de Aragón, 42 (Edificio Cervantes), 50009 Zaragoza, Spain; uribe@unizar.es
3 Department of Antiquity Sciences, ARAID—University of Zaragoza, c. Corona de Aragón, 42 (Edificio Cervantes), 50009 Zaragoza, Spain; j.angas@unizar.es
4 Department of Prehistory, Autonomous University of Barcelona, Edifici B Facultat de Filosofia i Lletres, 08193 Bellaterra, Spain; raquel.pique@uab.cat
5 ERAAUB, Department of Ancient History and Archaeology, University of Barcelona, Carrer de Montalegre, 6-8, 08001 Barcelona, Spain; vmartinez@ub.edu (V.M.-F.); jmgurt@ub.edu (J.M.G.)
* Correspondence: argil@usal.es

Abstract: The archaeological site of Ancient Termez is located in southern Uzbekistan. Despite the arid environment, the city benefited from its strategic position near two rivers, the Amu Darya and the Surkhan Darya. Its significance was mainly related to the expansion of trade routes connecting Eurasia. The city comprises several enclosures that attest long-term human-environment interactions. In order to identify the adaptive dynamics of the settlement models during an extended chronology covering the period from c. 300 BCE to c. 1220 CE (Greco-Bactrian/Yuezhi, Kushan, Kushano-Sasanian, and Islamic periods), a multidisciplinary study has been carried out, which includes: (1) archaeological excavations in several areas of the urban complex; (2) pedestrian surveying inside some enclosures and in the urban periphery; (3) an aerial survey based on high-resolution satellite imagery; (4) AMS dating of charcoal and bone samples; (5) archaeobotanical investigation through anthracological analysis; (6) zooarchaeological studies. The results point to variations in the development of the inhabited spaces, in which abandonment and occupation took place. The zooarchaeological and archaeobotanical data demonstrate the exploitation of natural resources in different environments (i.e., arid areas and irrigated land) and a certain evolution during the period considered.

Keywords: archaeological survey; remote sensing; Anthracology; Zooarchaeology; ancient Termez; central Asia; ancient period; medieval period

1. Introduction

Evolution of settlement models in the city of Termez (Uzbekistan) between c. 300 BCE and c. 1220 CE is studied through data obtained with archaeological methodology. The information from successive excavation seasons at the site (from the 1930s to the present) has been complemented with data obtained in pedestrian surveying. The overall interpretation of the different enclosures in the city of Termez has combined the data in the archaeological record with information provided by CORONA and WorldView3 satellite images. The systematic analysis of the satellite data has identified superficial traces (visible in the micro-relief or differential growth of vegetation) visible in the urban layout of the antique and medieval city. Research on urban construction, destruction, and rebuilding processes in Termez has been completed with a study of the settlement's natural environment by examining the remains of plants (charcoal) and animals (bones) found in the excavations. The research carried out aims to incorporate the techniques of landscape archaeology in the study of the urban complex of Termez. Remote sensing and surface surveys have been used to define the urban layout and the occupation chronologies of the different areas of the site. The results obtained with these techniques have been compared with the information provided by the archaeological excavations carried out at the site since 1936. A complete radiocarbon dating program has been used to obtain absolute dates in the use and reuse of the urban spaces. This has made it possible to address the dynamics of human occupation of the inhabited space between c. 300 BC and c. 1220 AD. The study also incorporates an analysis of the faunal remains and an important sample of charcoal recovered in the most recent excavation campaigns, in order to reconstruct the natural environment of Termez throughout the historical sequence. The final objective of the research is to determine how human activities and the natural environment were interrelated.

Termez was one of the most important urban centres of northern Bactria, a region of variable borders. It is generally accepted that during the Achaemenid Persian Empire (550–330 BC) its northern limit would have been marked by the course of the Amu Darya river, which would have delineated the border between the satrapies of Bactria and Sogdia. In the Seleucid period (323–238 BC), the border between the two satrapies would probably have shifted north of the Amu Darya, so that Bactria would have incorporated territories that, in the Achaemenid division, had been Sogdian. It is considered certain that at the time of the Greek kingdom (238–c. 140), Bactria would have incorporated into its dominions the region that today we conventionally call northern Bactria, which includes the lands north of the Amu Darya belonging to the republics of Uzbekistan, Tajikistan, and Turkmenistan. The northern boundary of the Greek kingdom would probably have been located in the gorge known as the Iron Gates, near Derbent (Uzbekistan). Little is known of the changes that could have been introduced by the nomadic Yuezhi Empire that ended the Greek kingdom of Bactria, but between the beginning of the 1st century AD and the middle of the 3rd century, the region was the central nucleus of the Kushan Empire, which far exceeded the limits of the ancient satrapy, especially to the east, as it included within its borders the city of Patna on the banks of the Ganges. The western part of the Kushan Empire—historical Bactria—was conquered by the Sassanian emperor Shapur I in military campaigns dated around 245–248. After the Islamic conquest of the region at the end of the 7th century, northern Bactria was included in the territory that Arab sources refer to as *Mawarannahr*, a term used to designate Transoxiana. Bactria is a geographical space with a dense urban occupation at least since the Achaemenid Empire. Classical sources refer to Bactria as the 'Land of a Thousand Cities' [1].

Old Termez was the predecessor of modern Termez. Both the old and new cities depend on the water in the Surkhan Darya, a tributary of the Amu Darya, which rises on the southern slopes of the Hisar mountain range (which reaches 4643 m in the summit of Khazret Sultan). In this way, irrigated crops can grow in an area characterised by its aridity. Old Termez was destroyed by Genghis Khan in 1220 [2] (p. 81) and [3]. The modern city is about ten kilometres to the south-east and was founded after the destruction of the old city. It was visited in the 1330s by Ibn Battuta, who described it as a prosperous town with numerous bazaars, irrigation canals, and gardens. Ruy González de Clavijo wrote about it in similar terms, after staying in the city in the autumn of 1404 [4] (pp. 8–11). Old Termez maintained a small population associated with craft workers until the 17th century [5] (pp. 438–439) and [6], but the place is now uninhabited.

The Termez archaeological site is at about 300 m above sea level in the Surkhan Darya Depression, which in turn is part of the Afghan-Tajik Depression (Figure 1). Some hills (Citadel, Kara Tepe, and Tchingiz Tepe) stand out on the plain with the ruins of the old city; these are the southernmost outcrops of the Cenozoic Baysuntau-Kugitangtau formation, a low mountain range (its highest peak is Khrebet Khaudat, 553 m) that delimits the Surkhan Darya basin on its western side [7] (pp. 92–93). The climate at Termez is characterised by a high evapotranspiration index (PETtho 1195.4) and marked aridity (Alu 0.12). The mean annual temperature is 16.5 °C. The warmest month is July (mean temperature 29.3 °C), and the coldest month is January (3 °C). The mean annual precipitation is 139 mm/y. Rainfall is distributed unequally, with most falling in spring, and varies greatly from year to year, from 78 mm/y to 223.4 mm/y. The lower valley of the Surkhan Darya is in the Irano-Turanian biogeographic region, with conditions suitable for low and not very dense xerophytic vegetation. The sudden temperature changes lead to freeze-thaw shattering of rocks and surface deposits. Winds are mainly from the south-east, with seasonal variations. South winds are most common in January, while east winds dominate in November [7]. The vegetation landscape of the region has been traditionally conditioned by human action. The lower Surkhan Darya valley is a traditional irrigated area, probably since prehistoric times. The Sherabad plain was not irrigated before the Soviet period. This area was put under irrigation in 1971 in order to develop large-scale cotton cultivation. The Sherabad plain is supplied with water by a large canal (the Sherabad canal) which takes water from the Surkhan Darya. Irrigation allows fruit tree crops, vineyards, and vegetables to thrive in the irrigated areas today, in addition to cotton, which still dominates large areas of cultivation [8,9]. The urban area of Termez is currently devoid of any tree cover, including the riparian vegetation that could thrive on the banks of the Amu Darya.

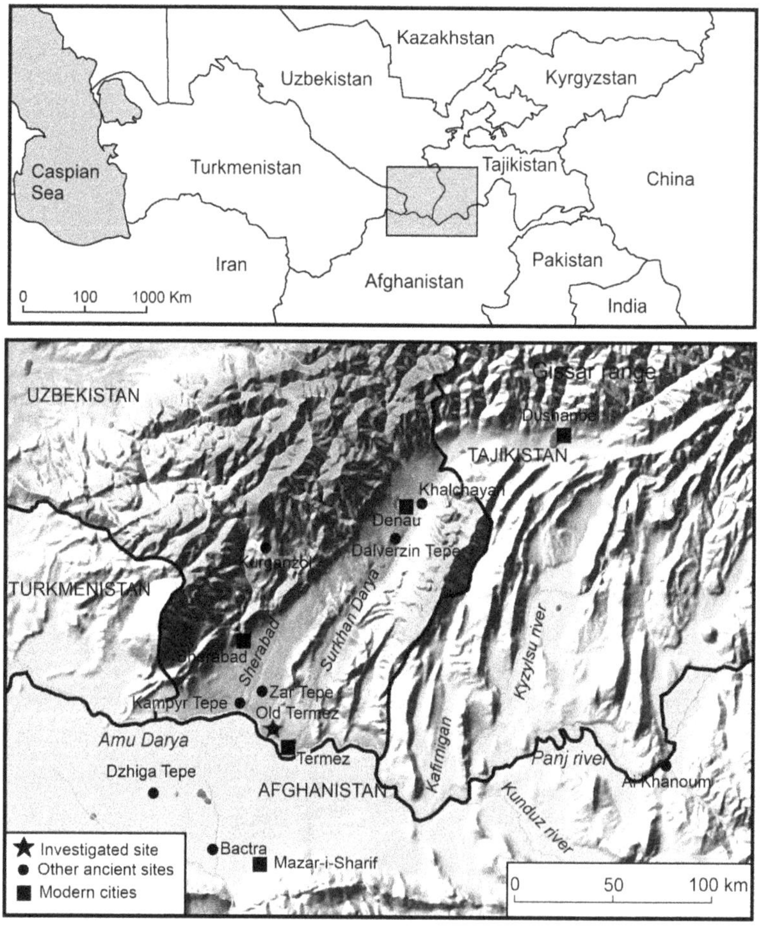

Figure 1. Map of Bactria with the location of Termez. Cartographic base 123RF (copyright: https://es.123rf.com/profile_mschmeling), accessed on 12 January 2022.

Wind and fluvial erosion are the two agents that have most affected the site's architectural structures and archaeological strata. Fluvial erosion has particularly damaged the Citadel and Tchingiz Tepe enclosures in the westernmost part of the urban site. The first plans of the site in 1936 show that most of the southern part of the Citadel was missing at that time, although its rectangular shape was still recognisable [10] (pp. 126–130). Since then, a large part of the Citadel has disappeared because of erosion. The comparison of the CORONA satellite images (DZB00403800056H011001, 20 October 1964) and modern pictures on Google Earth is indicative of the impact of fluvial erosion in this part of the site in the last fifty years [11] (Figure 2). Wind and rain have also contributed to the loss of archaeological structures in ancient Termez. The impact of those agents can be observed by comparing the state of buildings and walls in images taken by Masson's archaeological expedition in 1936 [12] with their present appearance. Aeolian deposits are found in several areas of the site, especially in Tchingiz Tepe [13]. A geoarchaeological survey in Tchingiz Tepe showed that the surface deposits, which are over two metres thick in some places (2 to 2.8 m), have formed in the last 2000 years through the action of two natural agents, gravity and wind, and an anthropic process (construction, abandonment, and rebuilding of dwellings) [7].

Figure 2. Comparison between satellite images of CORONA (DS1012-1039DA1963 20/10/1964) and WV3 (January 2017 012085673020).

The Termez archaeological complex occupies an area of about 550 ha. It comprises four fortified enclosures (Citadel, Tchingiz Tepe, Shahristan or Low Town, and Rabad). There was also human occupation in the suburbs, especially in the north (Potters' Quarter) and at the site's eastern end, with a palace complex dated to about the 11th century, which has now disappeared (Figure 3). The oldest part of the settlement was in the Citadel (*Phrourion*), where archaeological levels are dated to the late 4th–early 3rd centuries BCE. These levels are associated with colonisation by Greek populations who were present in the region following the conquests of Alexander the Great [14,15]. Although the Citadel must have been founded as a fortified settlement, the original Greek wall has not been located. In contrast, the wall from the Kushan period, made with adobe, and the Islamic wall (of brick) are known; the latter was attached to the Kushan wall on its outer face [2] (pp. 83–97). Two large adobe buildings located 600 m north-west of the Citadel (Buildings A and B) have been interpreted as part of a cult complex. Excavations in this area have suggested an age in the Greco-Bactrian period, with some remodelling in the Kushan period [16] (pp. 190–200) and [17] (pp. 34–36).

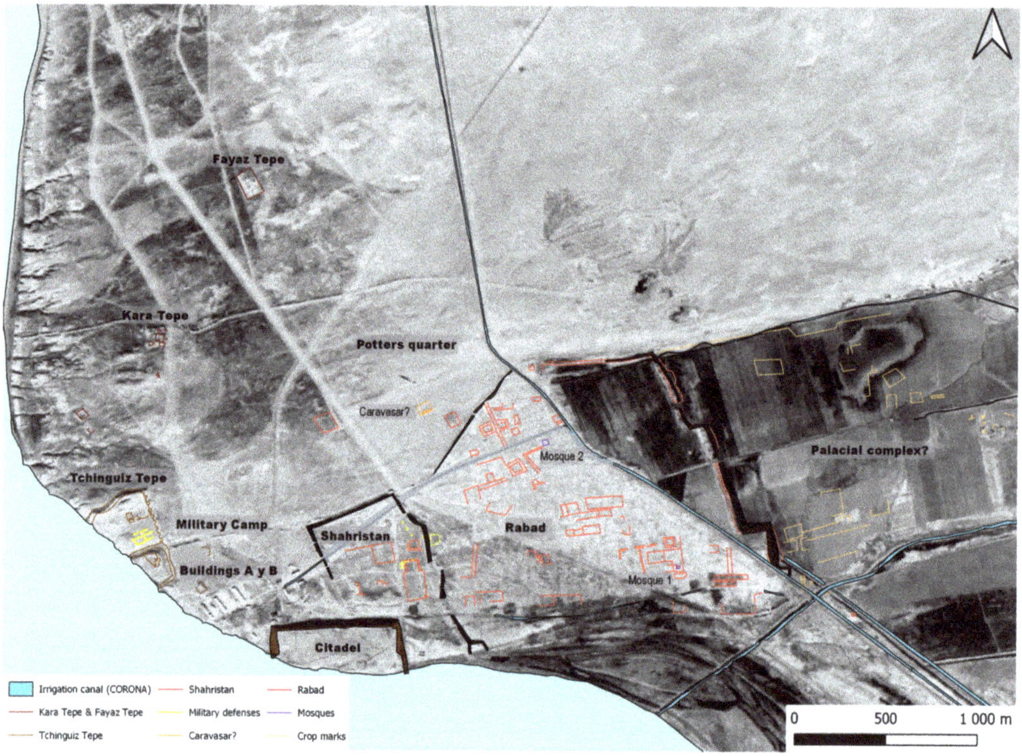

Figure 3. Image of Termez on the CORONA satellite (DS1029-2102DF114 09/02/1966) with the identification of the urban sectors mentioned in the text and the anomalies that may correspond to ancient urban structures.

Tchingiz Tepe's enclosure is located to the north of the cult complex. It has a trapezoidal shape, delimited by a wall of which only three sides have been preserved because of the action of the Amu Darya that has eroded the western wall. Tchingiz Tepe is the enclosure whose occupation sequence is best-known thanks to a full excavation programme accompanied by radiocarbon dating. The construction of the walls would have started in about the mid-second century BCE [13]. In the Kushan period, Tchingiz Tepe was the location of a Buddhist monastery [18]. The enclosure was no longer occupied after the 5th century, and it is the only fortified area in Termez with no evidence of medieval dwellings or activity.

The third walled enclosure at Termez is the Shahristan or Low Town. This district is located north of the Citadel and conserves the ruins of a thick wall of rammed earth (*pakhsa*). It is generally assumed that the wall was built in the Kushan period and rebuilt with the same technique in the Islamic period, although this hypothesis is based on field observation and the study of pottery in the rammed earth forming the wall [19,20]. Several parts of the Low Town have been excavated, but the results of this fieldwork have not been published completely. The only archaeological evidence supporting a Kushan presence in the Shahristan are some structures excavated in the bedrock, interpreted as Buddhist monastic cells, remains of a platform identified as part of a *stupa*, and a pillar capital decorated with acanthus leaves with an image of Buddha on each side [16] (pp. 189–190) [17] (pp. 50–55). The medieval Islamic sequence is much better documented and characterised due to the discovery of a group of houses made of stone and brick, many of them with blind wells used to drain wastewater. The foundation of these dwellings has been dated to about the 11th century, whereas their end is thought to be associated with the destruction of the city by Genghis Khan in 1220 [16] (188–189 and 210), Ref. [17] (pp. 44–45) and [21] (pp. 155–156). However, the Mongol conquest did not mark the definitive end of the occupation of this urban area, as several kilns used to fire pottery have been detected above the Islamic houses in the western sector of the Shahristan [22].

The Rabad, or Islamic City, is the largest of the walled enclosures at Termez, and it gives an idea of the vitality of the medieval city. The wall was built with *pakhsa* and was connected with the eastern part of the Shahristan wall. It enclosed a polygon with sides that contained defensive towers. Excavations have detected the ruins of at least two mosques. Mosque 1, in the south-east of Rabad, was founded in the 11th century and remodelled in the 12th–13th centuries. Mosque 2 was in the north of the Islamic city; its oldest phase is also dated to the 11th century, while it was remodelled on at least two occasions [23]. These remains documented by excavations in Islamic Termez are completed by a caravanserai, dated to the 12th century [24]. Outside the Rabad, at the site's eastern end, another walled enclosure held further important medieval remains. The TAKE expedition (Termez Archaeological Complex Expedition), directed by Masson in 1936–1937, found a palatial complex dated to the 11th century in this part of the city. No ruins of those buildings have been preserved because they were destroyed during the excavation of irrigation canals around Termez in the second half of the 20th century [4] (pp. 40–53).

Although the walled enclosures define the urban area, the city was also surrounded by its suburbs. On the plain between the enclosures of Tchingiz Tepe and Shahristan (Military Camp), our excavations have detected occupation levels whose oldest phase has been dated by radiocarbon to the Greek sequence (late 4th–early 3rd centuries BCE). The last phase in this sector is characterised by dwellings associated with the Islamic occupation. It does not display any evidence of destruction related to the conquest in 1220, although it should be borne in mind that only a small part of this area has been excavated [25,26] (p. 544).

To the north of the Shahristan and the Rabad, outside the walls, several potters' workshops have been documented in an inhabited area dated to the medieval period. For this reason, it is known as the Potters' Quarter [27]. Although this craft activity is undoubtedly well documented, the occupation sequence and urban structure in this suburb need to be defined more precisely. Finally, in addition to all these districts in the city, around Termez, the Buddhist monasteries of Kara Tepe [28,29], Fayaz Tepe [30], and Zurmala [31] (pp. 176–180), are all dated to the Kushan period.

In short, the historic layout of old Termez has been generally defined by archaeological fieldwork in several parts of the site, especially from detailed records of some of its sequences. Archaeological strata have been used to assign dates to the occupation of the walled enclosures, although the oldest phases in the different parts of the city are still largely undetermined. The functionality of the sites and the management of resources are equally pending study to a large extent.

2. Objectives, Materials and Methods

2.1. The Urban Dynamics in the Historical Sequence

One of the research objectives has been to analyse the dynamics of the construction, use, reuse, and abandonment of the different places in the city throughout its historical sequence. The study was designed to determine whether the urban expansion from the original centre of the Citadel really implied an increase in the inhabited area during the time in which the city was active (which would be indirect evidence of population growth) or if, on the contrary, the colonisation of new enclosures took place at the same time as the preceding inhabited areas were abandoned. The latter may have resulted from difficulties in maintaining the settlement areas due to the formation of natural aeolian deposits or anthropic causes (ruin and collapse of adobe architecture). Although archaeological fieldwork has been carried out in Termez since 1936, the publications are limited, and not all settlements can currently be identified in situ. Only some excavations in specific parts of the site have provided accurate stratigraphic records, while the date of the foundation of some enclosures still needs to be defined more precisely. Similarly, the date of the abandonment of the different parts of the urban site must also be better determined. Due to the large size of the inhabited area, intensive surveying and remote sensing were included in the study to combine the data provided by the archaeological excavations and surveys (surface finds of pottery) with the urban layout seen in satellite images.

2.1.1. Intensive Surveying

Intensive surveying has been carried out in three of the least-known areas at the site. The Shahristan has been surveyed in two grids: Transect Sha1 (20 × 30 m) and Transect Sha2 (15 × 60 m). A single area of 15 by 45 m was surveyed in the Rabad (Rabad1). The transect KBTK1 in the Potters' Quarter covered an area of 30 × 60 m (Figure 4). This work was performed with surveyors a metre apart and, therefore, it can be considered that the whole surface of the transects was examined. In each area, all the potsherds visible on the surface were collected. These remains were georeferenced by sub-dividing each transect into squares 5 × 5 m in size. The visibility conditions were good in all the surveyed areas, with the squares in the Shahristan completely free of vegetation, whereas some areas in the Rabad were covered by light shrub vegetation that did not affect the apparent visibility. The wind acts on the sand, covering the upper part of the soil at the site, removing it, and building up deposits. However, the accumulations of sand are not deep enough to affect visibility. Many of the potsherds were affected by the erosional action of the wind, indicating that they were exposed on the surface for the time needed to generate those alterations.

The pottery assemblage gathered in the surface sampling was classified into classes or families depending on their formal characteristics. Greek pottery represents the oldest occupation in the sequence (c. 300–c. 140 BCE). It is defined by at least one of its surfaces being covered by a black slip. This is well documented in Termez [5,6,14,26] and at sites in the vicinity [32–37]. A new production, known as grey clay ware, appears in the ceramic record from the mid-second century BCE onwards. This pottery characterises the occupation associated with nomadic Yuezhi or Tocharian populations (also known as Early Kushan) as documented at Termez [5,6] and Dalverzin Tepe [35]. However, Mirzakultepe [38], in the immediate surroundings of Termez, is the site of reference for the characterisation and dating of this ware. The change of era and first decades of the 1st century CE (c. 50) are usually linked to the beginning of the ceramic production characteristic of the central Kushan period.

Figure 4. Image of Termez on the WV3 (January 2017 012085673020) satellite showing the location of the survey transects and radiocarbon sampling areas. 1: Transect Shahristan 1 (Sha1); 2: Transect Shahristan 2 (Sha2); 3: Transect Rabad (Rabad1); Transect Ceramics Quarter (KBTK1).

The most notable novelty is the use of red slips to cover one or both surfaces of the recipients. Although the deep red slip is the most typical cover of Kushan ware, other vessels were covered in orange, brown, or black (Figure 5). Kushan ware has been found at Termez [5,6,39] and places near the site, like Akkurgan [40], Dalverzin Tepe [35], Kampyr Tepe [41,42], Dzhiga Tepe [36], Aktepe, Shodmon Kala, and other settlements in the lower Kofarnihon valley [43,44]. In the mid-third century, the Kushan kingdom was subjugated by the Sassanian Empire. However, the pottery in the Sassanian period maintained the main features of the Kushan products, especially the red slips [45]; therefore, both sequences are grouped in the classification of sherds found in the surface surveying since stratigraphic data is not available to establish the sequence.

At the end of the 7th century, the region was conquered by the Arabs, resulting in a complete change in the formal and decorative repertory of the pottery [5,6,22,27]. Glazes are frequent from the 9th and 10th centuries, especially in the case of tableware. They were often glazed with a single colour (white, green, yellow, or turquoise) but complex compositions with figurative or geometric motifs in different contrasting colours are quite common. The *sgrafiatto* technique consisted of scraping the surface with an awl before firing the vessel; it was combined with figurative or geometric decoration by applying layers of colour over the surface. Unglazed pottery is also common in Islamic pottery assemblage, usually decorated with various techniques, such as impression, incision, plastic applications, stamps, and moulds (Figure 6). Pots covered with a black slip or coating are also recognisable and exclusive to the Islamic period. This slip normally covers the interior of the vessel and part of the outer surface in a pattern of drip lines from the rim to the base. Other characteristic productions of the Islamic period are spherical-conical vessels, often made in moulds and usually decorated with complex patterns (sometimes with Arabic letters). They have thick walls and a hole with a very small diameter which would have been used to control the pouring out of the product they contained. Despite being very common in the archaeological record of ancient Termez, their purpose is still unknown.

Figure 5. Ceramic productions recovered in surface surveying. Ancient period. 1: Red-figure ware (Greek) (Sha1); 2: White-slipped ware (Greek) (Sha1); 3: Black-slipped tableware (Greek) (KBTK1); 4: Grey pottery (Yue-zhi) (Sha1); 5: Slipped tableware (Kushan/Sassanian) (Sha1); 6: Common ware (Kushan/Sassanian) (Sha1).

Common ware and coarse cooking ware appear in all the historical sequences. These pottery types (the most abundant in the surface record) are generally regarded as non-specific and cannot be used to determine the occupation period in the surveyed area. Some of the sherds belonging to those products can be assigned to the Kushan and Kushano-Sasanian sequence if their degree of conservation is satisfactory enough to identify a characteristic shape, which is well-known thanks to typological classifications.

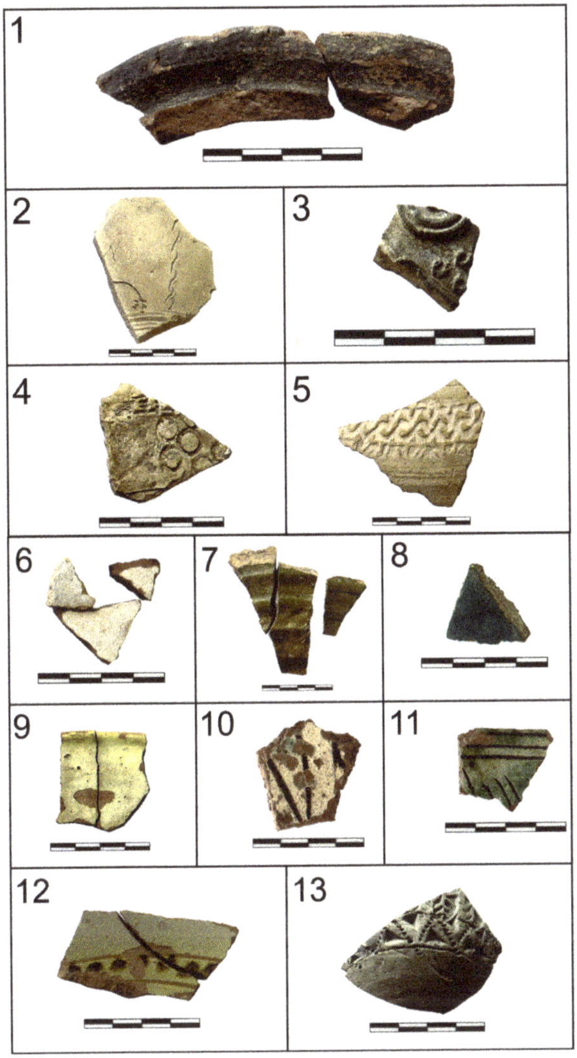

Figure 6. Ceramic productions recovered in surface surveying. Medieval period. 1: Common cooking ware, black slip partially covering the surface (Islamic) (Sha1); 2: Unglazed ware, comb-impressed/comb-incised decoration (Islamic) (Rabad1); 3: Unglazed ware, applied decoration (Islamic) (Sha1); 4: Unglazed ware, stamped relief decoration (Islamic) (Rabad1).; 5: Unglazed ware, moulded relief decoration (Islamic) (Sha2); 6: Glazed white ware (Islamic) (Rabad1); 7: Glazed green monochrome ware (Islamic) (Sha1); 8: Glazed turquoise monochrome ware (Islamic) (KBTK1); 9: Glazed yellow monochrome ware (Islamic) (Sha2); 10: Underglaze painted ware/slip painted (Islamic) (KBTK1); 11: Splashed *sgraffiato* ware (Islamic) (KBTK1); 12: Figured/geometric decoration glazed ware (Islamic) (Sha1); 13: Sphero-conical vessel (Islamic) (Sha1).

2.1.2. Radiocarbon Data

Samples of organic materials have been collected for radiocarbon dating in order to acquire new information about some of the less-known urban structures. A sample of bone was taken from the brick wall in the Citadel, and a second bone sample from the rammed earth wall in the Shahristan. Other samples were collected in the Islamic pottery kilns in the western sector of the Shahristan. The radiocarbon samples were completed with others obtained in the systematic archaeological excavations in Tchingiz Tepe, the Military Camp and the Potters' Quarter. The radiocarbon results, which mark the dates of the initial occupation and the abandonment of each area in Termez, are given in Table A1. All the dates were calibrated with OxCal v. 4.4.4 [46] and the IntCal20 calibration curve [47].

2.1.3. Remote Sensing

Satellite programmes with a chronological difference of nearly 50 years (1967–2016), including sets of panchromatic and multispectral images, were used to study the landscape by remote sensing. The full analysis of all the data will require further post-processing and laboratory research in the future. However, the preliminary results are useful to identify the most appropriate sectors to carry out future fieldwork [11,48]. The comparative analysis of the data provides significant information about the area of study, which has suffered major changes in the last fifty years due to both natural and anthropic processes. The study was carried out with high-resolution space images in the CORONA programme (1967) and the commercial WorldView3 satellite images (from 2016–2017) by interpreting the images (seven of CORONA and two of WV3) with the best visualisation of the remains. We have mainly worked with the images DZB00403800056H011001 in the CORONA programme and two in the WV3 (16AUG29063625-M2AS R1C2-012085673010 01 P001 from 2016 and 17JAN1305145-M2AS R1C2-012085673020 from 2017). Other CORONA images were examined to improve the visualisation of specific areas. The old city of Termez is in a natural and political border zone which is still of strategic interest, and therefore systems that might provide better resolution (such as drones) cannot be used. There is no automatic method for the pre-treatment and post-processing of the images and, therefore, the different processes used with the two satellite image programmes are described.

The first step is to determine both programme's world covers to purchase the images. Among the CORONA programmes, we selected images dating from 1960, known as DISP (Declassified Intelligence Satellite Photographs) [49], representing one of the oldest digitalised periods with a satellite sensor. In recent years, CORONA has been used by archaeologists in the Near East [50–52] because the various declassifications (1996, 2002, and 2013) offer a practically global cover with high-quality images and stereoscopy in some missions. In many cases, the images show the situation before the urban and agricultural changes and construction of infrastructures that took place from the late 1970s; therefore, they provide information of exceptional quality. In Termez, the images show the state of the landscape before the Afghan–Russian war and the anthropic changes that occurred in more recent times. After their purchase, the CORONA images were subject to a single treatment: georeferencing with the UTM coordinate system, Zone 42 (EPSG: 32642) and the WGS 84 datum, by referring to points in common with the WV3 images. The spectral resolution of the CORONA images varies, with different filters for the reduction of contrast [53], but it is always within the visible range for their panchromatic condition. As a result, it was not possible to generate neo-channels, which was possible with the WV3 images.

By complementing this CORONA programme with WV3 images, it was possible to extend the temporal framework to 2016–2017, increase the spatial resolution, and obtain multi-spectral information, which is not possible with the CORONA programme. The WV3 images offer greater spatial and spectral resolution, with panchromatic images and eight multi-spectral bands, with a pixel size of 0.31×1.24 m. The pre-treatment of these images is more complicated than in the case of the CORONA series. Once acquired, it is necessary to conduct an atmospheric correction, georeference them (UTM WGS 84, Zone 42, EPSG: 32642), and obtain a mosaic with a final pan-sharpening.

The atmospheric correction was achieved by rectification of the reflectivity values through a comparative statistical analysis of the results of the terrestrial cover with values of a Landsat-8 OLI image of a similar date. Homogeneous areas, such as bare land, pastures, shrubland, crops, water, and asphalt, were selected. The extraction of the mean value and the standard deviation of the WV3 pixels (the latter to verify the disparity in the values in a Landsat pixel) were performed with QGIS software. This process generated a database with information on 30 control points in the different land cover in the images.

The reflectivity values nearest to the true-earth values were identified with: (1) the Shapiro–Wilk test to determine the normality in the distribution of the variables; (2) ANOVA analysis; (3) the R2 determination coefficient; and (4) the root square mean error (RSME). The spectral signatures of each point were also analysed visually.

After obtaining the reflectivity image closest to reality, a pan-sharpening process was performed by combining the multi-spectral image with the panchromatic image obtained simultaneously to improve the visual interpretation of the photo-interpreted elements. The methods used were: Brovey Transform, Subtractive Resolution Merge (SRM), High Pass Filter Transformation (HPF), substitution based on the use of the ACP transformation, and substitution based on the HIS transformation [54]. These were all carried out with the Erdas Imagine 2015 software. Finally, the HPF method was selected to carry out the secondary magnitudes. Despite not being the image with the best colour, it achieves a very good spectral and spatial resolution and is a method with well-balanced results.

With the high-resolution spatial images from January and August 2016–2017, the pan-sharpening process was used to create secondary magnitudes. Neo-channels were created with the Erdas Imagine 2015 software for the most common indices, such as NDVI, SAVI and SR, and the ACP and edge detection techniques. The other indices (GRVI, DVI, RDVI, GDVI, GNDVI, IRG, MTVI, RDRER, GARI, OSAVI, GEMI, MSR, and RGRI) were obtained with the ArcMap 10.5 raster calculator to combine the bands resulting from the pan-sharpening process. The treatment of these indices has complemented the data identified with the CORONA programme. Finally, the structures visible in each index were drawn following the indicators of soil and crop marks.

2.2. Vegetal Landscape

Our excavations in Tchingiz Tepe and the Potters' Quarter have retrieved a significant number of charcoal fragments, which were examined to study the vegetal landscape in Termez and its surroundings at particular times in its historical sequence. A total of 220 fragments have been analysed from Tchingiz Tepe. These were selected by hand during the excavation of several structures in the sectors RB2 (SU 6, collapse level), RA (SUs 23 and 36, collapse levels), and RF (SU 26, the fuel used in a pottery kiln). The 57 samples from the Potters' Quarter were recovered during the excavation of Workshop 11 (the foundation level of the workshop on the bedrock, Kiln 1, and the waste dump that filled a *tannur* in one of the dwelling areas). The taxonomic classification of the archaeological charcoal was achieved by studying its microanatomy and comparing it with a reference collection. The anatomical analysis was performed with a reflected light microscope equipped with several lenses allowing 40× to 400× magnification. Identification of the taxon was based on observations of three sections of the wood (transversal, radial longitudinal, and tangential longitudinal) obtained by manually splitting the charcoal fragment. The keys used in the identification were published by Schweingruber [55], while the comparisons were made with the charred wood reference collection from the Archaeological Analysis Service at the Autonomous University of Barcelona. The charcoal corresponds to two sequences. The remains obtained in Tchingiz Tepe belong to Antiquity (2nd–5th centuries CE) and those from the Potters' Quarter to the Middle Ages (8th–10th centuries). These ages are based on the study of the stratigraphic record in those areas and supported by radiocarbon dates.

2.3. Fauna

Information about the fauna in old Termez has been obtained by studying skeletal remains found in the excavations conducted in six areas of the site: Tchingiz Tepe (sectors RA, RB1, RB2, RB3, and RC; Military Camp (sectors AC1 and AC2); Shahristan (Workshop 5); pottery kiln in Kara Tepe (Workshop 6); and Potters' Quarter (Workshop 11). The archaeozoological study has been carried out by grouping the finds in cultural sequences. The first is the Greek and Yuezhi periods (c. 300 BCE to 50 CE); the second covers nearly the whole Kushan and Kushano-Sasanian sequence (c. 150–600 CE), although data from the earliest phase (c. 50–150 CE) is missing. Practically all the information for the medieval period corresponds to the time after the Islamic conquest of the region (c. 800–1400).

The skeletal remains were studied with standard archaeozoology and taphonomy methodologies [56,57]. The anatomical and taxonomic identification of the remains was based on different osteological atlases for domestic and wild animals [58–61] and the reference collections of the Aranzadi Science Society in San Sebastián, Spain, and the Archaeoscience Laboratory in Lisbon. Several guidebooks on the fauna of Central Asia were also used to identify the species [62–65]. Specific studies were consulted to differentiate between species in a family or group of taxa [66–75]. Similarly, osteometric criteria were considered to discriminate between the domestic species represented [70,76–79]. To quantify the representation of skeletal elements, the Number of Remains (NR), Number of Identified Specimens (NISP), and Minimum Number of Individuals (MNI) were calculated by taking into account the most numerous anatomical part, the laterality of the bone, and the age and sex of the animal [56,57,80,81]. When the remains could not be identified to species or family level, they were grouped into three size categories, large (>200 kg), medium (100–200 kg), and small (<100 kg). The full data obtained in the faunal study at Termez can be consulted in Portero et al. [82,83].

3. Results

3.1. *Changes in the Use of Inhabited Areas*

3.1.1. Chronology and Occupation of Urban Spaces

The 11,899 potsherds found by surface surveying correspond to the following samples: Shahristan 1 (Sha1, n = 4674); Shahristan 2 (Sha2, n = 4014); Rabad (Rabad1, n = 1246); and Potters' Quarter (KBTK1, n = 1684). Table A2 quantifies the sherds in each of the transects of the survey and also their cultural ascription in order to allow a comparative analysis between the sequences. Common ware is the most frequent type in all the samples but cannot be used to identify the occupation as, in most cases, it is impossible to determine its cultural ascription.

Greek pottery appears in small quantities, only in Sha1 and KBTK1, and it can be considered off-site material. Human presence in the Seleucid and Greco-Bactrian periods is documented in the Citadel [14] and Tchingiz Tepe [13]. Therefore, the presence of Greek pottery in the Shahristan and Potters' Quarter may be attributed to activity in places close to the areas that were really inhabited. The small sample of Greek pottery includes a fragment with red figures and one with a white slip in its interior. The occupation sequence in the Tchingiz Tepe enclosure is well documented by a complete excavation programme and ^{14}C dates. At least three dates support human activity in the enclosure in the 2nd and 1st centuries BCE (Table A1: Tchingiz Tepe, Sector RB1, SU 5; Tchingiz Tepe, Sector RB2, SU10; Tchingiz Tepe, Sector RB2, SU11). However, the time of construction of the wall is still uncertain since the calibrated dates cannot discriminate between the time of the Greek kingdom of Bactria, or the Yuezhi period.

The surface survey results in the Shahristan indicate that the human occupation there can be dated to the Kushan period. Pottery associated with the Kushan and Kushano-Sasanian sequence is abundant in both sectors of the survey, especially in Sha2 (n = 382, 9.52% of the total), where it even outnumbers Islamic pottery (n = 346, 8.62%). In Sha1, Islamic pottery is more abundant (n = 655, 14.01%), but the number of sherds supporting the older period is still significant (n = 167, 3.57%). Excavations in the Shahristan have found remains consistent with a Kushan occupation [16] (pp. 189–190) and [17] (pp. 50–55). A radiocarbon date on a bone taken from the western part of the Shahristan wall corresponds to the time between the change of era and the early 2nd century CE (Table A1: Shahristan, Wall), which is further evidence for human occupation in that enclosure starting in the Kushan period. Moreover, data in other surveyed areas suggest that human activity expanded from the original centre in the Citadel over the whole urban area. Both the Rabad and the Potters' Quarter have provided a large number of potsherds dated to the Kushan and Kushano-Sasanian periods (n = 67, 4.39% in the Rabad; n = 39, 2.32% in the Potters' Quarter). In fact, the percentages of this type in the Rabad survey are higher than those in Sha1, so a Kushan population in that part of Termez cannot be ruled out even if no archaeological structure of this period has been detected to date through excavations [23,24]. The suburban area known as the Military Camp, located between Tchingiz Tepe and the Shahristan, was also inhabited in the Kushan period. Archaeological excavations have recovered pottery consistent with a stable occupation and determined radiocarbon dates that frame the oldest occupation phase in the 1st century CE, although an earlier occupation cannot be completely ruled out (Table A1: Military Camp, Sector AC1, SU12; Military Camp, Sector AC2, SU20; Military Camp, Sector AC2, SU33 [25] and [26] (p. 544).

The surface survey shows that the Islamic occupation was intense in all sectors, which is consistent with the excavation data. Two mosques and a caravanserai have been documented in the Rabad [23,24]. Similarly, the Potters' Quarter was used intensely in the Islamic period, with both potters' workshops and dwellings (Table A1: Potters' Quarter, Workshop 2, Kiln 2; Potters' Quarter, Workshop 11, *Tannur*; Potters' Quarter, Workshop 11, Kiln 1, SU1; Potters' Quarter, Workshop 11, Kiln 1, SU1; Potters' Quarter, Workshop 11, Kiln 3, SU21; Potters' Quarter, Workshop 11, SU32) [22,27,84]. The Shahristan was used for dwellings in the Islamic period [22,27]. In the Military Camp, the absolute dates and the stratigraphic record support an Islamic occupation (Table A1: Military Camp, Sector AC2, SU12; Military Camp, Sector AC2, SU18). Only Tchingiz Tepe seems to have been uninhabited in that period. Radiocarbon dates for Tchingiz Tepe indicate continuous occupation from its origin in the Greek or Yuezhi period (Table A1: Tchingiz Tepe, Sector RB3, SU2; Tchingiz Tepe, Sector RB3, SU6; Tchingiz Tepe, Sector RC, SU10; Tchingiz Tepe, Sector RF, SU20) but the most recent dates suggest that human settlement in the area did not continue after the mid-fifth century (Table A1: Tchingiz Tepe, Sector RA, SU17; Tchingiz Tepe, Sector RC, SU2; Tchingiz Tepe, Sector RC, SU3; Tchingiz Tepe, Sector RC, SU4; Tchingiz Tepe, Sector RC, SU5). The stratigraphic record in Tchingiz Tepe does not include a single level corresponding to the Islamic period [13]. The destruction of the city by the Mongol army of Genghis Khan affects the whole urban space. A ^{14}C date for the Citadel wall seems to be associated with building work or at least a repair in the late 11th or early 12th centuries (Table A1: Citadel/Tower 2). In the Shahristan, the pottery kilns have been dated to the late 14th or early 15th century (Table A1: Shahristan, Workshop 5, Kiln 3); however, this sequence seems to be residual as the potters' workshops are located in what had been a dwelling, which can be interpreted as evidence of the loss of the residential use of that part of the city.

3.1.2. Town Planning

Archaeological levels corresponding to the oldest phases in the Citadel are covered by thick medieval deposits, consisting of several metres in some areas. In this situation, the detection of evidence of structures belonging to the Greek, Kushan, and Sassanian periods through the analysis of satellite images is unlikely; however, it was able to provide important information about the Islamic occupation phase. An exception is the Citadel, where no type of urban structure appears, not even associated with its final occupation in the early 13th century. The 2017 image only reveals traces related to the use of the area for military training (Figure 3).

The information obtained by excavations, intensive surveying, and radiocarbon dating demonstrates that the Shahristan was occupied from the Kushan period to at least the 14th–15th centuries; therefore, the structures detected by remote sensing may belong to any of these periods. The study has detected traces that can be associated with the ruins of a structure of unusual size and characteristics (Figure 3). It is large and rectangular (206 × 111 m) and includes a rectangular head or annexe on its northern side (77 × 62 m). It is perfectly visible in the CORONA images 114_d/1966; 114 c/1966; 004c/1965; and 001c/1975. However, in the modern WV3 images (2016–2017), only the southern side of the structure can be observed in the January GARI vegetation index. The differences between the two satellite images show the importance of using multi-temporal data in remote sensing analysis.

It is hard to imagine the function and chronology of the structure visible in the CORONA image, but its shape, size, and apparent lack of internal divisions suggest that it might be interpreted as a religious or palatial space, perhaps associated with the Kushan period, a time when large monumental constructions with regular floor plans were built in Bactria. Some possible parallels are the Temples B and C at Surkh Kotal [85], the so-called temple of the Dioscuri in Dilberdzhin [86], and the palatial enclosure at Khalchayan [87]; the degree to which the structure in the Shahristan resembles the latter is much larger. An attribution to the Kushan period is also supported by the fact that the building does not seem to correspond to the typologies of the madrasahs or mosques detected in other parts of Termez associated with the Islamic occupation.

Many anomalies, visible in the old CORONA photographs (especially in the February 1966 image) and the more recent WV3 photos, are observed inside the Rabad enclosure. One that appears in the northern part stands out; in this area, there is now a small lake that formed barely fifty years ago, and which hinders archaeological exploration. The CORONA images provide information about how this northern part of the Rabad may have been organised. The 1966 image shows a large street in a north–south alignment. Two symmetrical spaces 50×55 m in size are associated with the middle part of the street. These two rectangular areas are identified as a topographic anomaly formed by the collapse of the perimeter walls. The relief that marks their position in the terrain is clearly visible in the 1964, 1966, and, above all, in the July 1975 photographs. The interpretation of these structures is difficult, although they would not correspond to propylaea, like those at Ai Khanoum, as the Termez structures are much larger (the structures at Ai Khanoum are 7×9 m in size) [88]. What is clear is that this street and the rectangular shapes annexed form a monumental complex around which the whole urban layout of the north sector of the Rabad is organised.

Judging by the traces visible in the satellite images, the layout of the Rabad was organised on a regular premeditated plan, although with variations in the directions of the streets. In the south, the alignment of the streets and buildings differs from those around the large street that articulates the north. The study of anomalies allows identifying the internal differences that might correspond to functional divisions. For example, in the north-east of the Rabad, some structures might be identified as large buildings with complex floor plans, probably residential structures. In contrast, the ruins in the south-west of the urban space are simpler and smaller, which might be indicative of an artisans' quarter in that sector. It is likely that the regular town planning discernible in satellite images corresponds to the Islamic occupations. Previous excavations in the Rabad identified two mosques, one of them built next to a dwelling [23], and they are visible in the satellite photographs. The southern mosque is perfectly visible in the WV3 panchromatic image from 2017 but not in the CORONA. It appears oriented in the same direction as the other traces in that part of the enclosure; in contrast, the mosque in the north is inserted in the regular layout around the monumental street in that area.

A caravanserai dated to the 12th century was documented in the Rabad [24] and excavated in 1986. This construction is no longer visible on the surface and does not appear in CORONA images. In contrast, the WV3 image exhibits some traces in the ground surface that may correspond to the ruins of this building. They are located north of the Rabad, outside the enclosure, near the north access. The palace to the east of the Rabad, excavated by the TAKE expedition in 1936 and 1937 [12], does not appear in the satellite images, although the area has been significantly altered since that time by irrigation channels for farming.

The TAKE expedition, directed by Masson in 1936–1937, excavated the palatial complex outside the Rabad, on the eastern side of the site, and dated it to the 11th century. This is now an area of crops, where the vegetation indices obtained with the WV3 images ought to detect traces of the ruins. However, the structures cannot be identified, perhaps because of the natural or anthropic sediment deposition [12].

3.2. Vegetal Landscape

The anthracological sample corresponding to the Kushan and Kushano-Sasanian sequence (2nd–5th centuries CE) comprises 220 pieces of charcoal. Their study reveals a predominance of taxa that thrive in riparian woodland and wetlands (Table 1, Figure 7). Trees and shrubs associated with riverbank vegetation are represented by willow and/or poplar (Salicaceae, n = 69) and tamarisk (*Tamarix* sp., n = 13). These species can be found in alluvial plains and in river valleys in Central Asia known as *tugai* [89,90]. The oleaster (*Elaeagnus* sp., n = 26) is native to the central Asian steppes and also flourishes in riparian sites, although it also tolerates moderately dry ground. Oleaster fruit can be consumed, and the tree might be encouraged or manged by traditional human societies. Monocotyledoneae (n = 32) represents many herbaceous species including steppe grasses and species that thrive in more humid conditions. Some charcoal was identified as juniper (*Juniperus* sp., n = 4), a genus with several slow-growing species that currently thrive in the region in dry and montane environments. Others, such as elm (*Ulmus* sp., n = 3) also grow in mountain areas in the region, which suggests that wood was acquired in more distant places. The plane tree (*Platanus orientalis*, n = 55) and nettle tree (*Celtis australis*, n = 1) are species that are thought to have been introduced into the region and therefore their presence in the site is indicative of a very early introduction and they may have been planted intentionally. Similarly, the common grapevine (*Vitis vinifera*, n = 5), the *Prunus* genus (n = 1), which may correspond to such fruit trees as cherry, plum, peach, apricot, and almond, and other remains identified as the Maloideae sub-family (Rosaceae/Maloideae, n = 1) suggest the possibility of cultivated species. Although it is not possible to discriminate between cultivated and wild varieties from the wood anatomy, other archaeobotanic studies in Central Asia, along the Silk Road, have documented the growth of grapevines, almonds, and peaches [91–93]. They might indicate the use of water resources to grow domestic species that would not flourish in the arid environment of Termez without a supply of water. They are all species grown today in the orchards of Termez thanks to irrigation with water from the Surkhan Darya.

Most of the pieces of charcoal associated with the Kushan and Kushano-Sasanian sequence (2nd–5th centuries CE) correspond to branches of a small calibre (up to 20 mm or a little larger), which seems to indicate selective gathering of wood, probably to be used as fuel, although those types of branches might be used for buildings, especially as roofing material. Pruning would have allowed the use of trees and shrubs for either of those two purposes. In the case of plane trees (*Platanus orientalis*), evidence has been found that the branches were cut in spring. The only taxon of which trunks have been identified is Salicaceae (willow or poplar), which suggests that the wood of this family was used as building material or sometimes to make furniture and tools.

The medieval sample (8th–10th centuries CE) was smaller, with 57 pieces. Nine riparian species were documented: willow or poplar (Salicaceae, n = 12), tamarisk (*Tamarix* sp., n = 11), and ash (*Fraxinus* sp., n = 7). Oleaster (*Elaeagnus* sp., n = 3) and *Prunus* sp. (n = 2) also appear. Nine pieces were identified as the Leguminosae family, which includes three sub-families: Caesalpinoioideae, Mimosoideae, and Papilionaceae. Their anatomic characteristics allow them to be attributed to the Papilionaceae sub-family, but the anatomical similarities hinder their identification at the taxonomical level. One piece of charcoal may be classified in the Chenopodiaceae family, which includes over 1500 herbaceous plants. Finally, as in the old phase, Monocotyledoneae has been documented (n = 3); it is a class that encompasses all herbaceous species. The lesser taxonomic diversity in this period might result from a smaller sample. However, continuity is seen in the use of riparian vegetation, especially Salicaceae and tamarisk, which are still predominant, together with ash. Again, most of the charcoal remains come from branches with a small calibre, which indicates a clear selection of that type of wood, possibly in connection with the function of the source of the sample, which is linked to a *tannur*.

Few anthracological studies have been carried out at archaeological sites in Central Asia to put the results of the present study in context. The use of Salicaceae as fuel was documented in Mesolithic and Bronze Age levels at Aigyrzhal-2 and the Naryn valley in Kyrgyzstan [94], and it has also been documented at 6th–4th century BCE sites in Kyzyltepa in Uzbekistan [95]. Despite the chronological and spatial distances, the continuous use of riparian woodland for firewood is worth noting.

Table 1. Frequencies of the different vegetal taxa documented in Termez. Indeterminate means that, although the anatomical features are well observed, we have not been able to identify it due to lack of reference materials. Not determinable means that the anatomical features have not been well preserved and do not provide enough information.

		Ancient Period (2nd–4th AD)			
Taxa	**Tchingiz Tepe/Sector RB2 SU 6**	**Tchingiz Tepe/Sector RA SU 23**	**Tchingiz Tepe/Sector RA SU 36**	**Tchingiz Tepe/Sector RF SU26**	**Total**
Celtis australis	1	-	-	-	1
Elaeagnus sp.	16	-	-	-	26
Juniperus sp.	-	1	3	-	4
Monocotyledoneae	-	11	1	20	32
Platanus orientalis	31	2	22	-	55
Prunus sp.	-	1	-	-	1
Rosaceae/Maloideae	-	-	1	-	1
Salicaceae	-	37	32	-	69
Tamarix sp.	-	5	8	-	13
Ulmus sp.	2	-	1	-	3
Vitis vinifera	-	5	-	-	5
Indeterminate	-	1	2	-	3
Not determinable	-	4	3	-	7
Total	50	75	75	20	220
		Medieval period (8th–10th AD)			
Taxa	Potters quarter/*tannur*	Potters quarter/Kiln 1 SU 1	Potters quarter/foundational level on bedrock		Total
Chenopodiaceae	-	-	1		1
Elaeagnus sp.	-	2	1		3
Fraxinus sp.	2	5	-		7
Leguminosae	7	1	1		9
Monocotyledoneae	3	-	-		3
Prunus sp.	-	1	1		2
Salicaceae	12	-	-		12
Tamarix sp.	11	-	-		11
Indeterminate	9	-	-		9
Total	44	9	4		57

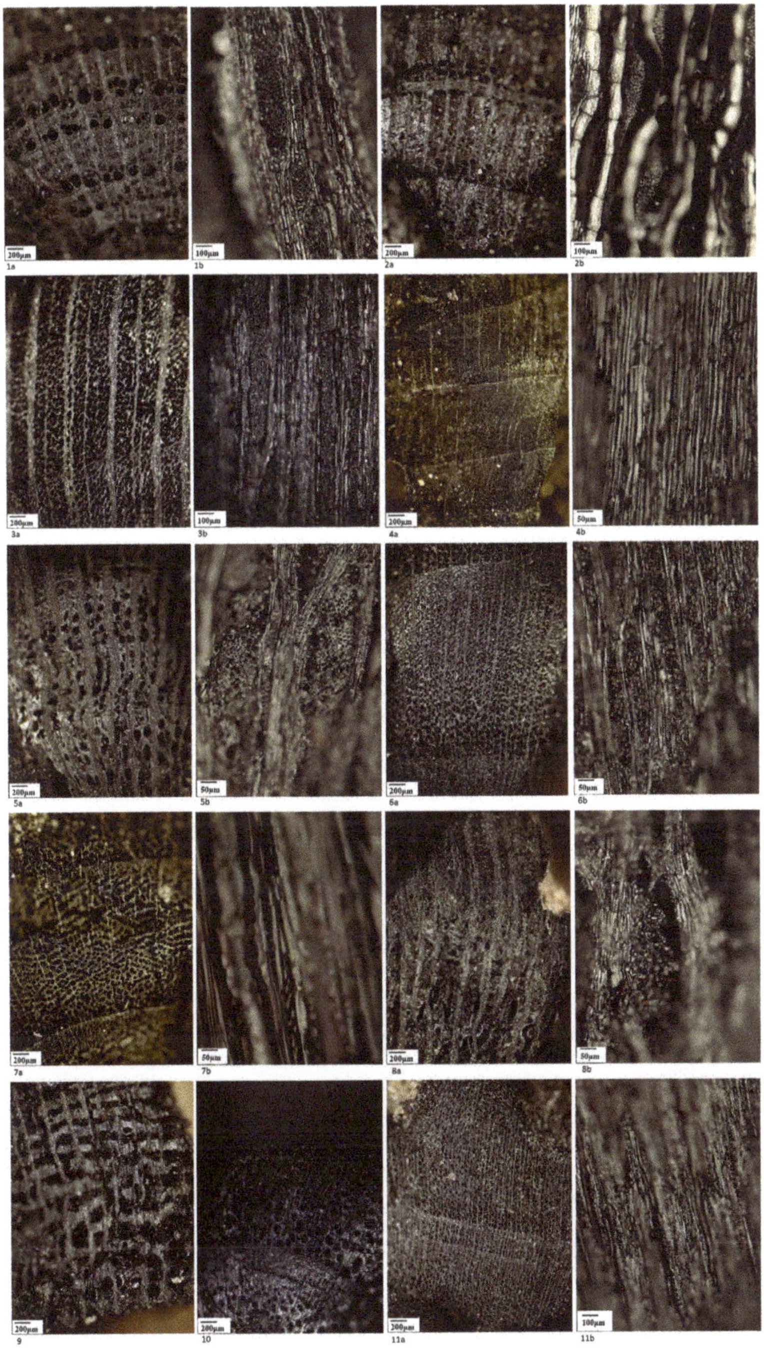

Figure 7. Vegetal taxa documented in Termez. 1: *Celtis australis*; 2: *Elaeagnus* sp.; 3: *Platanus orientalis*; 4: *Juniperus* sp.; 5: *Vitis vinífera*; 6: *Prunus* sp.; 7: Salicaceae; 8: *Tamarix* sp.; 9: *Ulmus* sp.; 10: *Fraxinus* sp.; 11: Rosaceae/Maloideae; a: cross section; b: tangential section.

3.3. Fauna

A number of 3762 faunal remains have been studied from Termez. Of these 1318 (35% of the NR) were determined to species or family levels. In general, the livestock at Termez was dominated by ovicaprids (>65% of the NISP), and these were the basis of the farming economy from the Greco-Bactrian period to the Islamic age (Table 2)

Table 2. Number of remains (NR), Number of Identified Specimens (NISP), and Minimum Number of Individuals (MNI) of the faunal remains in the different phases of occupation of Termez.

	Phase 1			Phase 2			Phase 3		
Specie	**NR**	**NISP**	**MNI**	**NR**	**NISP**	**MNI**	**NR**	**NISP**	**MNI**
Bos taurus	18	18	3	88	88	6	5	5	1
Ovis aries	31	31	6	38	38	8	52	52	5
Capra hircus	11	11	3	70	70	4	13	13	2
Caprinae	159	159	7	291	291	21	287	287	12
Gazella subgutturosa	2	2	1	5	5	2	1	1	1
Sus scrofa	5	5	2	46	46	5	3	3	2
Equus caballus	7	7	2	17	17	3	1	1	1
Equus asinus	1	1	1	6	6	2	1	1	1
Equus sp.	1	1	1	14	14	2	-	-	-
Camelidae	-	-	-	1	1	1	-	-	-
Saiga tatarica	1	1	1	-	-	-	-	-	-
Cervus elaphus	1	1	1	11	11	2	-	-	-
Canis l. familiaris	1	1	1	3	3	2	1	1	1
Vulpes corsac	-	-	-	1	1	1	-	-	-
Leporidae	-	-	-	1	1	1	15	15	1
Gallus g. domesticus	6	6	2	16	16	3	6	6	1
Columba livia/rupestris sp.	-	-	-	-	-	-	30	30	2
Corvus core/frugileus sp.	-	-	-	-	-	-	1	1	1
Anas/Mareca sp.	-	-	-	-	-	-	1	1	1
Passer sp.	-	-	-	-	-	-	18	18	2
Indeterminable bird	-	-	-	-	-	-	5	5	1
Cyprinidae	-	-	-	-	-	-	26	26	1
Large size mammal	52		2	138	-	4	43	-	3
Medium size mammal	16	-	2	102	-	4	4	-	2
Small size mammal	194	-	4	467	-	8	312	-	4
Indeterminable	377	-	-	534	-	-	205	-	-
Total	883	244	39	1849	608	79	1030	466	45

In the Greco-Bactrian and Yuezhi periods, sheep and goats were by far the most common domestic animals (82.37%), followed by cattle (7.38%) and equids (2.87%), which included both horses and donkeys. Poultry (2.46%) and, to a lesser extent, pigs (2.05%) were other sources of food. Dogs are represented by a single remain. Apart from the equids, all the domestic species display butchery marks indicating that their meat was consumed at the site. Wild species, like red deer, saiga antelope, and gazelle, appear in a very low proportion (<1%). These animals would live in the steppe areas near Termez, where they were hunted and consumed, as butchery marks have been observed on their bones, particularly in the case of the gazelle.

During the Kushan and Kushano-Sasanian periods of occupation, sheep and goats still dominated the livestock herds but in a smaller proportion than before (65.62%). Cattle and pigs increase their representation (14.47% and 7.57%, respectively), even if they require a larger amount of food and water than sheep and goats; therefore, their increasing presence in Termez may be associated with greater availability of vegetal and water resources in this period. Horses and donkeys continue appearing in low percentages, like poultry and dogs (<3%). In this period appears the only camelid remains documented at the site. Except for the donkey, all the other domestic species display cut marks related to the meat consumption of those animals. Hunting was still a source of food in the Kushan and Kushano-Sasanian periods, as attested by the remains of red deer, gazelles, and leporids, some of them with cut marks on their bones. A remain of corsac fox was also documented among the wild fauna.

In the Islamic period, sheep and goats again predominated (75.54%). The percentages of these animals increase compared to the previous period, while cattle and swine decrease (<2%). Equids, poultry, and dogs still appear in small percentages. As regards hunting, while no red deer or saiga antelopes have been documented in this period, gazelles were hunted, and their bones display cut marks. A wide range of birds include ducks, pigeons, rooks/crows, and sparrows, but no cut marks are seen on their skeletal remains. Fishing is also attested in this period by the presence of cyprinids caught in the Amu Darya.

In the sheep and goats' group, most specimens are young adults less than two years old in all the periods at the site. The consumption of animals of this age would be because the meat of young individuals is more tender. The importance of adult individuals in the record seems to indicate they were kept for their milk and wool. In contrast, cattle were consumed at an adult age, which may be because they were used for their milk and secondary products and as draught animals in agriculture and transport. The consumption of beef is attested in the whole occupation of Termez except in the Islamic period. Pigs, also present in all the periods, display an age of death with a predominance of infantile and juvenile individuals, which indicates consumption of their meat at those young ages. Swine were most common in the Kushan and Kushano-Sasanian periods. Equids, equally present throughout the occupation, are represented by horses and donkeys, the latter in low percentages. A series of remains could not be assigned to one or another animal and might correspond to hybrid animals. The mortality profiles indicate a more significant proportion of adult animals, so they may have been used as draught animals. However, some remains of immature individuals display butchery marks, so their meat may have been consumed. Dogs appear in the whole occupation sequence, which is common in a situation of animal husbandry. Anthropic processing of those animals has been documented in the Greco-Bactrian and Yuezhi periods, and therefore their meat or skins may have been used. This would not be unusual, as it has been documented at other sites dated in Antiquity [96,97].

Birds appear in all periods but never in large numbers. Cockerels and hens are represented the most and are the only ones with marks of anthropic processing. Their mortality profiles indicate the consumption of poultry at an old age, which can be related to their use to obtain eggs.

The economic exploitation of domestic livestock was complemented by hunting and fishing in the surroundings of Termez. Gazelles, red deer, and saiga antelopes are well-adapted to steppe conditions. Although anthropic marks have only been detected on the remains of gazelles, red deer, and saiga, antelopes were also consumed historically at different sites in central Asia [95,98–102]. Fishing is documented in the Islamic period by the remains of cyprinids, indicating the use of fluvial resources in the Amu Darya.

4. Discussion

The natural conditions in the lower Surkhan Darya valley made this area attractive for human occupation. The region's aridity was not an obstacle to the development and growth of Termez since its foundation in the late 4th century BCE. The urban surface area expanded considerably from the small enclosure of the Citadel, the centre of the Greek city, over the 1500 years of its existence. Archaeological data reveal that the greatest urban expansion in Termez occurred in two periods; the first between the change of era and the early 2nd century, under the rule of the Kushan kings; the second in the early Islamic period (in the 8th century). During the Kushan period, the Citadel, Tchingiz Tepe and Shahristan were urban centres, and the northern suburb (Military Camp) was also occupied. The activity was probably intense in other peripheral parts of Termez between the change of era and the late 6th century, especially in the area where the Rabad would grow later. In the Islamic period, the city's urban growth took place early. Radiocarbon dates obtained for the Potters' Quarter demonstrate that some workshops were active as early as the 8th–9th centuries. The Citadel, Shahristan, Rabad, and both northern suburbs were occupied, while Tchingiz Tepe was abandoned before the Arab conquest of the region.

Urban development at Termez gradually adapted to the constant topographic changes caused by the accumulation of deposits derived from both the collapse of buildings and the accumulation of geological materials from the frequent sandstorms in southern Uzbekistan [103]. In some parts of the city, the ground level rose because of the accumulation of (aeolian) sand brought by the wind and the periodic collapse and deposition of the building structures made of adobe and rammed earth. This is especially attested in the thick stratigraphy found in the Citadel, with levels over 14 m thick in some places [14]. The Citadel seems to have been the urban centre throughout the historical sequence, as it maintained a population from its foundation to its destruction in 1220. The urban evolution within the fortified enclosure of Tchingiz Tepe also entailed managing the sedimentation of aeolian sand deposits and the collapse of adobe buildings; this is evidenced in some areas within the fortress where depositions comprising stratigraphies 2.5 m thick were formed by those processes [7,13]. However, in this case, the enclosure was abandoned by its inhabitants during the 5th century CE, like the Kara Tepe and Fayaz Tepe Buddhist monasteries located in the westernmost sector of ancient Termez. The expansion of the city towards the east from the 8th century compensated for that loss of inhabited area by occupying the only free space: the interior alluvial plain.

Human occupation was compatible with the conservation of the natural environment, especially the riparian woodland. This area supplied fuel to the population of Termez for domestic purposes (heat and cooking), craft work (especially pottery but also glass and metals), or to obtain timber for buildings or other uses. The anthropisation of the environment is attested by the introduction of allochthonous woody species from at least the 2nd–5th centuries onwards. Part of the farmland near the city must have been irrigated, as some of the plant species would have been cultivated, such as the fruit trees and grapevines: species that can only be grown in an arid environment with a supplementary water supply. Domestic animals varied over time, although ovicaprids are predominant in all periods. Pigs, which are adapted to an orchard environment, appear throughout the sequence, but are scarce in the Middle Ages, when they would not be kept because of the prohibition on their consumption by Islam.

In Termez, judging from the anthracological and faunal record, the diversified agricultural exploitation, including species claiming irrigated crops, coexists with a livestock exploitation including all domestic species. From the data obtained in the study, there are hardly any significant variations in vegetation or livestock over the period studied. Or, in other words, the variations observed are less relevant than the continuity in the patterns of resource exploitation. In Termez, it has also not been possible to discern adaptations to the climatic variations detected in the historical sequences, unlike in Juuku Valley and Targaz, two areas around Lake Issyk-Kul (Kyrgyzstan) that have recently been the subject of detailed analyses [104–106]. In these regions, a system of agropastoralism (combined mountain agriculture) has been detected. This system is characterized by taking advantage of the variations in ecotones offered by the vertical gradient, an aspect that is especially important in the Juuku Valley.

However, the Termez records may not be incompatible with those obtained in the Issyk-Kul lake basin. The Issyk-Kul study areas have large variations due to the vertical gradient, resulting in diverse ecotones that allow for complementary economic use. In Juuku Valley, the vertical gradient makes the ecotones particularly sensitive to climatic variations that occurred during the Holocene. This probably means that the more visible variations in farming and herding strategies adapted to changes in vegetation at altitude. Termez is located in an oasis zone that benefits from the inputs of the Surkhan Darya to develop an irrigated crop agriculture. This irrigated agriculture seems to be an established fact, at least since Greek times. The faunal record of Termez presents some variations in the sequence, among which two stand out: the presence of wild species during the Greek-Yue zhi period (although in very low percentages) and the decline of the pig herd during the Islamic period (related to the religious prescription prohibiting the consumption of pork). However, judging from the data as a whole, these variations in the utilization of animal resources are less remarkable than the continuationist aspects.

The fauna and vegetation records detected in Termez are those of a large urban settlement in which agriculture and livestock coexist throughout the historical sequence. It would be necessary to carry out studies in the upper areas of the Surkhan Darya valley to detect which were the strategies that the human groups carried out in the peripheral spaces with respect to the settlement, in terms of agricultural and livestock use.

Termez was the most important city in northern Bactria (south of Uzbekistan) and prospered for longer than other cities founded at a similar time. About 20km to the west of Termez is Kampyr Tepe. According to the characteristics of the pottery found in the earliest levels, the foundation has been dated to about 300 BCE [107], or probably a little later, and posterior to the foundation of the Citadel in Termez [42,108]; however, some 14C chronologies suggest that a pottery workshop located within the Citadel of Kampyr Tepe was active in the late 4th century BCE [109]. The urban centre expanded to the north and east (low city) during the Greco-Bactrian, Yuezhi, and Kushan periods; it was abandoned in the late 2nd century CE [41], possibly because of a change in the course of the Amu Darya. Another relevant settlement in the Greek and Kushan periods was Dalverzin Tepe, 105km north of Termez, in the Shurkhan Darya valley. Its decline began at the end of the Kushan period (late 2nd century CE), although the population remained in specific areas of the urban layout until the 4th century. It was abandoned definitively in the 7th century; therefore, there is no evidence of Islamic occupation [110]. Khalchayan, located about 25km north of Dalverzin Tepe, was a centre founded in the Yuezhi period (mid-second century BCE) that became an important Kushan city; however, it was no longer inhabited in the Middle Ages as no Islamic levels have been documented [87,111]. Finally, Zar Tepe was a small city situated 15km north-west of Termez. It has provided a large pottery assemblage of the Kushan and Kushano-Sasanian periods. Its foundation has been dated to about the change of era, and it maintained a large population until the 4th century. Similar to other urban centres in the Surkhan Darya valley, the occupation seems to have been residual from that period, and it was depopulated in the Middle Ages [45].

The city of Termez, which was able to exist in a relatively hostile natural environment, was partially abandoned because of a catastrophic human action: the siege and destruction carried out by Genghis Khan in 1220. In fact, the Mongol conquest of central Asia had a powerful impact on the environment in the region, especially in the Amu Darya basin and the Aral Sea. The drop in the water level and the consequent regression detected in the 13th and 14th centuries might have resulted from the destruction of dams that controlled the tributaries of the Amu Darya by the Mongol armies. The impact would have been so significant as to change the course of the Amu Darya towards the Caspian Sea, leaving the Aral Sea without a supply of water [112]. Ancient Termez, which had outlived conquest by the Yuezhi nomads, the Sassanids and the Islamic forces, did not survive the destruction caused by the Mongol army. However, it was re-founded soon after, just 10km to the south of the old centre of the Citadel.

5. Conclusions

The populations that inhabited the lower valley of the Surkhan Darya benefited from the water supply by this tributary of the Amu Darya to prosper in a hostile environment characterised by extreme summer and winter temperatures and low rainfall. The remains of fauna and charred wood reveal the use of different species, including possible fruit trees and domestic livestock; they comprise practically all the biotopes around the city in both Antiquity and the Middle Ages. The botanical data allows a reconstruction of the landscape with quite extensive river-bank woodland until c. 1300, probably denser than at present, especially in the area around the city, which is now quite barren. The hunted wild animals are characteristic of a steppe environment, which in the antique and medieval periods was little altered by human action. Gazelles, saiga antelopes, and red deer were hunting prey for the inhabitants of Termez.

The city experienced continuous growth from its foundation c. 300 BC until its destruction in 1220. The urban expansion of Termez was mainly inland. The city was able to adapt to the topographic changes caused by the accumulation of deposits created by the collapse of structures and the aeolian contributions of sand. The two most expansive moments of urban growth are detected in the Kushan period (early 1st century AD and mid-third century) and in the early Islamic period (8th and 9th centuries). Only the urban enclosure of Tchinguiz Tepe was abandoned before the Islamic conquest, probably because of maintenance problems due to the accumulation of sand deposits.

Author Contributions: Conceptualization, E.A., P.U., J.A., V.M.-F. and J.M.G.; Formal analysis, E.A., P.U., J.A., R.P. (Raquel Piqué), R.P. (Rodrigo Portero) and J.M.G.; Investigation, E.A., P.U., J.A., R.P. (Raquel Piqué), R.P. (Rodrigo Portero), V.M.-F. and J.M.G.; Methodology, E.A., P.U., J.A., R.P. (Raquel Piqué), R.P. (Rodrigo Portero), V.M.-F. and J.M.G.; Software, P.U. and J.A.; Visualization, E.A., P.U., J.A., R.P. (Raquel Piqué), R.P. (Rodrigo Portero) and J.M.G.; Writing—original draft, E.A., P.U., J.A., R.P. (Raquel Piqué), R.P. (Rodrigo Portero), V.M.-F. and J.M.G.; Writing—review & editing, E.A., P.U., J.A., R.P. (Raquel Piqué), R.P. (Rodrigo Portero), V.M.-F. and J.M.G. All authors have read and agreed to the published version of the manuscript.

Funding: This research has been financed by three projects: 'Caracterización arqueológica y arqueométrica de cerámicas de Asia central. Producción, distribución y tecnología' ('Archaeological and archaeometric characterisation of ceramics from Central Asia. Production, distribution and technology'), funded by MCIN/AE (PID2020-114096GB-C21) (IP V. Martínez Ferreras), 'Arqueología de los centros de producción cerámica y de consumo en Asia central. Una aproximación desde la teledetección' ('Archaeology of ceramic production and consumption centrers in Central Asia. A remote sensing approach'), funded by MCIN/AE (PID2020-114096GA-C22) (IP Paula Uribe Agudo) and 'Termez en Bactriana (Uzbekistán): colonia griega, centro del budismo centroasiático y ciudad islámica' ('Termez in Bactriana (Uzbekistan): Greek colony, Buddhish centre and Islamic city'), funded by the Palarq Foundation (Paleontology and Archaeology) (IP J.M. Gurt Esparraguera).

Data Availability Statement: Not applicable.

Acknowledgments: Special thanks to S. Pidaev from the Academy of Sciences of the Republic of Uzbekistan for his great help during the campaign in Uzbekistan.

Conflicts of Interest: The authors declare no conflict.

Appendix A

Table A1. Analysed samples and ^{14}C results. Column A: Urban space, Excavation area, SU (Stratigraphic Unit); Colunm B: Dating material reference; Column C: Radiocarbon date code assigned by this Laboratory; Column D: Radiocarbon date with uncertainty expressed in terms of standard deviation; Column E: Calibrated date intervals centred on the modes of probability distribution for the true calibrated date corresponding to a total probability of 95.4% (range σσ) and the probability associated with each interval, respectively. In this set of intervals there is a probability of 95.4% that the true calibrated date will be found; Column F: Calibrated date intervals centred on the modes of probability distribution for the true calibrated date corresponding to a total probability of 68.3% (range σ) and the probability associated with each interval, respectively. In this set of intervals, there is a probability of 68.3% that the true calibrated date will be found.

A	B	C	D	E	F
Citadel, Tower 2	Charcoal	UBAR-1228	955 ± 30 BP	(95.4%) 1027–1160 cal AD	(11.2%) 1035–1049 cal AD (57.1%) 1081–1152 cal AD
Shahristan, Wall	Bone	Beta-504292	1960 ± 30 BP	(7.9%) 42 cal BC–8 cal BC (87.5%) 2 cal BC–130 Cal AD	(51.7%) 23–84 cal AD (16.6%) 95–116 cal AD
Shahristan, Workshop 5, Kiln 3	Charcoal	Beta-511510	560 ± 30 BP	(48.6%) 1312–1362 cal AD (46.9%) 1387–1428 cal AD	(34.9%) 1326–1351 cal AD (33.4%) 1394–1415 cal AD
Tchingiz Tepe, Sector RB1, SU 5	Charcoal	UBAR-989/ CNA-366	2130 ± 45 BP	(18%) 356–280 cal BC (0.2%) 253–250 cal BC (76.7%) 232–41 cal BC (0.5%) 9–2 cal BC	(6.6%) 340 – 325 cal BC (54.1%) 199 – 95 cal BC (7.6%) 73 – 56 cal BC
Tchingiz Tepe, Sector RB2, SU10	Bone	UBAR-1144/ CNA-1190	2075 ± 40 BP	(95.4%) 197 cal BC–23 cal AD	(66.4%)153–42 cal BC (1.9%) 8–4 cal BC
Tchingiz Tepe, Sector RB2, SU11	Charcoal	UBAR-1145/ CNA-1189	2095 ± 55 BP	(9%) 353–286 cal BC (0.7%) 228–218 cal BC (85.1%) 211 cal BC–28 AD (0.8%) 46–58 cal AD	(2.3%)195–187 cal BC (64.1%) 178–42 cal BC (1.8%) 8–2 cal BC
Tchingiz Tepe, Sector RB3, SU2	Bone	UBAR-1241/ CNA-2253	1905 ± 35 BP	(3.1%) 28–45 cal AD (92.4%) 58–225 cal AD	(8.5%) 84–96 cal AD (59.8%) 115–204 cal AD
Tchingiz Tepe, Sector RB3, SU6	Bone	UBAR-1240/ CNA-2252	1910 ± 35 BP	(95.4%) 26–219 cal AD	(12.2%) 80–99 cal AD (56%) 109–204 cal AD
Tchingiz Tepe, Sector RC, SU10	Charcoal	UBAR- 990	1940 ± 130 BP	(2.1%) 350–307 cal BC (93.4%) 208–AD 405 cal BC	(3%) 95 – 74 cal BC (65.2%) 56 cal BC–243 cal AD
Tchingiz Tepe, Sector RA, SU17	Charcoal	UBAR-1146	1670 ± 55 BP	(13.7%) 250–295 cal AD (81.7%) 311–540 cal AD	(9.5%) 258–281 cal AD (53.2%) 331–435 cal AD (3.3%) 518–529 cal AD
Tchingiz Tepe, Sector RC, SU2	Charcoal	UBAR-1043	1680 ± 35 BP	(15.1%) 252–290 cal AD (77.2%) 321–436 cal AD (1%) 465–475 cal AD (0.7%)500–509 cal AD (1.6%) 515–531 cal AD	(5.5%) 265–272 cal AD (62.8%) 351–417 cal AD
Tchingiz Tepe, Sector RC, SU3	Charcoal	UBAR-1044	1760 ± 35 BP	(95.4%) 232–401 cal AD	(14.2%) 244–262 cal AD (54%) 276–344 cal AD
Tchingiz Tepe, Sector RC, SU4	Charcoal	UBAR-1045	1740 ± 50 BP	(95.4%) 224–417 cal AD	(28.1%) 250–295 cal AD (40.2%) 310–376 cal AD
Tchingiz Tepe, Sector RC, SU5	Charcoal	UBAR-1046	1755 ± 45 BP	(95.4%) 221–408 cal AD	(15%) 242–366 cal AD (53.3%) 372–353 cal AD
Tchingiz Tepe, Sector RF, SU20	Charcoal	UBAR-1053	1690 ± 230 BP	(0.6%) 346–316 cal BC (94.3%) 204 cal BC–775 cal AD (0.4%) 791–821 cal AD	(68,3%) 120–602 cal AD
Military Camp, Sector AC1, SU12	Bone	UBAR-1242/ CNA-2254	1955 ± 35 BP	(95.4%) 51 cal BC–217 cal AD	(7.1%) 35–15 cal BC (61.2%) 5–127 cal AD

Table A1. *Cont.*

A	B	C	D	E	F
Military Camp, Sector AC2, SU20	Bone and Charcoal	UBAR-1125/ CNA–1121 UBAR-1126/ CNA 1122	1940 ± 25 BP	(92.9%) 12–169 cal AD (2.5%) 185–203 cal AD	(68.3%) 24–83 cal AD
Military Camp, Sector AC2, SU33	Bone	UBAR-1127/CNA-1123	2155 ± 40 BP	(30.6%) 359–276 cal BC (2.2%) 261–244 cal BC (62.7%) 235–52 cal BC	(21.7%) 350–307 cal BC (46.6%) 208–106 cal BC
Military Camp, Sector AC2, SU12	Bone	Beta-502182	1170 ± 30 BP	(73.9%) 772–901 cal AD (21.6%) 916–974 cal AD	(9.7%) 776–788 cal AD (49%) 825–894 cal AD (9.6%) 928–945 cal AD
Military Camp, Sector AC2, SU18	Bone	Beta-515844	1000 ± 30 BP	(57%) 992–1051 cal AD (38.5%) 1080–1154 cal AD	(51.3%) 994–1045 cal AD (5.2%) 1085–1093 cal AD (11.7%) 1105–1121 cal AD
Potters quarter, Workshop 2, Kiln 2	Charcoal	Beta-511511	1140 ± 30 BP	(4.9%) 774–787 cal AD (8.5%) 828–860 cal AD (82%) 870–992 cal AD	(16.1%) 883–903 cal AD (52.2%) 915–976 cal AD
Potters quarter, Workshop 11, Tannur	Charcoal	Beta-511509	1180 ± 30 BP	(82.7%) 771–900 cal AD (12.8%) 917–973 cal AD	(12.1%) 775–791 cal AD (56.2%) 821–891 cal AD (11.7%) 777–792 cal AD
Potters quarter, Workshop 11, Kiln 1, SU1	Charcoal	Beta-545671	1150 ± 30 BP	(7.5%) 773–789 cal AD (88%) 824–988 cal AD	(4.8%) 777–785 cal AD (5.4%) 835–846 cal AD (17.7%) 877–901 cal AD (40.3%) 916–974 cal AD
Potters quarter, Workshop 11, Kiln 1, SU1	Charcoal	Beta-502183	1190 ± 30 BP	(1.6%) 709–722 cal AD (88%) 771–897 cal AD (5.8%) 923–952 cal AD	(11.4%) 777–791 cal AD (1.6%) 805–807 cal AD (55.3%) 820–886 cal AD
Potters quarter, Workshop 11, Kiln 3, SU21	Charcoal	Beta-545670	1190 ± 30 BP	(1.6%) 709–722 cal AD (88%) 771–897 cal AD (5.8%) 923–952 cal AD	(11.4%) 777–791 cal AD (1.6%) 805–807 cal AD (55.3%) 820–886 cal AD
Potters quarter, Workshop 11, SU32	Charcoal	Beta-550228	1120 ± 30 BP	(1.8%) 774–785 cal AD (1.4%) 833–846 cal AD (92.3%) 876–995 cal AD	(34.1%) 893–933 cal AD (34.2%) 940–977 cal AD

Table A2. Quantification of ceramic productions recovered in surface surveying.

Square / Ware	Sha1	Sha1%	Sha2	Sha2%	Rabad1	Rabad1%	KBTK1	KBTK1%
Red-figure ware (Greek)	1	0.02						
White-slipped ware (Greek)	1	0.02						
Black-slipped tableware (Greek)	3	0.06					2	0.12
Total Greek pottery	5	0.11					2	0.12
Grey pottery (Yuezhi)	6	0.13	2	0.05				
Total grey pottery	6	0.13	2	0.05				
Slipped tableware (Kushan/Sassanian)	154	3.29	359	8.94	60	3.93	22	1.31
Common ware (Kushan/Sassanian)	13	0.28	12	0.30	5	0.33	16	0.95
Common cooking ware (Kushan/Sassanian)			11	0.27	2	0.13	1	0.06
Total Kushan/Sassanian pottery	167	3.57	382	9.52	67	4.39	39	2.32
Common cooking ware, black slip partially covering the surface (Islamic)	186	3.98	76	1.89	39	2,55	28	1.66

Table A2. *Cont.*

Ware \ Square	Sha1	Sha1%	Sha2	Sha2%	Rabad1	Rabad1%	KBTK1	KBTK1%
Unglazed ware, comb-impressed/comb-incised decoration (Islamic)	25	0.53	15	0.37	6	0.39	38	2.26
Unglazed ware, applied decoration (Islamic)	1	0.02					1	0.06
Unglazed ware, stamped relief decoration (Islamic)	10	0.21	6	0.15	3	0.20		
Unglazed ware, moulded relief decoration (Islamic)	11	0.24	5	0.12	2	0.13	5	0.30
Glazed white ware (Islamic)	96	2.05	88	2.19	62	4.06	13	0.77
Glazed green monochrome ware (Islamic)	18	0.39	6	0.15	11	0.72	6	0.36
Glazed turquoise monochrome ware (Islamic)	65	1.39	36	0.90	30	1.96	5	0.30
Glazed yellow monochrome ware (Islamic)	4	0.09	3	0.07	1	0.07		
Underglaze painted ware/slip painted (Islamic)	21	0.45	16	0.40	7	0.46	18	1.07
Splashed *sgraffiato* ware (Islamic)	50	1.07	37	0.92	20	1.31	24	1.43
Figured/geometric decoration glazed ware (Islamic)	122	2.60	34	0.84	20	1.31	5	0.30
Glazed overfired (not specific) (Islamic)	11	0.24	1	0.02	5	0.33		
Glazed altered (not specific) (Islamic)					1	0.07		
Sphero-conical vessel (Islamic)	35	0.75	23	0.57	7	0.46	38	2.26
Total Islamic ware	655	14.01	346	8.62	214	14.01	181	10.75
Common ware	3593	76.87	3187	79.40	1211	79.31	1420	84.32
Common cooking ware	248	5.31	97	2.42	35	2.29	42	2.49
Total undetermined pottery	3841	82.18	3284	81.81	1246	81.60	1462	86.82
Total	4674	100	4014	100	1527	100	1684	100

References

1. Leriche, P. Bactria, Land of Thousand Cities. In *After Alexander. Central Asia before Islam*; Cribb, J., Hermann, G., Eds.; Oxford University Press, Proceeding of the British Academy: Oxford, UK, 2007; pp. 121–153.
2. Leriche, P. Termez antique et médiévale. In *La Bactriane au Carrefour des Routes et des Civilisations de l'Asie Centrale*; Leriche, P., Pidaev, S., Gelin, M.K., Abdullaev, J., Eds.; with the Collaboration of Fourniau, V.; Maisonneuve and Larose: Paris, France, 2001; pp. 75–99.
3. Barthold, W. Tirmidh. In *Encyclopédie de l'Islam, Nouvelle Édition*; Bearman, P.J., Bianquis, T., Bosworth, C.E., van Donzel, E., Heinrichs, W.P., Eds.; T.-U. Brill: Leiden, The Netherlands, 2002; Volume X, pp. 582–584.
4. Masson, M.E. Gorodishcha ctarogo Termeza i ikh izuchyeniye. In *Termezskaya Arkhyeologichyeskaya Komplyeksnaya Ekspyeditsiya 1936 g.*; Academy of Sciences of the Soviet Socialist Republic of Uzbekistan, UzFAN: Tashkent, Uzbekistan, 1940; pp. 5–122.
5. Houal, J.-B. La céramique de Termez des époques antique et médiévale. In *L'Archéologie Française en Asie Centrale, Nouvelles Recherches et Enjeux Socioculturels*; Bendezu-Sarmiento, J., Ed.; with the Collaboración of Le Maguer, S.; Cahiers d'Asie Centrale: Paris, France; Bishkek, Kyrgyzstan; Kaboul, Afghanistan, 2013; Volume 21/22, pp. 423–442.
6. Houal, J.-B. Regards sur l'Asie centrale du IIIe siècle av. J.-C. au XVIIIe siècle. In *La Céramique Antique et Médievale de Termez et de Khaitabad (Ouzbékistan)*; Hermann: Paris, France, 2021.
7. Sánchez del Corral, A.; Thum, H. Geomorphology and Late Holocene morphogenesis of Tchingiz Tepe hill (Old Termez, Uzbekistan, Central Asia). *Quat. Int.* **2012**, *281*, 89–104. [CrossRef]

8. Stride, S. Regions and Territories in Southern Central Asia: What the Surkan Darya Province tells us about Bactria. In *After Alexander. Central Asia before Islam*; Cribb, J., Hermann, G., Eds.; Oxford University Press, Proceeding of the British Academy: Oxford, UK, 2007; pp. 99–117.
9. Stride, S. Géographie Archéologique de la Province du Surkhan Darya (Ouzbékistan du sud/Bactriane du Nord). Ph.D. Thesis, Université Paris I Panthéon-Sorbonne, Paris, France, 2004.
10. Shiskin, V.A. K istorichyeskoy topografii starogo Termeza. In *Termezskaya Arkhyeologichyeskaya Komplyeksnaya Ekspyeditsiya 1936 g.*; Academy of Sciences of the Soviet Socialist Republic of Uzbekistan, UzFAN: Tashkent, Uzbekistan, 1940; pp. 123–153.
11. Angás, J.; Uribe, P.; Ariño, E.; Gurt, J.M.; Martínez-Ferreras, V.; Pidaev, S. A multi-scalar photogrammetric recording approach in Termez (Uzbekistan). In *The International Archives of the Photogrammetry, Remote Sensing and Spatial Information Sciences, XLII (2), Proceedings of the 27th CIPA International Symposium, Documenting the Past for a Better Future, Avila, Spain, 5 September 2019*; CIPA-Icomos: Paris, France, 2019. [CrossRef]
12. *Termezskaya Arkhyeologichyeskaya Komplyeksnaya Ekspyeditsiya 1936 g.*; Academy of Sciences of the Soviet Socialist Republic of Uzbekistan, UzFAN: Tashkent, Uzbekistan, 1940.
13. Martínez-Ferreras, V.; Ariño, E.; Gurt, J.M.; Pidaev, S. The enclosure of Tchingiz Tepe (ancient Termez, Uzbekistan) during the Kushan and Kushan-Sassanian periods. Archaeological stratigraphy and 14C dating analyses. *Iran. Antiq.* **2014**, *49*, 413–474. [CrossRef]
14. Pidaev, S. Kyeramika gryeko-baktriyskogo vryemyeni s gorodishcha starogo Termeza. *Sov. Arkhyeologiya* **1991**, *1*, 210–224.
15. Pidaev, S. Contribution à l'histoire ancienne de Termez. In *La Bactriane au Carrefour des Routes et des Civilisations de l'Asie Centrale*; Leriche, P., Pidaev, S., Gelin, M.K., Abdullaev, J., Eds.; with the Collaboration of Fourniau, V.; Maisonneuve and Larose: Paris, France, 2001; pp. 47–57.
16. Leriche, P.; Pidaev, S. Termez in Antiquity. In *After Alexander. Central Asia before Islam*; Cribb, J., Hermann, G., Eds.; Oxford University Press, Proceeding of the British Academy: Oxford, UK, 2007; pp. 179–211.
17. Leriche, P.; Pidaev, S. *Termez sur Oxus: Cité-capitale d'Asie Centrale*; Maisonneuve and Larose: Paris, France, 2008.
18. Gurt, J.M.; Ariño, E.; Martínez-Ferreras, V.; Pidaev, S. The Buddhist occupation of Tchingiz Tepe (Termez, Uzbekistan) in the Kushan period through the ceramic contexts. *Archaeol. Res. Asia* **2015**, *8*, 19–33. [CrossRef]
19. Rakhmonov, U.V.A. Kukna Termiz sharistoni mudofaa devorlari. In *Uzbekiston Moddiy Madaniyati tarikhi/The History of Material Culture of Uzbekistan*; Shirinov, T.E., Ed.; Soghdian Publishing House: Samarkand, Uzbekistan, 1997; pp. 60–72.
20. Rakhmonov, U.V.A. Les fortifications de l'ancienne Termez. Antiquité et Moyen Âge. In *La Bactriane au Carrefour des Routes et des Civilisations de l'Asie Centrale*; Leriche, P., Pidaev, S., Gelin, M.K., Abdullaev, J., Eds.; with the Collaboration of Fourniau, V.; Maisonneuve and Larose: Paris, France, 2001; pp. 63–74.
21. Leriche, P. L'apport de la Mission Archéologique Franco-Ouzbèque (MAFOuz) de Bactriane du Nord à l'histoire de l'Asie Centrale. In *L'Archéologie Française en Asie Centrale, Nouvelles Recherches et Enjeux Socioculturels*; Bendezu-Sarmiento, J., Ed.; Cahiers d'Asie Centrale: Paris, France; Bishkek, Kyrgyzstan; Kaboul, Afghanistan, 2013; Volume 21/22, pp. 135–164.
22. Martínez-Ferreras, V.; Fusaro, A.; Gurt, J.M.; Ariño, E.; Pidaev, S.; Angourakis, A. The Islamic Ancient Termez through the lens of ceramics: A new archaeological and archaeometric study. *Iran* **2020**, *58*, 250–278. [CrossRef]
23. Pidaev, S. Mosquées de quartier dans l'ancienne Tirmidh (Ouzbékistan). *Archéologie Islam.* **2001**, *11*, 61–74.
24. Rakhmonov, U.V.; Pidaev, S. Raskopki karavan-saraya v rabadye sryednyevyekovogo Termeza. *IMKU* **1990**, *23*, 60–69.
25. Gurt, J.M. Archaeological record: Sector AC, military camps (former barracks). In *Preliminary Report of the Work of the International Pluridisciplinary Archaeological Expedition to Bactria, (IPAEB)*; Gurt, J.M., Pidaev, S., Rauret, A.M., Stride, S., Eds.; IPAEB: Samarkand, Uzbekistan; Barcelona, Spain, 2008; Volume 3, pp. 186–217. Available online: http://hdl.handle.net/2445/11504 (accessed on 1 August 2023).
26. Martínez-Ferreras, V.; Angourakis, A.; Hein, A.; Aulinas, M.; García-Vallés, M.; Gurt, J.M.; Ariño, E.; Sánchez del Corral, A.; Pidaev, S. Assessing Hellenistic to nomadic cultural patterns through pottery in ancient Termez, Uzbekistan. *Geoarchaeology* **2019**, *34*, 540–564. [CrossRef]
27. Fusaro, A.; Gurt, J.M.; Ariño, E.; Uribe, P.; Angás, J.; Pidaev, S. Islamic ancient Termez: An active and long-established ceramic manufacturing centre along the Silk Road. *Archaeol. Res. Asia* **2022**, *31*, 100375. [CrossRef]
28. Kato, K. The Excavation of the North Hill of Kara-Tepa (1985–1989). In Proceedings of the Sovyeshchaniye Ryegionalnykh Spyetsyalistov, Kulturnoye Naslyediye Tsyentralnoy Azii i Vklad Yaponii, Samarkand, Uzbekistan, 12–16 March 2008; National Institute for Research in Cultural Exchange Tokio, National Commission of Uzbekistan for UNESCO Affairs: Tashkent, Uzbekistan, 2010; pp. 153–155.
29. Pidaev, S.; Kato, K. The Excavation of the North and West (Central) Hill of Kara-Tepa (1998–2007). In Proceedings of the Sovyeshchaniye Ryegionalnykh Spyetsyalistov, Kulturnoye Naslyediye Tsyentralnoy Azii i Vklad Yaponii, Samarkand, Uzbekistan, 12–16 March 2008; National Institute for Research in Cultural Exchange Tokio, National Commission of Uzbekistan for UNESCO Affairs: Tashkent, Uzbekistán, 2010; pp. 157–185.
30. Mkrtychev, T.K. Buddiyskiy monastyr Fayaz-tepe (syevernaya Baktriya) po isslyedovaniyam poslyednikh let. *Ross. Arkheologiya* **2013**, *2*, 119–125.
31. Abdullaev, K. The Buddhist culture of Ancient Termez in old recent finds. *Parthica* **2013**, *15*, 157–187.

32. Bolelov, S.B. Goncharnaya mastyerskaya III-II vyekov do H.E. na Kampyrtepa. In *Matyerialy Tokharistanskoy Ekspyeditsii. Arkhyeologichyeskiye Isslyedovaniya Kampyrtepa*; Rtveladze, E.V., Ed.; Academy of Fine Arts of Uzbekistan, San'At: Tashkent, Uzbekistan, 2001; Volume 2, pp. 15–30.
33. Bolelov, S.B. Proizvodstvyennyy tsyentr epokhi Ellinizma na Tsitadyeli Kampyrtepa. In *Matyerialy Tokharistanskoy Ekspyeditsii. Arkhyeologichyeskiye Isslyedovaniya Kampyrtepa*; Rtveladze, E.V., Ed.; Academy of Fine Arts of Uzbekistan, San'At: Elets, Uzbekistan, 2011; Volume 8, pp. 48–79.
34. Lyonnet, B. La céramique hellénistique en Asie Centrale. In *Networks in the Hellenistic World. According to the Pottery in the Eastern Mediterranean and Beyond*; Fenn, N., Römer-Strehl, C., Eds.; BAR International Series; British Archaeological Reports: Oxford, UK, 2013; Volume 2539, pp. 351–368.
35. Nekrasova, E.F.; Pugachenkova, G.A. Kyeramika Dalverzintepe. In *Dalverzintepe. Kushanskiy Gorod na Yugye Uzbekistana*; Pugachenkova, G.A., Rtvladze, E.V., Eds.; Academy of Sciences of the Soviet Socialist Republic of Uzbekistan, Fan: Tashkent, Uzbekistan, 1978; pp. 143–161.
36. Pidaev, S. Kyeramika Dzhiga-tepe (iz raskopok 1976 g.). In *Dryevnyaya Baktriya, 3. Matyerialy Sovyetsko-Afganskoy Arkhyeo-logichyeskoy Ekspyeditsii/La Bactriane Ancienne, 3. Matériaux de l'Expédition Archéologique*; Afghano-Soviétique, I.T., Kruglikova, I.T., Eds.; Academy of Sciences of USSR, Nauka: Moscow, Russia, 1984; pp. 112–124.
37. Sverchkov, L.M. The Kurganzol Fortress (on the History of Central Asia in the Hellenistic Era). In *Ancient Civilizations from Scythia to Siberia*; Koninklijke Brill NV: Leiden, The Netherlands, 2008; Volume 14, pp. 123–191.
38. Pidaev, S. Syeroglinyanaya kyeramika Mirzakultepa. In *Istoriya Matyerialnoy Kultury Uzbyekistana*; Askarov, A.A., Ed.; Academy of Sciences of the Republic of Uzbekistan: Tashkent, Uzbekistan, 1991; Volume 25, pp. 88–98.
39. Pidaev, S. Stratigrafiya gorodishcha Starogo Termeza v svete novykh raskopok. Matyeryaly sovyetcko-frantsuzkogo kollkviyma. In *Gorodskaya cultura Baktrii-Tokharistana i Sogda: Antichnosct, Rannyeye Sryednyevekovye*; Fan: Taskhent, Uzbekistan, 1987; pp. 87–97.
40. Pidaev, S. *Posyelyeniya Kushanskogo Vryemyeni Syevyernoy Baktrii*; Academy of Sciences of the Soviet Socialist Republic of Uzbekistan, Fan: Tashkent, Uzbekistan, 1978.
41. Bolelov, S.B. Kyeramichyeskiy komplyeks pyerioda pravlyeniya Kanishki na Kampyrtepa (raskopki 2000–2001 godov). In *Matyerialy Tokharistanskoy Ekspyeditsii. Arkhyeologichyeskiye Isslyedovaniya Kampyrtepa*; Rtveladze, E.V., Ed.; Academy of Fine Arts of Uzbekistan, San'At: Tashkent, Uzbekistan, 2002; Volume 3, pp. 41–66.
42. Dvurechenskaia, N.D. *Itogi arkhyeologichyeskikh rabot 2004-2005 gg. v zhilom kvartalye-blokye 5 v syevyero-zapadnoy chasti Kampyrtepa In Matyerialy Tokharistanskoy Ekspyeditsii*; Rtveladze, E.V., Ed.; Academy of Fine Arts of Uzbekistan, San'At: Elets, Uzbekistan, 2006; Volume 6, pp. 110–127.
43. Mandelshtam, A.M. *Pamyatniki Kochyevnikov Kusanskogo Vryemyeni v Syevernoy Baktrii. Trudy Tadzhiskoy Arkhyelogichyeskoy Ekspyeditsii Instituta Arkheologii AN CCCP i Instituta Istorii*; VII, Academy of Sciences of USSR, Nauka: Leningrad, Russia, 1975.
44. Sedov, A.V. *Kobadian: Na porogye Rannyego Sryednyevyekovya*; Academy of Sciences of USSR, Academy of Sciences of the Soviet Socialist Republic of Tajikistan, Nauka: Moscow, Russia, 1987.
45. Zavyalov, V.A. *Kushanshakhr pri Sasanidakh (o Matyerialam Raskopok Gorodishcha Zartepa)/Kushanhahr under the Sasanian (on the Results of Excavations of the Zartepa Site)*; Russian Academy of Sciences, Institute of History of Material Culture Proceedings, St. Petersburg State University, Faculty of Philology and Arts: St. Petersburg, Russia, 2008; Volume 27.
46. Bronk Ramsey, C. Bayesian analysis of radiocarbon dates. *Radiocarbon* **2009**, *51*, 337–360. [CrossRef]
47. Reimer, P.; Austin, W.; Bard, E.; Bayliss, A.; Blackwell, P.; Bronk Ramsey, C.; Butzin, M.; Cheng, H.; Edwards, R.; Friedrich, M.; et al. The IntCal20 Northern Hemisphere radiocarbon age calibration curve (0–55 cal kBP). *Radiocarbon* **2020**, *62*, 725–757. [CrossRef]
48. Angás, J.; Uribe, P.; Bea, M.; Farjas, M.; Ariño, E.; Martínez-Ferreras, V.; Gurt, J.M. Potential of CORONA satellite imagery for 3D reconstruction of archaeological landscapes. In Proceedings of the 3rd Congress in Geomatics Engineering, Valencia, Spain, 7–8 July 2021; Polytechnic University of Valencia: Valencia, Spain, 2021. [CrossRef]
49. Schenk, T.; Csathó, B.; Shin, S. Rigorous panoramic camera model for DISP imagery. In *Proceedings of the ISPRS Workshop: High Resolution Mapping from Space*; Hannover University: Hannover, Germany, 2003.
50. Casana, J. Global-Scale Archaeological Prospection using CORONA Satellite Imagery: Automated, Crowd-Sourced, and Expert-led Approaches. *J. Field Archaeol.* **2020**, *45*, 89–100. [CrossRef]
51. Hammer, E.; Ur, J. Near Eastern Landscapes and Declassified U2 Aerial Imagery. *Adv. Archaeol. Pract.* **2019**, *7*, 107–126. [CrossRef]
52. Hammer, E.; Fitzpatrick, M.; Ur, J. Succeeding CORONA: Declassified HEXAGON intelligence imagery for archaeological and historical research. *Antiquity* **2022**, *96*, 679–695. [CrossRef]
53. Galiatsatos, N. Assessment of the CORONA Series of Satellite Imagery for Landscape Archaeology: A Case Study from the Orontes Valley, Syria. Ph.D. Thesis, Durham University, Durham, UK, 2004. Available online: http://etheses.dur.ac.uk/281/1/PhD_thesis.pdf?DDD14+ (accessed on 1 August 2023).
54. Ashraf, S.; Brabyn, L.; Hicks, B.J. Image data fusion for the remote sensing of freshwater environments. *Appl. Geogr.* **2012**, *32*, 619–628. [CrossRef]
55. Schweingruber, F. *Anatomy of European Woods*; Paul Haupt: Bern/Stuttgart, Germany, 1990.
56. Lyman, R.L. *Vertebrate Taphonomy*; Cambridge University Press: Cambridge, UK, 1994.
57. Reitz, E.; Wing, E. *Zooarchaeology (Cambridge Manuals in Archaeology)*, 2nd ed.; Cambridge University Press: Cambridge, UK, 2008.

58. Barone, R. *Anatomie Comparée des Mammifères Domestiques*; Vigot: Paris, France, 1976.
59. Cohen, A.; Serjeantson, D. *A Manual for the Identification of Bird Bones from Archaeological Sites*; Archetype Publications: London, UK, 2015.
60. López Plana, C.; Mayor, P.; Navarro, B.; Carretero, A.; Nacher, V.; Aige, G.; López, M.; Ruberte, J. *Atlas de Osteología de los Mamíferos Domésticos*; Universitat Autònoma de Barcelona: Barcelona, Spain, 2008; Available online: https://veterinariavirtual.uab.cat/anatomia/osteologia/Atlas_Virtual/Untitled2.html (accessed on 29 March 2020).
61. Pales, L.; Lambert, C. *Atlas Osteologiques des Mammiferes*; CNRS: Paris, France, 1972.
62. Castelló, J.R. *Bovids of the World: Antelopes, Gazelles, Cattle, Goats, Sheep, and Relatives*; Princeton Field Guides: Princeton, NJ, USA, 2016.
63. Huffman, B. Ungulates of the World. Species Fact Sheets. Available online: www.ultimateungulate.com (accessed on 29 March 2020).
64. Srinivasulu, C.; Srinivasulu, B. *South Asian Mammals: Their Diversity, Distribution, and Status*; Springer: New York, NY, USA, 2012.
65. Wilson, D.E.; Mittermeier, R.A. *Handbook of the Mammals of the World*; Lynx Edicions in association with Conservation International and IUCN: Barcelona, Spain, 2011; Volume 2.
66. Eisenmann, V. Comparative osteology of modem and fossil horses, half-asses and asses. In *Equids in the Ancient World*; Meadow, R.H., Uerpmann, H.P., Eds.; Dr. Ludwig Reichert Verlag: Weisbaden, Germany, 1986; pp. 67–116.
67. Fernandez, H. Ostéologie Comparée des Petits Ruminants Eurasiatiques Sauvages et Domestiques (Genres Rupicapra, Ovis, Capra et Capreolus): Diagnose Différentielle du Squelette Appendiculaire. Ph.D. Thesis, Université de Geneve, Geneva, Switzerland, 2001.
68. Halstead, P.; Collins, P. Sorting the Sheep from the Goats: Morphological Distinctions between the Mandibles and Mandibular Teeth of Adult *Ovis* and *Capra*. *J. Archaeol. Sci.* **2002**, *29*, 545–553. [CrossRef]
69. Helmer, D. Discrimination des genres *Ovis* et *Capra* à l'aide des prémolaires inférieures 3 et 4 et interprétation des âges d'abattage: L'exemple de Dikili Tash (Grece). *Anthropozoologica* **2000**, *31*, 29–38.
70. Jonhstone, C.J. A Biometric Study of Equids in the Roman World. Ph.D. Thesis, University of York, York, UK, 2004.
71. Payne, S. Morphological distinctions between the mandibular teeth of young sheep, and goats, capra. *J. Archaeol. Sci* **1985**, *12*, 139–147. [CrossRef]
72. Prummel, W.; Frisch, H.-J. A guide for the distinction of species, sex and body side in bones of sheep and goat. *J. Archaeol. Sci.* **1986**, *13*, 567–577. [CrossRef]
73. Uerpmann, H.-P. *Dental Morphology of Horses, Donkeys and Mules. Presented at the Horse*; Donkey and Co.: Basel, Germany, 2002.
74. Zeder, M.A.; Lapham, H.A. Assessing the reliability of criteria used to identify postcranial bones in sheep, *Ovis*, and goats, *Capra*. *J. Archaeol. Sci.* **2010**, *37*, 2887–2905. [CrossRef]
75. Zeder, M.A.; Pilaar, S.E. Assessing the reliability of criteria used to identify mandibles and mandibular teeth in sheep, *Ovis*, and goats, *Capra*. *J. Archaeol. Sci.* **2010**, *37*, 225–242. [CrossRef]
76. Davis, S.J.M. Measurements of a group of adult female shetland sheep skeletons from a single flock: A baseline for zooarchaeologists. *J. Archaeol. Sci.* **1996**, *23*, 593–612. [CrossRef]
77. Davis, S.J.M. Hacia una distinción métrica entre los astrágalos de oveja y cabra. In *What Bones Tell Us. el que Ens Expliquen els Ossos*; Lloveras, L., Rissech, C., Nadal, J., Fullola, J.M., Eds.; Universitat de Barcelona, Monografies del SERP 12: Barcelona, Spain, 2016; pp. 35–58.
78. Davis, S.J.M. Towards a metrical distinction between sheep and goat astragali. In *Economic Zooarchaeology. Studies in Hunting, Herding and Early Agriculture*; Rowley-Conwy, P., Serjeantson, D., Halstead, P., Eds.; Oxbow Books: Philadelphia, PA, USA, 2017; pp. 50–82.
79. Von den Driesch, A. *A Guide to the Measurement of Animal Bones from Archaeological Sites*; Harvard University, Peaboby Museum Bulletin I: Cambridge, MA, USA, 1976.
80. Binford, L.R. *Bones: Ancient Men and Modern Myths*; Academic Press: New York, NY, USA, 1981.
81. Brain, C.K. *The Hunters or the Hunted? an Introduction to African Cave Taphonomy*; University of Chicago Press: Chicago, IL, USA, 1981.
82. Portero, R.; Ariño, E.; Elorza, M.; Gurt, J.M.; Martínez-Ferreras, V.; Pidaev, S. Zooarchaeology of ancient Termez (Uzbekistan): From the Greco-Bactrian period to Islamic times. *Archaeol. Res. Asia* **2022**, *31*, 100384. [CrossRef]
83. Portero, R.; Fusaró, A.; Piqué, R.; Gurt, J.M.; Elorza, M.; Gabriel, S.; Pidaev, S. The Environment in the Islamic City of Termez (Uzbekistan): Zooarchaeology and Anthracology of a 9th-century *tannūr*. *J. Islam. Archaeol.* **2021**, *8*, 1–22. [CrossRef]
84. Lesguer, F. Les fours de potiers à Termez (Ouzbékistan) du IV siècle au XII après J.-C. In Proceedings of the Tourner autour du pot. Les ateliers de potiers médiévaux du Ve au XIIe siècle dans l'espace européen. Actes du colloque international de Douai, Caen, France, 5–8 October 2010; Thuillier, F., Louis, É., Eds.; Publications du Craham, Presses Universitaires de Caen: Caen, France, 2015; pp. 433–437.
85. Schlumberger, D.; Le Berre, M.; Fussman, G. *Surkh Kotal en Bactriane: Les Temples*; De Boccard, MDAFA XXV: Paris, France, 1983.
86. Kruglikova, I.; Pugachenkova, G.A. *Dilberdzhin (Raskopki 1970–1973)*; Nauka: Moscow, Russia, 1977.
87. Pugachenkova, G.A. *Khalchayan. K Problyemye Khudozhyestvyennoy Kulttury Syevernoy Bactrii*; Ministry of Culture of the Soviet Socialist Republic of Uzbekistan, Fan: Tashkent, Uzbekistan, 1966.
88. Guillaume, O.; Ligeron, J.-C.; Valence, R. *Fouilles d'Aï Khanoum II. Les Propylées de la rue Principale*; Mémoires de la Délégation Archéologique Française en Afghanistan, 26: Paris, France, 1983.
89. Thevs, N. *Ecology, Spatial Distribution, and Utilization of the Tugai Vegetation at the Middle Reaches of the Tarim River, Xinjiang, China*; Cuvillier Verlag: Göttingen, Germany, 2007.

90. Zeng, Y.; Zhao, C.; Kundzewicz, Z.W.; Lv, G. Distribution pattern of Tugai forests species diversity and their relationship to environmental factors in an arid area of China. *PLoS ONE* **2020**, *15*, e0232907. [CrossRef]
91. Nesbitt, M.; O'Hara, S. Irrigation agriculture in Central Asia: A long-term perspective from Turkmenistan. In *The Archaeology of Drylands. Living at the Margin*; Barker, G., Gilbertson, D., Eds.; Routledge: London, UK; New York, NY, USA, 2000; pp. 103–122.
92. Spengler, R.N.; Maksudov, F.; Bullion, E.; Merkle, A.; Hermes, T.; Frachetti, M. Arboreal crops on the medieval Silk Road. Archaeobotanical studies at Tashbulak. *PLoS ONE* **2018**, *13*, e0201409, Erratum in: *PLoS ONE* **2018**, *13*, e0204582. [CrossRef]
93. Tengberg, M. Économies végétales et environnement en Asie centrale du Néolithique à l'époque sassanide: La contribution des disciplines archéobotaniques. In *L'archéologie Française en Asie Centrale*; Bendezu-Sarmiento, J., Ed.; De Boccard, Cahiers d'Asie Centrale 21–22: Paris, France, 2013; pp. 545–558.
94. Motuzaite Matuzeviciute, G.; Preece, R.C.; Wang, S.; Colominas, L.; Ohnuma, K.; Kume, S.; Abdykanova, A.; Jones, M.K. Ecology and subsistence at the Mesolithic and Bronze Age site of Aigyrzhal-2, Naryn valley, Kyrgyzstan. *Quat. Int.* **2017**, *437*, 35–49. [CrossRef]
95. Wu, X.; Miller, N.F.; Crabtree, P. Agro-pastoral strategies and food production on the achaemenid frontier in Central Asia: A case study of Kyzyltepa in southern Uzbekistan. *Iran* **2015**, *53*, 93–117. [CrossRef]
96. Chenal-Velarde, I. Food, rituals?: The exploitation of dogs from Eretria (Greece) during the helladic and hellenistic periods. In *Dogs and People in Social, Working, Economic or Symbolic Interaction*; Snyder, L.M., Moore, E.A., Eds.; Oxbow Books: Havertown, PA, USA; Oxford, UK, 2006; pp. 24–31.
97. Snyder, L.M.; Klipper, W.E. From Lerna to Kastro: Further thoughts on dogs as food in ancient Greece; perceptions, prejudices and reinvestigations. *Br. Sch. Athens Stud.* **2003**, *9*, 221–231.
98. Frachetti, M.; Benecke, N. From sheep to (some) horses: 4500 years of herd structure at the pastoralist settlement of Begash (South-Eastern Kazakhstan). *Antiquity* **2009**, *83*, 1023–1037. [CrossRef]
99. Lhuillier, J.; Mashkour, M. Animal exploitation in the oases: Archaeozoological review of Iron Age sites in southern Central Asia. *Antiquity* **2018**, *91*, 655–673. [CrossRef]
100. Mashkour, M. Sociétés pastorales et économies de subsistance au nord-est de l'Iran et au sud du Turkménistan. In *L'archéologie Française en Asie Centrale*; Bendezu-Sarmiento, J., Ed.; De Boccard, Cahiers d'Asie Centrale 21–22: Paris, France, 2013; pp. 533–544.
101. Outram, A.K.; Kasparonv, A.; Stear, N.A.; Varfolomeev, V.; Usmanova, E.; Evershed, R.P. Patterns of pastoralism in later bronze age Kazakhstan: New evidence from faunal and lipid residue analyses. *J. Archaeol. Sci.* **2012**, *39*, 2424–2435. [CrossRef]
102. Taylor, W.T.T.; Shnaider, S.; Spengler, R.; Orlando, L.; Abykanova, A.; Krivoshapkin, A. Investigating ancient animal economies and exchange in Kyrgyzstan's Alay Valley. *Antiquity* **2018**, *93*, 1–5. [CrossRef]
103. Indoitu, R.; Orlovsky, L.L.; Orlovsky, N. Dust storms in Central Asia: Spatial and temporal variations. *J. Arid. Environ.* **2012**, *85*, 62–70. [CrossRef]
104. Chang, C.; Ivanov, S.S.; Tourtellotte, P.A. Landscape and Settlement over 4 Millennia on the South Side of lake Issiy Kul, Kyrgyzstan: Preliminary results of Survey Research in 2019–2012. *Land* **2022**, *11*, 456. [CrossRef]
105. Chang, C.; Ivanov, S.S.; Tourtellotte, P.A.; Spengler, R.N., III; Mir-Makhamad, B.; Kramar, D. Ancient Agricultural and Pastoral Lanscapes on the South Side of Lake Issik-Kul: Long-Term Diachronic Analysis of Changing Patterns of Land Use, Climate Change, and Ritual Use in the Juuku and Kizil Suu Valleys. *Land* **2022**, *11*, 902. [CrossRef]
106. Chang, C.; Ivanov, S.S.; Spengle, R.N., III; Tourtellotte, P.A.; Mir-Makhamad, B.; Tourtellotte, P.A. Montane Ecoclines in Ancient Central Asia: A Preliminary Study of Agropastoral Economies in Juuku, Kyrgyzstan. *Land* **2023**, *12*, 1406. [CrossRef]
107. Rtveladze, E. Kampyrtepa—Struktura. Pyeriodizatsyya. In *Matyerialy Tokharistanskoy Ekspyeditsii. Arkhyeologichyeskiye Isslyedovaniya Kampyrtepa*; Rtveladze, E.V., Ed.; Academy of Fine Arts of the Republic of Uzbekistan, San'At: Tashkent, Uzbekistan, 2000; Volume 1, pp. 5–18.
108. Sverchkov, L.M. Opyt sinkhronizatsii kyeramichyeskikh komplyesov epoki ellinizma (Kampyrtepa, Termez, Dzhigatepa, Kurganzol). In *Matyerialy Tokharistanskoy Ekspyeditsii. Arkhyeologichyeskiye Isslyedovaniya Kampyrtepa y Shortepa*; Rtveladze, E.V., Ed.; Academy of Fine Arts of the Republic of Uzbekistan: Tashkent, Uzbekistan, 2006; Volume 5, pp. 105–124.
109. Martínez-Ferreras, V.; Gurt, J.M.; Hein, A.; Pidaev, S.; Rtveladze, E.V.; Bolelov, S. Tableware in the Hellenistic tradition from the city of Kampyr Tepe in ancient Bactria (Uzbekistan). *Archaeometry* **2016**, *58*, 736–764. [CrossRef]
110. Pugachenkova, G.A.; Rtveladze, E.V. (Eds.) *Dalverzintepe. Kushanskiy Gorod na Yugye Uzbyekistana*; Academy of Sciences of the Soviet Socialist Republic of Uzbekistan, Fan: Tashkent, Uzbekistan, 1978.
111. Nehru, L. Khalchayan. *Encyclopædia Iranica*. Online Edition. 2017. Available online: http://www.iranicaonline.org/articles/khalchayan (accessed on 16 February 2023).
112. Boroffka, N.; Oberhänsli, H.; Sorrel, P.; Demory, F.; Reinhardt, C.; Wünnemann, B.; Alimov, K.; Baratov, S.; Rakhimov, K.; Saparov, N.; et al. Archaeology and climate: Settlement and lake-level changes at the Aral Sea. *Geoarchaeology* **2006**, *21*, 721–734. [CrossRef]

land

MDPI

Article

Resilience Thinking and Landscape Complexity in the Basentello Valley (BA, MT), c. AD 300–800

Matthew Munro

Department of Anthropology and Archaeology, The University of Calgary, 2500 University Drive NW, Calgary, AB T2N 1N4, Canada; matthew.munro@ucalgary.ca

Abstract: Archaeological data for the transformation of late Roman rural landscapes in Southern Italy over the sixth to eighth centuries AD are often meagre. This record often provides little explanatory power in the context of understanding the collapse of Roman political and economic hegemony and the framework for the regeneration of these relationships in the early medieval countryside. Resilience thinking offers a robust suite of heuristics to help guide both method and theory in understanding the key socio-environmental relationships involved in this transformative process based on limited material evidence. Through insights gained from developing a panarchic perspective of the Basentello landscape between AD 300 and 800, both capacities for and strategies of resilience to landscape-scale shocks and stressors emerge as key patterns in this collapse process. To explain how these patterns emerge, resilience thinking employs narratives from complexity science by framing landscapes as self-organizing complex adaptive systems. It is through appreciating this complexity that archaeologists can revolutionize how we understand landscape-scale transformations, the role of resilience in landscape history and, more broadly, the nature of societal collapse.

Keywords: archaeology; resilience; Late Antiquity; Early Middle Ages; panarchy; adaptive cycle; complexity; collapse; regeneration

1. Introduction

Research on the theory and insights of resilience is demonstrating the utility of 'resilience thinking' [1,2] when applied both to landscapes [3] and archaeological studies generally [4]. Resilience is still, however, a nebulous concept, which, similar to landscape, has shifting meanings and dimensions depending on the perspectives adopted. Studies in ecology suggest that when considered from the perspective of complexity science, resilience is not a measure of equilibrium-seeking systems bouncing back, but instead seeks to explain how complex systems develop by absorbing change [5,6]. Change can come in many forms, either from within or without, in the form of short-term shocks and long-term stressors, and the history of the landscape is derived from the decisions made by individuals and through institutions in the face of adversity [7].

Modern studies aim not to measure or quantify resilience but to understand its role in historical development qualitatively and, if there is sufficient evidence, to model capacities for resilience in specific landscapes and hypothesize strategies employed to respond to landscape-scale challenges. Such observations have potential to better explain patterns of landscape change and move debates beyond simplistic models of 'decline and fall' or 'continuity and discontinuity' with language and concepts derived from complexity science [8,9]. These heuristics are therefore of great importance to scholars of Late Antiquity and the Early Middle Ages since they provide another perspective through which to consider landscape developments associated with the transformation of the Roman landscape and the development of the medieval world.

To demonstrate the utility of resilience thinking in this historical and archaeological context, a case study in the Basentello Valley of central Southern Italy is illustrative. This region, the subject of several surveys and excavations [10–14], highlights the benefits and limitations of resilience thinking for landscape archaeologists and ancient historians by the nature of its location and archaeological record. Far removed from the extraordinary urban landscapes of Central and Northern Italy, the Basentello region provides a predominantly rural example where the contraction of social complexity following Roman political fragmentation appears more localized and less directly influenced by Lombard and Byzantine urban culture between c. AD 300 and 800.

Consequently, these inland regions of Southern Italy have long been considered a social backwater in the work of authors such as Carlo Levi [15] and historians such as Arnold Toynbee [16]. These early twentieth-century prejudices are being increasingly overturned for the Roman period [17], but, with few exceptions [18], they persist from the post-Roman world due to an impoverished archaeological and historical record for inland Basilicata and Southern Italy generally. While the heuristics of resilience thinking cannot replace historic evidence, they can hypothesize historic dynamics that fit and explain the little evidence available. As this research demonstrates, even this deeply rural and isolated population played a role in the collapse and regeneration of social complexity associated with the end of the Roman world and the development of the Middle Ages. Outlined here is a review of methods developed to investigate resilience in the Basentello Valley and a summary of results. A full description of the analysis and data is presented elsewhere [19].

2. Materials and Methods

2.1. Panarchy, Complexity, and Landscape

To demonstrate the role of resilience, it is necessary to view the landscape as a complex adaptive system composed of human–environmental relationships whose history is the result of how people handled shocks and stressors to their lives. These decisions had linear and nonlinear outcomes, which sometimes solved perceived problems and sometimes exacerbated them, dependent on the nature of the decisions and the scale at which they were made and perceived [20]. Resilience to these shocks and stressors can take many forms, ranging from perpetuating the system by coping with change, to adapting to a new social or environmental reality, or transforming social structures to enhance future well-being [21]. These decisions can ultimately change the identity of the system in question by, at one extreme, transforming its structures or, at the other, by collapsing into a simpler state.

Viewing the landscape from a complex systems perspective is vital because it provides the explanatory framework that gives resilience thinking its interpretive power. Emergence, nonlinearity, and edge of chaos dynamics have long been observed in complex systems such as human societies [22] and all influence the history and identity of landscapes. This complexity reinforces the observation that human systems are both robust and fragile to a wide suite of historically contingent shocks and stressors: sometimes, societies can weather the storm, while, at other times, quite unpredictably, collapse prevails.

This conflicting image of societies, one of robustness and one of fragility, is inherent to narratives of the rural Roman and early medieval landscapes [23]. Resilience thinking bypasses debates of continuity and discontinuity and allows for a systemic view on transformation and even the potential to explain it. One popular method derived from ecology uses an analytical framework called panarchy with a heuristic model called the adaptive cycle at its core [24]. The four-phase approach of the adaptive cycle is non-determinative and simply labels well-known periods of human societal development: growth and exploitation (r-phase), conservation (K-phase), release (Ω-phase), and reorganization (α-phase). These periods are assigned based on qualitative changes to the potential and connectedness of a given complex adaptive system, and thereby expose the capacity for resilience likely in each phase (Figure 1).

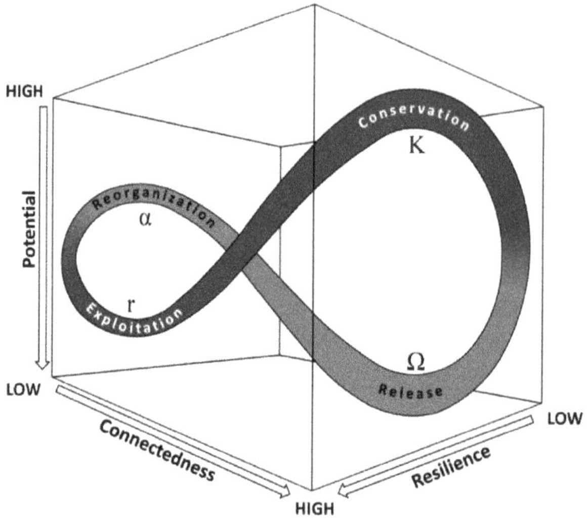

Figure 1. The adaptive cycle heuristic model.

This approach to modelling societies and societal change is familiar to historians. For Oswald Spengler, civilizations and cultures were biological entities with lifecycles of birth, life, and death [25] (p. 21). Arnold Toynbee, influenced by Spengler, adapted his organismic perspective less deterministically but still admitted to rhythms of growth and rhythms of disintegration resulting from the formula of challenge-and-response [26] (p. 548). Braudel's scalar approach, made famous by the longue durée, also recognized these rhythms [27]. More recently, Peter Turchin has applied this idea of societal lifecycles to what he calls 'Cliodynamics' in an effort to develop a predictive model of history [28,29].

These historical models are all problematic. Societies do not function as organisms for a very important reason: the relationships between parts of an organism are linear and reducible. An organism is the sum of its parts, whereas societies are not. The connections between relationships occurring at different historical scales were missing from Braudel's celebrated work on the sixteenth-century Mediterranean world [30]. Neither is history as predictable as Turchin implies, since human interactions and environmental relationships are linear and nonlinear: small changes in one part of society can have massive and unpredictable consequences [31].

The adaptive cycle itself is not a solution to these challenges. Given the non-reducibility of complex phenomena, such as those inherent in human social structures, modelling an entire society with a single adaptive cycle does not maximize the insights possible with panarchy (e.g., [32]). Panarchy provides conceptual tools to consider multiple adaptive cycles operating at different speeds and in different networks of relationships within a society. These conceptual tools include how to identify cross-scale interactions by comparing adaptive cycles or various speeds and at different scale and recognizing when resilience at one scale of the panarchy might serve to increase it or decrease it at another scale (i.e., the 'remember' and 'revolt' dynamics). Since the goal of panarchy is to 'be as simple as possible but no simpler than necessary', researchers suggest that at least three to five key interacting landscape components are included, and these should represent at minimum three qualitatively different speeds of adaptive cycles [33] (p. 102).

The landscape represents one such network of human–environmental relationships that can be heuristically modelled by the adaptive cycle and analyzed through panarchy. However, a landscape alone is insufficient to build a panarchy: a society must be analyzed from several scalar and functional perspectives. Political, economic, and urban cycles can be constructed based on available data, and so too can a wide range of possible shocks and stressors experienced by landscape inhabitants over time. All these elements are required to build a robust panarchy to facilitate analysis.

There is a further issue: neither the adaptive cycle nor panarchy can detail or explain any changes observed without considering how societies are structured or operate. The adaptive cycle can help to deduce when resilience is high or low, decreasing or increasing, in a network of relationships within a complex adaptive system, and only if we can determine relative changes in their potential and connectedness. However, these relationships, taking the form of an economy, political organization, or landscape, are between people and things that have a capacity for agency. In other words, people make decisions at various scales, and those decisions determine the outcome of history.

Research in social resilience has suggested that groups of people commonly act in the face of shocks and stressors to their lifeways to achieve three outcomes: to restore present well-being, to secure future well-being, or to enhance present and future well-being (Table 1). If these strategies for resilience can be identified in the archaeological record at the landscape scale, they will represent the foundational evidence for resilience at that scale.

Table 1. The matrix of social resilience, after Keck and Sakdapolrak [21] (p. 10).

	Coping Strategies	Adaptive Strategies	Transformative Strategies
Response to risk	Reactive	Proactive	Proactive
Temporal scope	Short-term	Long-term	Long-term
Degree of change	Low, seeking status quo	Medium, incremental change	High, radical change
Outcome	Restoration of present well-being	Security of future well-being	Enhancement of present and future well-being

Finally, for scholars of Late Antiquity and the Early Middle Ages, the utility of the resilience concept must focus on why resilience is seemingly reduced in a way that results in the collapse of Roman institutions and organization. The release phase (Ω) implies a loss of resilience and release of system resources, but it does not imply collapse. A collapse in any adaptive cycle is a special condition where the complexity of the relationships simplifies quickly [34,35]. In some cases, collapse in one level of the panarchy can send the other adaptive cycles into their own release phases, threatening to collapse the entire system [6,33].

From the perspective of the western empire, this process occurred over the fifth and sixth centuries due to a host of simultaneous shocks and long-term stressors to urban, political, and economic relationships at several scales and reflected in the archaeological record. The assumption here is that if the resilience of the western empire was sufficiently robust, then it would have survived the calamities of these centuries; as it did not, we must look to these centuries for proximate causation.

After choosing a defined landscape and establishing its identity, the method for landscape-scale analysis of resilience used here includes:

1. Establishing scales of inquiry where responses to adversity could be observed historically and archaeologically. These scales necessarily expand wider than the landscape under investigation to increase the likelihood of observing linear and nonlinear historical outcomes.
2. Constructing adaptive cycles from these scales and arranging them into a panarchy that also includes known shocks and stressors associated with the landscape under investigation.
3. Deducing periods with a heightened capacity for resilience in the landscape under investigation by analyzing relationships observed in the panarchy and explaining them through complexity science.
4. Examining the archaeological patterning of the landscape to hypothesize strategies for resilience that inhabitants may have used that resulted in the unfolding of landscape history.

2.2. Panarchy, Complexity, and Landscape

Resilience thinking can be applied to any landscape with evidence for human decision making. One value of this method is that we can say new and interesting things about the past using fragmented archaeological and historical records. This is valuable when studying landscape development from the Roman to medieval periods given the paucity of data for the latter in many regions of Mediterranean Europe, despite historical evidence for continuity of occupation, even if greatly reduced.

The 313-square-kilometer survey area formed by the Basentello Survey Project eastern zone [10] and the Basentello Valley Archaeological Research Project western zone [13,14] is a perfect case study: its data are robust, it spans the late Roman and early medieval periods, and it is relatively isolated from urban influence (Figure 2). Understanding resilience in this context is therefore specific to the rural world. Additionally, excavations at the Roman *vicus* of Vagnari [11,36] and the villa of San Felice [12,37] add tighter chronological control while providing detailed evidence through which to interpret strategies for resilience.

This evidence is situated within a geomorphological feature called the Fossa Bradanica, a geological rift of hills and flat-floored valleys separating the Apennines from the limestone plateau of the Murge (Figure 3). These surveys extend coverage between surveys and excavations in the Upper Bradano to the southwest [38], those in the *ager Venusinus* to the west [39], and an earlier survey to the north between the Via *Appia* and the Murge escarpment completed by Vinson in the late 1960s and early 1970s [10] (pp. 642–674). When considering the work done in the *chora* of Metaponto further southeast [40] and the Tavoliere further northwest [41], there is a robust landscape dataset from which to discuss similarities and differences between microregions across the wider Fossa Bradanica.

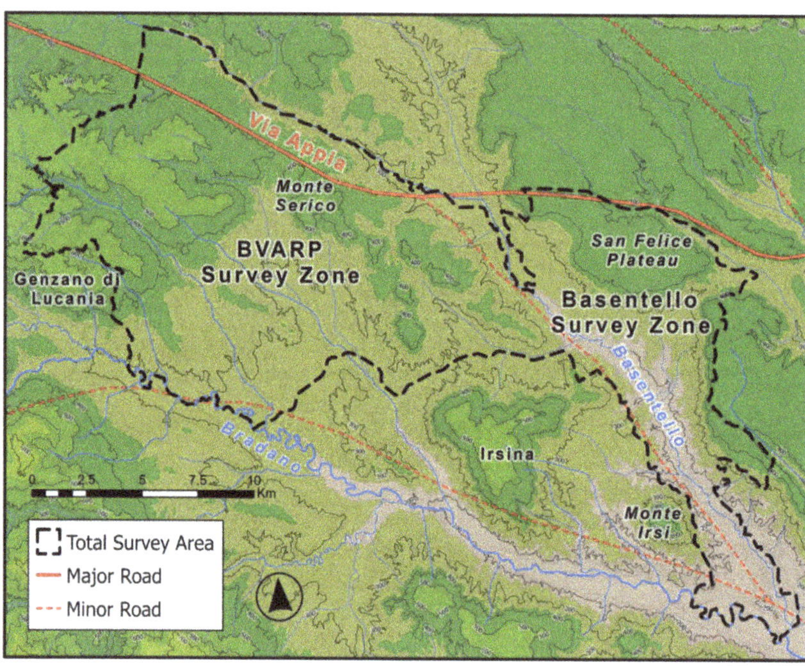

Figure 2. The Basentello Valley survey area.

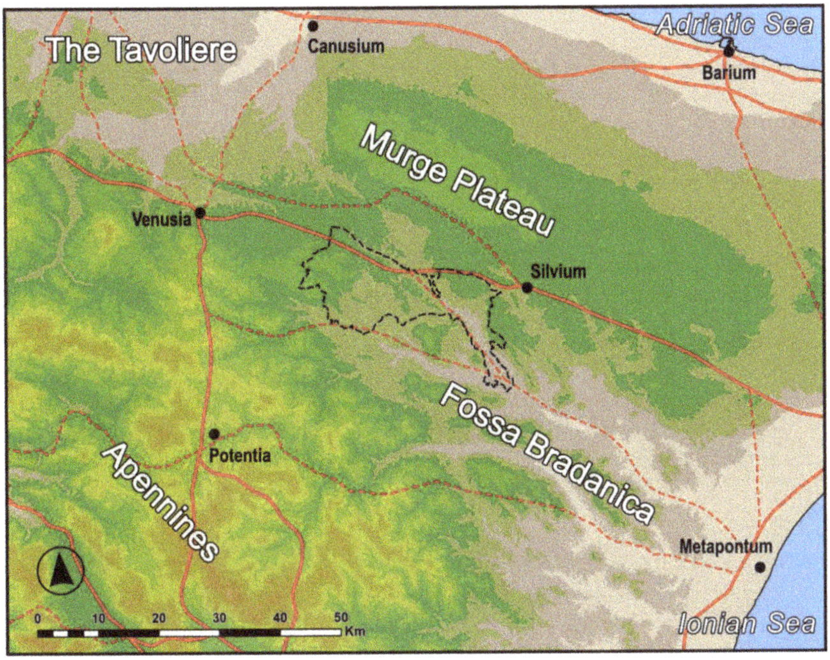

Figure 3. The survey area showing geographic features, the likely locations of major (solid) and minor (dashed) Roman roads, and nearby Roman towns.

On the Basentello Valley pedestrian survey [42], teams of 5 to 6 individuals walked approximately 15 m apart over all passable land (i.e., neither wet nor wooded) in the survey area. If more than five artifacts were identified within a 20 m transect, the area was selected for more intensive inspection through 10 × 10 m collection grids. After collection, the team classified artifact scatters as either 'findspots' (interpreted as casual or episodic frequentations of Paleolithic and Neolithic age) or 'sites' (interpreted as regular occupation, which is true of most Roman and medieval collections). McCallum and his team, on the other hand, conducted a more extensive survey, walking 15 to 25 m apart while focusing more closely on understanding changes within areas of defined cultural activity. Rather than identify 'sites' and attempt either 100% artifact collection or random samples of diagnostic sherds, they broke scatters into 0.25 ha 'collection units' and randomly sampled a 3-m-diameter circle within each unit systematically. While they later reassembled these collection units into 'sites' in a more traditional sense, this collection control allowed them to 'identify and study chronological, spatial, and functional variation within individual scatters or sites' [43] (p. 172).

The data generated from these projects are extensive and can be found in more complete forms in the survey volume by Small and Small [10] (pp. 321–395), in prepared articles by McCallum and his teams [13,14,43], and in a forthcoming doctoral thesis [19]. Sites dating to Late Antiquity and the Early Middle Ages include:

1. Agglomerated settlements (large scatters with evidence for diversified functional structures related to production, storage, and habitation for several families, i.e., agricultural villages or *vici*);
2. Villas (Roman residential structures of moderate size with evidence for agrarian production and upper-class habitation);
3. Farmhouses (scatters of moderate size with evidence for agrarian production and no elite material culture);

4. Small dwellings (of small scatter size with some domestic artifacts but undifferentiated function);
5. Working areas (scatters of different sizes that suggest agrarian or craft production with no evidence for sustained habitation).

Identified sites were assigned occupation date ranges based largely on ceramic evidence. African Red Slipware (ARS) in A (c. 100–200), C (c. 200–400), and D (c. 325–675) forms and Late Roman Painted Ware (LRPW) were the primary evidence for late antique occupations [44,45]. Early medieval occupation was identified through the presence of predominantly late Roman pottery associated with combed architectural tiles (used from the late sixth to late seventh centuries) [10] (pp. 627–629) and various locally produced medieval wares. If medieval pottery was associated with combed tiles, it was taken as a higher likelihood of some continuity of a site into the eighth century. Some medieval pottery, mostly plain wares and some glazed wares, is broadly dated from the eighth to tenth centuries. Future excavation is required to confirm or deny estimations made from the surface records in every case. This ambiguous evidence for sustained early medieval occupation after c. AD 800 will not be considered here.

An analysis of these Basentello Valley data [19] resulted in the identification of six landscape-scale patterns:

1. Agglomerated settlements became larger over the fourth and fifth centuries;
2. There was a shift over Late Antiquity from regional to local production;
3. Economic relations and land use suggest a shift from predominantly dry cereal agriculture to mixed agricultural and silvo-pastoral pursuits over the sixth century;
4. Settlement moved upland over the sixth to eighth centuries;
5. The early medieval landscape was one of sparsely occupied forests and pastures with significant settlement abandonment over the seventh and eighth centuries;
6. Agglomerated settlements (*vici*, hamlets, villages) were important articulating nodes in the production and settlement systems over both periods.

These patterns are familiar to scholars of late Roman and early medieval Italy. Other patterns, equally ubiquitous in other parts of Italy, are more ambiguous here. The question of late antique villa decline is unanswerable in this landscape since the excavations at San Felice, the only such excavation in the valley, indicate that its occupation ended in the mid-second century, perhaps due to an earthquake [37] (p. 127). Additionally, the monumentalization of surviving late antique villas observed at sites such as Faragola [46], San Giovanni di Ruoti [47], and Masseria Ciccotti [48] is not directly observed in the Basentello region; however, the growth observed in some site areas [13] (p. 342) might reflect this trend.

3. Results

A summation of the results is presented here. A full description of the data and discussion leading to these results, including the historical and archaeological data consulted to build the panarchy, is presented elsewhere [19].

3.1. Capacities for Resilience in the Basentello Landscape

The adaptive cycle helps to illustrate capacities for resilience if qualitative changes in connectedness and potential are known for a given landscape.

The connectedness of the Basentello landscape was a function of culturally mediated information flow, the availability and flow of landscape resources, and the infrastructure of connectedness [49,50]. Dimensions of connectedness include the ways in which information, mostly about land management, was disseminated through the settlement hierarchy.

Survey evidence suggests that this hierarchy grew rigid with top-down control over Late Antiquity judging by the imperial control observed at the *vicus* of Vagnari until c. AD 400 [51] (pp. 32–33). It appears that agglomerated settlements, some perhaps *vici* under imperial control, such as Vagnari, articulated this landscape thanks to their dominant positions on the Via *Appia* and the network of drove roads called *tratturi* (Figure 4).

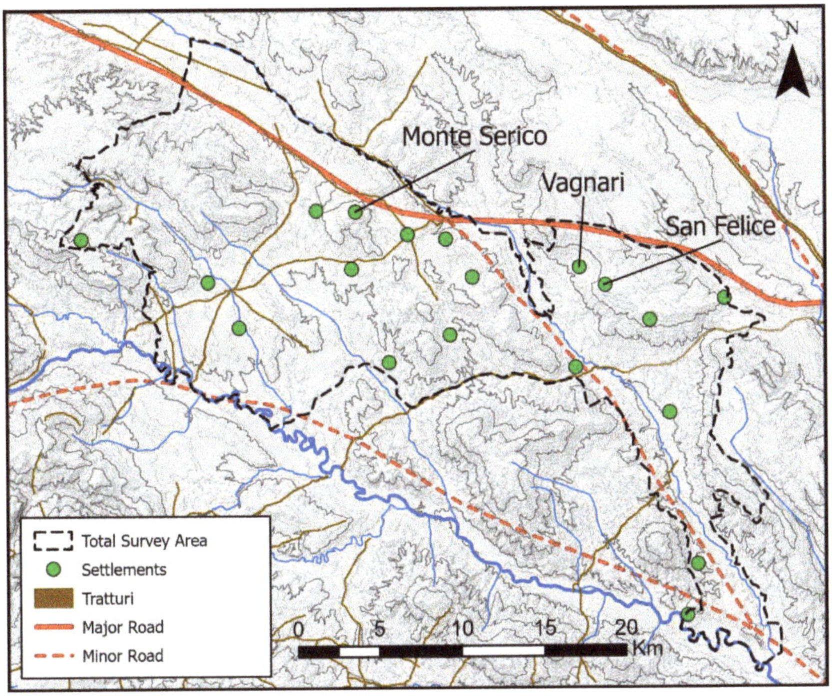

Figure 4. The Basentello Survey Area showing interaction between Roman roads, *tratturi*, watercourses, and agglomerated settlements (c. AD 200–1000) identified during surveys.

Overall, there is little sign of immediate change to the settlement system or land management in the sixth century. This is unexpected given the mid-century calamities, including the 536 dust veil event [52], the Ostrogothic War of 535–554 [53], and the spread of bubonic plague after c. 541 [54]. Some sites were abandoned over this century, but this likely resulted from dwindling landscape resources more than changes in connectivity (see below). It was over the seventh century that the settlement network collapsed into a simpler form, leaving mostly agglomerated settlements. Culturally mediated information flow in this landscape had effectively switched, over two centuries, from a rigid top-down control hierarchy to one that appears broadly heterarchical.

The history of culturally mediated information flow in the Basentello shares common elements with the changing availability and flow of landscape resources. Whereas changes to landscape potential are dictated in part by environmental and climatic considerations, in terms of connectedness, the flow and availability of resources is a social and economic consideration. Based on the survey evidence, inhabitants of the Basentello had a diversifying and localizing relationship with the available landscape resources. Broadly, in the Roman period, there appears a progressive move toward increasing grain cultivation at the expense of forests and pastures in this landscape. Imported pottery was still circulating, and this presumably meant that other goods such as textiles and foreign foodstuffs were at least as abundant. This reversed toward the end of Late Antiquity so that, by the Early Middle Ages, this landscape was providing nearly all the resources that its inhabitants required, however diminished in quantity and quality.

The flow of both information and landscape resources was dependent in large part on the infrastructure of connectedness. This infrastructure focused on rivers, roads, and *tratturi.* The rivers and streams of the Fossa Bradanica were too shallow for cargo vessels, but they could be followed on foot, on hoof, or by float. The Basentello flowed southeast into the Bradano river, which continues 50 km east until it empties into the Ionian near Metaponto, providing a corridor for travel and the shipping of goods, perhaps on donkeys or mules, or on small barges or logs, to the coast [42] (p. 367).

While these river routes did not change over our period, reasons for following them and their safety for travelers and inhabitants certainly did. The wetter weather of c. 400–550 and 650–750 likely saw a proliferation of the Anopheles mosquito in the Basentello region, making habitation and travel in the river valley more difficult. Overlaying the survey data onto Torelli's 1882 *Carta della Malaria dell'Italia* shows that the movement of sites uphill was partly a response to this malaria threat: the southern limit of settlement sites over the Early Middle Ages matches the area of light malaria associated with Central Basilicata (Figure 5). Malaria was a chronic disease that likely impacted populations more when environmental conditions were prime; the waves of bubonic plague that are registered in Italy over several waves were likely more deadly but are more difficult to identify historically for the Basentello region [55] (e.g., pp. 32–34).

The Roman road network was also fundamental in structuring settlement and economy in the Basentello region. Over the Roman period, sites were concentrated along or near the Via *Appia.* After the Via *Herculia* was built in the late third century, the prominence of the Via *Appia* in this part of Southern Italy declined. The road was likely still used regionally, but its interregional position was lessened. One long-theorized stressor in the late Roman and post-Roman landscape related to roads was the increased insecurity of travel caused by bandits or armies [56] (pp. 282–283). This situation is not obvious for the survey areas and, in fact, evidence suggests that as inland regions were abandoned, inhabitants moved closer to the Via *Appia.* The known shocks and stressors associated with living on or near roads during our period were not enough to trigger adaptive or transformative settlement strategies. People must have coped with the dangers, however infrequent.

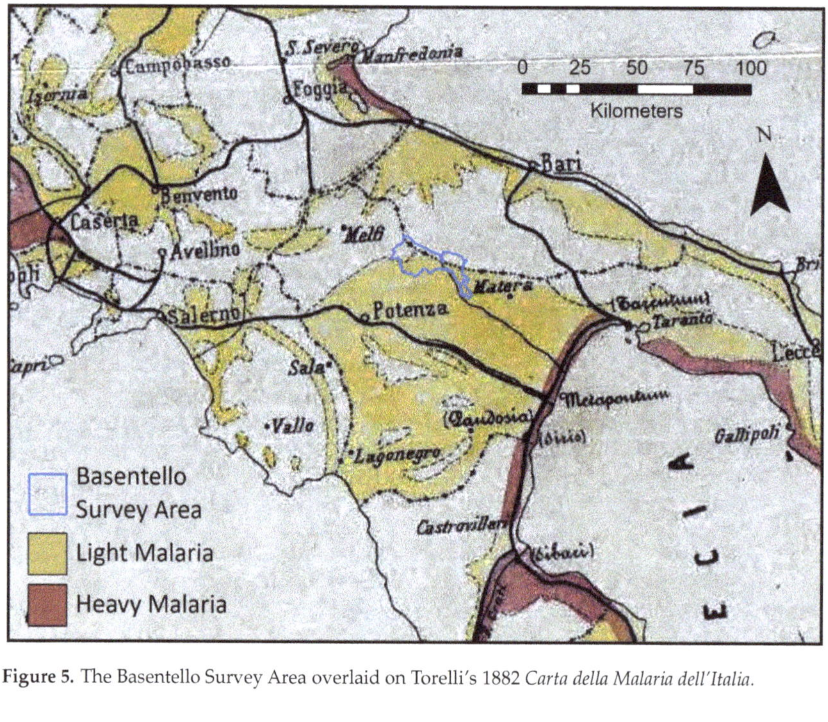

Figure 5. The Basentello Survey Area overlaid on Torelli's 1882 *Carta della Malaria dell'Italia.*

The connectedness of this landscape was dynamic over our period but certainly decreased qualitatively. Connectedness was high during c. 300–500, after which a slow drop started in the sixth century and continued until c. 700. This low level of landscape connectedness continued from c. 700 until c. 900, after which new sites started to appear and connectedness became more coherent. This increase continued into the Norman period and was strengthened with the appearance of castles, such as the re-founding of Monte Serico [57].

According to Holling [49] (pp. 393–394), potential represents what is available in the landscape to facilitate and undergo change. This can be summarized as forms of wealth, measured in terms of human-made, natural, sociocultural, and institutional capital [58]. These kinds of capital assume that there were resources that could be used to respond to shocks and stressors.

Human-made capital is largely a function of economic activity and labor. Judging by the pottery collected during a surface survey, economic potential weakened as sites were abandoned over the sixth and seventh centuries. Excavation at Vagnari suggests that this began during the first quarter of the fifth century [59]. The surface record suggests that imported items were nearly non-existent around 700 since very few date to after this period and those sites which still register imported items no longer accurately reflect their trade in the landscape.

In terms of natural capital, the landscape had good water sources but not soil [60]. Water sources included rivers, streams, and natural springs. Due to the geomorphology of the valley, a series of springs were regularly located at the c. 425 m contours of the San Felice plateau, for example [42] (pp. 357–358). These spots were unsurprisingly associated with sites, given that the environmental history of this region highlights its generally high aridity [61]. This association does not, however, extend beyond the San Felice plateau.

Over our period, and indeed for centuries before and after, this landscape appeared mostly silvo-pastoral. These vocations were partly influenced by connectivity and partly by the geomorphology of the region. The argillic, calcareous, and easily eroded slopes made cereal cultivation in this region difficult but not impossible given the economic advantages of intensification. In addition, the geomorphology of the region also indicates much environmental dynamism over the Holocene [60], resulting in at least 4–5 m of deposition in the main valleys suitable for dry cereal cultivation [42] (p. 349).

We have little evidence to discuss the sociocultural capital of the Basentello region after the decline of Vagnari. We may theorize aspects of social networks and the norms that connected and regulated decision making based only on patterns of settlement and economy. It is reasonable to suggest that as top-down control hierarchies weakened with disruptions to landownership in the later fifth and sixth centuries, the remaining population, however weakened by numbers or economic strength, had a larger measure of control over their daily practices. Relationships must have become more heterarchical and less rigidly hierarchical, with charismatic leaders facilitating some collective action amongst disparate peasant communities and responsible for the self-organizing nature of the complex landscape system. This model was previously proposed in the context of early medieval Italy by Wickham: growing peasant autonomy after c. 600–700 was a condition of weakening aristocratic authority and the inability of Lombard and Byzantine administrations to extract land tax from the countryside [62] (e.g., pp. 739–741). Complexity science suggests that the mechanism through which this autonomy was expressed was self-organization and heterarchical decision making.

Institutional capital is easier to understand than sociocultural capital based on the survey evidence. The main institutions operating in this landscape, namely the imperial government, the church, and the landowning class, all lost significant political control over this region during our period. The decline of Vagnari from c. 400 indicates that imperial patrimony waned or ceased at this time. The decline in villa sites over the fifth and sixth centuries shows that large-scale landowners were also losing authority in the landscape. More ambiguous are the agglomerated settlements that survived the sixth century and continued through the Early Middle Ages. There is no surface evidence to indicate social differentiation at these sites, and while some manner of individual or class leadership was likely, institutional capital must have been very low. Only the church had a form of hierarchical control over the lives of inhabitants in the survey region, and even with the founding of Belmonte in the mid-sixth century, this must have been tenuous until the High Middle Ages at the earliest [63].

The quality of the physical landscape potential alternated with various climatic and environmental changes. Socially mediated potential, however, appears to have declined over our period, similar to connectedness, with two noteworthy differences. As with connectivity, landscape potential, according to survey results, appears high from c. 300–400 but its decline began earlier, from c. 400–500, and remained very low from c. 500 until after the end of our period. The Basentello landscape did not experience a resurgence in landscape potential until the late tenth or early eleventh century at the earliest and due to the emergence of stable village communities throughout the *mezzogiorno* related to the Norman occupation [64]. Survey evidence suggests that the plateau of San Felice was reoccupied by the early twelfth century [65], and, while the evidence is mostly textual, settlement in the vicinity of Monte Serico appears to also have expanded during this period [66].

Overall, resilience thinking suggests that the period from c. 300 to 400 had a low capacity for resilience, whereas it increased from c. 400 to 700. The highest period of resilience in the landscape was from c. 700 to 900, after which resilience began to decrease slowly as landscape potential and connectedness was re-established, likely not until the eleventh century at the earliest (Figure 6). It is possible that these considerations for resilience had a part to play in the ease with which the Normans took control over Southern Italy in the mid to late eleventh century: the capacity for resilience in this landscape was growing but it was still low.

This landscape-scale release phase does not qualify as a collapse event, despite the observation that landscape organization of the seventh century was qualitatively simpler and more local compared to the fifth century. This qualitative interpretation is consistent with the coarse-grained reading of the survey data: instead of collapse, the Basentello Valley appears to have transformed over 150 to 200 years, with site losses in the sixth century due to the release phase and in the seventh century due to competition and attrition during the early medieval exploitation phase. The reorganization phase is absent because the landscape potential remained low following release, and surviving Roman landscape structures were reappropriated according to new medieval landscape logic. This all occurred over six to eight human generations.

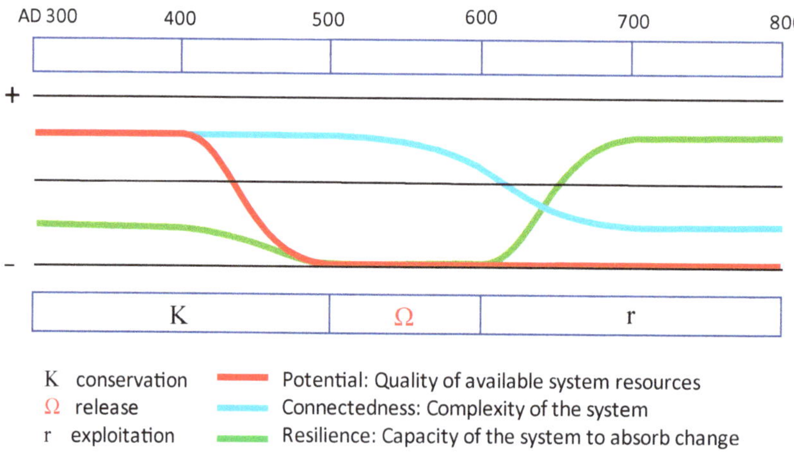

Figure 6. An adaptive cycle for the Basentello Valley landscape.

3.2. The Panarchy

Political, economic, and urban changes are well suited for heuristic modeling using the adaptive cycle and these form the basis for the panarchy. Rather than narrowing the analysis to the Basentello Valley itself, whose political, economic, and urban dimensions are largely unrecorded for our period, generalizing these dynamics for Southern Italy enhances future comparability while ensuring that cross-scale interactions are more visible. In terms of shocks and stressors, the most pertinent for Southern Italy include climatic and environmental dynamics, conflict (military, raiding, and border pressures), and mortality events (known human and animal disease outbreaks and food shortages). Estimated population figures for the Italian peninsula also help to gauge general landscape conditions for Southern Italy. The resulting panarchy is illustrated in Figure 7. This represents the panarchy through which the landscape dynamics can be understood in terms of resilience.

A full panarchic analysis requires reading the panarchy iteratively and considering the cross-scale dynamics that would have resulted in known historical and archaeological narratives. An appropriate resolution for the Basentello dataset is at the century scale, but this will become more refined as additional excavations are conducted.

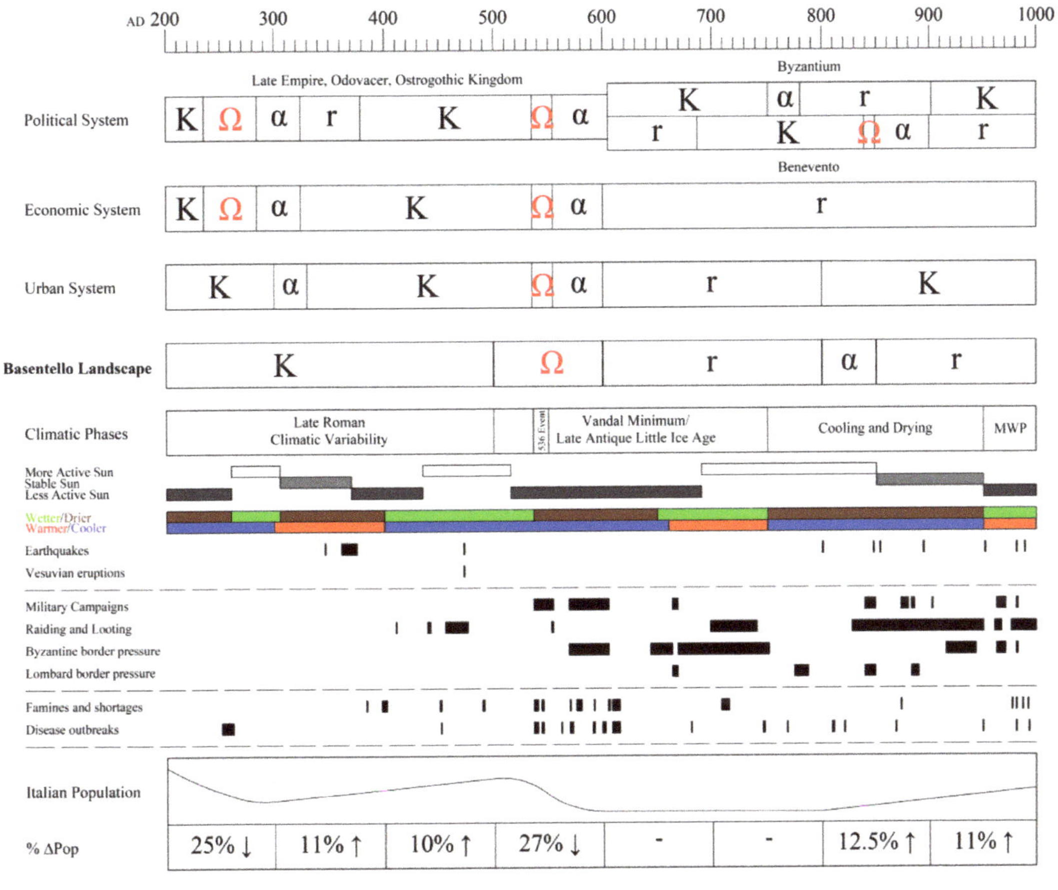

Figure 7. A panarchy for Southern Italy c. AD 200 to 1000.

Prior to discussing strategies for resilience, one observation about the panarchy is noteworthy. It is clear that a cross-scale cascading collapse occurred in the middle of the sixth century. The revolt dynamic was at work here as each level of the panarchy collapsed into a less complex state and shifted identities from Roman organizational logic to medieval. Our resolution for the Basentello Valley does not allow a more refined view of when this Ω phase occurred beyond the sixth century, but it is likely that if the political, economic, and urban systems were not all in the K phase, the landscape would not have experienced its own release, despite the mid-sixth century shocks and stressors. According to resilience thinking, the collapse of the western Roman empire was historically contingent on a nexus of randomly interacting phenomena entirely unpredictable from the perspective of late antique society.

3.3. *Strategies for Resilience*

Each of the six trends identified when reviewing the survey evidence for the Basentello Valley included several strategies that highlight the role of resilience in landscape development. Only two examples are presented here: a panarchic analysis of the observation that several sites grew larger over the fourth and fifth centuries, and an explanation of why agglomerated settlements were articulating nodes in the landscape during both Late Antiquity and the Early Middle Ages.

3.3.1. Sites Become Larger over the Fourth and Fifth Centuries

At the start of the fourth century, the political, economic, and urban reorganization of Southern Italy had no cross-scale impacts on the resilience of the Basentello landscape. Directed mainly from political top-down edicts by Diocletian and Constantine, this system continued benefiting the wealthiest landowners at the expense of the rural working populations [67]. These poorer classes had both legal rights and pathways to social advancement and wealth accumulation much reduced through taxation, which allowed wealthier landowners to further accumulate land at their expense [68]. This region remained in the K phase and any new social or economic opportunities afforded by these developments were absorbed by the landowning classes to increase landscape potential and connectedness, further benefiting the status quo.

Climatically, this century was warmer and drier than the preceding one, and the stable sun period suggests that warmth and precipitation for grain cultivation was predictable if not entirely optimal [69]. Environmentally, the only shocks included a major earthquake in 346 and a series of quakes from 361 to 375 [70]. It is likely that this tectonic activity was responsible for the shifting of the Vagnari settlement nucleus to the south side of the ravine [71]. Conflict in the landscape of Southern Italy was minimal apart from perhaps increased brigandage due to the new tax burden initiated by Diocletian [72], but we have no direct evidence. The food shortages of 383–384 [55] (pp. 207–208) and 397–402 [73] (pp. 134–135) likely affected this rural landscape given the biological and economic reliance of its inhabitants on agricultural resources during the fourth century. While it is unlikely that starvation was widespread since the population was relatively low, nonlinear effects would have stressed existing food resources as larger centers tried to extract more.

The capacity for resilience in this landscape continued to reduce over the fifth century. Landscape potential appears to have lessened, signaling that a shift into the release phase was likely and collapse was possible. Cross-scale analysis shows that this was unsurprising: the political system was in turmoil and the economic system was strained as the richest Italians continued to accumulate wealth while various barbarian groups confiscated lands all over Western Europe and the Mediterranean [74] (pp. 205–209). Since the imperial system was in the K phase, little could be done to face these challenges constructively. The cascading changes associated with these developments weakened the potential of Italian cities and, exemplifying its cascading nature, in turn weakened the potential of the landscape. Rather than enter a period of heterarchical peasant autonomy, the decreased power of the social hierarchy made this rural world more vulnerable to change since resilience remained very low. As the fifth century ended, any sufficient series of shocks and stressors could potentially send this entire complex adaptive system into the release phase.

There was no shortage of shocks and stressors during this period. The region appears wetter and cooler in the fifth than the fourth century [69], likely further reducing agricultural yields in this period of climatic variability. Incidentally, this is the century that Vagnari declined and the Basentello landscape likely began to shift from primarily cereal cultivation to mixed agrarian practices, but grain farming was still dominant [61]. The wetter, cooler climate would have also brought malarial mosquitoes closer to low-lying settlements while affecting farmers in the valley bottoms disproportionately more than the shepherds sticking to the uplands, as skeletal analysis from Vagnari [75] and early modern ethnographic comparison from Italy [76] both suggest.

Economic stressors were also part of this shift: cross-scale interaction between a contracting political and economic imperium lessened both hierarchical landscape oversight through financially weakened elites and the impetus for rural inhabitants to contribute to maintaining the status quo. There are few other socio-environmental explanations for the large-scale agrarian shifts interpreted for the Basentello valley; the Pollena Vesuvian eruption of 472 was far from this landscape and there is little evidence to suggest that it had any linear or obvious nonlinear cross-scale effects [77].

More conceivable are nonlinear effects derived from landscape insecurity. Coupled with economic strain, the Roman road network likely provided easy access for Visigoths to raid and loot the western half of the *mezzogiorno*. While there is no evidence that they entered the Basentello region, or indeed crossed far into the Apennines, tax levies for even *Apulia et Calabria* were much reduced at least into 413, presumably because local landowners and *coloni* did not have the resources to pay their share [78] (pp. 245–246).

The Visigothic attacks represent only the first wave of heightened conflict in the south. Piracy on the coasts and banditry on the road network increased over the fifth century, particularly after the fall of Carthage in 439 and then Sicily in 468 [74] (42–46). Legislation decreed during this period is informative: banditry was common enough along inland roads and near cities to warrant mention [79] (pp. 474–475). Direct archaeological evidence for this conflict is lacking in the Basentello region over the fifth century and the settlement system appears to have continued unhindered. Resilience to increasing landscape conflict might be found in the expansion of interregional connectivity experienced by the creation of the Via *Herculia* in the fourth century to service central *Lucania* [80]. The Via *Appia* appears to have been used predominantly for local travel from as early as the late fifth century [10] (p. 278), and, coupled with economic contraction, potential targets in the Basentello landscape were few.

There is little skeletal information that confirms incidences of food shortage or famine in the Basentello over the fifth century due to most interments at the Vagnari necropolis dating to the second and third centuries [81]. This rural landscape was more likely subject to nonlinear consequences of food shortages affecting the urban centers. The increasingly mixed agrarian economy of the region perhaps prevented the worst effects of shortages. However, food security strain on the inhabitants of the Basentello area was inevitable over this century thanks to economic contraction and political turmoil, along with their numerous cross-scale and nonlinear consequences.

This short panarchic analysis of the fourth and fifth centuries highlights why sites in the Basentello landscape may have grown in complexity and the role of resilience in this process. This growth was an adaptive strategy favoring rural elites looking to further extract resources from the countryside for economic gain to pay for the increased taxes of Diocletian and his successors. This was a proactive, incremental change to established sites, with the near-term goal of securing future well-being for the landowning class thanks to cross-scale interactions with political and economic cycles in the K phase. There was no centralized control mandating that landowners expand their operations or that settlements take in more inhabitants, and yet this self-organized patterning is evident in the archaeological record at several sites in the Basentello Valley.

This adaptive strategy would have created tension among landscape inhabitants. The growth in agglomerated settlements meant that more people were losing their land, likely through legal sale to avoid tax arrears, and being forced into more asymmetrical, quasi-indentured, agrarian service. This strategy for resilience benefited the landowning classes, at the expense of the social and political standing of the smallholders or *coloni*.

There was an unintended consequence to this accumulation of landed wealth: the resilience of the settlement network to shocks and stressors continued to decline. As fewer and fewer landowners acquired more and more resources, the ability for heterarchical information dissemination and control lessened. The system became even more hierarchically structured than before. Landowning elites were reducing peasant autonomy in exchange for continued economic benefit.

3.3.2. Agglomerated Settlements (*Vici*, Hamlets, Villages) Were Important Articulating Nodes in the Production and Settlement Systems over Both Periods

A total of 19 agglomerated settlements were identified in the survey area over the period c. AD 300 to 800 (Figure 4). Interestingly, it was during the sixth century that the highest proportion of sites were occupied (n = 18) (Table 2, Figure 8). The agglomerated settlements clearly contracted toward the Via *Appia* and the *tratturi*, and notably away from lower terrain in the river valleys, where malarial mosquitos were likely to frequent.

It has been argued that self-organization is a resilience-building component in complex adaptive systems [82]. It has also been observed that agglomerated settlements such as *vici* and villages have a lower degree of internal hierarchization than towns or cities [83] and they are found in landscapes adjacent to the Basentello Valley (e.g., the *ager Venusia* [84]. Importantly, their autonomy appears to have increased as Late Antiquity shifted into the Early Middle Ages, as Vagnari demonstrates: it was producing goods on an imperial estate and housing the local workforce, but there was no clear hierarchy in the settlement. As imperial control lessened over the fifth century, so too did control over sites such as Vagnari, and its residents recovered some autonomy and focused on local production strategies instead of exclusively farming grain, visible in the sharp increase in sheep and pig bones on the site [32,33,51]. Some smaller settlements continued to be occupied throughout the landscape over our period, but their material culture does not suggest that any social or economic differences existed between farmhouses, small dwellings, or agglomerated settlements.

The centrality of these settlements in the Basentello landscape, and the reason for their resilience over the transition between the Roman and medieval periods, resulted from several factors. First was the fact that late antique Italian elites increasingly lost wealth and power over the fifth to sixth centuries [85], weakening the hierarchical structure of the countryside and creating the possibility space for these settlements to maintain their structures. Second, urban sites were far removed from this landscape, meaning that it was more difficult for urban elites to exert direct control here. Third, heterarchical communal living made sense when accounting for the economic profile of the landscape: herding, small scale agriculture, and silviculture. Roman institutions of slavery and the *coloni* must have been weakened by the administrative decline of Italy, and while there is no reason to believe that social status vanished along with Roman oversight, we have no material evidence to conclude that traditional elites continued to exist in these settlements.

Table 2. Agglomerated settlement sites occupied during each century.

Century with Occupation	300–400	400–500	500–600	600–700	700–800
Number of agglomerated settlements	16	17	18	15	8

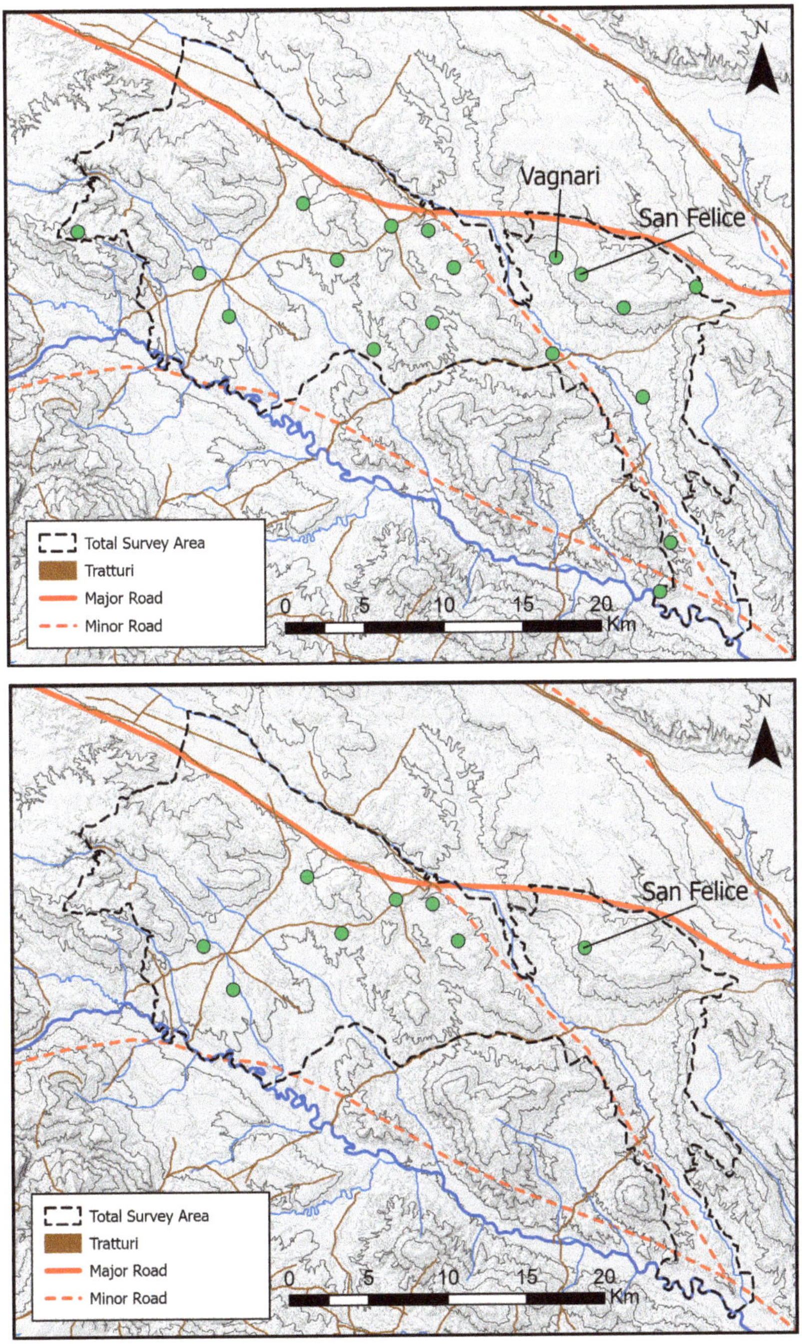

Figure 8. Agglomerated settlements occupied in the sixth (**top**) and the eighth centuries (**bottom**).

Why were agglomerated settlements so prolific in the Basentello landscape over our period? Self-organizing *rustici*, increasingly unburdened by Rome and only marginally influenced by the church, settled in ways that maintained this complex adaptive system at the edge of chaos: as complex as required but not any more or less. Hierarchical villa-style control over this landscape was unlikely because we have little evidence for such settlements, and economic and political potential was likely too low. Dispersed individual family farmsteads, a simpler social structure, was equally unlikely because it minimized the possibility for communal decision making and would have reduced connectedness and further localized economic relationships. If this situation persisted, a reduced state of potential, connectedness, and resilience would have resulted in what ecologists call a 'poverty trap' [33].

A poverty trap is likely responsible for the underdevelopment of this landscape in the early modern period, but it was not a reality for the Early Middle Ages thanks to these agglomerated settlements. They maximized the ability for rural inhabitants to absorb change from the fifth century on, and it was fortuitous that they were already part of the Roman settlement system. Thus, while the Basentello landscape experienced a collapse event in association with the collapse of Roman political power, Italian urban centrality, and the Mediterranean-wide economic system, the capacity for resilience of agglomerated settlement logic helped to avoid a poverty trap.

4. Discussion

Resilience thinking in the Basentello Valley has provided three core insights. First, the lack of stratigraphic excavations in this landscape means that panarchy provides a rigorous model through which to understand this region despite the paucity of detailed data. The survey data allow for a broad temporal understanding of settlement and interregional trade, but without excavations, it is impossible to argue for site-scale economic and social practices, nor is it possible to identify additional patterns resulting from strategies for resilience. This issue is compounded by the peculiarities of the late antique to early medieval transition since artifactual data become scarcer even in excavated contexts. The Basentello Valley data therefore highlight an important methodological insight: panarchy can help to theorize historic processes for landscapes where material and historic evidence are lacking or ambiguous. Panarchy cannot replace data, nor can it replicate the past [9], but it can outline a rigorous historical model open for testing by future excavations or, in extreme situations, stand as our single best understanding of a landscape with little unambiguous historic data available.

Another methodological insight is related to scales of inquiry. Using panarchy to understand resilience at a single site such as Vagnari is illustrative: for example, Building B, likely a market building for periodic fairs (*nundinae*) [51] (p. 30), was destroyed by a suspected earthquake immediately prior to the fifth century and it was not restored to its previous size or function. Instead, the inhabitants of Vagnari absorbed this change and modified their practices to reuse the structure as an animal pen. This is a resilient action and shows the capacity of these *rustici* to modify their practices in the face of changing conditions. However, as noted above, the rest of the Basentello Valley lacks this sort of detailed data for our period. Instead, a century-scale analysis is required to combine the adaptive cycle, panarchy, and the survey data, meaning that a longer-term perspective is productive for resilience research. The strategies for resilience identified above operated on the scale of two to three human generations at least and highlighted communal aspects of rural living. Future excavations in the Basentello Valley will add to this history of resilience by agreeing with, further refining, or rejecting the dynamics established here.

A third insight comes from the centrality of complexity concepts in resilience thinking and landscape dynamics. The panarchy could not exist without the adaptive cycle, and the adaptive cycle is impossible to explain unless the landscape is understood as a complex adaptive system. At its most basic, the existence of nonlinear interactions in the Basentello landscape and their unpredictability is assumed: it was impossible for *rustici* to predict that agglomerated settlements would replace villa agriculture and dispersed farmsteads after the sixth century. It was also impossible for them to predict that supplementing dwindling imports with local productions of pottery and tools in Late Antiquity would help to maintain local social and economic networks over the collapse of Roman hegemony and partially characterize the numerous regional economies of the Early Middle Ages [86,87].

The recognition of no central planning, and the inability for rural peoples to predict disastrous situations such as the mid-sixth century crises, helps to expose the core complexity concept that emerges from this case study: the resilience inherent to self-organization. As hierarchical control over this landscape diminished, the inhabitants made decisions locally and relied on social memory tied to the landscape to meet their social and economic needs. This capacity allowed them to persist over periods of great change, including the sixth century calamities. If connectedness over the Early Middle Ages was high, as during Late Antiquity, this landscape would have collapsed into a poverty trap. However, its long-term dynamics resulted in a landscape capable of persisting at the edge of chaos, balanced between total abandonment (banal simplicity) and collapse (unpredictable and impoverished chaos) over the sixth-century release phase.

With these insights in mind, we can now answer the core question posed earlier: what was the role of resilience in the landscape dynamics of the Basentello Valley over Late Antiquity and the Early Middle Ages?

The political connotations of resilience are made clear in Late Antiquity: landowners constrained the system for their own ends, and this endured until the calamities of the sixth century tipped it into the release phase and the landscape identity and organizational logic shifted from Roman to medieval. The capacity for resilience in this late antique landscape constrained the decision-making ability of the rural population to benefit the landowning aristocracy. As this class diminished and their control weakened, rural inhabitants turned to economic strategies for resilience to help perpetuate their social lives despite the shocks and stressors of the fifth and sixth centuries. Unpredictably, these strategies would become central in the subsequent early medieval landscape.

There is no evidence for hierarchical control in the early medieval landscape from either elites or the church. Instead, rural peoples living in agglomerated settlements were making local decisions, and the fact that other settlements made similar decisions is a sign of heterarchical relationships during a period of historically low political, economic, and urban potential. This change in landscape identity was political and economic, but it would have gone relatively unnoticed on the scale of human lives in the Basentello Valley.

While the resilience inherent to the early medieval rural populations causes them to appear more heterarchical, this state was not successful from the perspective of landscape longevity. Without cross-scale associations with imperial-scale politics, economics, and urbanism, this landscape could not endure external shocks or stressors through peasant autonomy alone. Early medieval landscape organization continued for a few centuries, but despite a capacity for resilience inherent to the exploitation phase, there were not enough resources to make early medieval settlement a worthy social investment beyond the ninth century. During this century, landscape logic was reorganized, and it appears nearly abandoned by start of the tenth century.

5. Conclusions

The Basentello example suggests that the role of resilience in landscape dynamics is to provide the capacity and tools for inhabitants to live in a manner that makes sense to them in their historical context considering the adversities to which they are exposed over time. As the western empire collapsed due to its low capacity for resilience during a period of impactful shocks and stressors, these cascading changes released the settlement logic of the Basentello Valley and changed its identity from Roman to medieval. This monumental historical phenomenon was less a 'decline and fall' and more a spectacular case of historical contingency. Agglomerated settlement and heterarchical relations endured this release phase and provided a blueprint for human–environment interactions for a further two centuries.

This case study is a detailed reminder that the nature of the Roman collapse in Western Europe was not a totalizing event for all associated landscapes. Much Roman infrastructure and settlement endured for centuries after the political, economic, and urban systems collapsed. This is why discussions about continuity and discontinuity mask much more complex and interesting historical realities. It is also why interest in the Roman collapse has endured since the time of Gibbon: an imperial system of robust connectivity and potential should logically have the capacity to endure a wide range of shocks and stressors, and we see this in the Eastern Empire [88].

Resilience thinking exposes a different story: the Western Empire was a complex adaptive system that unpredictably experienced a suite of shocks and stressors at a time when its components had reduced capacities for resilience. Despite its collapse, some landscapes endured, and their inhabitants developed strategies to adapt Roman landscape logic in a medieval social and economic context. In this way, the inhabitants of the Basentello Valley coped with and adapted to the collapse of the Roman world by transforming their landscape actively and without centralized control. Collapse is rarely the whole story, and landscape-scale investigations of resilience can rehabilitate narratives of peasantries in particular to highlight their agency when grand narratives seek to reduce them to, according to Braudel, 'crests of foam that the tides of history carry on their strong backs' [30] (pp. 21).

Funding: This research was conducted under a SSHRC Joseph-Armand Bombardier Canada Graduate Scholarship (767-2017-2697).

Data Availability Statement: Data from the Basentello Valley survey zone was compiled from published sources. Data from the Basentello Valley Archaeological Research Project (BVARP) survey zone included published and unpublished sources. Please visit the BVARP website at (https://www.smu.ca/projects/bvarp/, accessed on 14 February 2023) for contact information.

Acknowledgments: Thanks to Gerry Oetelaar for helping to focus my doctoral research topic and for thoughtfully questioning every interpretation. Thanks to Alistair and Carola Small for sharing with me an early draft of their comprehensive survey volume on the Basentello Valley. Finally, thanks to Myles McCallum for graciously providing me with the extensive and invaluable survey data collected during the BVARP project.

Conflicts of Interest: The author declares no conflict of interest.

References

1. Walker, B.; Salt, D. *Resilience Thinking: Sustaining Ecosystems and People in a Changing World*; Island Press: Washington, DC, USA, 2006.
2. Kinzig, A.P. Towards a deeper understanding of the social in resilience: The contributions of cultural landscapes. In *Resilience and the Cultural Landscape: Understanding and Managing Change in Human-Shaped Environments*; Plieninger, T., Bieling, C., Eds.; Cambridge University Press: Cambridge, UK, 2012; pp. 315–327.
3. Plieninger, T.; Bieling, C. (Eds.) *Resilience and the Cultural Landscape: Understanding and Managing Change in Human-Shaped Environments*; Cambridge University Press: Cambridge, UK, 2012.
4. Faulseit, R.K. (Ed.) *Beyond Collapse: Archaeological Perspectives on Resilience, Revitalization, and Transformation in Complex Societies*; Southern Illinois University Press: Carbondale, IL, USA, 2016.
5. Holling, C.S. Resilience and Stability of Ecological Systems. *Annu. Rev. Ecol. Evol. Syst.* **1973**, *4*, 1–23. [CrossRef]
6. Allen, C.R.; Angeler, D.G.; Garmestani, A.S.; Gunderson, L.H.; Holling, C.S. Panarchy: Theory and Application. *Ecosystems* **2014**, *17*, 578–589. [CrossRef]
7. McGlade, J. Ecohistorical Regimes and La Longue Durée: An Approach to Mapping Long-Term Societal Change. In *Complexity and Co-Evolution: Continuity and Change in Socio-Economic Systems*; Garnsey, E., McGlade, J., Eds.; Edward Elgar: Cheltenham, UK, 2006; pp. 77–114.
8. Daems, D. *Social Complexity and Complex Systems in Archaeology*; Routledge: London, UK; New York, NY, USA, 2021.
9. Gaddis, J.L. *The Landscape of History: How Historians Map the Past*; Oxford University Press: Oxford, UK, 2002.
10. Small, A.; Small, C. (Eds.) *Archaeology on the Apulian-Lucanian Border*; Archaeopress: Oxford, UK, 2022.
11. Carroll, M. (Ed.) *The Making of a Roman Imperial Estate: Archaeology in the Vicus at Vagnari, Puglia*; Archaeopress: Oxford, UK, 2022.
12. McCallum, M.; van der Leest, H.; Veal, R.; Taylor, A.; Cooney, L.; Brown, L.; Munro, M. The Roman Villa at San Felice: Investigations, 2004–2010. *Mouseion* **2011**, *11*, 25–108. [CrossRef]
13. McCallum, M.; van der Leest, H.; Hyatt, A. San Felice and the Basentello Valley Archaeological Research Project, July-August 2013 (Comune di Gravina in Puglia, Provincia di Bari, Regione Puglia). *Pap. Br. Sch.* **2014**, *82*, 343–383. [CrossRef]

14. McCallum, M.; Parsons, N.; van der Leest, H.; Garofalo, G.; Zotta, L. The Basentello Valley Archaeological Research Project, July–August 2014 (Comuni Di Genzano Di Lucania and Irsina, Province Di Potenza and Matera, Regione Basilicata). *Pap. Br. Sch.* **2015**, *83*, 319–323. [CrossRef]
15. Levi, C. *Christ Stopped at Eboli: The Story of a Year*; Farrar, Straus and Company: New York, NY, USA, 1947.
16. Toynbee, A.J. *Hannibal's Legacy: The Hannibalic War's Effects of Roman Life. Volume II: Rome and her Neighbours after Hannibal's Exit*; Oxford University Press: Oxford, UK, 1965.
17. Volpe, G. Introduzione. Documenti per la Storia e l'Archeologia dell'Italia Meridionale Tardoantica e Altomedievale. In *Paesaggi e Insediamenti Rurali in Italia Meridionale fra Tardoantico e Altomedioevo*; Volpe, G., Turchiano, M., Eds.; Edipuglia: Bari, Italy, 2005; pp. 11–22.
18. Volpe, G.; Turchiano, M. (Eds.) *Paesaggi e Insediamenti Rurali in Italia Meridionale fra Tardoantico e Altomedioevo*; Edipuglia: Bari, Italy, 2005.
19. Munro, M. An Archaeology of Resilience in Southern Italy from Late Antiquity to the Early Middle Ages (c. AD 300 to 800). Ph.D. Thesis, University of Calgary, Calgary, AB, Canada. manuscript in preparation, to be submitted.
20. Miller, J.; Page, S. *Complex Adaptive Systems: An Introduction to Computational Models of Social Life*; Princeton University Press: Princeton, NJ, USA, 2007.
21. Keck, M.; Sakdapolrak, P. What is social resilience? *Lessons learned and ways forward. Erdkunde* **2013**, *67*, 5–19.
22. Lansing, J.S. Complex Adaptive Systems. *Annu. Rev. Anthropol.* **2003**, *32*, 183–204. [CrossRef]
23. Bintliff, J. The Paradoxes of Late Antiquity: A Thermodynamic Solution. *Antiq. Tardive.* **2012**, *20*, 69–73. [CrossRef]
24. Holling, C.S.; Gunderson, L.H. Resilience and Adaptive Cycles. In *Panarchy: Understanding Transformations in Human and Natural Systems*; Gunderson, L.H., Holling, C.S., Eds.; Island Press: Washington, DC, USA, 2002; pp. 25–62.
25. Spengler, O. *The Decline of the West*; Alfred A. Knopf: New York, NY, USA, 1926.
26. Toynbee, A.J. *A Study of History. Abridgement of Volumes I-VI by D.C. Somervell*; Oxford University Press: Oxford, UK, 1946.
27. Braudel, F. Histoire et Sciences sociales: La longue durée. *Ann. Hist. Sci. Soc.* **1958**, *4*, 725–753. [CrossRef]
28. Turchin, P. *Historical Dynamics: Why States Rise and Fall*; Princeton University Press: Princeton, NJ, USA, 2003.
29. Turchin, P. Toward Cliodynamics–an Analytical, Predictive Science of History. *Cliodynamics* **2011**, 2, 167–186. [CrossRef]
30. Braudel, F. *The Mediterranean and the Mediterranean World in the Age of Philip II*; Harper & Row: New York, NY, USA, 1972.
31. Bentley, R. An Introduction to Complex Systems. In *Complex Systems and Archaeology*; Bentley, R., Maschner, H., Eds.; The University of Utah Press: Salt Lake City, UT, USA, 2003; pp. 9–23.
32. Storey, R.; Storey, G. Requestioning the Classic Maya Collapse and the Fall of the Roman Empire: Slow Collapse. In *Beyond Collapse: Archaeological Perspectives on Resilience, Revitalization, and Transformation in Complex Societies*; Faulseit, R., Ed.; Southern Illinois University Press: Carbondale, IL, USA, 2016; pp. 99–123.
33. Holling, C.S.; Gunderson, L.H.; Peterson, G. Sustainability and Panarchies. In *Panarchy: Understanding Transformations in Human and Natural Systems*; Gunderson, L.H., Holling, C.S., Eds.; Island Press: Washington, DC, USA, 2002; pp. 63–102.
34. Tainter, J. *The Collapse of Complex Societies*; Cambridge University Press: Cambridge, UK, 1988.
35. Bardi, U. *The Seneca Effect: Why Growth is Slow but Collapse is Rapid*; Springer: Cham, Switzerland, 2017.
36. Small, A. (Ed.) . *Vagnari. Il Villaggio, l'Artigianato, la Proprietà Imperial*; Edipuglia: Bari, Italy, 2011.
37. McCallum, M.; van der Leest, H. Research at San Felice: The Villa on the Imperial Estate. In *Beyond Vagnari. New Temes in the Study of Roman South Italy*; Small, A., Ed.; Edipuglia: Bari, Italy, 2014; pp. 123–134.
38. Fracchia, H. Rural agglomerations in the Upper Bradano Valley. In *Beyond Vagnari. New Themes in the Study of Roman South Italy*; Small, A., Ed.; Edipuglia: Bari, Italy, 2014; pp. 193–202.
39. Marchi, M.L. Ager Venusinus. Ville e Villaggi: Il Paesaggio Rural in Èta Tardoantica. In *Paesaggi e Insediamenti Rurali in Italia Meridionale fra Tardoantico e Altomedioevo*; Volpe, G., Turchiano, M., Eds.; Edipuglia: Bari, Italy, 2005; pp. 173–191.
40. Carter, J.C.; Prieto, A. (Eds.) . *The Chora of Metaponto 3: Archaeological Survey Bradano to Basento*; University of Texas Press: Austin, TX, USA, 2011; Volume 4.
41. Buglione, A.; De Venuto, G.; Goffredo, R.; Volpe, G. Dal Tavoliere alle Murge. Storie di Lana, di Grano e di Sale in Puglia Tra età Romana e Medioevo. In *Storia a Archeologia Globale 2: I Pascoli, i Campi, il Mare. Paesaggi d'Altura e di Pianura in Italia dall'Età del Bronzo al Medioevo*; Cambi, F., De Venuto, G., Goffredo, R., Eds.; Edipuglia: Bari, Italy, 2015; pp. 185–244.
42. Small, A.M.; Small, C.M.; Campbell, I.; MacKinnon, M.; Prowse, T.; Sipe, C. Field Survey in the Basentello Valley on the Basilicata-Puglia Border. *Class. Views* **1998**, *42*, 337–371.
43. McCallum, M.; Hyatt, A. A View of Vagnari from across the Basentello: Initial results from the BVARP survey, 2012. In *Beyond Vagnari. New Themes in the Study of Roman South Italy*; Small, A., Ed.; Edipuglia: Bari, Italy, 2014; pp. 169–180.
44. Reynolds, P. *Trade in the Western Mediterranean ad 400–700: The Ceramic Evidence*; B.A.R.: Oxford, UK, 1995.
45. Di Giuseppe, H. La fornace di Calle di Tricarico: Produzione e Diffusione. In *Ceramica in Italia: VI-VII Secolo*; Saguì, L., Ed.; All'Insegna del Giglio: Florence, Italy, 1998; pp. 735–752.
46. Volpe, G.; Turchiano, M. La villa tardoantica di Faragola (Ascoli Satriano) e oltre. In *La Villa del Casale e Oltre. Territorio, Popolamento, Economia Nella Sicilia Centrale Tra Tarda Antichità e alto Medioevo*; Rizzo, F., Ed.; Edizioni Università di Macerata: Macerata, Italy, 2013; pp. 305–352.
47. Small, A.; Tarlano, F. *La Villa Romana e Tardoantica di San Giovanni di Ruoti (Basilicata): Una Sintesi*; Pisani Teodosio Edizioni: Avigliano, Italy, 2016.

48. Gualtieri, M. La villa di Masseria Ciccotti di Oppido Lucano: Fasi edilize, architettura, mosaici. In *Felicitas Temporum. Dalla Terra Alle Genti: La Basilicata Settentrionale Tra Archeologia e Storia*; Russo Tagilente, A., Di Giuseppe, H., Eds.; Museo Arecheologico Nazionale di Muro Lucano: Potenza, Italy, 2008; pp. 265–287.
49. Holling, C.S. Understanding the Complexity of Economic, Ecological, and Social Systems. *Ecosystem* **2001**, *4*, 390–405. [CrossRef]
50. Redman, C.; Kinzig, A. Resilience of Past Landscapes: Resilience Theory, Society, and the Longue Durée. *Conserv. Ecol.* **2003**, *7*, 1–19. [CrossRef]
51. Small, A. Introduction. In *Vagnari. Il Villaggio, l'Artigianato, la Proprietà Imperial*; Small, A., Ed.; Edipuglia: Bari, Italy, 2011; pp. 11–36.
52. Newfield, T. Mysterious and Mortiferous Clouds: The Climate Cooling and Disease Burden of Late Antiquity. In *Environment and Society in the Long Late Antiquity*; Izdebski, A., Mulryan, M., Eds.; Brill: Leiden, The Netherlands, 2019; pp. 271–297.
53. Jacobson, T.C. *The Gothic War: Rome's Final Conflict in the West*; Westholme: Yardley, UK, 2009.
54. Harper, K. Invisible Environmental History: Infectious Disease in Late Antiquity. In *Environment and Society in the Long Late Antiquity*; Izdebski, A., Mulryan, M., Eds.; Brill: Leiden, The Netherlands, 2019; pp. 298–313.
55. Stathakopoulos, D. *Famine and Pestilence in the Late Roman and Early Byzantine Empire: A Systematic Survey of Subsistence Crises and Epidemics*; Ashgate: Aldershot, UK, 2004.
56. Christie, N. *From Constantine to Charlemagne: An Archaeology of Italy AD 300–800*; Ashgate: Aldershot, UK, 2006.
57. Ciriello, R.; Marchetta, I. Nuovi data da un sito fortificato della basilicata nord-orientale: Il caso di Monteserico e la questione delle fortificazione di confine sulla linea bradanica lucana. In *Atti. XLV Convengno Internazionale della Ceramica. Navi, Relitti e Porti: Il Commercio Marittimo della Ceramica Medievale e Postmedievale*; Tipolitografia Bacchetta: Savona, Italy, 2013; pp. 295–305.
58. Berkes, F.; Folke, C. Linking social and ecological systems for resilience and sustainability. In *Linking Social and Ecological Systems: Management Practices and Social Mechanisms for Building Resilience*; Berkes, F., Folke, C., Eds.; Cambridge University Press: Cambridge, UK, 1998; pp. 1–25.
59. Small, C.M. Vagnari the Basentello Survey. A brief summary. In *Beyond Vagnari. New Themes in the Study of Roman South Italy*; Small, A., Ed.; Edipuglia: Bari, Italy, 2014; pp. 65–72.
60. Wigand, P.; McCallum, M. The Varying Impact of Land Use and Climate in Holocene Landscape Dynamics in the Mezzogiorno. *Athens J. Mediterr. Stud.* **2017**, *2*, 121–150. [CrossRef]
61. Small, A. Grain from Apulia: The Changing Fortunes of Apulia as a Grain-Producing Area in the Hellenistic and Roman Periods. In *Tranquillitas: Mélanges en l'Honneur de Tran Tam Tinh*; Jentel, M.-O., Deschênes-Wagner, G., Eds.; Université de Laval: Laval, QC, Canada, 1994; pp. 543–555.
62. Wickham, C. *Framing the Early Middle Ages. Europe and the Mediterranean, 400–800*; Oxford University Press: Oxford, UK, 2005.
63. Ciminale, D.; Favia, P.; Giuliani, R. Nuove ricerche archeologiche nell'insediamento altomedievale di Belmonte (Altamura). *TARAS* **1994**, *14*, 339–440.
64. Loré, V. I villaggi nell'Italia meridionale (secoli IX-XI): Problemi di definizione. In *Paesaggi, Comunità, Villaggi Medievali. Atti del Convegno Internazionale di Studio, Bologna, 14–16 Gennaio 2010*; Galetti, P., Ed.; Centro Italiano di Studi Sull'Alto Medioevo: Spoleto, Italy, 2012.
65. Nanna, G.; Melillo, A.; Santovito, A. L'insediamento medievale di San Felice. In *Il Paesaggio Storico Ricostruito. L'insediamento di Piana San Felice a Gravina in Puglia*; Cossalter, L., Depalo, M., Eds.; Edipuglia: Bari, Italy, 2017.
66. Marchetta, I.; Ciriello, R. Il Castello di Monteserico a Genzano di Lucania. *Correva l'anno 1041 e arroccato e dominante controllava la battaglia? Mathera* **2017**, *1*, 36–42.
67. Sarris, P. Integration and Disintegration in the Late Roman Economy: The Role of Markets, Emperors, and Aristocrats. In *Local Economies? Production and Exchange of Inland Regions in Late Antiquity*; Lavan, L., Ed.; Brill: Leiden, Canada, 2015; pp. 167–188.
68. Métraux, G. Late Antique Villas: Themes. In *The Roman Villa in the Mediterranean Basin: Late Republic to Late Antiquity*; Marzano, A., Métraux, G., Eds.; Cambridge University Press: Cambridge, UK, 2018; pp. 401–425.
69. Sadori, L.; Giraudi, C.; Masi, A.; Magny, M.; Ortu, E.; Zanchetta, G.; Izdebski, A. Climate, environment and society in southern Italy during the last 2000 years. A review of the environmental, historical and archaeological evidence. *Quat. Sci. Rev.* **2016**, *136*, 173–188. [CrossRef]
70. Galadini, F.; Galli, P. The 346 A.D. earthquake (Central-Southern Italy): An archaeoseismological approach. *Ann. Geophys.* **2004**, *47*, 885–905. [CrossRef]
71. Favia, P.; Giuliani, R.V. Gli edifici nella parte meridionale del sito. In *Vagnari. Il Cillaggio, l'Artigianato, la Proprietà Imperial*; Small, A., Ed.; Edipuglia: Bari, Italy, 2011; pp. 143–230.
72. de Robertis, F.M. Prosperita e banditismo nella puglia e nell'Italia meridionale durante il basso imperio. In *Studi di Storia Pugliese in Onore di Giuseppe Chiarelli*; Paone, M., Ed.; Congedo Editore: Galatina, Italy, 1972; Volume 1, pp. 197–231.
73. Martin, J.-M. *La Pouille du VIe au XIIe Siècle*; École Française de Rome: Rome, Italy, 1993.
74. Christie, N. *The Fall of the Western Roman Empire. An Archaeological and Historical Perspective*; Bloomsbury Academic: London, UK, 2011.
75. Marciniak, S.; Herring, D.A.; Sperduti, A.; Poinar, H.N.; Prowse, T.L. A multi-faceted anthropological and genomic approach to framing Plasmodium falciparum malaria in Imperial period central-southern Italy (1st–4th c. CE). *J. Anthropol. Archaeol.* **2018**, *49*, 210–224. [CrossRef]
76. Sallares, R. *Malaria and Rome: A History of Malaria in Ancient Italy*; Oxford University Press: Oxford, UK, 2002.

77. De Simone, G.F.; Russell, B. The late-antique eruption of Vesuvius in A.D. 472 and its impact from the Bay of Naples to Aeclanum. *J. Rom. Archaeol.* **2019**, *32*, 359–388. [CrossRef]
78. Heather, P. *The Fall of the Roman Empire: A New History of Rome and the Barbarians*; Oxford University Press: Oxford, 2006.
79. Romer, F.E. Famine, pestilence and brigandage in Italy in the fifth century A.D. In *A Roman Villa and a Late Roman Infant Cemetery: Excavation at Poggio Gramignano, Lugnano in Teverina*; Soren, D., Soren, N., Eds.; L'Erma di Bretschneider: Rome, Italy, 1999; pp. 465–475.
80. Buck, R.J. The Via Herculia. *Pap. Br. Sch.* **1971**, *39*, 66–87. [CrossRef]
81. Prowse, T. *Lo studio antropologico dei resti scheletrici. Il Paesaggio Storico Ricostruito. L'insediamento di Piana San Felice a Gravina in Puglia*; Cossalter, L., Depalo, M., Eds.; Edipuglia: Bari, Italy, 2017; pp. 127–130.
82. Folke, C.; Colding, J.; Berkes, F. Synthesis: Building resilience and adaptive capacity in social–ecological systems. In *Navigating Social-Ecological Systems: Building Resilience for Complexity and Change*; Berkes, F., Colding, J., Folke, C., Eds.; Cambridge University Press: Cambridge, UK, 2003; pp. 352–387.
83. Arthur, P. From Vicus to Village: Italian Landscapes, AD 400-1000. In *Landscapes of Change: Rural Evolutions in Late Antiquity and the Early Middle Ages*; Christie, N., Ed.; Ashgate Publishing: Aldershot, UK, 2004; pp. 103–133.
84. Marchi, M.L. The landscape of Daunia: Ager Venusinus. In *Beyond Vagnari. New themes in the Study of Roman South Italy*; Small, A., Ed.; Edipuglia: Bari, Italy, 2014; pp. 181–192.
85. Arce, J. Otium et Negotium: The great estates, 4th-7th century. In *The Transformation of the Roman World AD 400–900*; Webster, L., Brown, M., Eds.; University of California Press: Berkeley, CA, USA, 1997; pp. 19–32.
86. Wickham, C. Overview: Production, Distribution, and Demand. In *The Sixth Century: Production, Distribution, and Demand*; Hodges, R., Bowden, W., Eds.; Brill: Leiden, The Netherlands, 1998; pp. 279–292.
87. Wickham, C. Overview: Production, Distribution and Demand, II. In *The Long Eighth Century*; Hansen, I., Wickham, C., Eds.; Brill: Leiden, The Netherlands, 2000; pp. 345–377.
88. Haldon, J. *The Empire that Would Not Die: The Paradox of Eastern Roman Survival, 640–740*; Harvard University Press: Cambridge, UK, 2016.

MDPI AG
Grosspeteranlage 5
4052 Basel
Switzerland
Tel.: +41 61 683 77 34

Land Editorial Office
E-mail: land@mdpi.com
www.mdpi.com/journal/land

www.ingramcontent.com/pod-product-compliance
Lightning Source LLC
LaVergne TN
LVHW070500061126
851858LV00011B/268
* 9 7 8 3 7 2 5 8 7 9 5 7 1 *